EXPLORATORY DATA ANALYSIS USING R

Chapman & Hall/CRC
Data Mining and Knowledge Series

Series Editor: Vipin Kumar

Computational Business Analytics
Subrata Das

Data Classification
Algorithms and Applications
Charu C. Aggarwal

Healthcare Data Analytics
Chandan K. Reddy and Charu C. Aggarwal

Accelerating Discovery
Mining Unstructured Information for Hypothesis Generation
Scott Spangler

Event Mining
Algorithms and Applications
Tao Li

Text Mining and Visualization
Case Studies Using Open-Source Tools
Markus Hofmann and Andrew Chisholm

Graph-Based Social Media Analysis
Ioannis Pitas

Data Mining
A Tutorial-Based Primer, Second Edition
Richard J. Roiger

Data Mining with R
Learning with Case Studies, Second Edition
Luís Torgo

Social Networks with Rich Edge Semantics
Quan Zheng and David Skillicorn

Large-Scale Machine Learning in the Earth Sciences
Ashok N. Srivastava, Ramakrishna Nemani, and Karsten Steinhaeuser

Data Science and Analytics with Python
Jesus Rogel-Salazar

Feature Engineering for Machine Learning and Data Analytics
Guozhu Dong and Huan Liu

Exploratory Data Analysis Using R
Ronald K. Pearson

For more information about this series please visit:
https://www.crcpress.com/Chapman--HallCRC-Data-Mining-and-Knowledge-Discovery-Series/book-series/CHDAMINODIS

EXPLORATORY DATA ANALYSIS USING R

Ronald K. Pearson

CRC Press
Taylor & Francis Group
Boca Raton   London   New York

CRC Press is an imprint of the
Taylor & Francis Group, an **informa** business
A CHAPMAN & HALL BOOK

CRC Press
Taylor & Francis Group
6000 Broken Sound Parkway NW, Suite 300
Boca Raton, FL 33487-2742

First issued in paperback 2020

© 2018 by Taylor & Francis Group, LLC
CRC Press is an imprint of Taylor & Francis Group, an Informa business

No claim to original U.S. Government works

ISBN-13: 978-0-367-57156-6 (pbk)
ISBN-13: 978-1-138-48060-5 (hbk)

Version Date: 20180312

Visit the Taylor & Francis Web site at
http://www.taylorandfrancis.com

and the CRC Press Web site at
http://www.crcpress.com

Contents

Preface

Much has been written about the abundance of data now available from the Internet and a great variety of other sources. In his aptly named 2007 book *Glut* [81], Alex Wright argued that the total quantity of data then being produced was approximately five *exabytes* per year (5×10^{18} bytes), more than the estimated total number of words spoken by human beings in our entire history. And that assessment was from a decade ago: increasingly, we find ourselves "drowning in a ocean of data," raising questions like "What do we do with it all?" and "How do we begin to make any sense of it?"

Fortunately, the open-source software movement has provided us with—at least partial—solutions like the *R* programming language. While *R* is not the only relevant software environment for analyzing data—*Python* is another option with a growing base of support—*R* probably represents the most flexible data analysis software platform that has ever been available. *R* is largely based on *S*, a software system developed by John Chambers, who was awarded the 1998 Software System Award by the Association for Computing Machinery (ACM) for its development; the award noted that *S* "has forever altered the way people analyze, visualize, and manipulate data."

The other side of this software coin is educational: given the availability and sophistication of *R*, the situation is analogous to someone giving you an F-15 fighter aircraft, fully fueled with its engines running. If you know how to fly it, this can be a great way to get from one place to another very quickly. But it is not enough to just have the plane: you also need to know how to take off in it, how to land it, and how to navigate from where you are to where you want to go. Also, you need to have an idea of where you do want to go. With *R*, the situation is analogous: the software can do a lot, but you need to know both how to use it and what you want to do with it.

The purpose of this book is to address the most important of these questions. Specifically, this book has three objectives:

1. To provide a basic introduction to *exploratory data analysis (EDA)*;

2. To introduce the range of "interesting"—good, bad, and ugly—features we can expect to find in data, and why it is important to find them;

3. To introduce the mechanics of using *R* to explore and explain data.

This book grew out of materials I developed for the course "Data Mining Using R" that I taught for the University of Connecticut Graduate School of Business. The students in this course typically had little or no prior exposure to data analysis, modeling, statistics, or programming. This was not universally true, but it was typical, so it was necessary to make minimal background assumptions, particularly with respect to programming. Further, it was also important to keep the treatment relatively non-mathematical: data analysis is an inherently mathematical subject, so it is not possible to avoid mathematics altogether, but for this audience it was necessary to assume no more than the minimum essential mathematical background.

The intended audience for this book is students—both advanced undergraduates and entry-level graduate students—along with working professionals who want a detailed but introductory treatment of the three topics listed in the book's title: data, exploratory analysis, and R. Exercises are included at the ends of most chapters, and an instructor's solution manual giving complete solutions to all of the exercises is available from the publisher.

Author

Ronald K. Pearson is a Senior Data Scientist with GeoVera Holdings, a property insurance company in Fairfield, California, involved primarily in the exploratory analysis of data, particularly text data. Previously, he held the position of Data Scientist with DataRobot in Boston, a software company whose products support large-scale predictive modeling for a wide range of business applications and are based on Python and R, where he was one of the authors of the `datarobot` R package. He is also the developer of the `GoodmanKruskal` R package and has held a variety of other industrial, business, and academic positions. These positions include both the DuPont Company and the Swiss Federal Institute of Technology (ETH Zürich), where he was an active researcher in the area of nonlinear dynamic modeling for industrial process control, the Tampere University of Technology where he was a visiting professor involved in teaching and research in nonlinear digital filters, and the Travelers Companies, where he was involved in predictive modeling for insurance applications. He holds a PhD in Electrical Engineering and Computer Science from the Massachusetts Institute of Technology and has published conference and journal papers on topics ranging from nonlinear dynamic model structure selection to the problems of disguised missing data in predictive modeling. Dr. Pearson has authored or co-authored five previous books, including *Exploring Data in Engineering, the Sciences, and Medicine* (Oxford University Press, 2011) and *Nonlinear Digital Filtering with Python*, co-authored with Moncef Gabbouj (CRC Press, 2016). He is also the developer of the *DataCamp* course on base *R* graphics.

Chapter 1

Data, Exploratory Analysis, and R

1.1 Why do we analyze data?

The basic subject of this book is data analysis, so it is useful to begin by addressing the question of why we might want to do this. There are at least three motivations for analyzing data:

1. to understand what has happened or what is happening;

2. to predict what is likely to happen, either in the future or in other circumstances we haven't seen yet;

3. to guide us in making decisions.

The primary focus of this book is on *exploratory data analysis*, discussed further in the next section and throughout the rest of this book, and this approach is most useful in addressing problems of the first type: understanding our data. That said, the predictions required in the second type of problem listed above are typically based on mathematical models like those discussed in Chapters 5 and 10, which are optimized to give reliable predictions for data we have available, in the hope and expectation that they will also give reliable predictions for cases we haven't yet considered. In building these models, it is important to use *representative, reliable data*, and the exploratory analysis techniques described in this book can be extremely useful in making certain this is the case. Similarly, in the third class of problems listed above—making decisions—it is important that we base them on an accurate understanding of the situation and/or accurate predictions of what is likely to happen next. Again, the techniques of exploratory data analysis described here can be extremely useful in verifying and/or improving the accuracy of our data and our predictions.

1.2   The view from 90,000 feet

This book is intended as an introduction to the three title subjects—data, its exploratory analysis, and the R programming language—and the following sections give high-level overviews of each, emphasizing key details and interrelationships.

1.2.1   Data

Loosely speaking, the term "data" refers to a collection of details, recorded to characterize a source like one of the following:

- an entity, e.g.: family history from a patient in a medical study; manufacturing lot information for a material sample in a physical testing application; or competing company characteristics in a marketing analysis;

- an event, e.g.: demographic characteristics of those who voted for different political candidates in a particular election;

- a process, e.g.: operating data from an industrial manufacturing process.

This book will generally use the term "data" to refer to a rectangular array of observed values, where each row refers to a different observation of entity, event, or process characteristics (e.g., distinct patients in a medical study), and each column represents a different characteristic (e.g., diastolic blood pressure) recorded—or at least potentially recorded—for each row. In R's terminology, this description defines a *data frame*, one of R's key data types.

The mtcars data frame is one of many built-in data examples in R. This data frame has 32 rows, each one corresponding to a different car. Each of these cars is characterized by 11 variables, which constitute the columns of the data frame. These variables include the car's mileage (in miles per gallon, mpg), the number of gears in its transmission, the transmission type (manual or automatic), the number of cylinders, the horsepower, and various other characteristics. The original source of this data was a comparison of 32 cars from model years 1973 and 1974 published in *Motor Trend Magazine*. The first six records of this data frame may be examined using the head command in R:

```
head(mtcars)

##                    mpg cyl disp  hp drat    wt  qsec vs am gear carb
## Mazda RX4         21.0   6  160 110 3.90 2.620 16.46  0  1    4    4
## Mazda RX4 Wag     21.0   6  160 110 3.90 2.875 17.02  0  1    4    4
## Datsun 710        22.8   4  108  93 3.85 2.320 18.61  1  1    4    1
## Hornet 4 Drive    21.4   6  258 110 3.08 3.215 19.44  1  0    3    1
## Hornet Sportabout 18.7   8  360 175 3.15 3.440 17.02  0  0    3    2
## Valiant           18.1   6  225 105 2.76 3.460 20.22  1  0    3    1
```

An important feature of data frames in R is that both rows and columns have names associated with them. In favorable cases, these names are informative, as they are here: the row names identify the particular cars being characterized, and the column names identify the characteristics recorded for each car.

A more complete description of this dataset is available through R's built-in help facility. Typing "help(mtcars)" at the R command prompt will bring up a help page that gives the original source of the data, cites a paper from the statistical literature that analyzes this dataset [39], and briefly describes the variables included. This information constitutes *metadata* for the mtcars data frame: metadata is "data about data," and it can vary widely in terms of its completeness, consistency, and general accuracy. Since metadata often provides much of our preliminary insight into the contents of a dataset, it is extremely important, and any limitations of this metadata—incompleteness, inconsistency, and/or inaccuracy—can cause serious problems in our subsequent analysis. For these reasons, discussions of metadata will recur frequently throughout this book. The key point here is that, potentially valuable as metadata is, we cannot afford to accept it uncritically: we should always cross-check the metadata with the actual data values, with our intuition and prior understanding of the subject matter, and with other sources of information that may be available.

As a specific illustration of this last point, a popular benchmark dataset for evaluating binary classification algorithms (i.e., computational procedures that attempt to predict a binary outcome from other variables) is the Pima Indians diabetes dataset, available from the UCI Machine Learning Repository, an important Internet data source discussed further in Chapter 4. In this particular case, the dataset characterizes female adult members of the Pima Indians tribe, giving a number of different medical status and history characteristics (e.g., diastolic blood pressure, age, and number of times pregnant), along with a binary diagnosis indicator with the value 1 if the patient had been diagnosed with diabetes and 0 if they had not. Several versions of this dataset are available: the one considered here was the UCI website on May 10, 2014, and it has 768 rows and 9 columns. In contrast, the data frame Pima.tr included in R's MASS package is a subset of this original, with 200 rows and 8 columns. The metadata available for this dataset from the UCI Machine Learning Repository now indicates that this dataset exhibits missing values, but there is also a note that prior to February 28, 2011 the metadata indicated that there were no missing values. In fact, the missing values in this dataset are not coded explicitly as missing with a special code (e.g., R's "NA" code), but are instead coded as zero. As a result, a number of studies characterizing binary classifiers have been published using this dataset as a benchmark where the authors were not aware that data values were missing, in some cases, quite a large fraction of the total observations. As a specific example, the serum insulin measurement included in the dataset is 48.7% missing.

Finally, it is important to recognize the essential role our *assumptions* about data can play in its subsequent analysis. As a simple and amusing example, consider the following "data analysis" question: how many planets are there orbiting the Sun? Until about 2006, the generally accepted answer was nine, with Pluto the outermost member of this set. Pluto was subsequently re-classified as a "dwarf planet," in part because a larger, more distant body was found in the Kuiper Belt and enough astronomers did not want to classify this object as the "tenth planet" that Pluto was demoted to dwarf planet status. In his book,

Is Pluto a Planet? [72], astronomer David Weintraub argues that Pluto should remain a planet, based on the following defining criteria for planethood:

1. the object must be too small to generate, or to have ever generated, energy through nuclear fusion;

2. the object must be big enough to be spherical;

3. the object must have a primary orbit around a star.

The first of these conditions excludes dwarf stars from being classed as planets, and the third excludes moons from being declared planets (since they orbit planets, not stars). Weintraub notes, however, that under this definition, there are at least 24 planets orbiting the Sun: the eight now generally regarded as planets, Pluto, and 15 of the largest objects from the asteroid belt between Mars and Jupiter and from the Kuiper Belt beyond Pluto. This example illustrates that definitions are both extremely important and not to be taken for granted: everyone knows what a planet is, don't they? In the broader context of data analysis, the key point is that unrecognized disagreements in the definition of a variable are possible between those who measure and record it, and those who subsequently use it in analysis; these discrepancies can lie at the heart of unexpected findings that turn out to be erroneous. For example, if we wish to combine two medical datasets, characterizing different groups of patients with "the same" disease, it is important that the same diagnostic criteria be used to declare patients "diseased" or "not diseased." For a more detailed discussion of the role of definitions in data analysis, refer to Sec. 2.4 of *Exploring Data in Engineering, the Sciences, and Medicine* [58]. (Although the book is generally quite mathematical, this is not true of the discussions of data characteristics presented in Chapter 2, which may be useful to readers of this book.)

1.2.2 Exploratory analysis

Roughly speaking, exploratory data analysis (EDA) may be defined as the art of looking at one or more datasets in an effort to understand the underlying structure of the data contained there. A useful description of how we might go about this is offered by Diaconis [21]:

> We look at numbers or graphs and try to find patterns. We pursue leads suggested by background information, imagination, patterns perceived, and experience with other data analyses.

Note that this quote suggests—although it does not strictly imply—that the data we are exploring consists of numbers. Indeed, even if our dataset contains nonnumerical data, our analysis of it is likely to be based largely on numerical characteristics computed from these nonnumerical values. As a specific example, categorical variables appearing in a dataset like "city," "political party affiliation," or "manufacturer" are typically tabulated, converted from discrete named values into counts or relative frequencies. These derived representations can be particularly useful in exploring data when the number of levels—i.e., the number of distinct values the original variable can exhibit—is relatively small. In such cases, many useful exploratory tools have been developed that allow us to examine the character of these nonnumeric variables and their relationship with other variables, whether categorical or numeric. Simple graphical examples include boxplots for looking at the distribution of numerical values across the different levels of a categorical variable, or mosaic plots for looking at the relationship between categorical variables; both of these plots and other, closely related ones are discussed further in Chapters 2 and 3.

Categorical variables with many levels pose more challenging problems, and these come in at least two varieties. One is represented by variables like U.S. postal zipcode, which identifies geographic locations at a much finer-grained level than state does and exhibits about 40,000 distinct levels. A detailed discussion of dealing with this type of categorical variable is beyond the scope of this book, although one possible approach is described briefly at the end of Chapter 10. The second type of many-level categorical variable arises in settings where the inherent structure of the variable can be exploited to develop specialized analysis techniques. Text data is a case in point: the number of distinct words in a document or a collection of documents can be enormous, but special techniques for analyzing text data have been developed. Chapter 8 introduces some of the methods available in R for analyzing text data.

The mention of "graphs" in the Diaconis quote is particularly important since humans are much better at seeing patterns in graphs than in large collections of numbers. This is one of the reasons R supports so many different graphical display methods (e.g., scatterplots, barplots, boxplots, quantile-quantile plots, histograms, mosaic plots, and many, many more), and one of the reasons this book places so much emphasis on them. That said, two points are important here. First, graphical techniques that are useful to the data analyst in finding important structure in a dataset are not necessarily useful in explaining those findings to others. For example, large arrays of two-variable scatterplots may be a useful screening tool for finding related variables or anomalous data subsets, but these are extremely poor ways of presenting results to others because they essentially require the viewer to repeat the analysis for themselves. Instead, results should be presented to others using displays that highlight and emphasize the analyst's findings to make sure that the intended message is received. This distinction between *exploratory* and *explanatory* displays is discussed further in Chapter 2 on graphics in R and in Chapter 6 on crafting data stories (i.e., explaining your findings), but most of the emphasis in this book is on exploratory graphical tools to help us obtain these results.

The second point to note here is that the utility of any graphical display can depend strongly on exactly what is plotted, as illustrated in Fig. 1.1. This issue has two components: the mechanics of how a subset of data is displayed, and the choice of what goes into that data subset. While both of these aspects are important, the second is far more important than the first. Specifically, it is important to note that the form in which data arrives may not be the most useful for analysis. To illustrate, Fig. 1.1 shows two sets of plots, both constructed

```
library(MASS)
library(car)
par(mfrow=c(2,2))
truehist(mammals$brain)
truehist(log(mammals$brain))
qqPlot(mammals$brain)
title("Normal QQ-plot")
qqPlot(log(mammals$brain))
title("Normal QQ-plot")
```

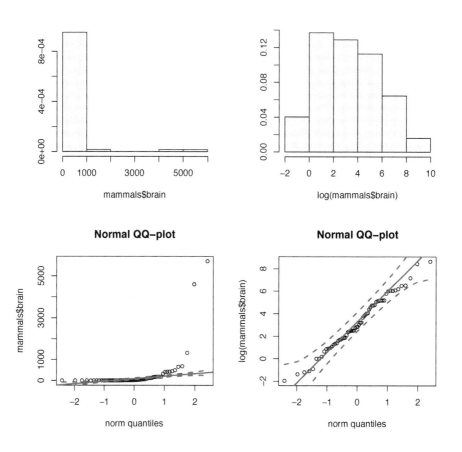

Figure 1.1: Two pairs of characterizations of the brain weight data from the `mammals` data frame: histograms and normal QQ-plots constructed from the raw data (left-hand plots), and from log-transformed data (right-hand plots).

from the `brain` element of the `mammals` dataset from the `MASS` package that lists body weights and brain weights for 62 different animals. This data frame is discussed further in Chapter 3, along with the characterizations presented here, which are histograms (top two plots) and normal QQ-plots (bottom two plots). In both cases, these plots are attempting to tell us something about the distribution of data values, and the point of this example is that the extent to which these plots are informative depends strongly on how we prepare the data from which they are constructed. Here, the left-hand pair of plots were generated from the raw data values and they are much less informative than the right-hand pair of plots, which were generated from log-transformed data. In particular, these plots suggest that the log-transformed data exhibits a roughly Gaussian distribution, further suggesting that working with the log of brain weight may be more useful than working with the raw data values. This example is revisited and discussed in much more detail in Chapter 3, but the point here is that exactly what we plot—e.g., raw data values vs. log-transformed data values—sometimes matters a lot more than how we plot it.

Since it is one of the main themes of this book, a much more extensive introduction to exploratory data analysis is given in Chapter 3. Three key points to note here are, first, that exploratory data analysis makes extensive use of graphical tools, for the reasons outlined above. Consequently, the wide and growing variety of graphical methods available in R makes it a particularly suitable environment for exploratory analysis. Second, exploratory analysis often involves characterizing many different variables and/or data sources, and comparing these characterizations. This motivates the widespread use of simple and well-known summary statistics like means, medians, and standard deviations, along with other, less well-known characterizations like the MAD scale estimate introduced in Chapter 3. Finally, third, an extremely important aspect of exploratory data analysis is the search for "unusual" or "anomalous" features in a dataset. The notion of an *outlier* is introduced briefly in Sec. 1.3, but a more detailed discussion of this and other data anomalies is deferred until Chapter 3, where techniques for detecting these anomalies are also discussed.

1.2.3 Computers, software, and R

To use R—or any other data analysis environment—involves three basic tasks:

1. Make the data you want to analyze available to the analysis software;

2. Perform the analysis;

3. Make the results of the analysis available to those who need them.

In this chapter, all of the data examples come from built-in data frames in R, which are extremely convenient for teaching or learning R, but in real data analysis applications, making the data available for analysis can require significant effort. Chapter 4 focuses on this problem, but to understand its nature and significance, it is necessary to understand something about how computer systems are organized, and this is the subject of the next section. Related issues arise when we attempt to make analysis results available for others, and these issues are also covered in Chapter 4. Most of the book is devoted to various aspects of step (2) above—performing the analysis—and the second section below briefly addresses the question of "why use R and not something else?" Finally, since this is a book about using R to analyze data, some key details about the structure of the R language are presented in the third section below.

General structure of a computing environment

In his book, *Introduction to Data Technologies* [56, pp. 211–214], Paul Murrell describes the general structure of a computing environment in terms of the following six components:

1. the *CPU* or *central processing unit* is the basic hardware that does all of the computing;

2. the *RAM* or *random access memory* is the *internal* memory where the CPU stores and retrieves results;

3. the *keyboard* is the standard interface that allows the user to submit requests to the computer system;

4. the *screen* is the graphical display terminal that allows the user to see the results generated by the computer system;

5. the *mass storage*, typically a "hard disk," is the *external* memory where data and results can be stored permanently;

6. the *network* is an external connection to the outside world, including the Internet but also possibly an *intranet* of other computers, along with peripheral devices like printers.

Three important distinctions between internal storage (i.e., RAM) and external storage (i.e., mass storage) are, first, that RAM is typically several orders of magnitude faster to access than mass storage; second, that RAM is *volatile*—i.e., the contents are lost when the power is turned off—while mass storage is not; and, third, that mass storage can accommodate much larger volumes of data than RAM can. (As a specific example, the computer being used to prepare this book has 4GB of installed RAM and just over 100 times as much disk storage.) A practical consequence is that both the data we want to analyze and any results we want to save need to end up in mass storage so they are not lost when the computer power is turned off. Chapter 4 is devoted to a detailed discussion of some of the ways we can move data into and out of mass storage.

These differences between RAM and mass storage are particularly relevant to R since most R functions require all data—both the raw data and the internal storage required to keep any temporary, intermediate results—to fit in RAM. This makes the computations faster, but it limits the size of the datasets you can work with in most cases to something less than the total installed RAM on your computer. *In some applications, this restriction represents a serious limitation on R's applicability.* This limitation is recognized within the R community and continuing efforts are being made to improve the situation.

Closely associated with the CPU is the *operating system*, which is the software that runs the computer system, making useful activity possible. That is, the operating system coordinates the different components, establishing and managing file systems that allow datasets to be stored, located, modified, or deleted; providing user access to programs like *R*; providing the support infrastructure required so these programs can interact with network resources, etc. In addition to the general computing infrastructure provided by the operating system, to analyze data it is necessary to have programs like *R* and possibly others (e.g., database programs). Further, these programs must be compatible with the operating system: on popular desktops and enterprise servers, this is usually not a problem, although it can become a problem for older operating systems. For example, Section 2.2 of the *R FAQ* document available from the *R* "Help" tab notes that "support for Mac OS Classic ended with R 1.7.1."

With the growth of the Internet as a data source, it is becoming increasingly important to be able to retrieve and process data from it. Unfortunately, this involves a number of issues that are well beyond the scope of this book (e.g., parsing HTML to extract data stored in web pages). A brief introduction to the key ideas with some simple examples is given in Chapter 4, but for those needing a more thorough treatment, Murrell's book is highly recommended [56].

Data analysis software

A key element of the data analysis chain (acquire → analyze → explain) described earlier is the choice of data analysis software. Since there are a number of possibilities here, why *R*? One reason is that *R* is a free, open-source language, available for most popular operating systems. In contrast, commercially supported packages must be purchased, in some cases for a lot of money.

Another reason to use *R* in preference to other data analysis platforms is the enormous range of analysis methods supported by *R*'s growing universe of add-on packages. These packages support analysis methods from many branches of statistics (e.g., traditional statistical methods like ANOVA, ordinary least squares regression, and *t*-tests, Bayesian methods, and robust statistical procedures), machine learning (e.g., random forests, neural networks, and boosted trees), and other applications like text analysis. This availability of methods is important because it greatly expands the range of data exploration and analysis approaches that can be considered. For example, if you wanted to use the multivariate outlier detection method described in Chapter 9 based on the MCD covariance estimator in another framework—e.g., Microsoft Excel—you would have to first build these analysis tools yourself, and then test them thoroughly to make sure they are really doing what you want. All of this takes time and effort just to be able to get to the point of actually analyzing your data.

Finally, a third reason to adopt *R* is its growing popularity, undoubtedly fueled by the reasons just described, but which is also likely to promote the continued growth of new capabilities. A survey of programming language popularity by the Institute of Electrical and Electronics Engineers (IEEE) has been taken for the last several years, and a summary of the results as of July 18,

2017, was available from the website:

```
http://spectrum.ieee.org/computing/software/
        the-2017-top-ten-programming-languages
```

The top six programming languages on this list were, in descending order: Python, C, Java, C++, C#, and R. Note that the top five of these are general-purpose languages, all suitable for at least two of the four programming environments considered in the survey: web, mobile, desktop/enterprise, and embedded. In contrast, R is a specialized data analysis language that is only suitable for the desktop/enterprise environment. The next data analysis language in this list was the commercial package MATLAB®, ranked 15th.

The structure of R

The R programming language basically consists of three components:

- a set of *base R packages*, a required collection of programs that support language infrastructure and basic statistics and data analysis functions;

- a set of *recommended packages*, automatically included in almost all R installations (the MASS package used in this chapter belongs to this set);

- a very large and growing set of *optional add-on packages*, available through the Comprehensive R Archive Network (CRAN).

Most R installations have all of the base and recommended packages, with at least a few selected add-on packages. The advantage of this language structure is that it allows extensive customization: as of February 3, 2018, there were 12,086 packages available from CRAN, and new ones are added every day. These packages provide support for everything from rough and fuzzy set theory to the analysis of twitter tweets, so it is an extremely rare organization that actually needs *everything* CRAN has to offer. Allowing users to install only what they need avoids massive waste of computer resources.

Installing packages from CRAN is easy: the R graphical user interface (GUI) has a tab labeled "Packages." Clicking on this tab brings up a menu, and selecting "Install packages" from this menu brings up one or two other menus. If you have not used the "Install packages" option previously in your current R session, a menu appears asking you to select a CRAN mirror; these sites are locations throughout the world with servers that support CRAN downloads, so you should select one near you. Once you have done this, a second menu appears that lists all of the R packages available for download. Simply scroll down this list until you find the package you want, select it, and click the "OK" button at the bottom of the menu. This will cause the package you have selected to be downloaded from the CRAN mirror and installed on your machine, along with all other packages that are required to make your selected package work. For example, the car package used to generate Fig. 1.1 requires a number of other packages, including the quantile regression packge quantreg, which is automatically downloaded and installed when you install the car package.

It is important to note that *installing* an *R* package makes it available for you to use, but this does *not* "load" the package into your current *R* session. To do this, you must use the `library()` function, which works in two different ways. First, if you enter this function without any parameters—i.e., type "library()" at the *R* prompt—it brings up a new window that lists all of the packages that have been installed on your machine. To use any of these packages, it is necessary to use the `library()` command again, this time specifying the name of the package you want to use as a parameter. This is shown in the code appearing at the top of Fig. 1.1, where the `MASS` and `car` packages are loaded:

```
library(MASS)
library(car)
```

The first of these commands loads the `MASS` package, which contains the `mammals` data frame and the `truehist` function to generate histograms, and the second loads the `car` package, which contains the `qqPlot` function used to generate the normal QQ-plots shown in Fig. 1.1.

1.3 A representative R session

To give a clear view of the essential material covered in this book, the following paragraphs describe a simple but representative *R* analysis session, providing a few specific illustrations of what *R* can do. The general task is a typical preliminary data exploration: we are given an unfamiliar dataset and we begin by attempting to understand what is in it. In this particular case, the dataset is a built-in data example from *R*—one of many such examples included in the language—but the preliminary questions explored here are analogous to those we would ask in characterizing a dataset obtained from the Internet, from a data warehouse of customer data in a business application, or from a computerized data collection system in a scientific experiment or an industrial process monitoring application. Useful preliminary questions include:

1. How many records does this dataset contain?

2. How many fields (i.e., variables) are included in each record?

3. What kinds of variables are these? (e.g., real numbers, integers, categorical variables like "city" or "type," or something else?)

4. Are these variables always observed? (i.e., is missing data an issue? If so, how are missing values represented?)

5. Are the variables included in the dataset the ones we were expecting?

6. Are the values of these variables consistent with what we expect?

7. Do the variables in the dataset seem to exhibit the kinds of relationships we expect? (Indeed, what relationships do we expect, and why?)

The example presented here does not address all of these questions, but it does consider some of them and it shows how the R programming environment can be useful in both answering and refining these questions.

Assuming R has been installed on your machine (if not, see the discussion of installing R in Chapter 11), you begin an interactive session by clicking on the R icon. This brings up a window where you enter commands at the ">" prompt to tell R what you want to do. There is a toolbar at the top of this display with a number of tabs, including "Help" which provides links to a number of useful documents that will be discussed further in later parts of this book. Also, when you want to end your R session, type the command "q()" at the ">" prompt: this is the "quit" command, which terminates your R session. Note that the parentheses after "q" are important here: this tells R that you are calling a *function* that, in general, does something to the argument or arguments you pass it. In this case, the command takes no arguments, but failing to include the parentheses will cause R to search for an object (e.g., a vector or data frame) named "q" and, if it fails to find this, display an error message. Also, note that when you end your R session, you will be asked whether you want to save your workspace image: if you answer "yes," R will save a copy of all of the commands you used in your interactive session in the file .Rhistory in the current working directory, making this command history—but not the R objects created from these commands—available for your next R session.

Also, in contrast to some other languages—$SAS^{®}$ is a specific example—it is important to recognize that R is *case-sensitive:* commands and variables in lower-case, upper-case, or mixed-case are *not* the same in R. Thus, while a SAS procedure like PROC FREQ may be equivalently invoked as proc freq or Proc Freq, the R commands qqplot and qqPlot are *not* the same: qqplot is a function in the stats package that generates quantile-quantile plots comparing two empirical distributions, while qqPlot is a function in the car package that generates quantile-quantile plots comparing a data distribution with a theoretical reference distribution. While the tasks performed by these two functions are closely related, the details of what they generate are different, as are the details of their syntax. As a more immediate illustration of R's case-sensitivity, recall that the function q() "quits" your R session; in contrast, unless you define it yourself or load an optional package that defines it, the function Q() does not exist, and invoking it will generate an error message, something like this:

```
Q()
```

```
## Error in Q(): could not find function "Q"
```

The specific dataset considered in the following example is the whiteside data frame from the MASS package, one of the *recommended* packages included with almost all R installations, as noted in Sec. 1.2.3. Typing "??whiteside" at the ">" prompt performs a fuzzy search through the documentation for all packages available to your R session, bringing up a page with all approximate matches on the term. Clicking on the link labeled MASS::whiteside takes us to a documentation page with the following description:

Mr Derek Whiteside of the UK Building Research Station recorded the weekly gas consumption and average external temperature at his own house in south-east England for two heating seasons, one of 26 weeks before, and one of 30 weeks after cavity-wall insulation was installed. The object of the exercise was to assess the effect of the insulation on gas consumption.

To analyze this dataset, it is necessary to first make it available by loading the MASS package with the `library()` function as described above:

```
library(MASS)
```

An *R* data frame is a rectangular array of *N* records—each represented as a row—with *M* fields per record, each representing a value of a particular variable for that record. This structure may be seen by applying the **head** function to the **whiteside** data frame, which displays its first few records:

```
head(whiteside)

##     Insul Temp Gas
## 1 Before -0.8 7.2
## 2 Before -0.7 6.9
## 3 Before  0.4 6.4
## 4 Before  2.5 6.0
## 5 Before  2.9 5.8
## 6 Before  3.2 5.8
```

More specifically, the first line lists the field names, while the next six lines show the values recorded in these fields for the first six records of the dataset. Recall from the discussion above that the **whiteside** data frame characterizes the weekly average heating gas consumption and the weekly average outside temperature for two successive winters, the first before Whiteside installed insulation in his house, and the second after. Thus, each record in this data frame represents one weekly observation, listing whether it was made before or after the insulation was installed (the **Insul** variable), the average outside temperature, and the average heating gas consumption.

A more detailed view of this data frame is provided by the **str** function, which returns structural characterizations of essentially any *R* object. Applied to the **whiteside** data frame, it returns the following information:

```
str(whiteside)

## 'data.frame': 56 obs. of  3 variables:
##  $ Insul: Factor w/ 2 levels "Before","After": 1 1 1 1 1 1 1 1 1 1 ...
##  $ Temp : num  -0.8 -0.7 0.4 2.5 2.9 3.2 3.6 3.9 4.2 4.3 ...
##  $ Gas  : num  7.2 6.9 6.4 6 5.8 5.8 5.6 4.7 5.8 5.2 ...
```

Here, the first line tells us that **whiteside** is a data frame, with 56 observations (rows or records) and 3 variables. The second line tells us that the first variable, **Insul**, is a *factor* variable with two levels: "Before" and "After." (Factors are an important *R* data type used to represent categorical data, introduced briefly in the next paragraph.) The third and fourth lines tell us that `Temp` and `Gas` are numeric variables. Further, all lines except the first provide summaries of the first few (here, 10) values observed for each variable. For the numeric variables, these values are the same as those shown with the `head` command presented above, while for factors, `str` displays a numerical index indicating which of the possible levels of the variable is represented in each of the first 10 records.

Because factor variables are both very useful and somewhat more complex in their representation than numeric variables, it is worth a brief digression here to say a bit more about them. Essentially, factor variables in *R* are special vectors used to represent categorical variables, encoding them with two components: a level, corresponding to the value we see (e.g., "Before" and "After" for the factor `Insul` in the `whiteside` data frame), and an index that maps each element of the vector into the appropriate level:

```
x <- whiteside$Insul
str(x)

##  Factor w/ 2 levels "Before","After": 1 1 1 1 1 1 1 1 1 1 ...

x[2]

## [1] Before
## Levels: Before After
```

Here, the `str` characterization tells us how many levels the factor has and what the names of those levels are (i.e., two levels, named "Before" and "After"), but the values `str` displays are the indices instead of the levels (i.e., the first 10 records list the the first value, which is "Before"). *R* also supports character vectors and these could be used to represent categorical variables, but an important difference is that the levels defined for a factor variable represent its only possible values: attempting to introduce a new value into a factor variable fails, generating a missing value instead, with a warning. For example, if we attempted to change the second element of this factor variable from "Before" to "Unknown," we would get a warning about an invalid factor level and that the attempted assignment resulted in this element having the missing value `NA`. In contrast, if we convert x in this example to a character vector, the new value assignment attempted above now works:

```
x <- as.character(whiteside$Insul)
str(x)

##  chr [1:56] "Before" "Before" "Before" "Before" "Before" "Before" ...

x[2]

## [1] "Before"

x[2] <- "Unknown"
str(x)

##  chr [1:56] "Before" "Unknown" "Before" "Before" "Before" "Before" ...
```

In addition to str and head, the summary function can also provide much useful information about data frames and other R objects. In fact, summary is an example of a *generic* function in R, that can do different things depending on the attributes of the object we apply it to. Generic functions are discussed further in Chapters 2 and 7, but when the generic summary function is applied to a data frame like whiteside, it returns a relatively simple characterization of the values each variable can assume:

```
summary(whiteside)
```

```
##     Insul           Temp             Gas
## Before:26    Min.   :-0.800    Min.   :1.300
## After :30    1st Qu.: 3.050    1st Qu.:3.500
##              Median : 4.900    Median :3.950
##              Mean   : 4.875    Mean   :4.071
##              3rd Qu.: 7.125    3rd Qu.:4.625
##              Max.   :10.200    Max.   :7.200
```

This result may be viewed as a table with one column for each variable in the whiteside data frame—Insul, Temp, and Gas—with a column format that depends on the type of variable being characterized. For the two-level factor Insul, the summary result gives the number of times each possible level occurs: 26 records list the value "Before," while 30 list the value "After." For the numeric variables, the result consists of two components: one is the mean value—i.e., the average of the variable over all records in the dataset—while the other is *Tukey's five-number summary*, consisting of these five numbers:

1. the *sample minimum*, defined as the smallest value of x in the dataset;

2. the *lower quartile*, defined as the value x_L for which 25% of the data satisfies $x \leq x_L$ and the other 75% of the data satisfies $x > x_L$;

3. the *sample median*, defined as the "middle value" in the dataset, the value that 50% of the data values do not exceed and 50% do exceed;

4. the *upper quartile*, defined as the value x_U for which 75% of the data satisfies $x \leq x_U$ and the other 25% of the data satisfies $x > x_U$;

5. the *sample maximum*, defined as the largest value of x in the dataset.

This characterization has the advantage that it can be defined for any sequence of numbers and its complexity does not depend on how many numbers are in the sequence. In contrast, the complete table of counts for an L-level categorical variable consists of L numbers: for variables like Insul in the whiteside data frame, $L = 2$, so this characterization is simple. For a variable like "State" with 50 distinct levels (i.e., one for each state in the U.S.), this table has 50 entries. For this reason, the characterization returned by the summary function for categorical variables consists of the complete table if $L \leq 6$, but if $L > 6$, it lists only the five most frequently occurring levels, lumping all remaining levels into a single "other" category.

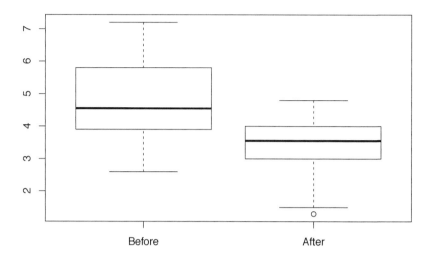

Figure 1.2: Side-by-side boxplot comparison of the "Before" and "After" subsets of the `Gas` values from the `whiteside` data frame.

An extremely useful graphical representation of Tukey's five-number summary is the *boxplot*, particularly useful in showing how the distribution of a numerical variable depends on subsets defined by the different levels of a factor. Fig. 1.2 shows a side-by-side boxplot summary of the `Gas` variable for subsets of the `whiteside` data frame defined by the `Insul` variable. This summary was generated by the following *R* command, which uses the *R formula interface* (i.e., `Gas ~ Insul`) to request boxplots of the ranges of variation of the `Gas` variable for each distinct level of the `Insul` factor:

```
boxplot(Gas ~ Insul, data = whiteside)
```

The left-hand plot—above the *x*-axis label "Before"—illustrates the boxplot in its simplest form: the short horizontal lines at the bottom and top of the plot correspond to the sample minimum and maximum, respectively; the wider, heavier line in the middle of the plot represents the median; and the lines at the top and bottom of the "box" in the plot correspond to the upper and lower quartiles. The "After" boxplot also illustrates a common variation on the "basic" boxplot based strictly on Tukey's five-number summary. Specifically, at the bottom of this boxplot—below the "sample minimum" horizontal line—is a single open circle, representing an *outlier*, a data value that appears inconsistent with the majority of the data (here, "unusually small"). In this boxplot, the

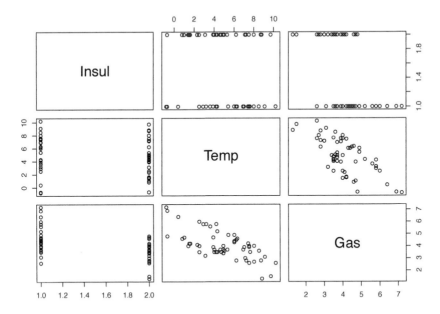

Figure 1.3: The 3 × 3 plot array generated by plot(whiteside).

bottom horizontal line does not represent the sample minimum, but the "small-est non-outlying value" where the determination of what values are "outlying" versus "non-outlying" is made using a simple rule discussed in Chapter 3.

Fig. 1.3 shows the results of applying the plot function to the whiteside data frame. Like summary, the plot function is also generic, producing a result that depends on the nature of the object to which it is applied. Applied to a data frame, plot generates a matrix of *scatterplots*, showing how each variable relates to the others. More specifically, the diagonal elements of this plot array identify the variable that defines the x-axis in all of the other plots in that column of the array and the y-axis in all of the other plots in that row of the array. Here, the two scatterplots involving Temp and Gas are simply plots of the numerical values of one variable against the other. The four plots involving the factor variable Insul have a very different appearance, however: in these plots, the two levels of this variable ("Before" and "After") are represented by their numerical codes, 1 and 2. Using these numerical codes provides a basis for including factor variables in a scatterplot array like the one shown here, although the result is often of limited utility. Here, one point worth noting is that the plots involving Insul and Gas do show that the Gas values are generally smaller when Insul has its second value. In fact, this level corresponds to "After" and this difference reflects the important detail that less heating gas was consumed after insulation was installed in the house than before.

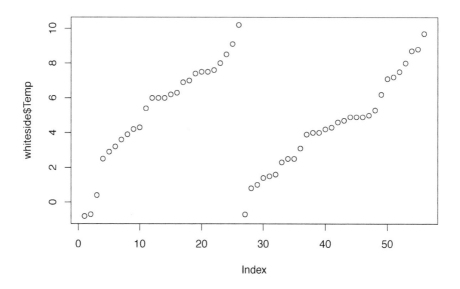

Figure 1.4: The result of plot(whiteside$Temp).

In Fig. 1.4, applying plot to the Temp variable from the whiteside data frame shows how Temp varies with its record number in the data frame. Here, these values appear in two groups—one of 26 points, followed by another of 30 points—but within each group, they appear in ascending order. From the data description presented earlier, we might expect these values to represent average weekly winter temperatures recorded in successive weeks during the two heating seasons characterized in the dataset. Instead, these observations have been ordered from coldest to warmest within each heating season. While such unexpected structure often makes no difference, it sometimes does; the key point here is that plotting the data can reveal it.

Fig. 1.5 shows the result of applying the plot function to the factor variable Insul, which gives us a *barplot*, showing how many times each possible value for this categorical variable appears in the data frame. In marked contrast to this plot, note that Fig. 1.3 used the numerical level representation for Insul: "Before" corresponds to the first level of the variable—represented as 1 in the plot—while "After" corresponds to the second level of the variable, represented as 2 in the plot. This was necessary so that the plot function could present scatterplots of the "value" of each variable against the corresponding "value" of every other variable. Again, these plots emphasize that plot is a generic function, whose result depends on the type of R object plotted.

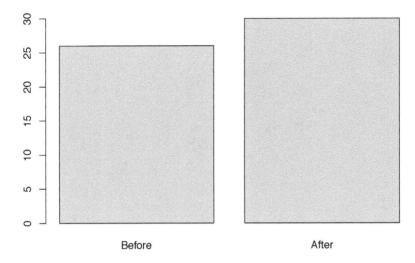

Figure 1.5: The result of plot(whiteside$Insul).

The rest of this section considers some refinements of the scatterplot between weekly average heating gas consumption and average outside temperature appearing in the three-by-three plot array in Fig. 1.3. The intent is to give a "preview of coming attractions," illustrating some of the ideas and techniques that will be discussed in detail in subsequent chapters.

The first of these extensions is Fig. 1.6, which plots Gas versus Temp with different symbols for the two heating seasons (i.e., "Before" and "After"). The following *R* code generates this plot, using open triangles for the "Before" data and solid circles for the "After" data:

```
plot(whiteside$Temp, whiteside$Gas, pch=c(6,16)[whiteside$Insul])
```

The approach used here to make the plotting symbol depend on the Insul value for each point is described in Chapter 2, which gives a detailed discussion of generating and refining graphical displays in *R*. Here, the key point is that using different plotting symbols for the "Before" and "After" points in this example highlights the fact that the relationship between heating gas consumption and outside temperature is substantially different for these two collections of points, as we would expect from the original description of the dataset. Another important point is that generating this plot with different symbols for the two sets of data points is not difficult.

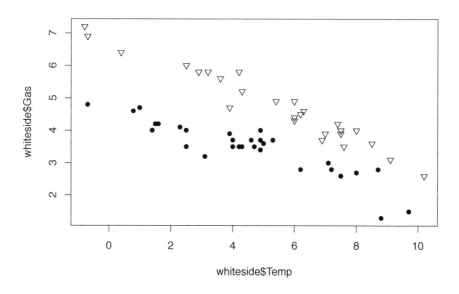

Figure 1.6: Scatterplot of Gas versus Temp from the whiteside data frame, with distinct point shapes for the "Before" and "After" data subsets.

Fig. 1.7 shows a simple but extremely useful modification of Fig. 1.6: the inclusion of a legend that tells us what the different point shapes mean. This is also quite easy to do, using the legend function, which can be used to put a box anywhere we like on the plot, displaying the point shapes we used together with descriptive text to tell us what each shape means. The R code used to add this legend is shown in Fig. 1.7.

The last example considered here adds two *reference lines* to the plot shown in Fig. 1.7. These lines are generated using the R function lm, which fits *linear regression models*, discussed in detail in Chapter 5. These models represent the simplest type of *predictive model*, a topic discussed more generally in Chapter 10 where other classes of predictive models are introduced. The basic idea is to construct a mathematical model that predicts a response variable from one or more other, related variables. In the whiteside data example considered here, these models predict the weekly average heating gas consumed as a linear function of the measured outside temperature. To obtain two reference lines, one model is fit for each of the data subsets defined by the two values of the Insul variable. Alternatively, we could obtain the same results by fitting a single linear regression model to the dataset, using both the Temp and Insul variables as predictors. This alternative approach is illustrated in Chapter 5 where this example is revisited.

```
plot(whiteside$Temp, whiteside$Gas, pch=c(6,16)[whiteside$Insul])
legend(x="topright",legend=c("Insul = Before","Insul = After"), pch=c(6,16))
```

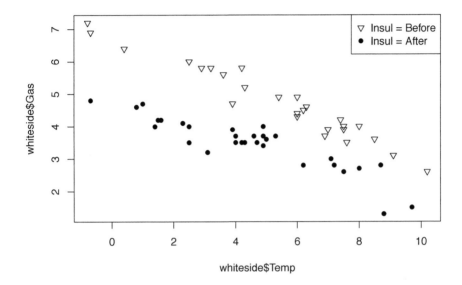

Figure 1.7: Scatterplot from Fig. 1.6 with a legend added to identify the two data subsets represented with different point shapes.

Fig. 1.8 is the same as Fig. 1.7, but with these reference lines added. As with the different plotting points, these lines are drawn with different line types. The R code listed at the top of Fig. 1.8 first re-generates the previous plot, then fits the two regression models just described, and finally draws in the lines determined by these two models. Specifically, the dashed "Before" line is obtained by fitting one model to only the "Before" points and the solid "After" line is obtained by fitting a second model to only the "After" points.

1.4 Organization of this book

This book is organized as two parts. The first focuses on analyzing data in an interactive R session, while the second introduces the fundamentals of R programming, emphasizing the development of custom functions since this is the aspect of programming that most R users find particularly useful. The second part also presents more advanced treatments of topics introduced in the first, including text analysis, a second look at exploratory data analysis, and an introduction to some more advanced aspects of predictive modeling.

```
plot(whiteside$Temp, whiteside$Gas, pch=c(6,16)[whiteside$Insul])
legend(x="topright",legend=c("Insul = Before","Insul = After"), pch=c(6,16))
Model1 <- lm(Gas ~ Temp, data = whiteside, subset = which(Insul == "Before"))
Model2 <- lm(Gas ~ Temp, data = whiteside, subset = which(Insul == "After"))
abline(Model1, lty=2)
abline(Model2)
```

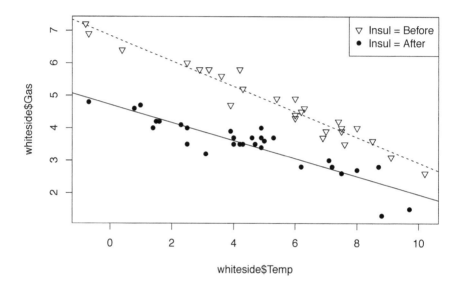

Figure 1.8: Scatterplot from Fig. 1.7 with linear regression lines added, representing the relationships between `Gas` and `Temp` for each data subset.

More specifically, the first part of this book consists of the first seven chapters, including this one. As noted, one of the great strengths of *R* is its variety of powerful data visualization procedures, and Chapter 2 provides a detailed introduction to several of these. This subject is introduced first because it provides those with little or no prior *R* experience a particularly useful set of tools that they can use right away. Specific topics include both basic plotting tools and some simple customizations that can make these plots much more effective. In fact, *R* supports several different graphics environments which, unfortunately, don't all play well together. The most important distinction is that between *base graphics*—the primary focus of Chapter 2—and the alternative *grid graphics* system, offering greater flexibility at the expense of being somewhat harder to use. While base graphics are used for most of the plots in this book, a number of important *R* packages use grid graphics, including the increasingly popular `ggplot2` package. As a consequence, some of the things we might want to do— e.g., add reference lines or put several different plots into a single array—can fail if we attempt to use base graphics constructs with plots generated by an R package based on grid graphics. For this reason, it is important to be aware of the different graphics systems available in R, even if we work primarily with base graphics as we do in this book. Since R supports color graphics, two sets of color figures are included in this book, the first collected as Chapter 2.8 and the second collected as Chapter 9.10 in the second part of the book.

Chapter 3 introduces the basic notions of exploratory data analysis (EDA), focusing on specific techniques and their implementation in R. Topics include descriptive statistics like the mean and standard deviation, essential graphical tools like scatterplots and histograms, an overview of data anomalies (including brief discussions of different types, why they are too important to ignore, and a few of the things we can do about them), techniques for assessing or visualizing relationships between variables, and some simple summaries that are useful in characterizing large datasets. This chapter is one of two devoted to EDA, the second being Chapter 9 in the second part of the book, which introduces some more advanced concepts and techniques.

The introductory R session example presented in Sec. 1.3 was based on the whiteside data frame, an internal R dataset included in the MASS package. One of the great conveniences in learning R is the fact that so many datasets are available as built-in data objects. Conversely, for R to be useful in real-world applications, it is obviously necessary to be able to bring the data we want to analyze into our interactive R session. This can be done in a number of different ways, and the focus of Chapter 4 is on the features available for bringing external data into our R session and writing it out to be available for other applications. This latter capability is crucial since, as emphasized in Sec. 1.2.3, everything within our active R session exists in RAM, which is volatile and disappears forever when we exit this session; to preserve our work, we need to save it to a file. Specific topics discussed in Chapter 4 include data file types, some of R's commands for managing external files (e.g., finding them, moving them, copying or deleting them), some of the built-in procedures R provides to help us find and import data from the Internet, and a brief introduction to the important topic of *databases*, the primary tool for storing and managing data in businesses and other large organizations.

Chapter 5 is the first of two chapters introducing the subject of *predictive modeling*, the other being Chapter 10 in the second part of the book. Predictive modeling is perhaps most simply described as the art of developing mathematical models—i.e., equations—that predict a *response variable* from one or more *covariates* or *predictor variables*. Applications of this idea are extremely widespread, ranging from the estimation of the probability that a college baseball player will go on to have a successful career in the major leagues described in Michael Lewis' popular book *Moneyball* [51], to the development of mathematical models for industrial process control to predict end-use properties that are difficult or impossible to measure directly from easily measured variables like temperatures and pressures. The simplest illustration of predictive modeling is the problem of fitting a straight line to the points in a two-dimensional scatterplot; both because it is relatively simple and because a number of important practical problems can be re-cast into exactly this form, Chapter 5 begins with a detailed treatment of this problem. From there, more general *linear regression* problems are discussed in detail, including the problem of *overfitting* and how to protect ourselves from it, the use of multiple predictors, the incorporation of categorical variables, how to include interactions and transformations in a linear regression model, and a brief introduction to robust techniques that are resistant to the potentially damaging effects of outliers.

When we analyze data, we are typically attempting to understand or predict something that is of interest to others, which means we need to show them what we have found. Chapter 6 is concerned with the art of crafting *data stories* to meet this need. Two key details are, first, that different audiences have different needs, and second, that most audiences want a summary of what we have done and found, and not a complete account with all details, including wrong turns and loose ends. The chapter concludes with three examples of moderate-length data stories that summarize what was analyzed and why, and what was found without going into all of the gory details of how we got there (some of these details are important for the readers of this book even if they don't belong in the data story; these details are covered in other chapters).

The second part of this book consists of Chapters 7 through 11, introducing the topics of R programming, the analysis of text data, second looks at exploratory data analysis and predictive modeling, and the challenges of organizing our work. Specifically, Chapter 7 introduces the topic of writing programs in R. Readers with programming experience in other languages may want to skip or skim the first part of this chapter, but the R-specific details should be useful to anyone without a lot of prior R programming experience. As noted in the Preface, this book assumes no prior programming experience, so this chapter starts simply and proceeds slowly. It begins with the question of why we should learn to program in R rather than just rely on canned procedures, and continues through essential details of both the structure of the language (e.g., data types like vectors, data frames, and lists; control structures like for loops and if statements; and functions in R), and the mechanics of developing programs (e.g., editing programs, the importance of comments, and the art of debugging). The chapter concludes with five programming examples, worked out in detail, based on the recognition that many of us learn much by studying and modifying code examples that are known to work.

Text data analysis requires specialized techniques, beyond those covered in most statistics and data analysis texts, which are designed to work with numerical or simple categorical variables. Most of this book is also concerned with these techniques, but Chapter 8 provides an introduction to the issues that arise in analyzing text data and some of the techniques developed to address them. One key issue is that, to serve as a basis for useful data analysis, our original text data must be converted into a relevant set of numbers, to which either general or highly text-specific quantitative analysis procedures may be applied. Typically, the analysis of text data involves first breaking it up into relevant chunks (e.g., words or short word sequences), which can then be counted, forming the basis for constructing specialized data structures like *term-document matrices*, to which various types of quantitative analysis procedures may then be applied. Many of the techniques required to do this type of analysis are provided by the *R* packages `tm` and `quanteda`, which are introduced and demonstrated in the discussion presented here. Another key issue in analyzing text data is the importance of *preprocessing* to address issues like inconsistencies in capitalization and punctuation, and the removal of numbers, special symbols, and non-informative *stopwords* like "a" or "the." Text analysis packages like `tm` and `quanteda` include functions to perform these operations, but many of them can also be handled using low-level string handling functions like `grep`, `gsub`, and `strsplit` that are available in base *R*. Both because these functions are often extremely useful adjuncts to specialized text analysis packages and because they represent an easy way of introducing some important text analysis concepts, these functions are also treated in some detail in Chapter 8. Also, these functions—along with a number of others in *R*—are based on *regular expressions*, which can be extremely useful but also extremely confusing to those who have not seen them before; Chapter 8 includes an introduction to regular expressions.

Chapter 9 provides a second look at exploratory data analysis, building on the ideas presented in Chapter 3 and providing more detailed discussions of some of the topics introduced there. For example, Chapter 3 introduces the idea of using random variables and probability distributions to model undertainy in data, along with some standard random variable characterizations like the mean and standard deviation. The basis for this discussion is the popular Gaussian distribution, but this distribution is only one of many and it is not always appropriate. Chapter 9 introduces some alternatives, with examples to show why they are sometimes necessary in practice. Other topics introduced in Chapter 9 include confidence intervals and statistical significance, association measures that summarize the relationship between variables of different types, multivariate outliers and their impact on standard association measures, and a number of useful graphical tools that build on these ideas. Since color greatly enhances the utility of some of these tools, the second group of color figures follows, as Chapter 9.10.

Following this second look at exploratory data analysis, Chapter 10 builds on the discussion of linear regression models presented in Chapter 5, introducing a range of extensions, including *logistic regression* for binary responses (e.g., the *Moneyball* problem: estimate the probability of having a successful major league career, given college baseball statistics), more general approaches to these *binary classification problems* like decision trees, and a gentle introduction to the increasingly popular arena of machine learning models like random forests and boosted trees. Because predictive modeling is a vast subject, the treatment presented here is by no means complete, but Chapters 5 and 10 should provide a useful introduction and serve as a practical starting point for those wishing to learn more.

Finally, Chapter 11 introduces the larger, broader issues of "managing stuff": data files, *R* code that we have developed, analysis results, and even the *R* packages we are using and their versions. Initially, this may not seem either very interesting or very important, but over time, our view is likely to change.

In particular, as we get further into a data analysis effort, the data sources we are working with change (e.g., we obtain newer data, better data, or simply additional data), our intermediate analysis results accumulate ("first, I looked at the relationship between Variable A and Variable B, which everybody said was critically important, but the results didn't seem to support that, so next I looked at the relationship between Variables A and C, which looked much more promising, and then somebody suggested I consider Variables D and E, ..."), and different people need different summaries of our results. Often, these components accumulate rapidly enough that it may take a significant amount of time and effort to dig up what we need to either explain exactly what we did before or to re-do our analysis with a "simple" modification ("Can you drop the records associated with the Florida stores from your analysis, and oh yeah, use the 2009 through 2012 data instead of the 2008 through 2013 data? Thanks."). The purpose of Chapter 11 is to introduce some simple ideas and tools available in *R* to help in dealing with these issues before our analytical life becomes complicated enough to make some of them extremely painful.

1.5   Exercises

1: Section 1.2.2 considered the `mammals` data frame from the `MASS` package, giving body weights and brain weights for 62 animals. Discussions in later chapters will consider the `Animals2` data frame from the `robustbase` package which gives the same characterizations for a slightly different set of animals. In both cases, the row names for these data frames identify these animals, and the objective of this exercise is to examine the differences between the animals characterized in these data frames:

   1a. The `rownames` function returns a vector of row names for a data frame, and the `intersect` function computes the intersection of two sets, returning a vector of their common elements. Using these functions, construct and display the vector `commonAnimals` of animal names common to both data frames. How many animals are included in this set?

   1b. The `setdiff` function returns a vector of elements contained in one set but not the other: `setdiff(A, B)` returns a vector of elements in set `A` that are not in set `B`. Use this function to display the animals present in `mammals` that are not present in `Animals2`.

   1c. Use the `setdiff` function to display the animals present in `Animals2` that are not present in `mammals`.

   1d. Can you give a simple characterization of these differences between these sets of animals?

2: Figure 1.1 in the text used the `qqPlot` function from the `car` package to show that the log of the `brain` variable (brain weights) from the `mammals` data frame in the `MASS` package was reasonably consistent with a Gaussian distribution. Generate the corresponding plot for the brain weights from the `Animals2` data frame from the `robustbase` package. Does the same conclusion hold for these brain weights?

3: As discussed at the end of Section 1.2.3, calling the `library` function with no arguments brings up a new window that displays a list of the *R* packages that have been previously installed and are thus available for our use by calling `library` again with one of these package names. Alternatively, the results returned by the `library` function when it is called without arguments can be assigned to an *R* data object. The purpose of this exercise is to explore the structure of this object:

  3a. Assign the return value from the `library()` call without arguments to the *R* object `libReturn`;

  3b. This *R* object is a named list: using the `str` function, determine how many elements this object has and the names of those elements;

  3c. One of these elements is a character array that provides the information normally displayed in the pop-up window: what are the names of the columns of this matrix, and how many rows does it have?

4: The beginning of Section 1.3 poses seven questions that are often useful to ask about a new dataset. The last three of these questions deal with our expectations and therefore cannot be answered by strictly computational methods, but the first four can be:

  4a. For the `cabbages` dataset from the `MASS` package, refer back to these questions and use the `str` function to answer the first three of them.

  4b. The combination of functions `length(which(is.na(x)))` returns the number of missing elements of the vector x. Use this combination to answer the fourth question: how many missing values does each variable in `cabbages` exhibit?

5: The generic `summary` function was introduced in Section 1.3, where it was applied to the `whiteside` data frame. While the results returned by this function do not directly address all of the first four preliminary exploration questions considered in Exercise 4, this function is extremely useful in cases where we do have missing data. One such example is the `Chile` data frame from the `car` package. Use this function to answer the following question: how many missing observations are associated with each variable in the `Chile` data frame?

6: As noted in the discussion in Section 1.2.2, the Gaussian distribution is often assumed as a reasonable approximation to describe how numerical variables are distributed over their ranges of possible values. This assumption is not always reasonable, but as illustrated in the lower plots in Figure 1.1, the `qqPlot` function from the `car` package can be used as an informal graphical test of the reasonableness of this assumption.

6a. Apply the `qqPlot` function to the `HeadWt` variable from the `cabbages` data frame: does the Gaussian assumption appear reasonable here?

6b. Does this assumption appear reasonable for the `VitC` variable?

7: The example presented in Section 1.3 used the `boxplot` function with the formula interface to compare the range of heating gas values (`Gas`) for the two different levels of the `Insul` variable. Use this function to answer the following two questions:

7a. The `Cult` variable exhibits two distinct values, representing different cabbage cultivars: does there appear to be a difference in cabbage head weights (`HeadWt`) between these cultivars?

7b. Does there appear to be a difference in vitamin C contents (`VitC`) between these cultivars?

8: One of the points emphasized throughout this book is the utility of *scatterplots*, i.e., plots of one variable against another. Using the `plot` function, generate a scatterplot of the vitamin C content (`VitC`) versus the head weight (`HeadWt`) from the `cabbages` dataset.

9: Another topic discussed in this book is *predictive modeling*, which uses mathematical models to predict one variable from another. The `lm` function was used to generate reference lines shown in Figure 1.8 for two subsets of the `whiteside` data from the `MASS` package. As a preview of the results discussed in Chapter 5, this problem asks you to use the `lm` function to build a model that predicts `VitC` from `HeadWt`. Refer back to the code included with Figure 1.8, noting that the `subset` argument is not needed here (i.e., you need only the formula expression and the `data` argument). Specifically:

9a. Use the `lm` function to build a model that predicts `VitC` from `HeadWt`, saving the result as `cabbageModel`.

9b. Apply the `summary` function to `cabbageModel` to obtain a detailed description of this predictive model. Don't worry for now about the details: the interpretation of these summary results will be discussed in Chapter 5.

10: Closely related to both scatterplots and linear regression analysis is the *product-moment correlation coefficient*, introduced in Chapter 9. This coefficient is a numerical measure of the tendency for the variations in one variable to track those of another variable: positive values indicate that increases in one variable are associated with increases in the other, while negative values indicate that increases in one variable are associated with decreases in the other. The correlation between x and y is computed using the `cor` function as `cor(x,y)`. Use this function to compute the correlation between `HeadWt` and `VitC` from the `cabbages` data frame: do these characteristics vary together or in opposite directions? Is this consistent with your results from Exercise 8?

Chapter 2

Graphics in R

It has been noted both that graphical data displays can be extremely useful in understanding what is in a dataset, and that one of R's strengths is its range of available graphical tools. Several of these tools were demonstrated in Chapter 1, but the focus there was on specific examples and what they can tell us. The focus of this chapter is on how to generate useful data displays, with the axes we want, the point sizes and shapes we want, the titles we want, and explanatory legends and other useful additions put where we want them.

2.1　Exploratory vs. explanatory graphics

In their book, Iliinsky and Steele [44] draw a distinction between *infograph-ics* and *data visualizations*, describing an infographic as an aesthetically rich, manually drawn representation of a specific data source, in contrast to a data visualization that is algorithmically drawn and "often aesthetically barren (i.e., data is not decorated)," but much more easily regenerated or modified, and much richer in data details. Here, we are concerned with data visualizations, which Iliinksy and Steele further divide into *exploratory graphics*, designed to help us understand what is in a dataset, and *explanatory graphics*, designed to convey our findings to others. Both exploratory and explanatory graphics are relevant here, since this book is concerned with both exploratory data analysis and conveying our results to others, which is the primary objective of Chapter 6 on data stories.

　　In describing exploratory visualizations, Iliinsky and Steele note that they are appropriate when you are attempting to understand what is in a large collection of data, and they offer two key observations [44, p. 7]. The first is reminiscent of the working definition of exploratory data analysis from Persi Diaconis [21] offered in Chapter 1:

> When you need to get a sense of what's inside your data set, translating it into a visual medium can help you quickly identify its features, including interesting curves, lines, trends, and anomalous outliers.

Their second observation concerns the level of detail that is appropriate to exploratory analysis:

> Exploration is generally best done at a high level of granularity. There may be a whole lot of noise in your data, but if you oversimplify or strip out too much information, you could end up missing something important.

In contrast, the purpose of explanatory visualizations is to convey the analyst's conclusions about what is in the data to others. Consequently, it is important in explanatory visualizations to remove or minimize data details that might obscure the message. Iliinsky and Steele make the point this way [44, p. 8]:

> Whoever your audience is, the story you are trying to tell (or the answer you are trying to share) is *known to you at the outset,* and therefore you can design to specifically accommodate and highlight that story. In other words, you'll need to make certain *editorial decisions* about which information stays in, and which is distracting or irrelevant and should come out. This is a process of selecting focused data that will support the story you are trying to tell.

The following two examples illustrate some of the key differences between exploratory and explanatory visualizations. Both are based on the `UScereal` data frame from the `MASS` package, which describes 11 characteristics of 65 breakfast cereals available for sale in the U.S., based mostly on information taken from the package label required by the U.S. Food and Drug Administration.

Fig. 2.1 represents a graphical data display that is best suited for exploration, constructed by applying the `plot` function to the entire data frame:

```
plot(UScereal, las = 2)
```

Specifically, this figure shows a plot array—a useful construct discussed further in Sec. 2.6—with one scatterplot for each pair of variables in the data frame. The diagonal elements of this array list the name of the variable appearing in the x-axis of all plots in that column, and the y-axis of all plots in that row. Since there are 11 variables in this data frame, the result is an array of 110 plots, making the result visually daunting at first glance. Further, because there are so many plots included in this array, each one is so small that it is impossible to see much detail in any individual plot. Nevertheless, this array represents a useful tool for preliminary data exploration because it allows us to quickly scan the plots to see whether any strong relationships appear to exist between any of the variable pairs. Here, there appear to be strong relationships between `fat` and `calories` (row 2, column 4 or vice versa: row 4, column 2), between `carbo` and `calories` (row 2, column 7 and vice versa), and between `potassium` and `fibre` (row 6, column 10 and vice versa). In addition, this display makes it clear that certain variables—e.g., `shelf` and `vitamins`—exhibit only a few distinct values. While this information can all be obtained using a combination of other displays

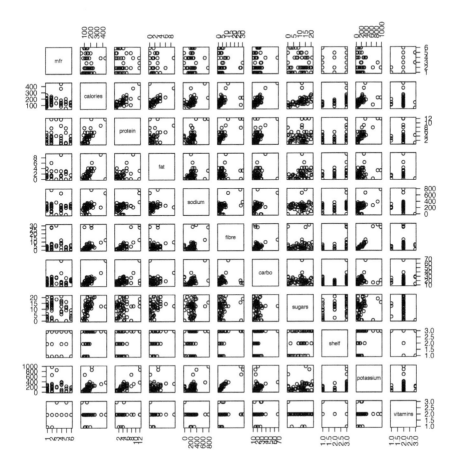

Figure 2.1: Array of pairwise scatterplots summarizing the UScereal data frame.

and/or nongraphical tools in *R*, this quick and simple plot array does provide a lot of useful preliminary information if we look at it carefully enough. That said, this plot array is *not* a good explanatory visualization because it contains far too much extraneous detail for any story we might wish to tell about any individual variable pair.

In contrast, Fig. 2.2 presents a much more detailed view of one of these scatterplots—that between the `calories` and `sugars` variables—augmented with a robust regression line emphasizing the general trend seen in most of this data, and with labels that explicitly identify two glaring outliers. Specifically, the dashed line in this plot represents the predictions of a robust linear regression model, generated using the `lmrob` function from the `robustbase` package discussed in Chapter 5. The key point here is that this dashed line highlights the trend our eye sees in the data if we ignore the two outlying points. These

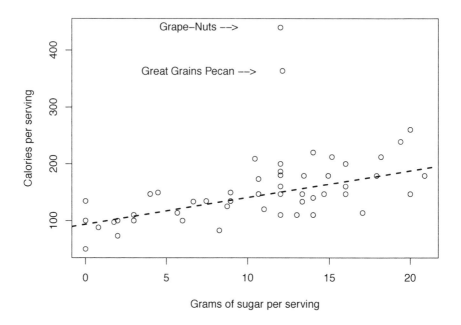

Figure 2.2: Annotated scatterplot showing the relationship between calories per serving and grams of sugar per serving from the UScereal data frame.

points correspond to the two cereals identified in the plot, representing much higher-calorie cereals than any of the others in the dataset. Fig. 2.2 tells a much more detailed story about the relationship between the variables `sugars` and `calories` from the `UScereal` data frame than we could infer from Fig. 2.1. Thus, Fig. 2.2 represents a much more effective way to describe or explain the relationship we have seen between `sugars` and `calories` to others.

2.2   Graphics systems in R

As noted, R supports many different graphical tools, but it is important to note that the underlying graphics systems on which these tools are built come in different flavors, and that tools built on different graphics systems generally don't play well together. The simplest of these systems is *base graphics*, described in Sec. 2.2.1 and used to create almost all of the graphical displays presented in this book. The other major graphics system in R is *grid graphics*, described in Sec. 2.2.2 and forming the basis for both *lattice graphics*, described in Sec. 2.2.3, and the `ggplot2` graphics package, described very briefly in Sec. 2.2.4.

Function	Object type	Nature of plot generated
`plot`	Many	Depends on the object type
`barplot`	Numeric	Bar plot (Sec. 2.5.2)
`boxplot`	Formula, numeric, or list	Boxplot summary (Chapter 3)
`hist`	Numeric	Histogram (Chapter 3)
`sunflowerplot`	Numeric + Numeric	Sunflower plot (Chapter 3)
`mosaicplot`	Formula or table	Mosaic plot (Chapter 3)
`symbols`	Multiple numeric	Bubbleplots, etc. (Sec. 2.5.3)

Table 2.1: A few of the more common base graphics functions.

2.2.1 Base graphics

The terms *base graphics* or *traditional graphics* [57] refer to the graphics system originally built into the *R* language. Because it is typically the default graphics system and is, in many respects, the easiest to learn, it is the primary graphics system used in this book, with exceptions noted when they arise. Probably the most common base graphics function is `plot`, described in detail in Sec. 2.3. As discussed in Chapter 1, this is a *generic function*, whose results depend on the type of *R* object we ask it to plot, an important concept discussed further in Sec. 2.3.2. Also, the detailed appearance of base graphics displays is partially controlled by a collection of 72 graphics parameters discussed further in Sec. 2.3.3. In addition, these displays can be further customized by using what Murrell calls *low-level* plotting functions [57] that add lines, points, text, and other details to an existing plot, discussed in Sec. 2.4.

Table 2.1 lists a few of the more common base graphics functions, along with the types of *R* objects they can accept as plot data arguments and the types of plot they generate. As noted, the basic `plot` function listed there is generic, accepting many different *R* object types and generating many different types of plots as a result; examples illustrating the range of plots possible with this function are presented in Sec. 2.3. The other functions listed in Table 2.1 are less flexible in the range of object types they accept, but as examples presented in Sec. 2.5 and in Chapter 3 illustrate, these functions can be extremely useful in generating both exploratory and explanatory data visualizations.

2.2.2 Grid graphics

As Murrell notes [57], almost all *R* graphics functions are ultimately based on the *graphics engine* represented by the `grDevices` package, which supports the lowest-level interface between *R* and the devices that display our graphics, handling details like fonts, colors, and display formats. Traditional or base graphics functions are based on the `graphics` package, while *grid graphics* is based on the `grid` package (which Murrell developed). Murrell gives the following description of this package [57, p. 18]:

> The `grid` package provides a separate set of basic tools. It does not provide functions for drawing complete plots, so it is not often used directly to produce statistical plots. It is more common to use one of the graphics packages that are built on top of `grid`, especially either the `lattice` package or the `ggplot2` package.

In addition to these large graphics systems—described briefly in the following sections—it is important to note that certain data analysis packages in *R* are also built on grid graphics. A specific example is the `vcd` package, which provides a number of extremely useful visualization tools for categorical data. If we want to modify these plots—e.g., add annotations, construct multiple plot arrays, etc.—it is necessary to use grid graphics to do this.

While a detailed introduction to grid graphics is beyond the scope of this book, it is important to be aware of its existence, its general incompatibility with base graphics, and its basic structure. A key component of grid graphics is the *viewport*, which Murrell defines as "a power facility for defining regions" [57, p. 174]. To construct a graphical display using the `grid` package, the basic steps are these:

- create a viewport;

- put a collection of graphic objects in the viewport;

- render the viewport to obtain a graphical display.

A useful introduction to the `grid` package, with a number of very good examples, is Murrell's package vignette "grid Graphics." The package is also supported by about a dozen other vignettes, giving more detailed discussions of different aspects of working with grid graphics, and additional details are given in Part II of his book [57].

Finally, it is important to note two other points. First, a potential source of confusion is the existence of the `grid` *function*, which is part of the *base graphics system* and *not* related to the `grid` package. The base graphics function `grid` adds a rectangular grid to an existing base graphics plot; refer to the results from `help(grid)` for details. The second important point is that Murrell has also developed the `gridBase` package which allows both grid and base graphics to be used together. For an introduction to what is possible and some preliminary ideas on how to use this package, refer to the `gridBase` package vignette, "Integrating Grid Graphics Output with Base Graphics Output."

2.2.3   Lattice graphics

As noted, one of the complete graphics systems in *R* that is based on the `grid` package is *lattice graphics*, implemented in the `lattice` package. This package provides an alternative implementation of many of the standard plotting functions available in base graphics, including scatterplots, bar charts, boxplots, histograms, and QQ-plots. Two of the primary advantages of this package over base graphics are, first, that many prefer the lattice default options (e.g., colors, point shapes, spacing, and labels) over the corresponding base defaults [57, p. 123], and, second, that lattice graphics provides simple implementations of certain additional features. One example, illustrated in Fig. 2.3, is the *multipanel conditioning plot*, which shows how the relationship between two variables depends on a third categorical *conditioning variable*. Specifically, Fig. 2.3 shows six interrelated scatterplots, each describing the relationship between the variables `Horsepower` and `MPG.city` from the `Cars93` data frame in the `MASS` package, but only for a single level of the categorical conditioning variable `Cylinders`. The code to generate this plot is extremely simple:

```
library(lattice)
xyplot(MPG.city ~ Horsepower | Cylinders, data = Cars93)
```

The first line here loads the `lattice` package, which makes the scatterplot function `xyplot` available for use. The second line applies this function, which supports R's standard formula interface. Specifically, the first argument represents a three-component formula: the variable `MPG.city`, appearing to the left of the ~ symbol, represents the response variable to be plotted on the y-axes of all plots; the variable `Horsepower`, appearing to the right of this symbol but to the left of the symbol |, defines the x-axis in all of these plots; and the variable `Cylinders` that appears to the right of this symbol is the categorical conditioning variable. Thus, the `xyplot` function constructs one scatterplot of `MPG.city` versus `Horsepower` for each distinct value of `Cylinders` and displays them in the format shown in Fig. 2.3. The `data` parameter in the above function call specifies the data frame containing these variables.

Another useful feature of the `lattice` package is the `group` argument, which allows different groups within a dataset (e.g., distinct `Cylinders` values in the previous example) to be represented by different point shapes in a single scatterplot. In addition, a legend is automatically generated that identifies the groups and their associated plotting symbols. Nevertheless, these capabilities come at a price [57, p. 135]:

> One advantage of the lattice graphics system is that it can produce extremely sophisticated plots from relatively simple expressions, especially with its multipanel conditioning feature. However, the cost of this is that the task of adding simple annotations to a lattice plot, such as adding extra lines or text, is more complex compared to the same task in traditional graphics.

It is for this reason that this book focuses on traditional (i.e., base) graphics.

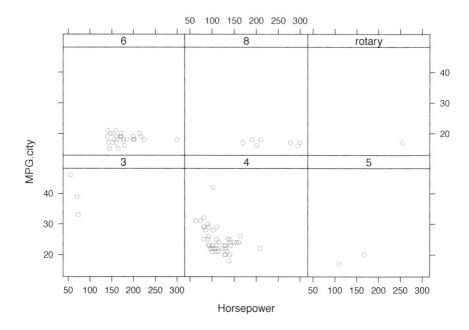

Figure 2.3: Lattice conditioning plot, showing the relationship between Horse-power and MPG.city, conditional on Cylinders, all from the Cars93 data frame.

2.2.4   The ggplot2 package

As noted, another *R* graphics system based on the `grid` package is `ggplot2` by Hadley Wickham, similar in some respects to lattice graphics, but with a fundamentally different basis and structure. Specifically, `ggplot2` is based on the *grammar of graphics*, a systematic approach to constructing graphical objects described in a book of the same title by Wilkinson [76]. Like lattice graphics, many of the default options for plots generated by `ggplot2` are based on research in human perception and are therefore preferred by many to the base graphics defaults [57, p. 21]. Also like lattice graphics, `ggplot2` provides better support for multipanel conditioning plots, and because this package is highly extensible and has become extremely popular, many extensions are available in the form of other *R* packages that provide significant additional capabilities beyond the `ggplot2` package itself. As with lattice graphics, however, one price paid for this additional flexibility is a steeper learning curve; another is the greater complexity of generating multiple plot arrays, which requires explicitly working with viewports in grid graphics.

Both because the learning curve is steeper and because some extremely use-ful tools like the `qqPlot` function from the `car` package, based on traditional graphics, are not available as built-in functions in `ggplot2`, this book uses base graphics instead of the more flexible `ggplot2` package. (Note that the basic

QQ-plot is available in `ggplot2` via the `qq` stat, but this option only displays the points in the plot: addition of the reference line, confidence intervals, and options for non-Gaussian distributions is all possible, but requires additional programming; in the `qqPlot` function, this is all included.) For a detailed introduction to the `ggplot2` package, good references are Hadley Wickham's book [73] and the chapter on grammar of graphics in Murrell's book [57, Ch. 5].

2.3 The plot function

Probably the most commonly used base graphics function is `plot`, which is a *generic* function, meaning that the nature of the plot it generates depends on the type of R object we pass to it. Most of the plots in Chapter 1 were generated using the `plot` function, usually augmented with some of the added details described in Sec. 2.4. Sec. 2.3.1 presents a collection of examples that illustrate the range of capabilities of the `plot` function, and Sec. 2.3.2 presents a brief but broader discussion of the concept of generic functions, giving some typical examples in R and their relationship to *S3 objects*. The essential idea is that an S3 object has certain defining characteristics, and generic functions with methods defined for a specific S3 object class can exploit those characteristics to return class-specific results. In the case of the generic `plot` function, this means that a command like "plot(x, y)" can generate a scatterplot if x and y are both numeric, a boxplot summary if x is categorical and y is numeric, or a mosaic plot if both variables are factors.

2.3.1 The flexibility of the plot function

The flexibility of the `plot` function was illustrated in the sample R session presented in Chapter 1, where results were shown for this function applied to a complete data frame, a numeric vector, a factor, and a pair of numeric variables: the same function returned an array of scatterplots, a plot of the numerical values in their order of appearance, a bar chart, and a scatterplot. In addition, this sample R session began by using the `boxplot` function to generate a boxplot summary of heating gas consumption both before and after the installation; this boxplot summary can also be generated by using the `plot` function:

```
plot(whiteside$Insul, whiteside$Gas)
```

Many of the modeling functions in R return an object of the type discussed in Sec. 2.3.2 (i.e., an S3 object), and special plot methods have frequently been developed for these objects. Fig. 2.4 provides an example, based on the class of *decision tree models* discussed in Chapter 10. This model predicts the average value of the heating gas consumption `Gas` in the `whiteside` data frame from the values of the other two variables, `Temp` and `Insul`. It is easily generated using the `rpart` package, with the following code:

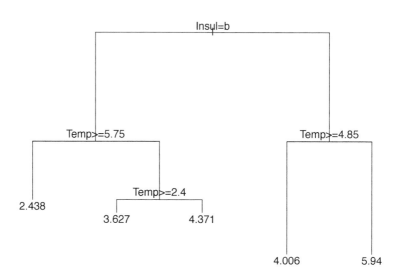

Figure 2.4: Plot of an rpart model built from the whiteside data frame.

```
library(rpart)
rpartModel <- rpart(Gas ~ ., data = whiteside)
```

For now, don't worry about the details of this model, its interpretation, or the **rpart** function used to obtain it: this example is discussed in detail in Chapter 10. The key point here is that the **rpart** function returns an S3 object of class "rpart," which the **plot** function has a method to support. Thus, we can execute the command **plot(rpartModel)** to obtain the plot shown in Fig. 2.4. Actually, the **plot** function only displays the tree structure of the model, without labels; to obtain the labels, we must also use **text**, another generic function with a method for rpart objects:

```
plot(rpartModel)
text(rpartModel)
```

The second model-based example is shown in Fig. 2.5 and it belongs to the class of *MOB models*, also discussed in Chapter 10. Like the **rpart** model just described, this model has a tree-based structure, but rather than generating a single numerical predicted value to each terminal node of the tree (i.e., each "leaf"), each terminal node contains a linear regression model that generates

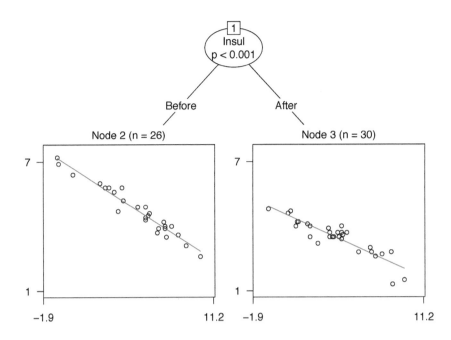

Figure 2.5: Plot of an MOB model built from the whiteside data frame.

predictions from other covariates. These models may be fit using the `lmtree` function from the `partykit` package, using very similar code to that used to generate the `rpart` model discussed above:

```
library(partykit)
MOBmodel <- lmtree(Gas ~ Temp | Insul, data = whiteside)
```

The resulting model object, `MOBmodel`, is an S3 object of class "lmtree" and when the `plot` function is applied to this object, we obtain the result shown in Fig. 2.5. In this case, the model is created using a three-part formula structure: the variable `Gas` appears to the left of the ~ symbol to indicate it is the response variable to be predicted; the variable `Temp` appears between this symbol and the | symbol to indicate it is the covariate used to predict the response variable in the models that appear at the terminal nodes of the tree; and the variable `Insul` that appears to the right of the | symbol is the partitioning variable used to build the tree. Since this partitioning variable is binary in this example, the resulting tree has two nodes, one corresponding to the "Before" data and the other corresponding to the "After" data. The structure of this model is clear from the plot: all records are assigned to one of these nodes, and a separate linear regression model that predicts `Gas` from `Temp` is built for each node. In favorable cases—like this one—the MOB model class can be extremely effective in finding and exploiting strong heterogeneity in the underlying data, a point discussed further in Chapter 10 where this example is revisited.

The primary point of this discussion has been to illustrate the vast range of graphical results that can be generated using the generic `plot` function. The following section introduces the concepts of S3 object classes and their associated methods, the concepts that make this behavior possible. In fact, it is easy to define our own S3 object classes and construct methods for generic functions like `plot` or `summary` that make them generate specialized results for our object classes. This idea is discussed in detail in Chapter 7.

2.3.2   S3 classes and generic functions

The ability of functions like `plot` to behave very differently depending on the *type* of the *R* object we give it is a consequence of the language's *object-oriented* structure. Many programming languages are object-oriented, and in his book *Advanced R*, Hadley Wickham offers the following useful description of the minimal requirements for an object-oriented language [74, p. 99]:

> Central to any object-oriented system are the concepts of class and method. A **c**lass defines the behavior of **o**bjects by describing their attributes and their relationship to other classes. The class is also used when selecting **m**ethods, functions that behave differently depending on the class of their input. Classes are usually organized in a hierarchy: if a method does not exist for a child, then the parent's method is used instead; the child **i**nherits behavior from the parent.

Wickham further notes that *R* has three distinct object-oriented systems, based on three different object types: *S3 objects*, *S4 objects*, and *reference classes*. Since the S3 system is both the simplest of these three and the one we encounter the most often, it is useful to know something about S3 classes, particularly in understanding the behavior of *generic functions* like `plot`. Also, it is easy—and sometimes extremely useful—to define our own S3 objects and their associated methods, an important idea discussed further in Chapter 7.

The key feature of the S3 system is that methods belong to functions—specifically, generic functions—which is different from the object-oriented systems encountered in languages like *Python*, *Ruby*, or *Java*, where methods belong to objects. The newest object-oriented system in *R*—reference classes—has this more traditional structure, making it very different from the S3 system considered here. The S4 system is also significantly different from the S3 system—methods still belong to generic functions, but the structure is more formal; like the `lattice` and `ggplot2` graphics systems, the S4 system has greater flexibility, but a correspondingly steeper learning curve. Consequently, only S3 objects and their associated methods are considered here.

To see how this system works, consider the decision tree model, `rpartModel`, discussed in Sec. 2.3.1, built using the `rpart` package. The `class` function shows that the result is an S3 object of class "rpart":

```
class(rpartModel)

## [1] "rpart"
```

Fig. 2.4 was generated from this S3 object with the generic functions `plot` and `text`. In each case, the behavior of the function depends on the class of the object: for example, `text` is the same generic function as that used in Sec. 2.4.2 below to add annotation to scatterplots, but the two *methods* used with this generic function are different. Specifically, if we type "text" without trailing parentheses, we are asking R to display the function's code, which is:

```
text

## function (x, ...)
## UseMethod("text")
## <bytecode: 0x000000001599f1f8>
## <environment: namespace:graphics>
```

This result tells us that `text` is a function, taking one required argument (x) and allowing an unspecified number of optional arguments, via "..." (see Chapter 7 for a detailed discussion). The second line of this result tells us that this function is generic, having different methods associated with different object types. To see a list of these methods, use the `methods` function with the name of the generic function, i.e.:

```
methods("text")

## [1] text.default  text.formula* text.rpart*
## see '?methods' for accessing help and source code
```

In the examples presented in Sec. 2.4.2, the `text.default` method is used to add text to a scatterplot, while the `text.rpart` method was used to add the text labels to the `rpart` plot. The methods not marked with the asterisk (*) can be displayed directly as we did with the `text` function above, while those marked with the asterisk cannot; also, note that additional information is available if we type "?methods" and follow the instructions given there.

It is also possible to ask what methods are available for a given S3 object class. Again, we use the `class` function, but now we specify the desired class:

```
methods(class = "rpart")

##  [1] as.party    labels      meanvar     model.frame plot
##  [6] post        predict     print       prune       residuals
## [11] summary     text
## see '?methods' for accessing help and source code
```

We can see from this result that—based on the packages we currently have loaded in our R session—we have 12 methods available for S3 objects of class "rpart," including both the `plot` and `text` methods used above.

Finally, it is worth noting that some generic functions have *many* associated methods, and this number depends on what packages are loaded into our *R* session (specifically, many packages define new S3 objects and methods). For example, in the *R* environment used to develop this book, the `plot` function has 133 methods, and the `summary` function has 156.

2.3.3   Optional parameters for base graphics

It was noted earlier that there are 72 optional base graphics parameters that affect many of the base graphics plot functions. These parameters are set by the `par` function, which can also be called to return a named list with the current values for these parameters. The names are:

```
names(par())
```

```
##   [1] "xlog"      "ylog"      "adj"       "ann"       "ask"
##   [6] "bg"        "bty"       "cex"       "cex.axis"  "cex.lab"
##  [11] "cex.main"  "cex.sub"   "cin"       "col"       "col.axis"
##  [16] "col.lab"   "col.main"  "col.sub"   "cra"       "crt"
##  [21] "csi"       "cxy"       "din"       "err"       "family"
##  [26] "fg"        "fig"       "fin"       "font"      "font.axis"
##  [31] "font.lab"  "font.main" "font.sub"  "lab"       "las"
##  [36] "lend"      "lheight"   "ljoin"     "lmitre"    "lty"
##  [41] "lwd"       "mai"       "mar"       "mex"       "mfcol"
##  [46] "mfg"       "mfrow"     "mgp"       "mkh"       "new"
##  [51] "oma"       "omd"       "omi"       "page"      "pch"
##  [56] "pin"       "plt"       "ps"        "pty"       "smo"
##  [61] "srt"       "tck"       "tcl"       "usr"       "xaxp"
##  [66] "xaxs"      "xaxt"      "xpd"       "yaxp"      "yaxs"
##  [71] "yaxt"      "ylbias"
```

Detailed descriptions of these parameters are available via the `help(par)` command, which notes that some of them are *read-only*, meaning that their values are fixed and cannot be modified (an example is `cin`, the default character size in inches). The following discussion does not attempt to discuss all of these parameters, only a few that are particularly useful. It is also worth noting that some of these parameters can be set in calls to certain base graphics functions (e.g., `plot`), while others can only be set through a call to the `par` function.

One of the most useful of the graphics parameters set by `par` is `mfrow`, a two-dimensional vector that sets up an array of plots; discussion of this parameter is deferred to Sec. 2.6 where the generation of plot arrays is covered in detail.

Several of these parameters come in closely related groups. One is the "cex-family" that specifies the extent to which text and symbols should be magnified relative to their default size. These parameters include:

- `cex` specifies the values for text and plotting symbols in the next plot generated, serving as a base for all of the other parameters in this group;

- `cex.axis` specifies the scaling of the axis annotations, relative to `cex`;

- `cex.lab` specifies the scaling of the axis labels, relative to `cex`;

- `cex.main` specifies the scaling of the main plot title, relative to `cex`;

- `cex.sub` specifies the scaling of the plot subtitle, relative to `cex`.

It is important to note that some functions, like `points` and `text`, allow the `cex` parameter to be specified in the function call as a *vector*, allowing individual points in the plot to have different sizes.

Two other, similarly structured parameter families are the "col-family" that specifies colors for points, lines, and text that is discussed in detail in Sec. 2.7.2, and the "font-family," discussed next. This family specifies font options for text appearing in plots, in axis notations, in labels, and in titles:

- `font` specifies one of the following four font types for text in the plot:

  - `font` = 1 specifies plain text;
  - `font` = 2 specifies bold face;
  - `font` = 3 specifies italic;
  - `font` = 4 specifies bold italic.

- `font.axis` specifies the font to be used for axis notations;

- `font.lab` specifies the font to be used for axis labels;

- `font.main` specifies the font to be used for the main plot title;

- `font.sub` specifies the font to be used for the plot subtitle.

The use of these optional parameters is illustrated in Sec. 2.4.2, as is the `adj` parameter, which specifies the *justification* of the text. Specifically, `adj` is a numerical parameter, whose value typically lies between 0 (corresponding to left-justification) and 1 (corresponding to right-justification), with the default value 0.5 (corresponding to centered text). Another useful text parameter is `srt`, which specifies the rotation of the text string in degrees: the default value of 0 corresponds to standard horizontal positioning of the text, while 90 rotates it 90 degrees so it reads from bottom to top, and a value of 180 rotates it 180 degrees, making it appear upside down.

The `lty` and `lwd` parameters take integer values and specify the *line type* and *line width*, respectively. Both of these parameters are illustrated in Sec. 2.4.1, as is the `pch` parameter, which can be used to specify point shapes in base graphics plots in a variety of different ways. The `las` parameter is used to specify the orientation of axis labels, and it is discussed further in Sec. 2.4.4.

The `ask` parameter is a logical flag (the default value is **FALSE**) that specifies whether the graphics system should halt and prompt the user for a response before displaying the next plot. This option is useful if we want to look through a sequence of individual plots—e.g., if we create a loop that generates a number of scatterplots between different variables and we want to have time to carefully look at each one before going on to the next. This option is commonly set by the built-in *R* examples that show many plots, and it is important to be aware of this use because the `ask` option is not always reset correctly after these example plots have all been displayed. To reset the graphics system to display plot results immediately, it is then necessary to execute the following code:

```
par(ask = FALSE)
```

*Finally, it is important to note that some of these parameters can be set as passing parameters when we call base graphics functions like **plot**, while others cannot.* As a specific example, the `las` parameter is ignored—without warning—if we pass it to the `plot` function when creating a mosaic plot; there, we must use the `par` command to set this parameter globally *before* we call the `plot` function. Then, if we want to return to the default option, we must explicitly call `par(las = 0)` after creating the mosaic plot.

2.4 Adding details to plots

Using only their default options, the basic plot functions in *R* generally provide useful displays, especially for exploratory analysis purposes, but they typically lack important details that we may want to add, particularly if we plan to share the results with others. Much improvement is possible using the base graphics parameters described in Sec. 2.3.3, but we often want to add other details that cannot be obtained this way. Specific examples include trend lines, text on the plot itself to draw the viewer's attention to key details, or legends to succinctly describe what the plot is showing us. The following sections describe some simple ways of accomplishing these objectives.

2.4.1 Adding points and lines to a scatterplot

This section and the next two describe the *R* base graphics options used to construct plots like the explanatory visualization of the relationship between `sugars` and `calories` from the `UScereal` data frame shown in Fig. 2.2. The starting point for these examples is Fig. 2.6, obtained using the `plot` function with only default options, which represents each data point as an open circle, with axis labels constructed from the names of the variables passed to the `plot` function. Subsequent examples will show, first, how to change these axis labels and the point shapes and sizes, and, second, how to add reference lines. Sec. 2.4.2 then shows how to add text to this plot, annotating the outliers as done in Fig. 2.2, and Sec. 2.4.3 will show how to add a legend.

Fig. 2.7 shows the results of adding axis labels to Fig. 2.2, varying the point shapes and sizes, and adding two linear regression reference lines that will be discussed in detail in Chapter 5. This plot was generated using the code listed below. The first line sets the `mfrow` parameter discussed in Sec. 2.6.1 so that only a single plot is generated in the plot window; this is not necessary in general, but the `mfrow` value persists, so it is necessary to do this if the last plots generated were in a multiple plot array (see Sec. 2.6.1 for a discussion of this parameter).

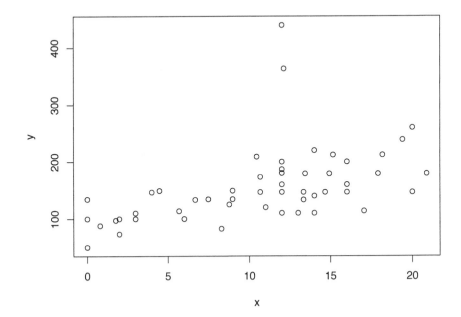

Figure 2.6: Default plot of `calories` vs. `sugars` from the `UScereal` data frame.

The following two lines define the variables x and y for convenience, since they are re-used in many places in the code that follows:

```
par(mfrow=c(1,1))
x <- UScereal$sugars
y <- UScereal$calories
```

The next line invokes the `plot` function, specifying three optional arguments:

- `xlab` is a character string specifying the text for the x-axis label;
- `ylab` is a character string specifying the text for the y-axis label;
- `type = "n"` specifies that the basic plot is *constructed, but not displayed.*

This last option is useful in cases like this one where we wish to display different subsets of the data using different point sizes or shapes. Specifically, calling the `plot` function with `type = "n"` sets up the basic framework for the plot, including limits of the x- and y-axes, label names, and any other options specified in the `plot` command. Here, the code is:

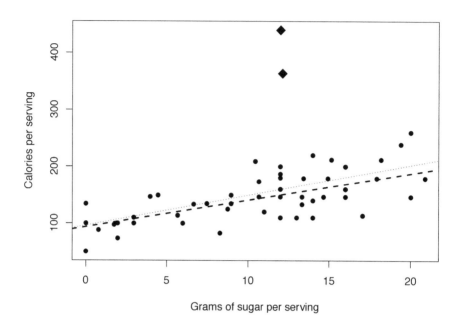

Figure 2.7: Enhanced scatterplot with different point sizes and shapes, and two added reference lines.

```
plot(x, y, xlab = "Grams of sugar per serving",
       ylab = "Calories per serving", type = "n")
```

The next three lines first construct a pointer to the outliers in the `calories` variable (`index`), and then add points to the plot using the `points` function:

```
index <- which(y > 300)
points(x[-index], y[-index], pch = 16)
points(x[index], y[index], pch = 18, cex = 2)
```

The `points` function behaves much like `plot`, except that it adds points to an existing plot instead of creating a new plot. In the first case here, the point shape is specified to be a solid circle via the `pch = 16` optional parameter setting, and the points added to the plot are the *non-outliers*, selected as `x[-index]` and `y[-index]`; that is, the `-index` specification keeps all points *except* those selected by `index`. The second call to the `points` function adds the outlying points, selected as `x[index]` and `y[index]`, drawing them as solid diamonds (via the specification `pch = 18`) that are twice as large as the other points (via the specification `cex = 2`).

The final four lines of code add the two reference lines shown in Fig. 2.7:

```
olsModel <- lm(y ~ x)
abline(olsModel, lty = 3)
robustModel <- lmrob(y ~ x)
abline(robustModel, lty = 2, lwd = 2)
```

Specifically, the `lm` function is first called to construct a linear regression model (`olsModel`) via the method of ordinary least squares (refer to Chapter 5 for a discussion of this, most popular, linear regression modeling approach), and the function `abline` is then called to display the predictions of this model as the thin dotted line seen in the plot. Lines can be added to an existing plot either with the `lines` function analogous to the `points` function, or with the `abline` function used here. This function is typically called with an intercept parameter `a` and a slope parameter `b` that specify a line (see Chapter 5 for a further discusson of how lines are represented), but this function will also accept an S3 object of class `lm` like `olsModel`. The other `abline` argument is `lty`, which specifies the line type. The default line type is `lty = 1`, giving a solid line, while `lty = 3` specifies a dotted line.

Finally, the last two lines of code listed above create and display a second linear regression model, this one using the alternative linear model fitting procedure `lmrob` from the `robustbase` package; this function is an outlier-resistant alternative to ordinary least squares regression, also discussed in Chapter 5. Specifying `lty = 2` gives a dashed line, while specifying `lwd = 2` makes this line twice as thick as normal.

Before leaving this discussion, it is important to return to the last figure presented in the sample R session in Chapter 1 (i.e., to Fig. 1.6 in Sec. 1.3). Recall that this example used the `plot` function to construct a scatterplot of `Gas` vs. `Temp` data from the `whiteside` data frame, with different plotting symbols used for the "Before" and "After" data, determined by the `Insul` variable. The specific R code used to generate this plot was:

```
plot(whiteside$Temp, whiteside$Gas, pch=c(6,16)[whiteside$Insul])
```

The novel part of this code is the specification of the `pch` parameter, which works as follows. First, note that `c(6, 16)` defines a two-dimensional vector with components 6 and 16. The following code fragment—`[whiteside$Insul]`—converts the factor variable `Insul` into its numerical representation, giving a numerical vector of the same length as `Insul`, but having the values 1 or 2, indexing the two possible values of `Insul`. This numeric vector is then used as an index into the vector $(6, 16)$, selecting the first value when `Insul` takes its first value ("Before") and selecting the second value when `Insul` takes its second value ("After"). Thus, the final plot generated uses `pch = 6` for the "Before" data points, giving a downward-point open triangle, and `pch = 16` for the "After" points, giving a solid circle. More generally, note that the `pch` parameter will accept a vector argument of the same length as the x- and y-vectors used to construct the scatterplot. Point sizes and colors can also be specified this way using the `cex` and `col` parameters.

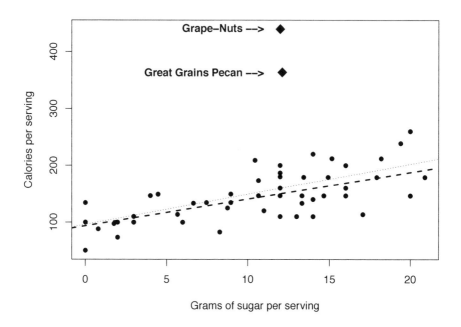

Figure 2.8: Further enhanced scatterplot with outliers labelled.

2.4.2 Adding text to a plot

Fig. 2.8 is the same plot as Fig. 2.7, except that the `text` function has been used to add identifying labels to the two outliers. Like the `points` and `lines` functions, `text` adds details to the current plot. Although this function supports a number of other options (see `help(text)` for details), the basic three are: (1) the x-position of the text, (2) the y-position of the text, and (3) the text to be displayed. The left-to-right alignment of the text is determined by the `adj` parameter: by default, this parameter has the value 0.5, which causes the text to be centered. In this case, the x-position in the `text` function call specifies the location of the center of the text; the most useful alternatives are `adj = 0`, which causes the text to be left-justified (i.e., the x-position specified in the function call defines the left end of the text string in the plot), and `adj = 1`, which causes the text to be right-justified (i.e., the x-position specified in the function call defines the right end of the text string). The code used to generate this plot is identical to that used to generate the previous plot, except that these two lines were added at the end:

```
pointLabels <- paste(rownames(UScereal)[index], "-->")
text(11, y[index], pointLabels, adj = 1, font = 2)
```

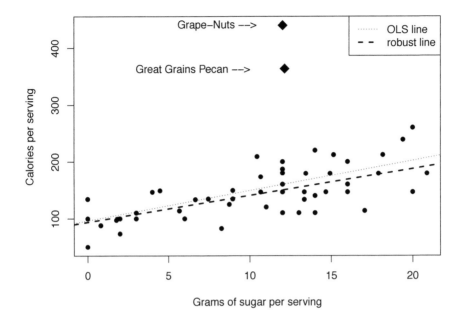

Figure 2.9: Previous scatterplot with legend added to identify reference lines.

Here, the `paste` function, discussed further in Chapter 8, has been used to append the trailing arrow to the names of the two outlying cereals, given by the row names of the `UScereal` data frame. The result is a character vector with two elements, and the `text` function adds these two text strings to the plot, with the right end of each text string at position $x = 11$ (note the use of `adj = 1` here, modifying the `adj` parameter for this plot only), and the y-axis positions corresponding to the outlying values of `calories` for these two outliers. Also, note the use of `font = 2` here to make this text appear in boldface.

This example does not illustrate the full range of optional parameters available with the `text` function. Particularly useful parameters include `cex`, which scales the size of the text string added to the plot, relative to the default size, and `srt`, the angle in degrees to rotate the text string around the position specified by the `adj` parameter. For further details, refer to the results of the `help(text)` and `example(text)` commands.

2.4.3   Adding a legend to a plot

As a final embellishment of the previous examples, Fig. 2.9 adds a *legend* that tells us something about the two lines that appear in the plot. This is accomplished with the `legend` function, which puts a boxed explanatory text display at a specified location in the current plot. In its simplest form, this function can be used like the `text` function, specifying x- and y-positions and the text string to be added, but the `legend` function has many more optional parameters, allowing considerable flexibility in what appears in the legend box. The legend in Fig. 2.9 was added to the plot in Fig. 2.8 with this additional R code:

```
legend(x = "topright", lty = c(3,2), lwd = c(1,2),
           legend = c("OLS line", "robust line"))
```

Here, the x- and y-positions are specified by the single text string "topright" for the x parameter, which places the legend box and its contents in the top right corner of the plot; other possible text strings accepted for this parameter are "bottomright," "bottom," "bottomleft," "left," "right," "center," "topleft," and "top," which have analogous effects. The text to be included in the legend box is defined by the `legend` parameter: here, this is a two-component character vector, so two lines of text are included in the legend. In addition, the `lty` and `lwd` parameters are specified as two-element numerical vectors, causing a short line segment of type specified by the `lty` element and width specified by the `lwd` element to be included to the left of the text. Similarly, points could be placed before the legend text using `pch` and related parameters.

2.4.4   Customizing axes

Considerable flexibility is available in base R graphics in customizing the axes. We have already seen how axis labels are easily specified via the `xlab` and `ylab` parameters, and how the limits of the x- and y-axes can be set using the `xlim` and `ylim` parameters. A number of the other base graphics parameters that can be set via the `par` function can also be used to customize the axes in a plot. As noted in Sec. 2.3.3, one is the `las` parameter, used to specify the orientation of the axis labels. This parameter accepts these integer values:

- `las = 0`—labels are displayed parallel to the axes (default option);

- `las = 1`—labels are always horizontal;

- `las = 2`—labels are always perpendicular to the axes;

- `las = 3`—labels are always vertical.

Since horizontal labels are easier to read, some journals require this label orientation (i.e., `las = 1`). Conversely, long horizontal labels on the x-axis of a plot will often "collide," causing the R graphics system to omit some of them, displaying only a legible subset. Specifying `las = 2` or `las = 3` can be used to overcome this problem, since these options force the x-axis labels to be vertical. In the case of very long x-axis text labels, it may also be necessary to specify the parameter `cex.lab` to have a value smaller than 1 to shrink these labels enough to fit within the graphical display area.

It is also possible to specify your own axes. This is done in two steps:

1. Execute the desired base graphics function with `axes = FALSE`;

2. Use the `axis` function to specify your own axes.

Specifying `axes = FALSE` suppresses the default axes normally generated with the plot, and the `axis` function allows you to add any of the following four axes:

- `side = 1` creates the (default) lower x-axis, below the plot;

- `side = 2` creates the (default) left y-axis, to the left of the plot;

- `side = 3` creates an upper x-axis, above the plot;

- `side = 4` creates a right y-axis, to the right of the plot.

The only required parameter for the `axis` function is `side`, but useful optional parameters include `at`, which specifies the locations of the tick-marks on the axis, `labels`, which specifies the text labels attached to these tick-marks, `col`, which specifies the color of the axis line, and `col.ticks`, which specifies the color of the tick-marks. For a more complete discussion of the `axis` function and its parameters, refer to the results from `help(axis)`.

Fig. 2.10 provides a simple illustration of how the `axis` function can be used to create useful custom axes. The basic plot here is a boxplot of the range of `sugars` values for each of the three levels of the `shelf` variable, corresponding to the positions of three grocery store shelves: `shelf = 1` is at the floor level, `shelf = 2` is the middle shelf (about eye-level for a child riding in a shopping cart), and `shelf = 3` is the top shelf (about eye-level for the adult pushing the shopping cart). The code used to generate this plot was:

```
boxplot(sugars ~ shelf, data = UScereal, axes = FALSE,
              xlab = "Shelf", ylab = "Grams of sugar per serving",
              varwidth = TRUE)
axis(side = 1, at = c(1, 2, 3), labels = c(1, 2, 3))
yRange <- seq(0, max(UScereal$sugars), 5)
axis(side = 2, at = yRange, labels = yRange)
axis(side = 3, at = c(1, 2, 3), labels = c("Floor", "Mid", "Top"))
```

The first line constructs the basic boxplot display—in this case, a variable-width boxplot, so that the width of each individual boxplot reflects the number of different cereals on each shelf. Note that the `xlab` and `ylab` parameters define the x- and y-axis labels, but since we have specified `axes = FALSE`, the axes themselves are not drawn, only their labels. The second line invokes the `axis` function to create the lower x-axis, which puts tick-marks at the numerical values 1, 2, and 3, corresponding to the three levels of the `shelf` variable, and labels the tick-marks with these same numbers. The next two lines construct the y-axis to the left of the plot: `yRange` is a sequence of values from 0 to the maximum `sugars` value from the `UScereal` data frame in steps of size 5, which is used for both the `at` and `labels` parameters for the `axis` function call with `side = 2` to construct the y-axis. Finally, the last line constructs a top x-axis above the plot, with tick-marks at the same location as those in the lower x-axis, but with different labels, giving the actual shelf positions.

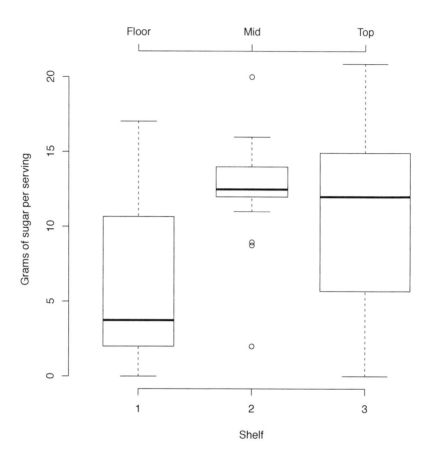

Figure 2.10: Boxplot example with custom axes.

2.5   A few different plot types

It has been noted repeatedly that one of R's great strengths is its range of graphical visualization tools. The following sections describe a few specific examples, some like the bar chart that have been discussed previously and others like the `symbols` function that are introduced here. It is not reasonable to attempt to cover the complete range of graphical tools available in R, nor are the graphics tools discussed in this chapter the only ones introduced in this book: highly specialized graphics functions like `corrplot` are introduced as they are needed in subsequent chapters (the `corrplot` function is useful for visualizing correlation matrices, which are discussed in Chapter 9). The following discussions focus on graphical functions that are either particularly useful, or best avoided, like the pie charts discussed next.

2.5.1   Pie charts and why they should be avoided

Pie charts represent non-negative numerical data vectors in the form of a circular "pie" with one "slice" for each element of the vector, whose size is proportional to its relative value. Pie charts are implemented in R by the `pie` function, but the help files associated with this function strongly discourage its use:

> Pie charts are a very bad way of displaying information. The eye is good at judging linear measures and bad at judging relative areas. A bar chart or dot chart is a preferable way of displaying this type of data.

This point is illustrated in Fig. 2.11, which shows two representations of the `veh_body` variable in the `dataCar` data frame from the `insuranceData` package. This categorical variable has 13 distinct levels indicating different classes of vehicle, and both of the plots shown in Fig. 2.11 were generated from a tabulation of the number of times each level occurs, obtained with the `table` function:

```
xTab <- table(dataCar$veh_body)
```

The plot shown on the left is the pie chart constructed from `xTab`, while the right-hand plot is the corresponding bar chart representation, generated by the `barplot` function discussed further in Sec. 2.5.2. Clearly, the bar chart on the right summarizes the relative frequencies of the vehicle body types much more effectively here than the pie chart on the left. One reason is that many of the labels on the pie chart overlap badly enough that they cannot be read at all, but even ignoring this difficulty, the bar chart on the right gives us a much clearer picture of the magnitude of the differences in the relative frequencies of the different vehicle types in the dataset.

In their discussion of pie charts, Iliinsky and Steele [44] note, first, that this graphical representation is highly controversial with both strong opponents and strong supporters, but, second, that it is easier for us to compare lengths than to compare angles. Overall, they offer the following recommendations [44, pp. 61–62]:

> The bottom line is that pie graphs are best used when precision isn't particularly important, and when there are relatively few wedges to compare (or few that matter). ...

> While there are some valid uses of pie graphs, in most cases, there's a rectangular solution that is as good or better. For comparing fractions of a whole, stacked bar graphs are the most common alternative. For comparing fractions to each other, standard bar graphs are preferred. That's not to say, 'Never use pie graphs'—just use them selectively.

Given the limitations of pie charts cited and illustrated here, the only compelling case for their use is that "we have to": a boss or a customer insists, or pie charts are required in keeping with the format of a report or other document that includes our analysis results.

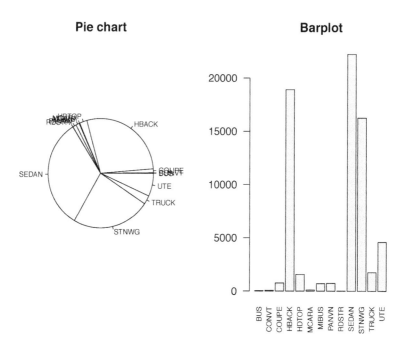

Figure 2.11: Two summaries of veh_body from the dataCar data frame.

2.5.2   Barplot summaries

As Fig. 2.11 illustrates, bar charts can be effective in displaying integer-valued numerical data like that obtained with the `table` function. As shown in Sec. 2.3.1, we can obtain a bar chart by applying the `plot` function to a categorical variable, but the following discussion focuses on the `barplot` function, which is extremely flexible, capable of generating both vertical bar charts like those seen in the previous examples and horizontal bar charts like the one described next, along with other variations like the stacked bar chart discussed in Sec. 2.7.2. For a more complete illustration of the range of plots possible with the `barplot` function, execute `example(barplot)`.

Fig. 2.12 is a bar chart representation of the same `veh_body` data summarized in the pie chart and bar chart shown in Fig. 2.11, but with two important differences. First, Fig. 2.12 is a *horizontal* bar chart, and, second, the data used to generate this plot has been sorted so that the longest bar, corresponding to the most frequently occurring vehicle type in the dataset, appears at the top of the plot and the shortest bar appears at the bottom. The R code used to generate this plot was:

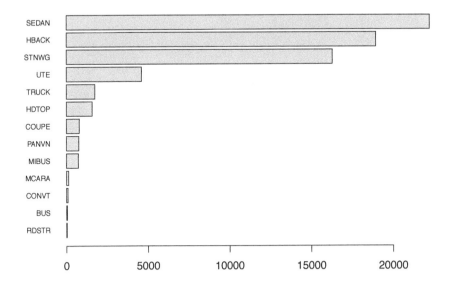

Figure 2.12: Horizontal barplot summary of veh_body values from the dataCar data frame, in decreasing order of frequency.

```
barplot(sort(xTab), cex.names = 0.7, las = 1, horiz = TRUE)
```

The **sort** function re-orders the tabulation vector **xTab** so that the smallest count (27) appears first and the largest count (22,233) appears last. In a standard vertical bar chart (the default option for **barplot**), the first data value (i.e., the smallest count) would appear at the left and the last data value would appear at the right of the plot; by specifying **horiz = TRUE**, we request a horizontal bar chart, constructed from the bottom up, so the smallest count determines the length of the bottom bar in the plot and the largest count determines the length of the top bar. The parameter **cex.names** specifies the size of the names associated with the bars, and the parameter **las = 1** specifies horizontal labeling for both axes, as discussed in Sec. 2.4.4.

2.5.3 The symbols function

The **symbols** function is an extremely flexible base graphics function that supports the generation of plots that show the relationship between more than two numerical variables. A specific example is the *bubbleplot*, a scatterplot of two

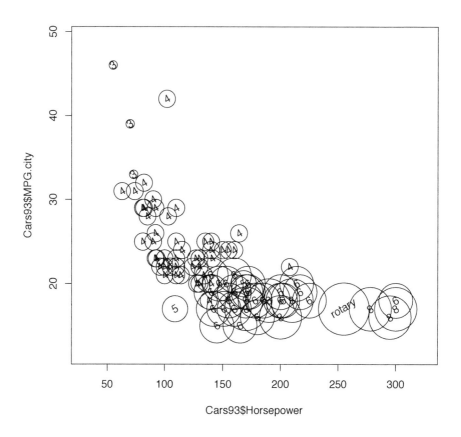

Figure 2.13: Bubble plot of MPG.city vs. Horsepower, using the symbols function, with overlaid text.

numeric variables, with points represented by circles where the size of each circle is determined by a third numeric variable. These plots can be generated using the symbols function by specifying the usual scatterplot arguments x and y, along with a third optional parameter circles as a numeric vector z, of the same length as x and y. Other options available in the symbols function include squares, rectangles, stars, thermometers, and boxplots.

A simple example of a bubble plot constructed using the symbols function is shown in Fig. 2.13. As noted, a bubble plot is a scatterplot—in this case, of MPG.city vs. Horsepower from the Cars93 data frame from the MASS package— where the points are represented by circles, whose radius is determined by a third variable. Here, this third variable is the factor Cylinders from the Cars93 data frame, converted to the numeric index of each distinct value. The inches parameter sets the size of the largest symbol appearing in the plot, with a default value of 1; the smaller value used here was determined by trial and error to make the plot easier to interpret. In addition, the plot in Fig. 2.13 has overlaid the `Cylinders` value of each point, rotated by 30 degrees to improve readability. The *R* code used to construct this plot is:

```
symbols(Cars93$Horsepower, Cars93$MPG.city,
        circles = as.numeric(Cars93$Cylinders),
        inches = 0.4)
text(Cars93$Horsepower, Cars93$MPG.city,
     Cars93$Cylinders, srt = 30)
```

Several points are clear from this plot that are not apparent from a simple scatterplot of `MPG.city` vs. `Horsepower`. First, note that the smallest circles correspond to the three records with `Cylinders = 3`, and that these cars generally exhibit the highest city gas mileage, but the lowest horsepower. The sole exception is the one vehicle with `Cylinders = 4` that exhibits a gas mileage over 40 miles per gallon—second-best overall—with a horsepower of 100, substantially larger than all of the three-cylinder cars. Similarly, the one car with `Cylinders = 5` exhibits about the same horsepower—i.e., approximately 100—but a much lower gas mileage value, comparable with the majority of the six- and eight-cylinder engines. Finally, note that the single rotary engine exhibits much larger horsepower than most others, but again among the lowest gas mileage values seen. The key point here is that the bubble plot allows us to see these details, whereas a simple scatterplot does not.

More generally, the `symbols` function allows us to construct plots that attempt to show relationships between three or more numerical variables. The `circles` parameter used above generates bubble plots for three variables, while the `squares` parameter generates analogous plots with points represented as squares instead of circles. The `rectangles` parameter allows the generation of plots involving four numerical variables, with two specifying the lengths and widths of the rectangles. More variables can be represented using the `thermometers`, `boxplots`, or `stars` parameters, but with range restrictions in some cases. Used with sufficient care, it is possible to generate highly informative explanatory data visualizations that emphasize important details in the data, but this can involve substantial design effort with much trial-and-error selection of variables, scaling parameters, and other details. Plots that attempt to show relationships between too many variables can be difficult enough to interpret that they are not useful as either explanatory or exploratory data visualizations.

2.6 Multiple plot arrays

Often, we want to place multiple plots together, either to facilitate comparisons between plots of similar things or to emphasize differences between plots of distinct things. Base graphics provides two ways of doing this: the simpler way is to use the `mfrow` plot parameter as described in Sec. 2.6.1, and while this approach is extremely convenient, it is of limited flexibility. The slightly more complicated but more flexible approach is to use the `layout` function as described in Sec. 2.6.2. It is important not to get carried away with multiple plots, however, and this is one of the points of the following discussion.

2.6.1   Setting up simple arrays with mfrow

The plot parameter `mfrow` is a two-dimensional vector that defines the number of elements in a rectangular array of plots. The first element of this vector specifies the number of rows in this plot array, while the second element specifies the number of columns. To create an array with R rows and C columns, we first specify `par(mfrow = c(R, C))` and then use any of the base graphics functions to create each individual plot. In creating these plots, we proceed from the top left corner of the array, creating the plots in the top row from left to right, then we create the second row from left to right, repeating until we reach the bottom row, from left to right. In principle, R and C can be any positive integers, but if we make either of them too large, the individual plots become very small and difficult to read. This point was illustrated in the exploratory data analysis plot presented in Sec. 2.1 for the `UScereal` data frame (Fig. 2.1), which corresponded to an 11×11 plot array. Recall that while gross features were visible in the individual scatterplots, nothing more was apparent.

The basic recipe just described is illustrated in Fig. 2.14, which shows an array with one row of plots and two columns. This structure is extremely useful when we wish to compare two similar plots, as in this example: the left-hand plot in this array shows the city gas mileage (`MPG.city`) from the `Cars93` data frame, plotted against the `Horsepower` variable. As we saw in the previous example, there is a roughly monotone decreasing trend in this data: the greater the horsepower, in general the lower the gas mileage. The right-hand plot shows the corresponding plot for the highway gas mileage (`MPG.highway`) against the horsepower. These variables are seen to exhibit a generally similar relationship, although the highway mileage curve appears shifted upwards relative to the city mileage curve, reflecting the fact that the highway mileage is generally greater. Note that to see this difference clearly, it is necessary to specify the same ranges for both plots; since the x-variable is identical in these plots, this means that we need to specify the same y-axis limits for both plots using the `ylim` parameter. The R code used to generate these plots is:

```
par(mfrow=c(1,2))
plot(Cars93$Horsepower, Cars93$MPG.city, ylim = c(15, 50))
title("Plot no. 1")
plot(Cars93$Horsepower, Cars93$MPG.highway, ylim = c(15, 50))
title("Plot no. 2")
```

The key points here are, first, that the specification `par(mfrow = c(1,2))` appears before either plot is generated, in order to set up the one-row, two-column plot array we want. Next, the left-hand plot is generated by calling the `plot`

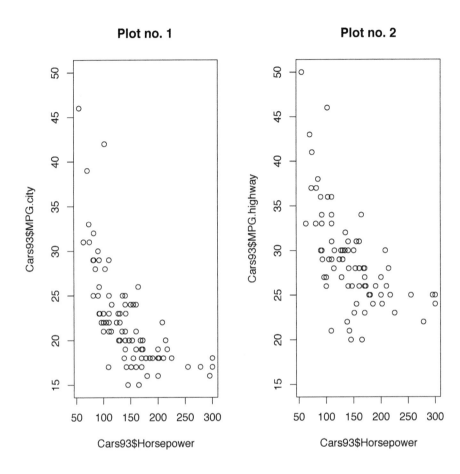

Figure 2.14: Side-by-side comparison: city (Plot no. 1) and highway (Plot no. 2) gas mileage vs. horsepower.

function with the two variables to be plotted and ylim = c(15, 50) to set the y-axis scaling. The plot title is then added, after which the right-hand plot and its title are generated, using the same y-axis scaling.

One potential disadvantage of the plots in Fig. 2.14 is that the y-axis is greatly elongated relative to the x-axis, unlike the square plots obtained with par(mfrow = c(1,1)), which put both variables on a more equal footing visually. This elongation is a consequence of the default value for the graphics parameter pty, which is 'm' and means "use the maximum available space." When the number of rows and columns are the same, this gives equal spacing, but when these numbers are different, we get rectangular plots, as in this example; similarly, the two resulting plots for par(mfrow = c(2,1)) would each be much wider than they are tall. We can override this default and obtain square

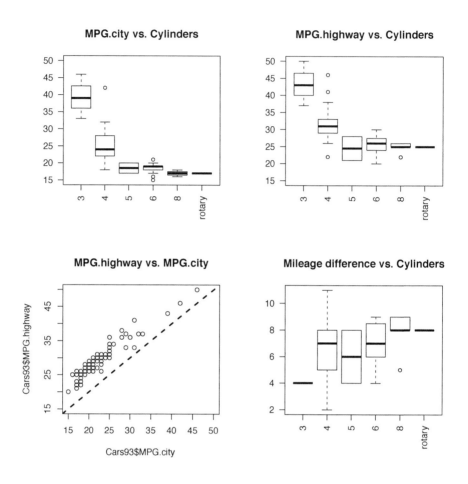

Figure 2.15:   Relationships between mileages and cylinders: city mileage vs. cylinders (upper left), highway mileage vs. cylinders (upper right), highway vs. city mileage (lower left), and mileage difference vs. cylinders (lower right).

plots by specifying par(pty = 's') when we set up the plot array. One advantage of the 2 × 2 plot array is that it creates an array of four square plots, all the same size and typically large enough to see useful details.

This point is illustrated in Fig. 2.15, which presents four different views of the relationships between highway gas mileage, city gas mileage, and the number of cylinders. Specifically, the upper two plots show boxplots of MPG.city and MPG.highway vs Cylinders, illustrating that both of these mileage values generally decline as the number of cylinders increases, with the rotary engine behaving essentially the same as the 8-cylinder engines. The lower left plot shows the highway mileage versus the city mileage, with an equality reference line to emphasize that the highway mileage is always greater than the city mileage. Finally, the lower right plot is a boxplot summary of the difference between these mileages (highway mileage minus city mileage) versus the number of cylinders. While not uniformly so, it appears generally true that the difference between highway and city mileage increases as the number of cylinders increases, with the rotary engine again behaving essentially like the 8-cylinder engines. The code used to generate these plots was:

```
par(mfrow=c(2,2))
plot(Cars93$Cylinders, Cars93$MPG.city, las = 2, ylim = c(15, 50))
title("MPG.city vs. Cylinders")
plot(Cars93$Cylinders, Cars93$MPG.highway, las = 2, ylim = c(15, 50))
title("MPG.highway vs. Cylinders")
plot(Cars93$MPG.city, Cars93$MPG.highway, xlim = c(15, 50),
        ylim = c(15, 50))
title("MPG.highway vs. MPG.city")
abline(a = 0, b = 1, lty = 2, lwd = 2)
delta <- Cars93$MPG.highway - Cars93$MPG.city
plot(Cars93$Cylinders, delta, las = 2)
title("Mileage difference vs. Cylinders")
```

Key points to note here are that whenever either MPG.city or MPG.highway appears in a plot, the same values for the x- or y-axis limits are used, i.e., xlim = c(15, 50) or ylim = c(15, 50), as appropriate. This scaling choice facilitates comparison between the upper two boxplots, and it causes the equality line to appear as a diagonal line in the lower left plot, helping to emphasize that the highway mileage is always larger than the city mileage. Also, note the use of the las = 2 parameter specification in all of the boxplots, making the **rotary** label visible on the x-axes of these plots: adopting the default orientation (las = 0) causes this label to disappear from all of the plots.

2.6.2 Using the layout function

The layout function provides a more flexible approach to creating plot arrays. Although using layout is a little more complicated than using the mfrow parameter as described in Sec. 2.6.1, it allows greater flexibility in specifying the sizes, shapes, and positions of plots. The basis for this flexibility is a *layout matrix*, which has integer elements specifying plot positions, and constructing this matrix correctly is probably the most challenging aspect of using the layout function.

To see how this works, consider the following simple example:

```
vectorOfNumbers <- c(rep(0, 4), rep(c(1,1,2,2), 2), rep(0, 4))
layoutMatrix <- matrix(vectorOfNumbers, nrow = 4, byrow = TRUE)
layoutMatrix

##      [,1] [,2] [,3] [,4]
## [1,]   0    0    0    0
## [2,]   1    1    2    2
## [3,]   1    1    2    2
## [4,]   0    0    0    0
```

The first line of code constructs a vector with 16 elements, using the c function to gather its arguments into a vector, each generated with the rep function to generate repeated subsequences. For example, rep(0, 4) creates a sequence with its first argument (0) repeated the number of times specified by its second argument (4). The matrix function then constructs a matrix from the vector specified in its first argument, with nrow rows. The number of columns is determined by the number of rows and the length of the vector: here, the vector length is 16, so there are 4 columns (16 divided by 4), giving a 4×4 square matrix. Specifying byrow = TRUE builds this matrix by rows, taking the first four elements of the vector as the first row, the next four elements as the second row, and so forth. The result is the 4×4 matrix shown, with the element values 0, 1, or 2.

The layout function takes a matrix like the one constructed above as an argument and uses it to set up a plot array, much like the mfrow parameter, but with greater flexibility, as the following examples demonstrate. *To set up this array correctly, the matrix must be structured into* blocks, *with all elements having the same value in each block.* Given such a matrix, after the layout function is executed, the first plot is put in the array position corresponding to the block with the value 1, the second plot is put in positions corresponding to the block with the value 2, etc. Blocks with the value 0 specify whitespace where no plot appears.

This layout is illustrated in Fig. 2.16, which contains the same two plots as in Fig. 2.14, but here the two plots are both square, in contrast to the elongated rectangular plots shown in Fig. 2.14. The R code used to generate this figure was essentially identical to that used before, but with layout(layoutMatrix) substituted for par(mfrow = c(1, 2)):

```
layout(layoutMatrix)
plot(Cars93$Horsepower, Cars93$MPG.city, ylim = c(15, 50))
title("Plot no. 1")
plot(Cars93$Horsepower, Cars93$MPG.highway, ylim = c(15, 50))
title("Plot no. 2")
```

Here, because the first and last rows of layoutMatrix constitute blocks of contiguous zeros, they define strips of whitespace at the top and bottom of the plot array. The 2×2 block of contiguous 1's in this matrix specifies the position and size of the first plot, while the block of 2's defines the second plot. Since both of these blocks are square and of the same size, we obtain side-by-side square plots of the same size to facilitate visual comparison.

The primary advantage of using layout is the ability to construct arrays of plots with different sizes and/or shapes, with greater flexibility in their placement. The following example illustrates this point:

```
layoutVector <- c(rep(c(1,1,0,0,0,0),2), rep(c(0,0,2,2,2,2),4))
layoutMatrix <- matrix(layoutVector, nrow = 6, byrow = TRUE)
layoutMatrix

##       [,1] [,2] [,3] [,4] [,5] [,6]
```

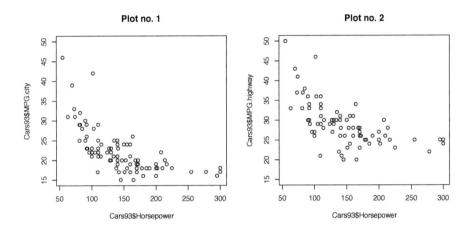

Figure 2.16: City (Plot 1) and highway (Plot 2) gas mileage vs. horsepower.

```
## [1,]     1     1     0     0     0     0
## [2,]     1     1     0     0     0     0
## [3,]     0     0     2     2     2     2
## [4,]     0     0     2     2     2     2
## [5,]     0     0     2     2     2     2
## [6,]     0     0     2     2     2     2
```

Using this matrix, the `layout` function sets up an array with a small plot
(the first) in the upper left corner, and a larger plot (the second) in the lower
right. The data for this example consists of the `insulin` and `triceps` variables
from the `PimaIndiansDiabetes` dataset in the `mlbench` package of benchmark
datasets. The *R* code used to generate the plot in Fig. 2.17, based on the layout
matrix constructed above, was:

```
layout(layoutMatrix)
plot(PimaIndiansDiabetes$insulin, PimaIndiansDiabetes$triceps)
title("Scatterplot")
sunflowerplot(PimaIndiansDiabetes$insulin, PimaIndiansDiabetes$triceps)
title("Sunflowerplot")
```

This combination gives two different views of the same variables—triceps skin-fold thickness versus the serum insulin—each with a different emphasis. Specifically, the scatterplot in the upper left shows how the two variables cluster, with no indication of repeated values. In contrast, the larger sunflower plot in the lower right allows us to see clearly the extent to which records with physically impossible zero serum insulin values tend to have repeated triceps skinfold thickness values, especially when this variable also has the physically impossible value zero. Note that this use of different plot sizes to give different emphasis to the individual plots is easy with the `layout` function, but impossible with the simpler `mfrow` parameter specification discussed in Sec. 2.6.1.

2.7   Color graphics

Used well, color can add greatly to a graphical data display, helping us to see important details, but used badly, color can obscure these details completely. Both good and bad examples are discussed in Sec. 2.7.1, which offers general guidelines for using color effectively. Conversely, color is not always available and, even worse, plots that begin life in color and are subsequently rendered in black and white completely lose those details we were using color to emphasize. That said, there are cases where color is extremely beneficial, allowing us to see differences that are much less evident in a black and white plot with differing shades of gray: the stacked barplot discussed at the end of Sec. 2.7.2 and the `tableplot` function described in Sec. 2.7.3 are cases in point.

2.7.1   A few general guidelines

As a number of authors have noted, "color is tricky" [44, 65, 83]. For one thing, color has three distinct attributes that we can discern: *hue*, which represents the color itself (e.g., "red," "green," or "purple"), *saturation* or "intensity," and *brightness*, also called *luminence* or "tint." Further, it is important to note that while saturation and brightness are *ordered*, color itself (i.e., hue) *is not*. This means that the different aspects of color have different uses in formulating effective displays. Tufte has a marvelous illustration of this point in his book *Visual Explanations* [65, pp. 76–77], where he presents two maps that attempt to show both land elevations and ocean depths in the area of the Japan Sea. The first is highly effective, encoding differences in elevation and depth with different tints of two colors: tan for elevation, and blue for depth. Also, the map adds gray ocean depth contour lines with labels to present additional detail, and black text labels to identify major land masses and bodies of water. Tufte describes his second map with these words: "In ghastly contrast below, a rainbow encodes

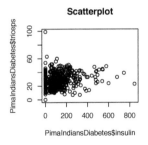

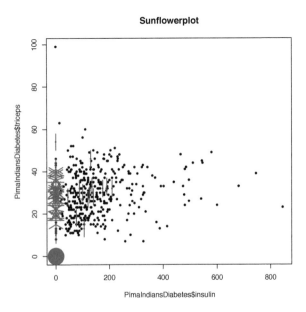

Figure 2.17: Scatterplot (upper left) and sunflower plot (lower right) of triceps vs. insulin from the PimaIndiansDiabetes data frame.

depth." There, he uses 21 different colors to encode the same information, and the result is indeed ghastly: differences that are clearly evident in the first map at a single glance are difficult to see at all in the second.

To make effective use of the different aspects of color, Yau proposes three different types of color scale [83, p. 312]:

- *sequential:* increasing saturation levels of the same hue are used to represent increasing values of a non-negative variable;

- *diverging:* two contrasting hues are used, one for negative values and the other for positive values, with increasing saturation levels used to indicate "more negative" or "more positive," with a third, neutral color used to indicate zero;

- *qualitative:* contrasting hues are used to represent the different levels of an un-ordered categorical variable.

Tufte's example described above shows that it is also possible to construct sequential or diverging color scales based on brightness (i.e., luminance or tint) instead of saturation (i.e., intensity), since both of these color attributes are naturally ordered.

When hue is used to represent different levels of a categorical variable (e.g., different subgroups of a dataset), it is important not to use too many colors, or to make them too similar. Iliinsky and Steele offer this advice [44, p. 66]:

> The standard advice for using color to encode categories is to limit your selection to ideally about six—hopefully no more than 12, and absolutely no more than 20—colors and corresponding categories. This will allow you to select colors that are different enough that they can easily be differentiated and clearly named.

The authors then present a list of 12 colors, recommending the first six be used in preference to the others. These colors are shown in the horizontal bar chart in Fig. 2.18, with longer bars corresponding to the preferred six colors. *This figure appears as part of a group at the end of this chapter, along with all of the other color figures discussed here.* Fig. 2.18 was generated with this R code:

```
Top12Colors <- c("red", "green", "yellow", "blue", "black", "white",
           "pink", "cyan", "gray", "orange", "brown", "purple")
colorVector <- rev(Top12Colors)
barLengths <- c(rep(1,6), rep(2,6))
yvec <- barplot(barLengths, col = colorVector, horiz = TRUE, axes = FALSE)
axis(2, yvec, colorVector, las = 2)
```

Note the use of the function `rev` in the second line here, which reverses the order of the vector `Top12Colors`; this is necessary since the `barplot` function constructs the plot from the bottom up, and we want the top-ranked colors to appear at the top of the plot. Also, note the use of the assignment statement in the next-to-last line here: this retrieves the positions of the center of each bar in the plot in the vector `yvec`, which is then used in the `axis` function call to put the *y*-axis labels in the appropriate places.

Finally, many authors have noted the difficulties posed by color coding for those who suffer from color blindness. For example, Yau notes that approximately 8% of men and 0.5% of women suffer from a red-green color deficiency (the most common form of color blindness), and he includes a figure to illustrate how someone with red- and/or green-deficient vision sees a sequence of eight differently colored squares [83, p. 103]. His example clearly illustrates that using hue to convey the different levels of a categorical variable will be completely ineffective for those with red-green color deficiency.

2.7.2   Color options in R

The base graphics system in R supports 657 colors, specified with the appropriate graphics parameters. Specifically, the family of `col` parameters is structured the same as the `cex` family discussed in Sec. 2.3.3:

- `col` specifies the color of text and plotting symbols;
- `col.axis` specifies the color of the axes;
- `col.lab` specifies the color of the axis labels;
- `col.main` specifies the color of the main plot title;
- `col.sub` specifies the color of the plot subtitle.

An important aspect of color is that it can be specified in several different ways. Probably the simplest is to use a character string like those used to specify the colors of the bars in Fig. 2.18. The names and colors of 50 randomly selected colors available in R are shown in Fig. 2.19, displayed in alphabetical order from the bottom left to the top right. The R code used to generate this plot was:

```
set.seed(3)
colorNames <- sort(sample(colors(), size = 50))
plot(seq(1, 10, 1), rep(5, 10), ylim = c(0, 6), xlab = "", ylab = "",
     axes = FALSE, type = "n")
for (i in 1:5){
    index <- seq(1, 10, 1) + (i-1) * 10
    angle <- (-1)^(i+1) * 45
    text(seq(1, 10, 1), i, colorNames[index], col = colorNames[index],
         srt = angle)
}
```

In this plot, alternate rows are oriented at either 45 degrees or -45 degrees to minimize the overlap of the text. Also, note that several of these colors are not distinguishable from the background, and some are barely distinguishable from each other (e.g., "gray11," "gray21," "gray26," and "grey22").

Colors in R can also be specified with a character string of the form "#RRGGBB," where the pairs "RR," "GG," and "BB" are hexidecimal codes for a Red-Green-Blue representation. The `rainbow` function is called with a number n and returns a vector with n color specifications of this form, representing colors spaced along the rainbow from red, through orange, yellow, green, and blue, to violet. Fig. 2.20 shows the colors generated by this function for $n = 10$, along with their character string representations, generated with the following R code:

```
x <- seq(1, 10, 1)
xColors <- rainbow(10)
plot(x, x, xlim = c(0,11), ylim = c(0, 11), xlab = "",
     ylab = "Element of rainbow(10)", type = "n",
     axes = FALSE)
text(x, x, xColors, col = xColors)
axis(2, x, x)
```

For more on R color specifications, refer to Murrell [57, Sec. 10.1].

A *stacked barplot* is an extension of the bar charts discussed in Sec. 2.5.2, which can be used to represent the results of an arbitrary two-variable contingency table. Fig. 2.21 shows two examples, one in black-and-white and one in color, both constructed from the contingency table describing the relationship between the variables `region` and `vote` in the `Chile` data frame:

```
table(Chile$region, Chile$vote)

##
##        A   N   U   Y
##   C   44 210 141 174
##   M    2  18  23  38
##   N   30 102  46 135
##   S   42 214 148 275
##   SA  69 345 230 246
```

The upper plot in Fig. 2.21 is the result of applying the `barplot` function to this contingency table, using only default options except for the axis label orientation parameter, which is specified as `las = 1`. Each bar in this plot corresponds to one value of the `vote` variable, with the total bar height representing the number of records listing each of these four possible values. Within each bar, the display is partitioned into rectangles that each represent the total number of records that list the corresponding `vote` value but also list one of the five possible `region` values. The bottom plot shows the same results as the top, but color-coded using the first five of the twelve colors recommended by Iliinsky and Steele, as discussed in Sec. 2.7.1. In addition, the lower plot also includes a legend, which gives the region designation for each color used in constructing the plot. Comparing these two plots, it is clear that the relationship between the variables `vote` and `region` is much easier to see in the colored plot than in the gray-scale plot. For example, note that the "M" region is the smallest and appears to be essentially absent from the "A" barplot representing those voters intending to abstain in the election, and it appears to be most prevalent in the "Y" barplot representing those who intend to vote "Yes." Similarly, note that the color-coded plot makes it clear that "SA" (city of Santiago) is generally the largest region among all voters except those intending to vote "Yes." These details are difficult if not impossible to see from the gray-scale plot.

2.7.3   The tableplot function

The `tableplot` function from the `tabplot` package provides graphical characterizations of datasets with an arbitrarily large number of records, giving a quick and useful view of how the different variables relate. Further, this tool is applicable to mixed datasets with both numerical and categorical variables, but to represent categorical variables, it makes extensive use of color.

Fig. 2.22 shows the result of applying the `tableplot` function to the data frame `PimaIndiansDiabetes` from the `mlbench` package, which has 768 records and 9 variables in each record, all but one of which is numeric. Specifically, this plot was constructed using the following R code:

```
library(tabplot)
library(mlbench)
```

```
data(PimaIndiansDiabetes)
tableplot(PimaIndiansDiabetes)
```

To construct this plot, the `tableplot` function first sorts all records in descending order of the variable specified by the `sortCol` parameter; by default, this is the first variable in the data frame, here `pregnant`. Records are then grouped into subsets or "bins"—by default, `tableplot` constructs 100 bins, each with approximately $N/100$ of the N data records. In this case, each bin contains 8 records, as indicated in the bottom left corner of the plot. The tableplot consists of one horizontal barplot for each bin, where the length of each bar is the average of the variable over all records in the bin. Since the records are sorted in descending order of the `pregnant` variable, the plot for this variable decreases monotonically. The other columns of the tableplot display are constructed analogously: each bin is represented by a horizontal bar whose length represents the average value of the corresponding variable for that bin. One advantage of this construction is that it highlights relationships that may exist between the reference variable used to order the data and other variables in the dataset. In particular, if another variable increases or decreases in an approximately monotonic fashion, this implies a systematic variation of this variable with changes in the reference variable. In the plot shown in Fig. 2.22, there does appear to be a weak but somewhat systematic relationship between the reference variable `pregnant` and the variable `age`. Finally, note that `diabetes`—the right-most column in this display—is a two-level categorical variable taking on the values "pos" or "neg." In this horizontal barplot display, the length cannot be represented by an "average value" over each bin, so it is represented as a stacked barplot, with orange representing the fraction of "neg" values and light blue representing the fraction of "pos" values.

If we specify `sortCol` explicitly, either as a column number or a variable name, we can construct a tableplot based on whatever reference variable we choose. Fig. 2.23 shows the tableplot that results when we select `glucose` as our reference variable instead of `pregnant`, constructed as:

```
tableplot(PimaIndiansDiabetes, sortCol = "glucose")
```

In this tableplot summary, we see that the `glucose` barplots are decreasing monotonically, reflecting the fact that it is our reference variable. In addition, although it is not monotonic in its variation, the `insulin` variable appears to generally decline as `glucose` decreases, and the fraction of positive diabetes diagnoses appears to decrease almost monotonically with decreasing `glucose`. Since elevated glucose levels are the key diagnostic for diabetes, this last association is not surprising.

As a final tableplot example, Fig. 2.24 shows the results obtained for the `Chile` data frame from the `car` package. Recall that this dataset characterizes voter attitudes and stated intentions prior to an election in Chile. This data frame contains 2700 records, with 8 variables, including both numerical and categorical variables, with a small fraction of missing values, making it ideally suited to characterization using the `tableplot` function. Since we have not specified the `sortCol` parameter explicitly, the records are sorted in descending order by the first column in the data frame, with missing values put last. In this case, this first column corresponds to the categorical variable `region`, which has five levels, none of which are missing. For a categorical variable, "descending order" corresponds to *reverse alphabetical order*, so the tableplot in Fig. 2.24 is constructed by first sorting all records so that the level "SA" (city of Santiago) appears first, followed by "S" (south), then "N" (north), "M" (metropolitan Santiago area), and finally "C" (central). The second column of the tableplot characterizes the numerical variable `population`, which is seen to be strongly associated with `region`: the population value for the "SA" region is constant and *much* larger than that for any other region. Most of the other regions exhibit bin-to-bin variations, reflecting the fact that they are large enough geographically to exhibit significant local variation in population. A particularly interesting exception is the "M" region—described as "metropolitan Santiago area"—which has the smallest population of all regions. The only other systematic variation across regions that is evident in this plot is `income`, which appears to be larger in Santiago (region "SA") than elsewhere, although the relationship is not nearly as strong as that with `population`. The key point of this example is that since 4 of the 8 variables in this data frame are categorical, color-coding is almost essential here to see the relationships, both between different categorical variables and between categorical and numerical variables.

2.8 Exercises

1: The `fgl` data frame from the `MASS` package characterizes 214 forensic glass samples in terms of their refractive index (`RI`), a type designation (`type`), and percentages by weight of eight different elemental oxides. Using the options discussed in the text, generate a plot of the magnesium oxide concentration (`Mg`) versus record number, with these features:

   – *x*-axis label: "Record number"
   – *y*-axis label: "Mg concentration"
   – use the `las` parameter to make the labels horizontal for both axes

2: It was noted in Section 2.3 that the generic function `plot(x, y)` generates a boxplot when `x` is a categorical variable and `y` is a numerical variable. Boxplots will be discussed further in Chapter 3, but this problem asks you to use this observation to create a boxplot summary showing how the magnesium concentration in the `fgl` dataset considered in Exercise 1 varies with the different values of the categorical `type` variable. Specify the *x*-axis label as "Forensic glass type" and the *y*-axis label as "Mg concentration", and make the labels horizontal for both axes.

3: A useful feature of the `plot` function is that it accepts the *formula interface*, commonly used in modeling functions like `lm` for linear regression

models or **rpart** for decision tree models. That is, if x and y are variables
in the data frame **dataFrame**, the following two plot function calls give
the same results:

```
plot(y ~ x, data = dataFrame)
   and
plot(dataFrame$x, dataFrame$y)
```

Use the formula interface to generate a plot of refractive index versus
calcium oxide concentration (**Ca**) values from the **fgl** data frame. Specify
the x-axis label as "Ca concentration" and the y-axis label as "Refractive
index" and make the labels horizontal for both axes.

4: Section 2.5.2 introduced barplot summaries of categorical variables. Using the **barplot** function, construct the following summary for the **type**
   variable from the **fgl** data frame in the **MASS** package:

   4a. Using the **sort** and **table** functions as in the R code that created
       Figure 2.12, create a horizontal barplot of the **type** variable record
       frequencies, with x-axis label "Records listing glass type" and no y-
       axis label. Use the **font.lab** parameter discussed in Section 2.3.3 to
       make this label boldface. Use the **las** parameter to make the type
       name labels horizontal.

   4b. The **paste** function can be used to combine several elements into a
       single text string; here, you are asked to use this function to create
       a title string **tString** containing three components:

       * first, the text string "Horizontal barplot of the"
       * second, the number of levels of the **type** variable
       * third, the text string "glass types"

   4c. Use the **title** function with the **tString** character vector from (4b)
       to add a title in italics to the plot.

5: One of the optional graphics parameters introduced in Section 2.3.3 and
   discussed further in Section 2.6 is the **mfrow** parameter, which provides
   a simple way of generating multiple plot arrays. Using this parameter,
   construct a two-by-two plot array showing the concentrations of the following four oxides versus the record number in the dataset: (1) magnesium
   (chemical symbol **Mg**), top left; (2) calcium (chemical symbol **Ca**), top right;
   (3) potassium (chemical symbol **K**), lower left; and (4) barium (chemical
   symbol **Ba**), lower right. In all cases, the x-axis label should read "Record
   number in dataset" and the y-axis should read "Xx oxide concentration"
   where "Xx" is the appropriate chemical symbol. Each plot should have
   a title spelling out the name of the element on which the oxide is based
   (e.g., "Magnesium" for the top-left plot).

6: The UScereal data frame from the MASS package characterizes 65 break-fast cereals sold in the U.S., from information on their FDA-mandated labels. Three of the variables included in this data frame are the calories per serving (calories), the grams of fat per serving (fat), and a one-character manufacturer designation mfr. Using the text function discussed in Section 2.4.2, create a plot of fat versus calories where the data points are represented using the single-letter manufacturer designations. To improve readability, use the srt argument to tilt these text strings −30 degrees with respect to the horizontal axis. Set the x-axis label to "Calories per serving" and the y-axis label to "Grams of fat per serving". (Hint: the type = ''n'' option is useful here.)

7: As noted in the text and several of the previous exercises, the form of the display generated by the generic plot function depends on the types of the R objects passed to it. This exercise asks you to create and compare side-by-side plots that attempt to characterize the relationship between the variables mfr and shelf from the UScereal data frame in the MASS package. The first of these variables is a factor variable with 6 distinct levels, but the second is represented as an integer with 3 distinct levels.

7a. Using the mfrow parameter, set up a side-by-side plot array; set the pty graphics parameter to obtain square plots.

7b. Using the plot function and the natural representations for these variables, construct a plot that attempts to show the relationship between these variables. Label the x- and y-axes "mfr" and "shelf" and give the plot the title "shelf as numeric".

7c. Using the plot function with UScereal$shelf converted to a factor, re-construct this plot with the same x- and y-axis labels and the title "shelf as factor".

8: It was noted in Section 2.4.1 that the type = ''n'' option can be extremely useful in cases where we want to first specify key plot details (e.g., x- and y-axis limits and labels), and then display different data subsets in slightly different formats (e.g., point shapes or line types) to highlight these subset differences. The following exercise asks you to do this; specifically:

8a. Using the type = ''n'' option, set up the axes and labels for a plot of VitC versus HeadWt from the cabbages data frame in the MASS package. Use the default axis scalings, but specify the x-axis label as "Head weight" and the y-axis label as "Vitamin C".

8b. Construct the vectors indexC39 and indexC52 that point to records in the cabbages data frame for which Cult has the values "c39" and "c52", respectively.

8c. Using the points function, include the scatterplot points for VitC versus HeadWt, restricted to the Cult "c39" subset, representing these points as open triangles (pch = 6).

8d. Using the `points` function, include the scatterplot points for `VitC` versus `HeadWt`, restricted to the `Cult` "c52" subset, representing these points as solid triangles (`pch = 17`).

8e. Using the `legend` function, add a legend to the upper right corner of the plot, including the point shapes and the text "Cultivar c39" and "Cultivar c52".

9: This example uses the `layout` function discussed in Section 2.6.2 to provide a somewhat more informative view of the relationship between the variables `Cult`, `vitC`, and `HeadWt` in the `cabbages` data frame from the `MASS` package considered in Exercise 8.

9a. Using the `matrix` function, construct the 2×2 matrix `layoutMatrix` with plot designations 1 and 2 in the first row, and 3 and 3 in the second, giving a single wide bottom plot. Display `layoutMatrix` and use the `layout` function to set up the plot array.

9b. Construct the vectors `indexC39` and `indexC52` that point to records in the `cabbages` data frame for which `Cult` has the values "c39" and "c52", respectively.

9c. In the upper left position of the array, generate a plot of `VitC` versus `HeadWt` for those records with `Cult` equal to "c39". Use the `ylim` parameter to make the y-axis in this plot span the complete range of the `VitC` data for all records in the dataset. Specify the x-axis label as "Head weight" and the y-axis label as "Vitamin C" and give the plot the title "Cultivar c39".

9d. In the upper right position of the array, generate a plot of `VitC` versus `HeadWt` for those records with `Cult` equal to "c52". Use the `ylim` parameter to make the y-axis in this plot span the complete range of the `VitC` data for all records in the dataset. Specify the x-axis label as "Head weight" and the y-axis label as "Vitamin C" and give the plot the title "Cultivar c52".

9e. In the bottom plot, put a boxplot summary of `VitC` by `Cult` value, with no x- or y-axis labels and the title "Boxplot summary of vitamin C by cultivar".

10: As a further illustration of the flexibility of the `layout` function in configuring plot arrays, this problem asks you to construct an array of three plots: two small plots to the left, one over the other, showing a random data sample and its estimated density, with an elongated plot to its right, with its normal QQ-plot, a characterization discussed in more detail in Chapter 3.

10a. Using the `matrix` function, construct the 2×2 matrix `layoutMatrix` that specifies the plot designations 1 and 3 in the first row, and 2 and 3 in the second row. Display `layoutMatrix` and use the `layout` function to set up the plot array.

10b. Using the `rnorm` function, generate a vector `x` of 100 zero-mean, unit-variance Gaussian random variables. For consistency, use the `set.seed` function with required argument `seed` equal to 3. Plot `x` in the upper small plot in the array with the x-axis label "Sample number" and the y-axis label "Random value". Give the plot the title "Data sample".

10c. Using the `density` function, display the estimated density in the lower left plot in this array. Use the default x and y labels, but use the `main` argument to give the plot the title "Density estimate".

10d. Using the `qqPlot` function from the `car` package, display the normal QQ-plot for `x` in the larger right-hand plot. Using the `main` argument for this function, give this plot the title "Normal QQ-plot".

Color Figures, Group 1

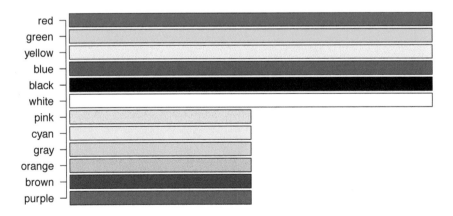

Figure 2.18: Horizontal bar chart of Illinsky and Steele's recommended colors: longer bars are the first six (preferred), shorter bars are the last six.

Figure 2.19: Random sample of 50 colors available in R.

Figure 2.20: Color names returned by rainbow(10).

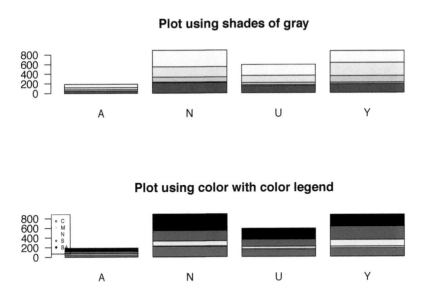

Figure 2.21: Stacked bar chart of voter intention by region from the Chile data frame: using shades of gray (top), and using color (bottom).

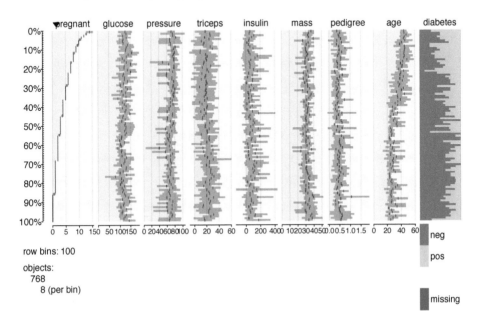

Figure 2.22: Default tableplot of the PimaIndiansDiabetes data frame.

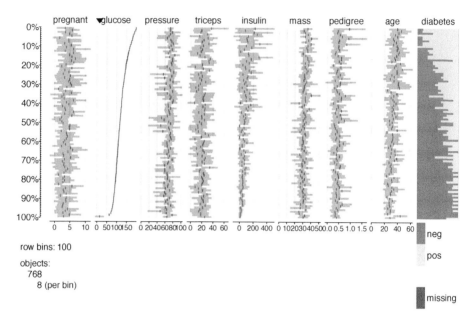

Figure 2.23: Summary of the PimaIndiansDiabetes data frame with the table-plot function, ordered by the glucose variable.

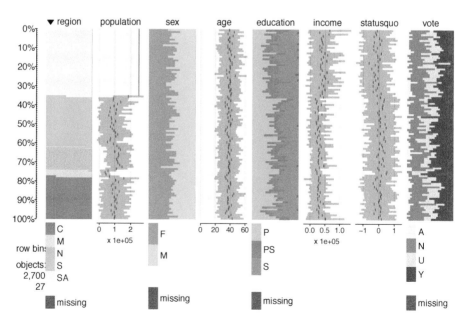

Figure 2.24: Summary of the Chile data frame with the tableplot function.

Chapter 3

Exploratory Data Analysis: A First Look

The basic notion of exploratory data analysis was introduced in Chapter 1, starting with the informal description by Persi Diaconis [21] that may be very roughly paraphrased as "exploratory data analysis is the art of looking at data in a careful and structured way." This initial description was followed by a very high-level overview of some of the techniques used in doing exploratory data analysis, noting that graphical characterizations play a central role in this process. This chapter is the first of two included in this book to give a more detailed description of the objectives of exploratory data analysis, the reasons for its importance in practice, and some particularly useful tools and techniques that are either available as built-in procedures or can be easily implemented in R. This chapter includes sections on exploring a new dataset (Sec. 3.1), summarizing numerical variables (Sec. 3.2), anomalies in numerical data, including outliers and their detection (Sec. 3.3), and visualizing relations between variables (Sec. 3.4). More ideas and techniques for exploratory data analysis are presented in Chapter 9.

Four key concepts in exploring data

Many of the ideas presented here and in Chapter 9 are closely related to "the four R's of exploratory data analysis" proposed by Velleman and Hoaglin [69]:

1. revelation;

2. residuals;

3. re-expression;

4. resistance.

Here, the term *revelation* refers to data visualization, previously noted as an important part of exploratory data analysis and emphasized throughout this chapter and the rest of this book. The term *residuals* refers to the set of differences between the observed values of a variable and its predictions from some mathematical model. A key difference between the models considered for exploratory analysis and those considered in Chapters 5 and 10 on predictive modeling is that the models used in exploratory analysis are often very simple, sometimes trivially so. In particular, looking at the differences between individual data values and a mean or median can be useful in exploratory data analysis, even though such "intercept-only models" are seldom useful in a predictive modeling application except as a "minimum performance benchmark" for comparison. The third R listed above is *re-expression*, which refers to the application of mathematical transformations to one or more variables. The utility of this idea is a consequence of the fact that the data values we are given to analyze are not always in their most informative representation. This point is illustrated with some examples in Chapters 5 and 9, and Chapter 9 also introduces a simple *correlation trick* that is sometimes very useful in suggesting when transformations may be beneficial. Finally, the fourth R—*resistance*—refers to the ability of a data characterization to avoid the undue influence of outliers or other data anomalies, an extremely important topic discussed further in Sec. 3.3. A careful exploratory analysis before we attempt to use data in explaining phenomena, building predictive models, or making decisions is one of the best ways of finding these anomalies before they can adversely impact our results. The following section gives a high-level overview of how to apply these ideas to the preliminary characterization of a new dataset, and the rest of this chapter is devoted to detailed discussions of a few of the ideas and tools used to carry out this task.

3.1   Exploring a new dataset

In the era of Gauss—who lived from 1777 to 1855, who developed the method of ordinary least squares, and for whom the ubiquitous Gaussian distribution is named—about the only way to have data to analyze was to either collect it yourself or get it from a friend or colleague. In either case, the result was a relatively small dataset, and you had a pretty good idea of what it contained. Today, the typical datasets we encounter are both much larger and collected by other people or organizations with whom we have little or no direct connection. Even in cases where a researcher is analyzing their own data, the dataset is often collected with the aid of electronic data acquisition systems or Internet web scrapers. As a consequence, a very useful first step when we obtain a dataset is to look at it carefully to see what it contains. This is even true if we have collected the data ourselves, since this preliminary exploration allows us to determine—sooner rather than later—whether the data we have is what we were expecting to have. In particular, even the best laid data collection plans, equipment, and software—especially software—can fail.

3.1.1 A general strategy

In exploring a new dataset, the following basic sequence is often useful:

1. Assess the general characteristics of the dataset, e.g.:

    a. How many records do we have? How many variables?

    b. What are the variable names? Are they meaningful?

    c. What type is each variable—e.g., numeric, categorical, logical?

    d. How many unique values does each variable have?

    e. What value occurs most frequently, and how often does it occur?

    f. Are there missing observations? If so, how frequently does this occur?

2. Examine *descriptive statistics* for each variable (Secs. 3.2.1 and 3.2.2);

3. Where possible—certainly for any variable of particular interest—examine exploratory visualizations like those described in Sec. 3.2.5;

4. Again, where possible, apply the procedures described in Sec. 3.3 to look for data anomalies;

5. Look at the relations between key variables using the ideas described in Sec. 3.4;

6. Finally, summarize these results in the form of a *data dictionary*, to serve as a basis for subsequent analysis and explanation of the results.

Step (1) in this sequence was mentioned at the beginning of the sample R session presented in Chapter 1 as a good starting point in analyzing data. Some—but not all—of these characteristics can be obtained with built-in R functions like `summary` or `str`. The simple custom function `BasicSummary` described in Sec. 3.1.2 provides all but the first two characteristics listed above, which are easily obtained with the built-in `dim` function giving the numbers of rows (i.e., records in the data frame) and columns (i.e., variables in the data frame). The `BasicSummary` function returns a summary in the form of another data frame, with one row for each variable in the original data frame, giving the rest of the characteristics listed in Step (1) above. The simplest illustration of Step (2) in this sequence is the use of the mean discussed in Sec. 3.2.1 and the standard deviation discussed in Sec. 3.2.2 to characterize numerical variables; other descriptive statistics for numerical variables and alternatives for non-numeric variables are discussed in Chapter 9. As emphasized repeatedly, looking at the data—Step (3)—can be an extremely effective way of both confirming that certain key features are present in the data and discovering unexpected features that may be important in your subsequent analysis. In addition, the anomaly detection methods described in Sec. 3.3 can enhance this visual inspection, drawing our attention to anomalies that might have gone unnoticed; this fact—together with the potentially disastrous consequences of some data anomalies, discussed in Secs. 3.3.1, 3.3.3, and 3.3.5—motivates Step (4) above.

So far, all of these steps have focused on *univariate* views of the data: i.e., looking at each variable by itself. Usually, however, the motivation for our data analysis is to understand and quantify relationships between variables, which motivates Step (5) above. Since bivariate—i.e., two-variable—relationships are the easiest to visualize and characterize, it is best to start with this case, first considering the type-appropriate plots of one variable against another described in Sec. 3.4. An alternative approach that is particularly useful in cases where there are many possible variable pairs to consider is the use of numerical *association measures* that attempt to characterize the strength of the relationship—if any—between two variables. These measures are introduced in Chapter 9, starting with the best known case, the *correlation coefficient*. An important point, discussed further in Sec. 3.4, is that for large datasets with many variables, it may not be practical to consider all possible variable pairs. Specifically, the number of pairs of m variables is $m(m-1)/2$, which grows rapidly with increasing m: this number is 10 when $m = 5$, but it is 45 for $m = 10$ and 4950 for $m = 100$. This fact motivates the restriction to "key variable pairs" in Step (5) above. Characterization of relationships between more than two variables is more difficult, both because we can't directly visualize plots in more than three dimensions and because the number of higher-order combinations grows even more rapidly than the number of pairs. For example, the number of three-way combinations of 10 variables is 120, while that for 100 variables is $161,700$; for four-way combinations, these numbers grow to 210 for 10 variables and just over 3.9 million for 100 variables. Still, useful techniques are available for examining multivariate relationships and Chapter 9 describes a few of them.

Finally, although it is often more work than it seems like it should be, the construction of a data dictionary in Step (6) is time well spent. At a minimum, this document—often in the form of a spreadsheet—should include the summary information from Step (1). In addition, however, a good data dictionary should also include at least pointers to, if not details from, other sources of information about the dataset and its contents. The objective is to capture the information and insights you wish you would have had before you began the analysis of the data, including notes about anything unexpected or ambiguous. One reason a good data dictionary is important is that, very often, one analysis leads to another: "Oh, that's an interesting finding. I wonder if the same thing holds for this other case?" A good data dictionary can save a lot of time in either addressing concerns ("No, those anomalous records were excluded: the range of X values was ...") or in setting up the next analysis, especially if the request comes six months or a year later, with different people involved.

3.1.2   Examining the basic data characteristics

As noted, the `BasicSummary` procedure generates all of the data characteristics listed in Step (1) above, except for the numbers of records and variables. A detailed discussion of the R code for this function is given in Chapter 7, but the following example illustrates its utility. The `Chile` data frame from the `car` package describes voter intentions prior to an election in Chile in 1988. This data frame has 2700 records and 8 variables, characteristics provided by the `dim` function as noted earlier:

```
dim(Chile)
```

```
## [1] 2700    8
```

The `BasicSummary` function returns the following summary data frame:

```
BasicSummary(Chile)
```

```
##       variable       type levels topLevel topCount topFrac missFreq missFrac
## 1       region     factor      5       SA      960   0.356        0    0.000
## 2   population    integer     10   250000     1300   0.481        0    0.000
## 3          sex     factor      2        F     1379   0.511        0    0.000
## 4          age    integer     54       21       96   0.036        1    0.000
## 5    education     factor      4        S     1120   0.415       11    0.004
## 6       income    integer      8    15000      768   0.284       98    0.036
## 7    statusquo    numeric   2093 -1.29617      201   0.074       17    0.006
## 8         vote     factor      5        N      889   0.329      168    0.062
```

It is clear from this summary that most of the variables in this data frame have reasonably explanatory names, with the possible exception of `statusquo`, discussed further in Sec. 3.3.6. This point is important, both because good variable names make our life easier ("Wait—was the quarterly sales volume for the northeast region X27 or X28?") and because some *R* functions automatically change variable names that exhibit certain pathologies (e.g., spaces or special characters), a problem discussed further in Chapter 4.

Of the eight variables in the `Chile` data frame, four are factors, three are integers, and one is numeric. It is important to note that *R* distinguishes between integers (i.e., "whole numbers") and numeric variables (i.e., numbers with decimal points). Although most mathematical operations work equally well with both types, there are some important differences between these two species of "numerical" variables, a point that will be discussed further in Chapter 9. One important difference between integer and numeric data types is that we expect numeric data types to have many unique levels, ideally as many levels as there are data records. The reason for this expectation is discussed in Chapter 9, but the point here is that the numeric variable `statusquo` is fairly consistent with these expectations, while the integer variables all have many fewer levels. In the case of the variable `age`, this characteristic is not surprising: each record corresponds to one voting-age adult, and adult ages are almost always recorded in integer years. In fact, the range of this variable is from 18 to 70. The number of levels for the other two integer variables, `population` and `income`, is more surprising and is highly suggestive of coarse rounding. For example, the `income` variable has the following distinct levels:

```
table(Chile$income)
```

```
##
##   2500   7500  15000  35000  75000 125000 200000
##    160    494    768    747    269     88     76
```

For factor variables, we generally expect a relatively small number of unique levels, as in the four factor variables included here, which exhibit between 2 and 5 unique values. As a specific example, the factor variable `education` exhibits four levels, the most common of which is "S" (secondary school education), which occurs 41.5% of the time. *It is important to note that missing values are counted as a single level: the fact that `missFreq` is not zero for this variable means that the variable `education` exhibits three non-missing levels and one missing level, for a total of four distinct values.* The reason for including both the absolute count and the fraction of records exhibiting the most frequent value is that, when these numbers are small, the raw counts are generally easier to interpret, while if they are large, the fractions are generally easier to interpret. The same reasoning applies to the missing record counts: there is only one record with a missing `age` value, which is reported as 0.000% due to the finite (3-digit) precision used to record the fraction. In computations like the age range given above, it is important to know that the number of missing values is not zero, even if the fraction is very small, so we know to remove these records when computing minima, maxima, or averages.

3.1.3   Variable types in practice

One of the key data characteristics included in the preliminary data summary just described is *variable type*. Like most other data analysis platforms, *R* supports a set of predefined variable types, including numeric, character, logical, and factors. Unfortunately, these basic variable types don't always fully reflect the variable's "actual character." An important example is date variables, which can be represented in more than one way, each supporting different kinds of computations: dates represented as character strings like "15-Nov-2015" are useful for human readers, but they do not provide the basis for computing the number of days between this date and another. That requires a numerical representation, and such representations are available, but they don't provide the same utility for human readers. Thus, special date formats are often developed that allow both human interpretation and simple computations, a topic discussed at length in Chapter 9. Unfortunately, these representations are often lost in transferring data from their original source into an *R* session, requiring us to explicitly recognize date variables represented as character strings and re-encode them into a special date format. More generally, it is usually possible and sometimes extremely useful to convert variables from one type to another, as subsequent examples will demonstrate. The following discussion gives a brief introduction to some key ideas related to variable type, many of which will be revisited in more detail in later sections of this book.

Numeric vs. ordinal vs. nominal variables

Many data analysis problems are concerned with numerical data. The analysis of the `whiteside` dataset introduced in Chapter 1 is a case in point: there, the primary variables of interest were the weekly heating gas consumption and the outside temperature, both numbers. An important characteristic of numerical data is that we can apply many mathematical operations to it, computing sums, differences, products, quotients, averages, square roots, and many other combinations and/or transformations. This possibility forms the basis for descriptive statistics like the mean and standard deviation discussed in Secs. 3.2.1 and 3.2.2, together with many of the other data characterization tools discussed in this chapter and used throughout this book (e.g., the scatterplots discussed in Sec. 3.4.1). Not all variables are numeric, however, and most of these mathematical operations are not applicable to non-numeric variables. Again, the whiteside dataset provides an illustration: the Insul variable is an example of a *categorical, nominal,* or *factor* variable that can assume either of two values, each represented by a character string (i.e., Before and After).

In the case of nominal variables like Insul, none of the mathematical operations mentioned above apply: we cannot compute sums, differences, products, quotients, averages, or square roots of these data values. Indeed, in working with nominal variables, about all we can do is count and compare, asking questions like the following—given a categorical variable C:

- How many distinct values or "levels" does the variable exhibit?

- How often does each of these levels occur in the dataset?

- How does the behavior of another variable X vary over the levels of C?

These questions form the basis for the categorical variable characterizations described in Chapter 9 and some of the visualization tools described in Sec. 3.4. These questions also form the basis for the techniques described in Chapters 5 and 10 for including categorical variables in predictive models.

Another class of variables that is in many respects intermediate between numerical and nominal variables is the class of *ordinal variables,* also referred to as *ordered categorical variables* or *ordered factors.* These variables assume non-numeric values so the full range of mathematical operations is not available to us in working with them, but they do possess an inherent *order,* so we can say that one value of the variable is "smaller than" or "precedes" another value. Examples here might include variables coded as "low, medium, and high" or survey results recorded on an ordered scale like "strongly agree, agree, no opinion, disagree, strongly disagree." While we cannot compute sums or averages for these variables, we may be able to compute medians or other characterizations that depend strictly on ordering. For example, character vectors can be sorted alphabetically in R, so if we have a character vector of odd length, it is possible to compute the median. As a specific example, the variable Make in the Cars93 data frame from the MASS package is included as a factor variable, but it can be converted to a character vector of length 93, from which the median can be computed as:

```
median(as.character(Cars93$Make))
```

```
## [1] "Hyundai Sonata"
```

It is also possible to extend the Spearman rank correlation described in Chapter 9 to ordered categorical variables. If the ordering of these variables is of particular interest, this characterization may have practical utility, but in other cases—like the alphabetical ordering of vehicle makes—this utility is not obvious, serving more as an illustration that strange computational results are sometimes possible.

Before leaving this discussion of ordered categorical variables, it is worth noting that variables can be *partially ordered*, meaning that some levels can be compared but others cannot. As a simple example, consider the distribution of levels for the variable `Cylinders` from the `Cars93` data frame:

```
table(Cars93$Cylinder)

##
##      3      4      5      6      8 rotary
##      3     49      2     31      7      1
```

If we exclude the single record with the troublesome level `rotary` from this data frame, we obtain a numerical variable with a clear ordering: the smallest value is "3," the largest value is "8," and all other values fall between these extremes. This possible ordering may be important since cars with fewer cylinders generally exhibit lower horsepower and higher gas mileage. If we retain the non-numeric value `rotary` for this variable, this simple summary statement is no longer adequate: we must somehow treat "rotary" as a special case.

There is a well-developed mathematical theory of partially ordered sets or *POSETS*, and while their potential utility for data mining has been discussed [63, Part II], they have not yet worked their way into mainstream data analysis. Thus, it is important to be aware, first, that partially ordered variables can arise in practice, and second, that the simplest way of dealing with them is to either treat them as nominal (i.e., un-ordered categorical) variables, or exclude "troublesome" levels so that we are left with a strictly ordered variable.

Finally, note that a fourth important class of variables—*binary variables*—can also be viewed as intermediate between the numerical and categorical variable classes. Specifically, binary variables take only two values, like the `Insul` variable from the `whiteside` data frame, and these values can be encoded either as character strings (e.g., `Before` and `After`) or as numbers (typically 0 or 1). In fact, in developing predictive models, categorical prediction variables are usually re-coded as a collection of binary indicators for each level, with each indicator taking the value 1 if the variable assumes the value under consideration and the value 0 otherwise (refer to Chapter 5 for a detailed discussion of this idea). In practical terms, this "chameleon character" of binary variables means that we can adopt whichever representation suits our needs: character-coded nominal, or numerical. Nevertheless, because binary variables do have certain unique characteristics, a number of specialized techniques have been developed for working with them, and these techniques usually give better results. This point is discussed further in Chapter 9 where the *odds ratio* is introduced as a tool for characterizing the relationship between two different binary variables.

Text data vs. character strings

Categorical variables—either nominal or ordinal—are commonly represented as character strings. A key requirement for this representation is that each distinct level of the variable have a unique character string representation, and these designations are often—in favorable cases—chosen to provide some degree of human interpretation. Thus, there is generally no reason to "parse" these character strings into substrings and use these substrings as variables in subsequent analysis or modeling efforts. In contrast, *text data* consists of character strings—often, very long character strings—that convey more complex information (e.g., multiple data characteristics, information about relations between different entities of interest, etc.). There, it *is* advantageous to parse these character strings into components, either to detect new features in the data (e.g., similarities or differences between records that were not obvious from the other variables) or to use the components as new covariates in building predictive models.

This second case corresponds to the problem of *text analysis* or *text mining* introduced in Chapter 8. Specialized techniques have been developed to address critical issues that arise when doing text analysis but which are not needed in dealing with simpler categorical variables. For example, the phrases "High-strength material" and "Material strength: high" convey the same meaning, but are distinct as character strings, differing in capitalization, punctuation, and word order. Recognizing these phrases as the same so they can be grouped together in text analysis requires specialized pre-processing techniques that are not required in dealing with simple categorical variables where the text represents a (typically small) set of fixed levels. A complete treatment of the techniques required for text analysis is beyond the scope of this book, but Chapter 8 gives an introduction to some of the main ideas, preprocessing techiques, and supporting R procedures.

3.2 Summarizing numerical data

The term "descriptive statistics" refers to simple characterizations like the mean of a numerical data variable discussed in Sec. 3.2.1, or the standard deviation discussed in Sec. 3.2.2. In fact, these two characterizations are probably the most common and, in some respects, the most important descriptive statistics: the mean attempts to give us a "typical value" for a numerical variable, while the standard deviation attempts to convey an idea of the "spread" or "scatter" of the individual data observations around this typical value. More generally, descriptive statistics are usually size-independent characterizations, having the same format and interpretation for small datasets as they do for large datasets. Because they are simple summaries, however, descriptive statistics are limited in their power to describe data, and in unfavorable cases they can be misleading, a point illustrated in Sec. 3.2.3. For this reason, it is important to consider a range of different descriptive statistics, an idea discussed further in Chapter 9.

3.2.1   "Typical" values: the mean

The term *location estimator* refers to a class of statistical characterizations of "the typical value" of a sequence $\{x_k\}$ of N numerical data values. As noted in the previous paragraph, probably the best known of these location estimators is the *mean* or *average,* defined as:

$$\bar{x} = \frac{1}{N} \sum_{k=1}^{N} x_k. \tag{3.1}$$

The mean is extremely easy to compute, and because it is so widely used, it is built into almost any software environment that has even rudimentary data analysis capabilities, including probably all spreadsheet packages and most or all database packages. In *R*, it is implemented as the built-in `mean` function, which belongs to the class of *generic functions* discussed in Chapters 2 and 7. Here, it is enough to note three things: first, applied without any optional parameters, this function simply computes the mean defined in Eq. (3.1); second, the `mean` function supports optional parameters that give it additional flexibility, discussed briefly in the following paragraphs; and third, this function even gives sensible results when applied to the *date variables* discussed in Chapter 9.

One of the optional parameters of *R's* `mean` function is `na.rm`, a logical variable that specifies how missing values (coded with *R's* default representation `NA`) should be handled. The default for `na.rm` is `FALSE`, which causes the `mean` function to return the result `NA` in the face of missing data, often not what we want. To obtain the average of the non-missing values, we must specify `na.rm` as `TRUE`. This point is illustrated in the following example for the variable `income` from the `Chile` data frame, which has missing values:

```
mean(Chile$income)

## [1] NA

mean(Chile$income, na.rm=TRUE)

## [1] 33875.86
```

Fig. 3.1 is a plot of the fiber content in grams per serving for the 65 breakfast cereals characterized in the `UScereal` data frame from the `MASS` package in *R*. The dashed horizontal line corresponds to the mean value 3.870844 computed from these numbers. (The dotted line lies one standard deviation above the mean, discussed in Sec. 3.2.2.)

3.2.2   "Spread": the standard deviation

The term *scale estimator* refers to a class of statistical characterizations of the "spread" or "scatter" of a sequence $\{x_k\}$ of N numerical data values. Like the mean, the best-known scale estimator is unquestionably the *standard deviation,*

```
library(MASS)
x <- UScereal$fibre
plot(x, xlab="Record number, k",
     ylab="Grams of fiber per serving, kth cereal")
abline(h = mean(x), lty=2, lwd=2)
abline(h = mean(x) + sd(x), lty = 3, lwd = 3)
```

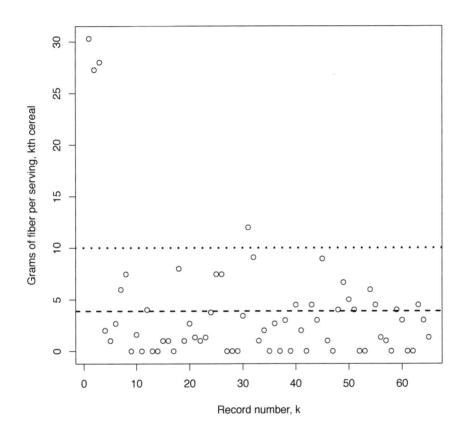

Figure 3.1: Plot of the variable `fibre` from the `UScereal` data frame.

defined as:

$$\sigma_x = \sqrt{\frac{1}{N-1}\sum_{k=1}^{N}(x_k - \bar{x})^2}, \tag{3.2}$$

where $\bar{x}$ is the mean defined in Eq. (3.1). Also like the mean, the standard deviation is easy to compute and is built into almost all data analysis platforms, spreadsheet packages, or databases. One reason the standard deviation is so popular is its strong connection with the Gaussian distribution, an important point discussed in Sec. 3.2.4. Unfortunately, the standard deviation is even more sensitive to outliers than the mean is, a fact that has important consequences for automated outlier detection procedures. This crucial topic is discussed at some length in Sec. 3.3.1.

In R, the standard deviation is implemented with the function sd, which has only one optional parameter: na.rm, with the same interpretation and default value (i.e., FALSE) as for the mean function. Applying the sd function to the fibre variable from the UScereal data frame used to illustrate the mean function gives the standard deviation as 6.1334039. As noted, the dotted line in Fig. 3.1 lies one standard deviation above the mean; the fact that all but four of the 65 points in this data sequence lie within one standard deviation of the mean gives a rough, informal illustration of the interpretation of the standard deviation as a "typical measure of spread" in the data. We can make this interpretation more precise by introducing some simple ideas from statistics, a point discussed further in Sec. 3.2.4, but this precision comes at the expense of assumptions about how the data values are distributed. As with any working assumption, these assumptions can be wrong, sometimes causing our interpretations to be very wrong. This possibility is particularly important in the automated detection of outliers, disussed in Sec. 3.3.1.

3.2.3   Limitations of simple summary statistics

Fig. 3.2 shows plots of four small datasets, collectively known as "the Anscombe quartet" after the author of the paper that used them to illustrate the limitations of simple summary statistics like the mean and standard deviation [3]. These plots were generated from the anscombe data frame in the datasets package in R, which includes the 11 pairs of x and y observations for each of these datasets. The apply function discussed in Chapter 7 can be used to apply a single function to all rows or columns of a data frame. Using it here with the mean function gives the following summary of the mean values for the x and y variables in all four datasets, corresponding to the eight columns of the anscombe data frame:

```
apply(anscombe, MARGIN=2, mean)

##       x1       x2       x3       x4       y1       y2       y3       y4
## 9.000000 9.000000 9.000000 9.000000 7.500909 7.500909 7.500000 7.500909
```

Note that the means of the four x values are identical to six decimal places, while the means for the four y values differ by approximately 0.01%. The agreement between the standard deviations is comparably impressive—again, the values are identical to six decimal places for the x values, and the y values differ by less than 0.06%:

```
apply(anscombe, MARGIN=2, sd)

##       x1       x2       x3       x4       y1       y2       y3       y4
## 3.316625 3.316625 3.316625 3.316625 2.031568 2.031657 2.030424 2.030579
```

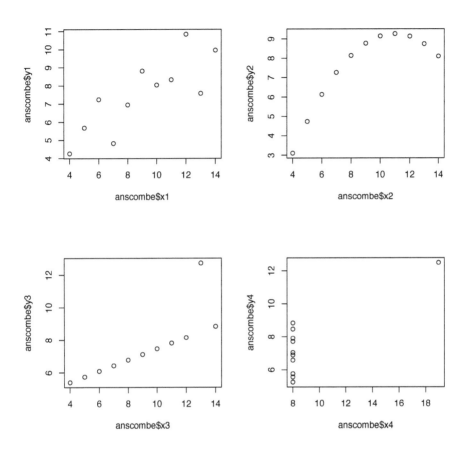

Figure 3.2: Plots of the four "Anscombe quartet" data examples.

Characterizing these four datasets in terms of their means and standard deviations alone, we might be tempted to declare them identical or nearly so, but the plots shown in Fig. 3.2 make it clear that this is not the case at all.

If we apply different characterizations, we may be able to see evidence of the differences between these four datasets. As a specific example, comparing the medians introduced in Sec. 3.3.2 for these variables yields results that are no longer identical for all of the x values and that show differences between all of the y sequences:

```
apply(anscombe, MARGIN=2, median)

##    x1    x2    x3    x4    y1    y2    y3    y4
## 9.00  9.00  9.00  8.00  7.58  8.14  7.11  7.04
```

We might be tempted to conjecture on the basis of this example that the difficulty somehow lies with the mean, that if we always used the median—or, if not the median, some other "yet better" descriptive statistic—we could avoid the difficulty illustrated here. Unfortunately, this is not the case: even though the median gave more informative results here, we could construct other examples where datasets differed in important ways but their median values were identical. *The real point of this discussion is that, because of their simplicity, descriptive statistics cannot give us a complete view of an entire dataset, even a small one like those considered here.* In particular, the four datasets considered here have only two variables with 11 observations each: we can expect summary statistics to be still less complete characterizations of datasets with thousands or millions of rows and dozens or hundreds of variables. That said, the partial characterizations provided by summary statistics can still be useful, especially when two descriptive statistics that should give "the same" results are computed and compared, an important idea discussed further in Chapter 9.

3.2.4   The Gaussian assumption

One of the key characteristics of numerical data—whether in the form of variables that are originally numeric, or numeric characterizations of non-numeric data—is *imprecision*. Alternatively known as *uncertainty* or *inaccuracy*, imprecision is an essential characteristic of real data: even under pristine measurement conditions, it is not possible to obtain *exact* measurements of physical variables, and *typical* measurements are at best accurate to a few percent [58, Sec. 2.4]. Similarly, numerical data describing demographics or other social or business characteristics may be deliberately rounded off [45] or imprecise due to limitations in our knowledge, proxy estimates of quantities that are not measurable directly (e.g., sales volumes of specific products sold by privately-held competitors), or a variety of other causes. For all of these reasons, it is important to acknowledge the presence of uncertainty in numerical data, and to develop practical strategies for dealing with it. This is most commonly done using the mathematical machinery of statistics. While this book does *not* provide a comprehensive introduction to this important subject, the following paragraphs provide a brief introduction to probability distributions, which form the basis for the statistical approach to treating uncertainty in numerical data.

Without question, the most common description of uncertainty in numerical data is the *Gaussian* or *normal* distribution that is assumed to describe the behavior of many physical variables, corresponding to the "bell-shaped curve" shown in Fig. 3.3. One specific example we will consider in detail in Sec. 3.2.5 is the weights of chicks at various ages; for this example, we will see that the Gaussian distribution appears to be a fairly reasonable data description, while in other examples, this is not the case at all. When our data does approximately conform to the Gaussian distribution, we expect to see most values near the mean of the distribution—which corresponds to the peak in the bell curve—with values far from the mean increasingly rare as the distance increases. One advantage of assuming a Gaussian data distribution is that it allows us to make

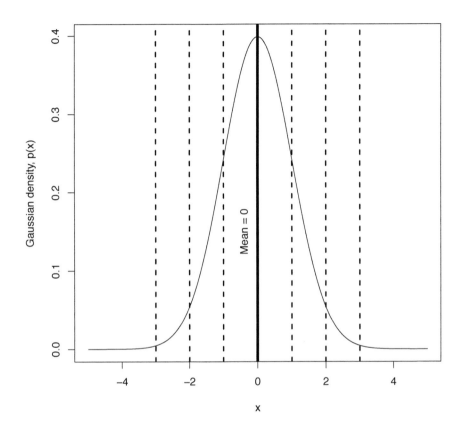

Figure 3.3: Gaussian probability density function, with lines at the mean (solid line), mean $\pm\sigma$, mean $\pm2\sigma$, and mean $\pm3\sigma$ (dashed lines).

these statements more precise. For example, under the Gaussian distribution, approximately 68.3% of all data values lie within one standard deviation of the mean, and 99.7% lie within three standard deviations. This last observation forms the basis for the "three sigma edit rule," a popular outlier detection strategy described in Sec. 3.3.2; unfortunately, this is a case where Gaussian intuition often does not serve us well, a point that is also discussed in Sec. 3.3.2.

Mathematically, the density function defining the Gaussian distribution— i.e., the "bell-shaped curve" shown in Fig. 3.3—has the following form:

$$p(x) = \frac{1}{\sqrt{2\pi\sigma}} \exp\left\{ -\frac{1}{2} \left(\frac{x - \mu}{\sigma} \right)^2 \right\}, \tag{3.3}$$

where μ is the mean of the distribution and σ is the standard deviation. *An* important point here is that the Gaussian distribution is completely defined by these two constants, μ and σ. In practical terms, if our data is well approximated by a Gaussian distribution, knowing the mean and standard deviation allows us to compute many other data characteristics. As one specific example, we can compute the probability of observing values more extreme than $\mu \pm t\sigma$ for various values of t, as described below.

The Gaussian probability density function $p(x)$ defined in Eq. (3.3) is probably the best-known characterization of the distribution among non-statisticians, but other closely related characterizations exist and are useful in various circumstances. A particularly important characterization is the *cumulative distribution function (CDF)* $P(z)$, which describes the probability that the random variable x exhibits a value no larger than z. This function may also be viewed as the area under the density function $p(x)$ from $x = -\infty$ to $x = z$; mathematically, these relationships are expressed as:

$$P(z) = \text{Probability } \{x \leq z\} = \int_{-\infty}^{z} p(x)dx. \tag{3.4}$$

(For those with little or no prior exposure to calculus, don't worry about the last part of this equation: it is included as a conceptual aid for those who find it helpful. It is also used to derive the following result, which simplifies to something that can be used without any knowledge of calculus.) Unfortunately, the function $P(z)$ does not have a simple representation like Eq. (3.3), but its values can be computed numerically, and they are provided by the R function pnorm discussed below. From Eq. (3.4), it follows that the probability that a Gaussian variable x lies in an interval $a \leq x \leq b$ is given by:

$$\begin{aligned} \text{Probability}\{a \leq x \leq b\} &= \int_{a}^{b} p(x)dx, \\ &= \int_{-\infty}^{b} p(x)dx - \int_{-\infty}^{a} p(x)dx, \\ &= P(b) - P(a). \end{aligned} \tag{3.5}$$

As a specific application of this result, note that we can use it to compute the probability of observing values more extreme than $\mu \pm t\sigma$ for any $t > 0$. Specifically, note that the probability that x lies between $\mu - t\sigma$ and $\mu + t\sigma$ is:

$$\text{Probability}\{\mu - t\sigma \leq x \leq \mu + t\sigma\} = P(\mu + t\sigma) - P(\mu - t\sigma). \tag{3.6}$$

In R, the two probabilities on the right-hand side of Eq. (3.6) are easily computed using the pnorm function, leading to the specific results cited above (e.g., the 99.7% probability of falling within three standard deviations of the mean). To calculate the probability that an observation lies *outside* these limits, note that:

$$\text{Probability}\{x < a \text{ or } x > b\} = 1 - \text{Probability}\{a \leq x \leq b\}. \tag{3.7}$$

From this, it follows that the probability of observing a Gaussian data value more than three standard deviations from the mean is only 0.3%.

An important feature of the cumulative distribution function is that it is both monotonically increasing and continuous, implying it is invertible [48, p. 181]. In practical terms, this means that the following defining equation for the *distribution quantile* x_q associated with a probability q has a unique solution:

$$P(x_q) = q \;\Rightarrow\; x_q = P^{-1}(q), \tag{3.8}$$

where $P^{-1}(q)$ represents the inverse of the cumulative distribution function. In R, the cumulative distribution function for the Gaussian distribution can be computed with the **pnorm** function, and its inverse—called the *quantile function*—can be computed with the **qnorm** function.

An important practical application of the quantile function is that it allows us to ask the opposite question to that considered above: if we want to capture, say 90% of the data range, how far out must we go into the tails of the distribution? For this specific case, the answer is given by:

```
qnorm(c(0.05, 0.95))

## [1] -1.644854  1.644854
```

This result says that, for a Gaussian distribution with zero mean and unit standard deviation, there is a 5% probability of observing data samples more than 1.64 standard deviations below the mean, and a 95% probability of observing data samples less than 1.64 standard deviations above the mean. Thus, from Eq. (3.8), there is a 10% probability of observing values outside the range $-1.64 \le x \le 1.64$. Results like this provide a practical basis for constructing *confidence intervals*, an important topic discussed in Chapter 9.

3.2.5 Is the Gaussian assumption reasonable?

Despite its overwhelming historical popularity as a description for the behavior of numerical data, the Gaussian distribution is not always a reasonable approximation, for a variety of different reasons. Some of these reasons and some of the important ways the Gaussian distribution can be inadequate are discussed in Sec. 3.3.1 and in Chapter 9, but the following paragraphs first address the practical question, "how can we tell whether this assumption is reasonable or not?" Specifically, the following discussion introduces and illustrates three graphical tools that can be used to give approximate answers to this question:

1. *histograms*, probably most familiar but least effective;

2. *density plots*, essentially smoothed histograms;

3. *quantile-quantile (QQ) plots*, probably least familiar but most effective.

Since the histogram is the best known of these graphical characterization tools, it represents the most reasonable place to start. Essentially, this characterization is constructed by first grouping the data values into a finite collection of "bins" or "buckets," and then plotting either the number data observations

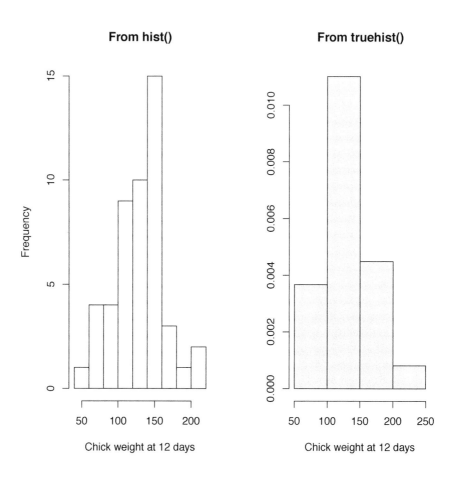

Figure 3.4: Two default histograms of measured chick weights at 12 days.

falling into each bin, or a scaled version of these counts that may be interpreted as a probability density estimate. In R, there are at least two ways of constructing histograms: the hist function that is part of the graphics R package, and the truehist function from the MASS package, both of which are available in essentially all R installations. Example histograms generated using both of these functions are shown in Fig. 3.4, generated from 49 chick weights measured 12 days after birth, from the ChickWeight data frame in the datasets R package. The R code used to generate these plots was:

```
par(mfrow=c(1,2))
index12 <- which(ChickWeight$Time == 12)
wts12 <- ChickWeight$weight[index12]
hist(wts12, main = "From hist()", xlab="Chick weight at 12 days")
truehist(wts12, main = "From truehist()", xlab="Chick weight at 12 days")
```

Thus, the plot on the left was generated with the `hist` function and it illustrates the first histogram description given above: each vertical bar in the plot represents the count of records whose `weight` value falls into the indicated bin. The plot on the right was generated using the `truehist` function from the `MASS` package, and this plot is different in two important aspects. First, it is based on different bins, so the plot consists of fewer and wider vertical bars, and second, it has been scaled so the vertical axis represents an estimated probability density instead of the number of records in each bin. As the following example illustrates, the advantage of this scaling is that it allows us to compare the histogram with estimated probability density functions. These estimates may be constructed either by assuming a Gaussian distribution and plotting the Gaussian density with the same mean and standard deviation as the data, or by using the density estimation function described next. The key point here, however, is that both of the histograms in Fig. 3.4 give only very crude assessments of the data distribution. In particular, the reasonableness of the Gaussian assumption for the data is not obvious from either of these plots.

The second graphical tool listed above—the *nonparametric density estimate*—represents a *smoothed* histogram and is shown in Fig. 3.5. Detailed descriptions of how either histograms or density estimates are constructed are beyond the scope of this book (for a discussion of both topics, refer to Chapter 8 of *Exploring Data* [58]), but these plots are easily generated in *R* with the `density` function. The adjective "nonparametric" here refers to the fact that no specific (i.e., parametric) distributional form is assumed in constructing this density estimate. Conversely, another way of constructing a probability density estimate would be, as mentioned above, to first *assume* a Gaussian distribution and then construct the "best approximating" Gaussian density by matching its mean and standard deviation to the corresponding values computed from the data. For this example, the following computations accomplish this task:

```
mu <- mean(wts12)
sigma <- sd(wts12)
x <- seq(0,250,2)
px <- dnorm(x, mean = mu, sd = sigma)
```

The object `px` constructed here is the Gaussian probability density function $p(x)$ that best approximates the 12 day chick weights.

This density is shown as the dashed line overlaid in Fig. 3.5 on top of the `truehist` histogram from Fig. 3.4; note that the difference in appearance between the two versions of this histogram is primarily due to the wider range of x-values in Fig. 3.5. The solid line in this figure represents the nonparametric density estimate generated using the `density` function applied to the chick weight data vector `wts12`. Comparing the two curves in this plot, the shape of the (solid) nonparametric density estimate does appear similar enough to the (dashed) parametric Gaussian density curve to suggest that the Gaussian distribution may be a reasonable approximation here. Still, comparisons like this one can be somewhat misleading, motivating the alternative described next.

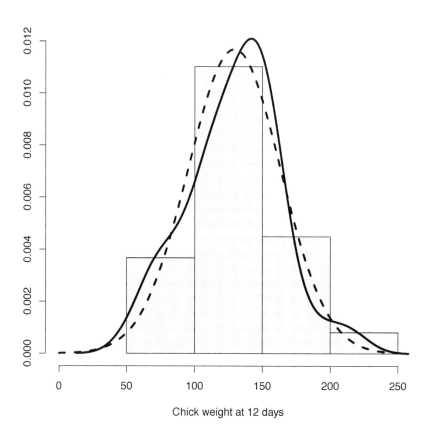

Figure 3.5: Histogram, Gaussian density, and nonparametric density estimate for the 12 day chick weight data.

A *quantile-quantile plot*, or more simply a *QQ-plot*, is a plot like that shown in Fig. 3.6 that gives an informal graphical assessment of the reasonableness of a specified data distribution. While a detailed discussion of the construction of these plots is beyond the scope of this book, the basic idea is to, first, sort the observed data values from smallest to largest, and then construct a "warped" x-axis, such that if the data values are drawn from the chosen reference distribution, a plot of the sorted data values against this warped x-axis should fall approximately on a straight line. The example in Fig. 3.6 shows a normal QQ-plot for 49 measured chick weights, obtained from the ChickWeight data frame from the R datasets package. This data frame gives the age in days for 45 to 50 chicks, along with their measured weight in grams, a unique identifier for each chick, and a number indicating which of four diets they were raised

```
qqPlot(wts12, xlab="Chick weight at 12 days")
```

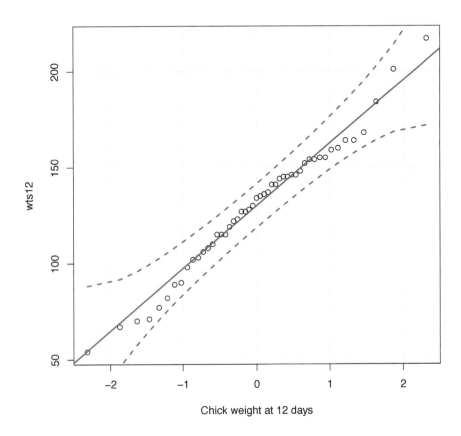

Figure 3.6: Normal quantile-quantile plot of chick weights at 12 days.

on. The QQ-plot shown in Fig. 3.6 was constructed from the chick weights measured at 12 days, irrespective of their diet. A useful ancillary feature of the qqPlot function from the **car** package used to construct Fig. 3.6 is the inclusion of both a reference line and 95% confidence limits around this line (for an introductory discussion of confidence intervals, refer to Chapter 9). If the points fall within these reference lines—as they do here—this indicates that the observed data values conform reasonably well to the assumed data distribution. Here, this reference distribution is Gaussian, the default for the qqPlot function; as discussed in Chapter 9, other distributions are easily specified for this function.

While a detailed discussion of the math behind QQ-plots is well beyond the scope of this book, they are included here because they are an extremely useful diagnostic tool for characterizing numerical variables. They can help us identify cases where classical analysis tools like means and standard deviations that are based on Gaussian working assumptions are probably safe to use, but more importantly, they can help us identify important anomalies in a dataset, regardless of the data distribution. For a discussion of the theory behind QQ-plots—both for the most common Gaussian case and for other continuous data distributions—refer to Chapter 8 of *Exploring Data* [58, Sec. 8.3].

Returning to the title question posed by this section—"Is the Gaussian assumption reasonable?"—the general answer is "sometimes, but not always." The key point of this discussion and of many others, both later in this chapter and throughout the rest of this book, is that "standard" working assumptions like "the data has an approximately Gaussian distribution" should *not* be taken for granted, especially for important numerical variables. Sometimes, "standard" data analysis and modeling techniques are reasonably forgiving of even fairly significant departures from this and other working assumptions, but often these departures can cause serious damage, rendering results that were developed by "standard, well-accepted methods" uselessly incorrect or even damaging if accepted without question. The following section introduces a number of important anomalies that can occur in numerical data, with examples of some of their consequences.

3.3   Anomalies in numerical data

Fig. 3.7 shows a normal QQ-plot for the variable `fibre` from the `UScereal` data frame in the `MASS` package that was plotted in Sec. 3.2.1 in connection with the discussion of the mean. This QQ-plot highlights two important data anomalies examined in the following sections. First, the upper tail of this distribution—in particular, the three cereals with the largest `fibre` values—emphasizes the presence of outliers in this data sequence, a topic discussed in detail in Secs. 3.3.1 and 3.3.2. Second, the flat lower tail in this QQ-plot suggests the presence of *inliers*, a topic discussed in Sec. 3.3.3. Another variable included in this data frame—`potassium`—illustrates the problem of *metadata errors*, a potentially catastrophic problem discussed further in Sec. 3.3.4. Various aspects of the problem of missing data have been discussed in the earlier sections of this chapter and in previous chapters, but Sec. 3.3.5 presents a more focused discussion of this topic, introducing the important ideas of *systematic missing data* and *disguised missing data*, this latter notion representing a particular type of metadata error. Finally, Sec. 3.3.6 provides further illustrations of the use of normal QQ-plots in detecting data anomalies of different types.

3.3.1   Outliers and their influence

The term "outlier" has been mentioned several times, and it is in sufficiently common use to serve as the title of a popular book [30]. Precisely because this term is so widely used, it is important to be explicit about what it means in

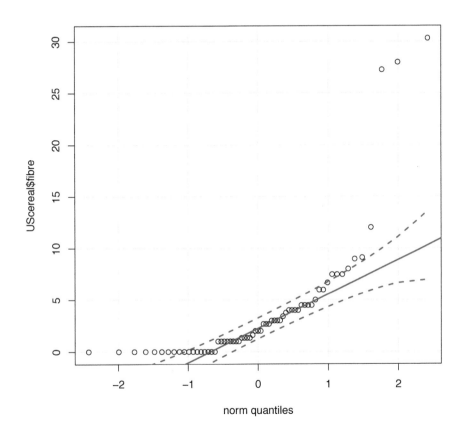

Figure 3.7: Normal quantile-quantile plot of the `fibre` variable from the `UScereal` data frame.

this book. The working definition adopted here is that of Barnett and Lewis, who define an outlier as [5, p. 4]:

> an observation (or subset of observations) which appears to be inconsistent with the remainder of that set of data.

The two plots shown in Fig. 3.8 illustrate both the notion and impact of outliers. These plots show the `fibre` values characterized in Fig. 3.7, with the first three points representing cereals whose `fibre` values exceed 24 grams per serving. These points lie well outside the range seen for the other cereals, which vary from 0 to approximately 12, so these three cereals meet the defining criteria for outliers offered by Barnett and Lewis cited above. In addition to these

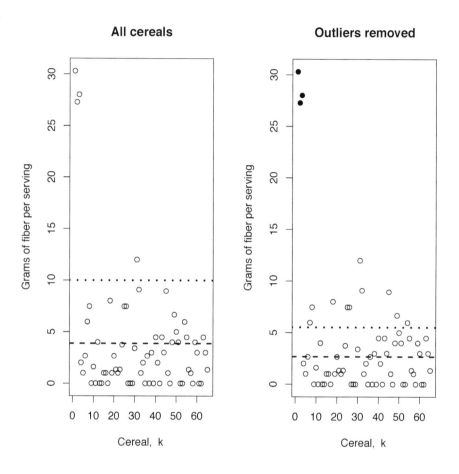

Figure 3.8: Mean (dashed lines) and mean plus one standard devation (dotted lines) overlaid with fibre values, for all cereals (left plot) and excluding the three outliers (right plot).

data points, both plots in Fig. 3.8 include two horizontal lines, representing analogous reference values, but computed in two different ways. In the left-hand plot, the dashed line corresponds to the mean `fibre` value computed from all 65 cereals in the dataset, and the dotted line above corresponds to this mean plus one standard deviation. In the right-hand plot, the three outlying cereals are marked as solid circles while the rest are marked as open circles. Here, the dashed line represents the mean, not of all `fibre` values, but only of the non-outlying values. That is, the first three elements of the `fibre` data sequence have been removed and the mean has been computed from the remaining elements. Comparing dashed lines in the left- and right-hand plots, we see that the mean computed without these outlying records is slightly smaller than for the original

Case	Mean	Standard Deviation	Median	MADM Scale
All records	3.87	6.13	2.00	2.97
Outliers removed	2.68	2.84	1.80	2.67
Percent change	−30.8%	−53.8%	−10.0%	−10.0%

Table 3.1: Influence of outliers in the fibre variable on the mean, standard deviation, median, and MADM scale.

data sequence. Similarly, the dotted line in the right-hand plot represents the mean plus one standard deviation for the data sequence with the three outliers removed. Comparing the dotted lines in the left- and right-hand plots, we see a much larger difference, reflecting the fact that the standard deviation in the absence of the outliers is *substantially* smaller than the value computed from all of the data records.

To make this comparison more explicit, Table 3.1 compares the values of the mean and standard deviation computed from the complete fibre record against the corresponding values computed after removing the three outlying cereals. In addition, the plot also shows the corresponding values for two other descriptive statistics—the median and the MADM scale estimator, both discussed in Sec. 3.3.2—computed for the same two cases. Finally, the last row of the table lists the percentage change that results in each computed value when the three outlying points are removed. For the mean, we see that removing the three outlying cereals causes the mean fibre value to decline by 30.8%, even though the points removed represent less than 5% of the data records. The change in standard deviation is even greater: removing the three outlying cereals causes a decrease of 53.8% in the standard deviation. The point of both of these examples is that the presence of outliers in a data sequence can profoundly influence even simple data characterizations; as we will see in Chapters 5 and 9, the impact of outliers can be quite profound for most "standard" analysis and modeling procedures. This fact motivates the fairly detailed discussion of outlier detection given in Sec. 3.3.2.

Before proceeding to that discussion, however, it is important to briefly discuss the "meaning" of outliers: how do they arise in our data and how should we view them? One common source of outliers is *gross measurement errors:* the recorded value is simply wrong, and wrong by a large amount. *Indeed, some people view the terms "outliers" and "gross data errors" as synonymous, but this is a serious conceptual error.* For example, the three prominent outliers in the fibre data correspond to the following breakfast cereals:

```
outlierIndex <- which(UScereal$fibre > 25)
rownames(UScereal)[outlierIndex]

## [1] "100% Bran"                  "All-Bran"
## [3] "All-Bran with Extra Fiber"
```

From these names—especially the last one—it is clear that these three cereals are special, developed to have a substantially higher fiber content than the other cereals in this collection. Thus, it is important to emphasize that outliers do not *necessarily* arise from gross data errors, although they can and do *sometimes* arise this way. In particular, outliers can arise from large data recording errors (especially when working with manually recorded data), inadvertent unit changes (e.g., most temperatures recorded in degrees Celsius but a few recorded in degrees Kelvin), or data acquisition system failures (e.g., missing data values recorded as zero for a variable that is large and positive). Conversely, non-erroneous outliers commonly arise in situations like the UScereal dataset where a few data records are obtained accurately from sources that are significantly different from the majority of those represented (e.g., special high-fiber cereals versus "ordinary" breakfast cereals, or gas mileages recorded for traditional vehicles with a few gas/electric hybrids included in the collection of vehicles). The practical importance of these outliers lies in the fact that these "special" data subsets may be present without our knowledge in the data we need to analyze: even if they represent a very small minority of the data records, their influence may be large enough to give us misleading or meaningless results if we do not recognize their presence and account for them appropriately.

3.3.2   Detecting univariate outliers

The term *univariate outlier detection* refers to the process of identifying outliers in a single variable. In this case, "inconsistent with the remainder of the data" is typically interpreted to mean "unusually far above or below the 'typical' data values." This idea can be turned into an automated outlier detection procedure by defining a computable "typical" data value, a measure of "data spread," and a threshold for "how many 'data spreads' from the 'typical value' constitutes 'unusually far' from the 'typical' value." By taking different measures for "typical," "data spread," and "threshold for unusualness," we obtain different automated univariate outlier detection procedures, with different strengths and weaknesses. The following discussions present three alternatives and briefly describe their strengths and weaknesses.

The three-sigma edit rule

Probably the best-known automatic outlier detection procedure is the "three-sigma edit rule," known in the statistics literature by various names including the *extreme Studentized deviation (ESD) identifier* [18]. The basis is the observation that, for approximately Gaussian data, points lying more than three standard deviations from the mean should occur with probability less than about

0.3%, based on the tail probability results presented in Sec. 3.2.4. This suggests the following choices for detecting outliers in a sequence $\{x_k\}$ of N numbers:

- the "typical" data value is the mean $\bar{x}$;

- the "data spread" is the standard deviation σ_x;

- the "threshold for unusualness" is $t = 3$ standard deviations.

These choices lead to the following classification rule, where $t = 3$:

$$\text{classification of } x_k = \begin{cases} \text{outlier} & \text{if } |x_k - \bar{x}| > t\sigma, \\ \text{nominal} & \text{otherwise.} \end{cases} \tag{3.9}$$

Despite its simplicity, popularity, and intuitive appeal, this outlier detection procedure often performs poorly in practice, as the following example illustrates. Following this example, two alternative outlier detection procedures are described and illustrated for the same example, and this section concludes with a brief summary of general recommendations.

Fig. 3.9 shows a plot of the `potassium` values for each cereal in the `UScereal` data frame in the `MASS` package. As discussed in Sec. 3.3.4, there is strong reason to believe the units given for this variable in the `help` file associated with the `UScereal` data frame are incorrect, but the following discussion assumes that, aside from this probable unit error, the data values are correct. As with the `fibre` variable considered earlier from this data frame, it appears that the first three cereals are outliers: these values all exceed 600, while all of the other 62 values for this variable lie between 0 and 400. The mean of the `potassium` values is approximately 159 and is indicated in the plot as the dashed line. The dotted line at approximately 700 corresponds to the upper outlier detection limit under the three-sigma edit rule. Note that while this outlier detection rule identifies the first two cereals as outliers, it misses the third cereal: although this point is clearly less extreme than the first two, it still lies "unusually far from the majority of the data values," and should be declared an outlier.

The failure of the three-sigma rule to detect this "least extreme outlier" is a consequence of the *inflation* that outliers cause for the usual standard deviation estimator. This effect was illustrated in Sec. 3.3.1 in connection with the `fibre` data example: recall that removing the three most extreme `fibre` values—visually obvious outliers in that data sequence—caused a reduction of 53.8% in the standard deviation. As a consequence, the three-sigma rule often fails to detect outliers that are present in the data sequence, a phenomenon called *masking*. In extreme cases, this effect can cause the three-sigma rule to fail completely, detecting no outliers at all in a data sequence, regardless of their magnitude [58, Sec. 7.1.2]. One way of overcoming this problem is to replace the outlier-sensitive mean and standard deviation on which the three-sigma rule is based with estimators that are less sensitive to the presence of outliers; this is the idea behind the outlier detection procedure discussed next.

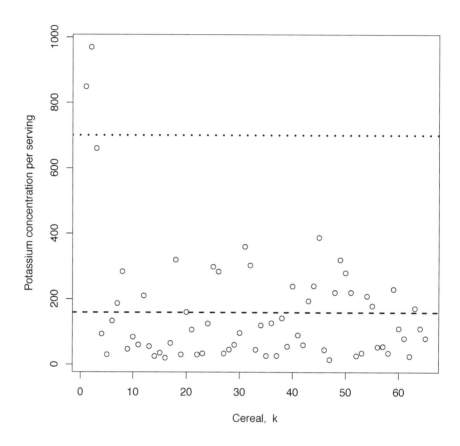

Figure 3.9: Potassium per serving from the UScereal data frame, with the mean value (dashed line) and the upper outlier detection limit from the three-sigma rule (dotted line).

The Hampel identifier

Table 3.1 in Sec. 3.3.1 showed the impact of removing three outlying cereals from the **fibre** data sequence, resulting in a decrease of 30.8% for the mean and a decrease of 53.8% for the standard deviation, as noted in the preceding discussion. This table also showed the impact of removing these outliers on two other summary statistics—the *median* and the *MADM scale estimator*—both of which showed a decrease of only 10%. The median discussed next is reasonably well known and may be viewed as an outlier-resistant alternative to the mean, while the MADM scale estimator represents an outlier-resistant alternative to the standard deviation. Replacing the mean with the median and the standard deviation with the MADM scale estimator gives rise to the *Hampel identifier*

[18], which is much less sensitive to masking effects than the three-sigma rule [58, Sec. 7.5]. For reference, this outlier detection procedure corresponds to the following choices:

- the "typical" data value is the median $x^\dagger$, defined below;

- the "data spread" is the MADM scale estimator S, defined below;

- the "threshold for unusualness" is $t = 3$ MADM scale estimates.

As we will see, this outlier detection rule gives much better results than the three-sigma edit rule for the `potassium` data, although this procedure can also fail badly, a point discussed further at the end of this section.

As noted, the median $x^\dagger$ is a reasonably well-known characterization of a numerical data sequence $\{x_k\}$, computed as follows. First, we sort the original data sequence $\{x_k\}$ from smallest value to largest to obtain the new, ordered sequence denoted $\{x_{(j)}\}$. That is, $x_{(1)}$ is the smallest value in the original data sequence (note that if the sequence includes both positive and negative values, this is the most negative value), and $x_{(N)}$ is the largest value. More generally, these values satisfy the ordering:

$$x_{(1)} \leq x_{(2)} \leq \cdots \leq x_{(N-1)} \leq x_{(N)}. \qquad (3.10)$$

Now, if N is an odd number, there is a unique "middle element" $x_{(j)}$ where $j = (N + 1)/2$, and the median of the original data sequence is this value:

$$\text{median}\{x_k\} = x_{((N+1)/2)} \text{ if } N \text{ is odd.} \qquad (3.11)$$

Conversely, if N is an even number, there is no single "middle element" in this sequence: both $x_{(N/2)}$ and $x_{((N/2)+1)}$ lie squarely in the middle of this sequence. In this case, the median is defined as the average of these two values:

$$\text{median}\{x_k\} = [x_{(N/2)} + x_{((N/2)+1)}]/2 \text{ if } N \text{ is even.} \qquad (3.12)$$

Like the mean, the median is also very widely available as a built-in function. In R, it is computed with the generic function `median`, which returns "reasonable" values for cases where the defining conditions in Eq. (3.11) or Eq. (3.12) can be computed. Also like the `mean` function, `median` has an optional parameter `na.rm`, with the same interpretation and default value (i.e., `FALSE`), so the function returns `NA` if there are missing data values unless we specify `TRUE` for this parameter.

Although not nearly as well known as the median, the scale estimator that is in many respects the median's analog is the *MADM* or *MAD* scale estimator, an abbreviation for *median absolute deviation from the median*. If we let $x^\dagger$ denote the median of the data sequence $\{x_k\}$, the MADM scale estimator is defined as:

$$\text{MADM}(x) = 1.4826 \times \text{median}\{|x_k - x^\dagger|\}. \qquad (3.13)$$

To understand the rationale for this estimator, proceed as follows. First, note that the sequence $\{x_k - x^\dagger\}$ represents the deviation of each data observation

x_k from the median reference value $x^\dagger$. Thus, we can interpret the sequence $\{|x_k - x^\dagger|\}$ of absolute values of these deviations as a sequence of distances that each point lies from the reference value. Taking the median of this sequence of distances then tells us how far a "typical" observation lies from the median. By itself, this number gives us a measure of the "scatter" or "spread" of the data around the median, but it tends to (usually) be smaller than the standard deviation. The correction factor 1.4826 accounts for this difference, adjusting it so that on average, the MADM scale estimate should be equal to the standard deviation for Gaussian data. (In statistical terms, this makes $MADM(x)$ an unbiased estimate of the standard deviation for Gaussian data; for those interested, a detailed derivation is given in *Exploring Data* [58, Sec. 7.4.2].) In R, the MADM scale estimator is implemented as the built-in function `mad`, which has the optional parameter `na.rm` with the default value `FALSE`.

Fig. 3.10 again shows the potassium per serving values from the `UScereal` data frame, in the same format as Fig. 3.9. Here, however, the dashed line corresponds to the median instead of the mean, and the dotted line represents the upper outlier detection limit of the Hampel identifier instead of the three-sigma edit rule. For this example, the Hampel identifier gives dramatically better results: the upper outlier detection limit corresponds almost exactly with the largest of the points our eye classifies as "nominal" or "not-outlying," and all three of the visually obvious outliers lie well above this limit. *As popular automobile ads used to emphasize, however, "actual mileage may vary": there are some datasets where the Hampel identifier performs very badly.* This point is discussed further and illustrated in Chapter 9; here, it is enough to note that, while the three-sigma edit rule is often not aggressive enough in finding outliers in the data, the Hampel identifier is sometimes too aggressive in finding them, declaring "nominal" data points to be outliers. Conversely, there are cases where the full aggressiveness of the Hampel identifier is needed, a point illustrated at the end of this section.

The boxplot outlier rule

In many cases, a useful "intermediate" outlier detection procedure—more aggressive than the three-sigma edit rule but less aggressive than the Hampel identifier—is the boxplot rule, described next. The formulation of this outlier detection procedure is slightly different than the previous two, in that we define *upper outliers* as points that are "unusually far above" a "typical upper value" and *lower outliers* as points that are "unusually far below" a "typical lower value." More specifically, the boxplot rule is based on these four numbers, all discussed further in the following paragraphs:

- the "typical upper value" is the *upper quartile* x_U, discussed below;

- the "typical lower value" is the *lower quartile* x_L, discussed below;

- the "data spread" is the *interquartile distance* Q, defined below;

- the "threshold for unusualness" is $t = 1.5Q$.

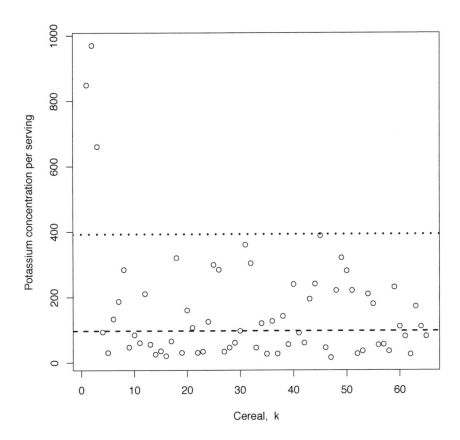

Figure 3.10: Potassium per serving from the UScereal data frame, with the median value (dashed line) and the upper outlier detection limit from the Hampel identifier (dotted line).

Thus, the boxplot outlier detection rule yields the following three classifications:

$$\text{classification of } x_k = \begin{cases} \text{upper outlier} & \text{if } x_k > x_U + 1.5Q, \\ \text{lower outlier} & \text{if } x_k < x_L - 1.5Q, \\ \text{nominal} & \text{otherwise.} \end{cases} \quad (3.14)$$

In Chapter 1, Tukey's five-number summary was presented both as a simple way of characterizing a variable and as the basis for the boxplots that were also introduced there and which are discussed further in Sec. 3.4.2. This summary is based on quantiles, computed by the quantile function. Like the mean and median functions, the quantile function returns the value NA if the sequence $\{x_k\}$ contains missing values, but this may be suppressed by specifying na.rm

= TRUE. By default, this function returns the quantiles defining Tukey's five-number summary, which are:

1. the $q = 0$ quantile, $x_{0.0}$, is the sample minimum: 0% of the data observations fall below this value;

2. the $q = 0.25$ quantile, $x_{0.25}$, is the lower quartile: 25% of the data observations fall below this value;

3. the $q = 0.5$ quantile, $x_{0.5}$, is the sample median: 50% of the data observations fall below this value;

4. the $q = 0.75$ quantile, $x_{0.75}$, is the upper quartile: 75% of the data observations fall below this value;

5. the $q = 1$ quantile, $x_{1.0}$, is the sample maximum: 100% of the data observations fall on or below this value.

Thus, the "typical" upper and lower data values in the boxplot outlier rule are $x_U = x_{0.75}$ and $x_L = x_{0.25}$, both of which are easily computed with the quantile function by specifying the optional argument probs:

```
xU <- quantile(x, probs = 0.75)
xL <- quantile(x, probs = 0.25)
```

The *interquartile distance (IQD)* or *interquartile range (IQR)* Q is defined as the difference between these quartiles:

$$Q = x_U - x_L. \qquad (3.15)$$

The following custom R function, discussed in Chapter 7, computes this value for any numerical data vector x:

```
ComputeIQD <- function(x){
  #
  quartiles <- quantile(x, probs = c(0.25, 0.75))
  IQD <- as.numeric(quartiles[2] - quartiles[1])
  return(IQD)
}
```

(The use of the built-in function as.numeric here suppresses the names attached to the return value from the quantile function; failure to do so causes the ComputeIQD function's return value delta to have the name "75%" attached to it, a potential source of confusion.)

Fig. 3.11 shows the results of applying the boxplot rule to the potassium data sequence from the UScereal data frame. Here, the dashed line represents the upper quartile, and the fact that it runs through the "upper portion" of the *nominal* data supports the interpretation of this value as a "typical upper value." The dotted line represents the upper detection limit for the boxplot rule: comparing this line with the dotted lines in Figs. 3.9 and 3.10 shows

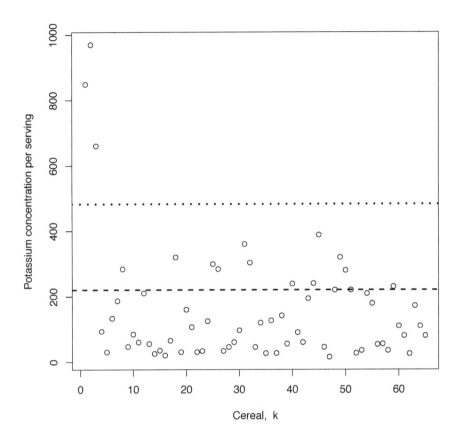

Figure 3.11: Potassium per serving from the UScereal data frame, with the upper quartile (dashed line) and the upper outlier detection limit from the boxplot rule (dotted line).

the intermediate behavior of the boxplot rule. Specifically, note that the upper detection limit for the boxplot rule is approximately 482, smaller than the three-sigma limit of 700 and larger than the Hampel limit of 393. Like the Hampel identifier, note that the boxplot rule clearly separates the potassium data values into "nominal" and "outlying" groups that correspond to what our eye expects: the first three cereals are declared outliers, while the others are declared "nominal." Like the Hampel identifier, the boxplot rule can also fail, a point discussed further in Chapter 9.

Summary: some practical advice on outlier detection

Outlier detection is discussed further in Chapter 9, but a few practical recommendations are offered here, to facilitate the use of the procedures just described.

First, it is important to note that, while outlier *detection* can be automated as a mathematical procedure—the three-sigma edit rule, the Hampel identifier, and the boxplot rule represent three specific examples—the *interpretation* of outliers cannot be automated, as it is *not* a mathematical problem. Further, as the preceding examples have demonstrated, these different procedures frequently identify different sets of points as outliers. For these reasons, the following general strategy is recommended:

1. Apply all three of these outlier detection procedures to your data sequence and carefully examine the results, comparing: (1), the number of outliers detected by each procedure; (2), the data values declared to be outliers; and (3), the range of data values *not* declared to be outliers;

2. If at all possible, apply application-specific "reasonableness tests" to both the outlier limits and the ranges of non-outlying values: does this "nominal range" seem reasonable? And do the outlying values seem extreme enough to be suspicious?

3. Examine plots of the data, either with the outlier detection thresholds indicated on the plot, or with the outlying points marked with different point shapes or colors to highlight which ones have been deemed "suspicous" by these outlier detection procedures.

Generally, as in the `potassium` variable in the `UScereal` data frame considered here, the three-sigma edit rule is the least aggressive, finding the fewest outliers, the Hampel identifier is the most aggressive, finding the most outliers, and the boxplot rule yields results that are intermediate between these two extremes. There are cases, however, where this ordering is not observed, as in the example considered next. To facilitate these comparisons, a simple function `FindOutliers` is described in detail in Chapter 7. Briefly, this function applies all three of the outlier detction procedures just described and returns a list that contains the following four data frames as elements:

1. the `summary` element is a three-row data frame, with one row summarizing the results of each outlier detection procedure;

2. the `threeSigma` element is a three-column data frame, with one row for each outlier identified, giving the record index k, the outlying value x_k, and the designation as either an upper ("U") or lower ("L") outlier;

3. the `Hampel` element is a three-column data frame, in the same format as the `threeSigma` element;

4. the `boxplotRule` element is also a three-column data frame, in the same format as the `threeSigma` element.

Applying this function to the **potassium** variable from the **UScereal** data frame gives the following results. The **summary** element of the list returned when the **FindOutliers** function is applied to the **potassium** variable contains the following information:

```
fullSummary <- FindOutliers(UScereal$potassium)
fullSummary$summary

##            method  n nMiss nOut     lowLim    upLim minNom    maxNom
## 1      ThreeSigma 65     0    2 -381.7460 699.9855     15  660.0000
## 2          Hampel 65     0    3 -199.5921 392.7740     15  388.0597
## 3     BoxplotRule 65     0    3  -42.5000 482.5000     15  388.0597
```

The first column gives the name of the outlier detection rule applied, the second and third give the total number of observations and the number of missing observations (i.e., data records coded as "NA"), and the fourth column gives the number of outliers detected by each rule. Here, we see that the three-sigma edit rule identifies only two outliers, while the other two each identify three outliers. The next two columns in the data frame give the lower and upper outlier detection limits, and the final two columns give the lower and upper limits of the non-outlying data values.

The **threeSigma** element of the list returned by **FindOutliers** contains one row for each of the two points identified as outliers by the three-sigma edit rule, giving the row number of each observation in the data frame (**index**), the outlying value, and its designation as either an upper ("U") or a lower ("L") outlier:

```
fullSummary$threeSigma

##    index   values type
## 1      1 848.4849    U
## 2      2 969.6970    U
```

Similarly, the **Hampel** and **boxplotRule** elements give the corresponding results for these other two outlier identification rules:

```
fullSummary$Hampel

##    index   values type
## 1      1 848.4849    U
## 2      2 969.6970    U
## 3      3 660.0000    U

fullSummary$boxplotRule

##    index   values type
## 1      1 848.4849    U
## 2      2 969.6970    U
## 3      3 660.0000    U
```

Finally, it is worth emphasizing the extent to which looking at plots of data can be useful in determining which—if any—of these outlier identifications to

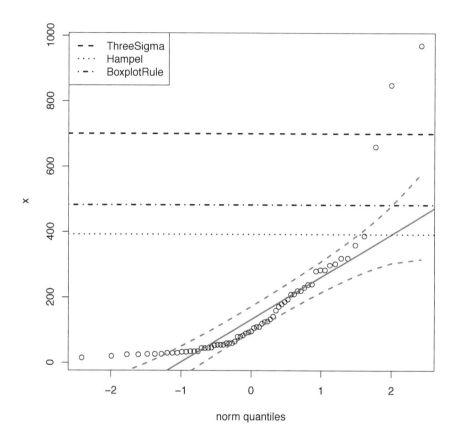

Figure 3.12: Normal QQ-plot of the UScereal potassium variable, with three upper outlier detection limits.

accept as a basis for either data editing or setting observations aside for special scrutiny. The plots shown in Figs. 3.9 through 3.11 illustrate the utility of including outlier detection limits on simple scatterplots of the data values versus their record index, but these reference lines can also be included in QQ-plots for an even more complete view of the data distribution. The next two examples show how these reference lines can help us see the differences between these outlier detection strategies and decide which of their results to accept.

The plot in Fig. 3.12 consists of the normal QQ-plot generated by the qqPlot function from the car package, overlaid with the upper outlier detection limits of the three outlier detection rules considered here. It is clear from this plot that the three-sigma edit rule is the least aggressive, failing to detect the third outlier that is obvious from the QQ-plot, while both of the other outlier detection rules

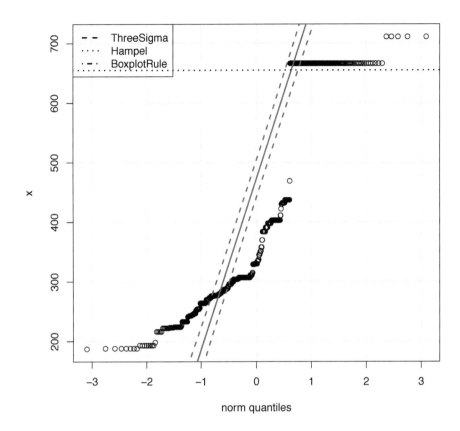

Figure 3.13: Normal QQ-plot of the tax rate data from the Boston data frame, with the upper outlier detection limits for the Hampel identifier: the three-sigma and boxplot limits lie above the upper limit of the plot and are not shown.

correctly identify this point as unusual. The lower outlier detection limits are not shown in this plot because these limits are all negative, and it is not possible to modify the y-axis limits of the qqPlot function, so these negative detection limits cannot be shown.

Fig. 3.13 shows a normal QQ-plot of the tax variable from the Boston data frame in the MASS package, again with upper outlier detection limits marked as horizontal lines. As before, those reference lines that lie beyond the range of the data cannot be shown, so the only line that actually appears in this plot is the dotted line for the Hampel identifier's upper detection limit. This example is particularly interesting, because there is a large collection of *repeated* outliers in this data sequence, as indicated by the "flat stretch" in the upper tail of the QQ-plot. The fact that these points—together with five more points that are even more extreme, but again, all sharing the same value—are well separated from the majority of the data values makes them outliers under the working definition considered here, even though neither the three-sigma edit rule nor the boxplot rule identify them as such. The `summary` element of the results returned by the `FindOutliers` function discussed above gives us a clearer picture of what is happening here:

```
taxSummary <- FindOutliers(Boston$tax)
taxSummary$summary

##       method   n nMiss nOut    lowLim    upLim minNom maxNom
## 1  ThreeSigma 506     0    0 -97.37419  913.8485    187    711
## 2      Hampel 506     0  137   5.31060  654.6894    187    469
## 3 BoxplotRule 506     0    0  85.50000 1246.5000    187    711
```

The `tax` data values range from 187 to 711, which corresponds to the "nominal" data range for both the three-sigma edit rule and the boxplot rule, since neither of these procedures identify any points as outliers in this data sequence. In contrast, the Hampel identifier declares 137 points to be outliers, corresponding to the 5 points with the value 711 and the "plateau" of 132 points, all with the value 666. The range of the "nominal" data values under this outlier designation is from 187 to 469, well below these outlying values. Note that the presence of these repeated outliers has inflated both the standard deviation and the interquartile range relative to this nominal data subset, to the extent that the upper three-sigma detection limit is almost 50% greater than the maximum data value, and the boxplot rule detection limit is even larger.

The key points of example are, first, the utility QQ-plots in visualizing data distributions, and second, the fact that simple rules of thumb like "the boxplot rule is a safe intermediate choice between the over-aggressive Hampel identifier and the overly conservative three-sigma edit rule" do not always hold.

3.3.3   Inliers and their detection

The term *inliers* is sometimes used to refer to the "nominal" or "non-outlying" points in a data sequence, but this book defines this term two different ways, both distinct from this meaning. The first defines an inlier as *"a data value that lies in the interior of a statistical distribution and is in error"* [20, 78]. One important source of inliers is *disguised missing data*, a problem discussed further in Sec. 3.3.5, but another common source is data collected from business transaction systems that require an entry in order to complete the transaction, applied in cases where no acceptable entry is correct. The following example, described by Adriaans and Zantinge, is typical [1, p. 84]:

> Recently a colleague rented a car in the U.S.A. Since he was Dutch, his post code did not fit into the fields of the computer program. The car hire representative suggested that she use the zip code of the rental office instead.

By itself, this definition of inliers refers to a situation that is potentially serious, but it does not provide the basis for any mathematical detection procedure, in contrast to the case of outliers just discussed. Fortunately, a common characteristic of inliers in numerical data is that they often represent *the same value, repeated unusually frequently.* If we take this as our working definition of an inlier, we now have a basis for detecting them: count the number of times each value occurs, and look for outliers in this count sequence.

One of the important points discussed in Chapter 9 is that, in continuously distributed random data (e.g., Gaussian data sequences), repeated values cannot occur. For real numerical data, repeated values are likely to occur at least occasionally due to finite precision effects. For example, if we have a variable that lies in the range between 0 and 1 and is recorded to only two decimal places, any collection of more than 101 observations necessarily has at least one repeated value. Typically, however, the number of repetitions is small in numerical data. As a specific example, consider the `statusquo` variable in the `Chile` data frame from the `car` package. This data frame has 2700 records and this variable assumes 2093 unique values. Using the `table` function to count the number of times each value occurs, we obtain the following range of counts:

```
Tbl <- table(Chile$statusquo, useNA = "ifany")
range(Tbl)

## [1]   1 201
```

We can apply the `table` function again to tabulate these counts:

```
table(Tbl)

## Tbl
##    1    2    3    4    5    6    8    9   13   17   18   21   61  201
## 1955   72   22   19    8    5    4    1    1    2    1    1    1    1
```

From this table, we see that the overwhelming majority of values occur only once, but a few values appear much more frequently. *This example also illustrates an important practical point: since the vast majority of numerical values occur only once, the median count is 1 and the MAD scale estimate is zero:*

```
median(as.numeric(Tbl))

## [1] 1

mad(as.numeric(Tbl))

## [1] 0
```

This behavior of the MAD scale estimate is called *implosion* and it represents an important failure mechanism for both the MAD scale estimate and the Hampel identifier, a problem discussed further in Chapter 9. In fact, the boxplot outlier detection rule suffers from the same problem for this case; specifically, both the upper and lower quartiles are 1, so the interquartile distance is zero:

```
quantile(Tbl)
```

```
##    0%   25%   50%   75%  100%
##     1     1     1     1   201
```

As a consequence of these scale implosions, both the Hampel identifier and the boxplot rule will declare *every data point that occurs more than once to be an outlier.* Since we expect a few data values to occcur two or three times by the finite precision argument noted above, this result seems too extreme.

Despite its ineffectiveness in many outlier detection applications, the three-sigma edit rule does not exhibit implosion: the standard deviation can only be zero if all of the data values are identical. *Thus, for inlier detection, the three-sigma edit rule should be applied to the count data sequence, and not the Hampel identifier or the boxplot outlier rule.*

As an illustration of how to use these ideas for inlier detection, consider the example of the `tax` variable from the `Boston` data frame in the `MASS` package. A plot of these data values is shown in Fig. 3.14, and it is clear that the second-largest value in this data sequence (666) is "repeated unusually frequently," qualifying it as an inlier under the definition given above. In fact, its value makes it appear to be an outlier, although only the Hampel identifier described earlier flags either of the largest values as such.

If we count the number of times each value occurs in this sequence and then tabulate these counts, we obtain:

```
Tbl <- table(Boston$tax)
table(Tbl)
```

```
## Tbl
##    1    2    3    4    5    6    7    8    9   10   11   12   14   15   30   40  132
##   13   12    5    5    6    1    6    4    3    2    2    2    1    1    1    1    1
```

Applying the three-sigma edit rule to this count sequence identifies the following values as occurring "unusually frequently":

```
mu <- mean(Tbl)
sigma <- sd(Tbl)
Tbl[which(Tbl > mu + 3 * sigma)]
```

```
## 666
## 132
```

The question of why this value is repeated so frequently is revisited in the Boston Housing data story presented in Chapter 6.

3.3.4   Metadata errors

As noted in Chapter 1, *metadata* refers to "data about data." Ideally, metadata includes detailed definitions of the variables, ranges of admissible values, number of missing observations and the notation used to indicate them, along with any

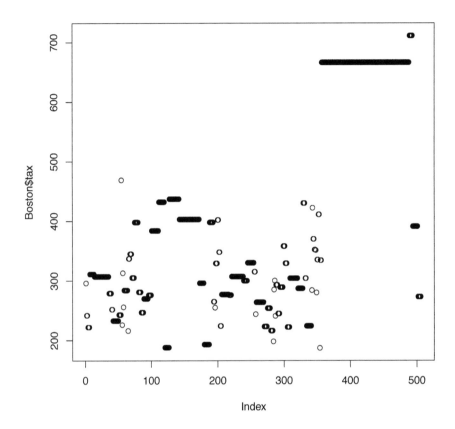

Figure 3.14: Plot of the tax values from the Boston data frame.

other noteworthy features or peculiarities. Typically, however, metadata is much less complete than we would like, and it is difficult to maintain quality control because metadata is highly unstructured. As the following examples illustrate, errors in metadata can be both extremely serious and difficult to detect.

The first example is the most extreme. On December 11, 1998, NASA launched the Mars Climate Orbiter, a robotic space probe designed to collect and send back detailed information about the Martian atmosphere, climate, and surface changes. On September 23, 1999, radio contact with the satellite was lost and never recovered. It was ultimately determined that, due to final course correction errors, the $125 million satellite burned up in the Martian atmosphere. In the end, the cause of this failure was determined to be a metadata error: values that were supposed to be expressed in units of newton-seconds were provided in pound-seconds, too large by a factor of 4.45.

The second example of a metadata error is the Pima Indians diabetes dataset from the UCI Machine Learning Repository, introduced in Chapter 1. As noted in that discussion, the metadata given on the associated website now includes the following statement:

> UPDATE: Until 02/28/2011 this web page indicated that there were no missing values in the dataset. As pointed out by a repository user, this cannot be true: there are zeros in places where they are biologically impossible, such as the blood pressure attribute. It seems very likely that zero values encode missing data. However, since the dataset donors made no such statement we encourage you to use your best judgement and state your assumptions.

As a result of the original metadata error, a number of studies characterizing binary classifiers have been published using this dataset as a benchmark where the authors were not aware that data values were missing. In some cases, quite a large fraction of the total observations are missing: as a specific example, 48.7% of the serum insulin values are missing. This example illustrates the problem of *disguised missing data*, discussed further in Sec. 3.3.5.

As a final example, consider the `potassium` variable in the `UScereal` data frame from the `MASS` package. The metadata from the `help` file indicates the units of this variable to be "grams per cup," but the maximum reported value is 969.7. If the units really are in grams, since one kilogram is equivalent to 2.2 pounds, the potassium content of one cup of this cereal would be 1.93 *pounds*. This seems highly unlikely, since the weight of one cup of granulated sugar is aproximately 0.5 pounds [41, p. 771]. Further, for a "typical" 14 ounce box of cereal, this reported weight of potassium in one cup of this "most potassium rich" cereal would correspond to the weight of 2.44 boxes. Given that values for the `sodium` variable in the `UScereal` data frame are reported to be in milligrams, it seems most likely that there is a metadata error for the `potassium` variable: the weight should be in milligrams instead of grams. While this unit error is unlikely to have the catastrophic consequences of the Mars Climate Orbiter unit error, it does illustrate the importance of asking the question, "do these data values seem reasonable?"

3.3.5   Missing data, possibly disguised

Various aspects of the problem of missing data have been discussed in earlier chapters and in previous sections of this chapter, and other aspects will be discussed in later chapters (e.g., the impact of missing data on program logic in certain cases, an issue discussed in Chapter 7). One important aspect of missing data is that there is no single, universal representation for it, a point discussed in some detail in the unclaimed bank account data example introduced in Chapter 4 and revisited in subsequent chapters. The following two sections briefly discuss two other aspects of missing data: the problem of *disguised missing data*, and the problem of *systematic missing data*.

Disguised missing data

An excellent example of *disguised missing data* is the Pima Indians diabetes dataset that has been mentioned several times previously, where the metadata originally indicated there were no missing values, but physically impossible zeros appear in certain variables (e.g., diastolic blood pressure). This dataset is available from several different sources: the original is available from the UCI Machine Learing Repository, as noted in Chapter 1, and different versions are available in various *R* packages, including a corrected version in the MASS package and a version in the mlbench package that has these zeros replaced with the missing value indicator "NA." The dataset considered here is PimaIndiansDiabetes from the mlbench package, which corresponds to the original available from the UCI Machine Learning Repository, with the unphysical zeros. The dataset is a popular benchmark for classification algorithms like those discussed in Chapter 10, where the objective is to predict the probability of a diabetic diagnosis from the other variables included in the dataset.

Analyzing this dataset without recognizing these zero values as missing data can have serious consequences. This point was emphasized by Breault, who examined the results of approximately 70 published analyses of this data, most of which appear not to have recognized these zero values as codes for missing data [8]. In these analyses, the objective is to predict the diabetic diagnosis of the patients from the other characteristics (like diastolic blood pressure) included in the dataset. Since 500 of the 768 patients characterized in the dataset are diabetic, simply declaring *everyone* diabetic results in a prediction accuracy of 65.1%, and some of the published models do little better than this. The problem of building binary classification models is discussed in detail in Chapter 10 where this example is revisited, but the point here is that failure to recognize the missing data from this dataset—and treating it appropriately— can result in significant degradation of model accuracy.

If a single numeric code is used to represent missing data—e.g., the zeros used to represent missing data in this example—these values typically appear as "plateaus" in normal QQ-plots, which serve as an excellent screening tool for detecting disguised missing data. This point is illustrated in Fig. 3.15, which shows normal QQ-plots of four variables from the Pima Indians diabetes dataset. The upper left plot characterizes the variable pressure, representing the patient's diastolic blood pressure: the pronounced "flat lower tail" in this plot represents repeated zero values in this dataset, which are biologically impossible. The upper right plot shows the normal QQ-plot constructed from the triceps skinfold thickness variable, triceps, an alternative obesity measure to the more familiar body mass index (BMI) characterized by the QQ-plot in the lower right; again, the flat lower tails in both of these plots represent zero values, which are not biologically possible. Finally, the lower left plot shows a QQ-plot constructed from the serum insulin values included in the dataset, again showing physically impossible zero values as a flat lower tail in the data distribution.

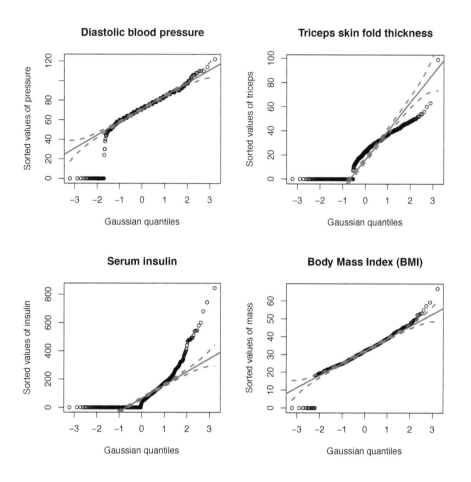

Figure 3.15: Four normal QQ-plots highlighting the zeros in four variables from the PimaIndiansDiabetes data frame.

Systematic missing data

Missing data is often assumed to be a random phenomenon. The most popular working assumption is that missing values occur *completely at random*, meaning that the probability a specific record x_k is missing from a dataset containing N observations of the variable x, along with a number of other variables, is independent of the values of any of these variables. This working assumption—referred to as the *missing completely at random (MCAR)* missing data model [53, p. 12]—is extremely popular in part because it is simple and in part because it does not introduce biases into our results. Specifically, if we omit missing data observations under this working assumption, the effect is to reduce the size of our dataset, which does reduce the precision of our results—an important idea discussed further in Chapter 9—but it does not cause our results to be

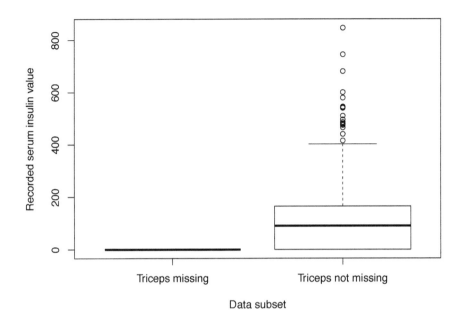

Figure 3.16: Boxplot comparisons of serum insulin values for records with triceps missing and not missing.

systematically wrong (i.e., biased). A weaker assumption is the *missing at random (MAR)* missing data model [53, p. 12], which allows the probability that x_k is missing to depend on the other non-missing values in the dataset. This situation is one we can attempt to detect and develop ways of addressing, by looking at the distributions of missing records as a function of the other variables in the dataset. For example, Fig. 3.16 presents a boxplot comparing the distribution of `insulin` values for records with missing (i.e., zero) `triceps` values and non-missing (i.e., non-zero) `triceps` values. It is clear from this plot that the `insulin` value is missing (i.e., recorded as zero) whenever the `triceps` value is missing; the extent to which `insulin` values are zero when `triceps` is not missing is not clear from this plot; a more detailed picture may be obtained using the `sunflowerplot` function described in Sec. 3.4.1.

The most difficult case is the *missing not at random (MNAR)* data model, where the probability that x_k is missing depends on its actual (i.e., unobserved) value. This case is difficult both because it is impossible to detect from the available data, and because it can cause large biases in our results. To illustrate, the following example compares three data sequences: (1) a completely observed sequence of $N = 200$ samples; (2) a variation where 25% of the data values are missing completely at random; and (3) a second variation where the largest 25% of the data values are missing, representing a typical MNAR scenario. Here, the complete data sequence is a zero-mean, unit variance Gaussian random number sequence, and the computed mean and standard deviation are:

```
set.seed(333)
x <- sort(rnorm(200))
mean(x)

## [1] 0.07512683

sd(x)

## [1] 0.9577256
```

Both of these values are in reasonable agreement with the "true" values of 0 and 1, respectively. (Note that this data sequence has been sorted, to simplify the task of excluding the top 25% of the data; neither the mean nor the standard deviation depend on the order of the data values, so this sorting has no impact on these computed values.) The easiest way of simulating 25% randomly missing data is to draw a random sample from the original data sequence that contains 75% of the original data values; in this case, since $N = 200$, this retained sample has $M = 150$ observations:

```
set.seed(5)
xMCAR <- sample(x, size = 150, replace = FALSE)
mean(xMCAR)

## [1] 0.09788207

sd(xMCAR)

## [1] 0.9519862
```

Not surprisingly, the results we obtain under this missing data model are slightly different from those we obtained from the complete data sequence, but not radically so. A different picture emerges if we *systematically* omit the top 25% of the data values, corresponding to a MNAR missing data setting:

```
xMNAR <- x[1:150]
mean(xMNAR)

## [1] -0.3247635

sd(xMNAR)

## [1] 0.7200448
```

In this last case, both the mean and the standard deviation are substantially smaller than the "true" values, reflecting the fact that we have systematically reduced the range of the data.

The primary points of this discussion have been, first, to highlight the often implicit assumption that missing data values occur randomly throughout a dataset, and, second, the fact that this assumption may not be valid. Further, as the example just presented demonstrates, the problem of *systematic* missing data can lead to biased—i.e., incorrect—analysis results. Since the probability that an observation x_k is missing depends on its true but unknown value under the MNAR assumption, this potentially very damaging case is essentially impossible to detect from the data alone. An intermediate approach that is sometimes helpful in detecting MNAR missing data is to examine the MAR missingness assumption, as in the Pima Indians diabetes example presented above. Significant associations between missing values in distinct variables, or systematic differences in the distributions of other covariates for the missing and non-missing values of some variable x, rule out the simple MCAR assumption, suggesting that we do whatever we can to learn about the missing x values. That is, is there evidence to suggest that the missing observations represent a significantly different segment of the population from which the data collection was drawn, and is there reason to expect that the distribution of x-values for this segment may be substantially different?

3.3.6   QQ-plots revisited

Sec. 3.2.5 presented QQ-plots as an informal graphical tool to evaluate specific distributional assumptions, identify outliers, detect asymmetry, highlight multimodality, and identify inliers. The following examples show how combining these plots with other simple data views like scatterplots, histograms, and nonparametric density estimates can supplement these insights.

The UScereal potassium data

Fig. 3.17 presents four views of the UScereal variable potassium. The upper left is the scatterplot from Sec. 3.3.2, highlighting the three glaring outliers discussed there. These outliers are also evident in the histogram in the upper right, but the primary impression conveyed there is the asymmetry of the data distribution: small potassium values are most prevalent, and the distribution decays fairly rapidly as potassium increases. The nonparametric density estimate in the lower left conveys this asymmetry more clearly but obscures the outliers. The normal QQ-plot in the lower right has the advantage of highlighting both of these features: the outliers appear as isolated points in the upper tail, while the general curvature of the data points and the fact that the lower tail lies above the 95% confidence limits for the reference line suggest distributional asymmetry. Finally, it might appear from this plot that the lower tail consists of repeated values like those seen in the Pima Indians diabetes dataset discussed in Sec. 3.3.3, but this is not the case, as may be seen by tabulating these data values and looking at the range of repetition frequencies:

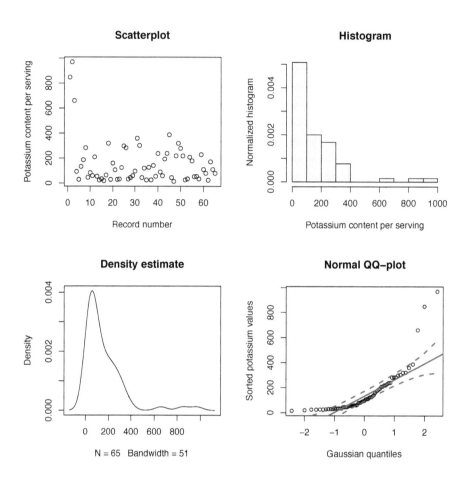

Figure 3.17: Four-plot summary of the US cereal potassium data.

```
tabulation <- table(UScereal$potassium)
table(tabulation)

## tabulation
##  1  2  3
## 38  9  3
```

Thus, while there are a few values repeated two or three times, there is no frequent repetition indicative of inliers.

The Cars93 Price data

Fig. 3.18 gives the analogous four-plot summary for the `Cars93` average sales price variable `Price` from the `MASS` package. Again, the scatterplot in the upper left suggests one or possibly two outliers, although the evidence here is not quite

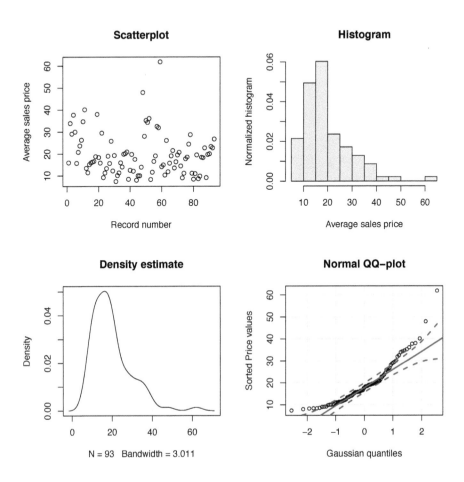

Figure 3.18: Four-plot summary of the Cars93 Price data.

as compelling as for the **potassium** data (especially for the second, less extreme point), partly because of the greater asymmetry of the main data distribution, seen clearly in both the histogram and the density plot. The normal QQ-plot in the lower right is very similar in appearance to that for the **potassium** data, identifying both the potential outliers and the distributional asymmetry.

One of the useful features of the **qqPlot** function from the **car** package is its flexibility in supporting many reference distributions, a point discussed further in Chapter 9. Fig. 3.19 shows the *gamma QQ-plot* for the **Price** data, providing a nice illustration of this capability. In particular, the nonparametric density estimate in the lower left plot in Fig. 3.18 strongly resembles the shape of the gamma density discussed in Chapter 9, appropriate to variables that can only assume positive values. This distribution is characterized by a parameter **shape** that defines the overall shape of the density function, and the gamma QQ-plot in

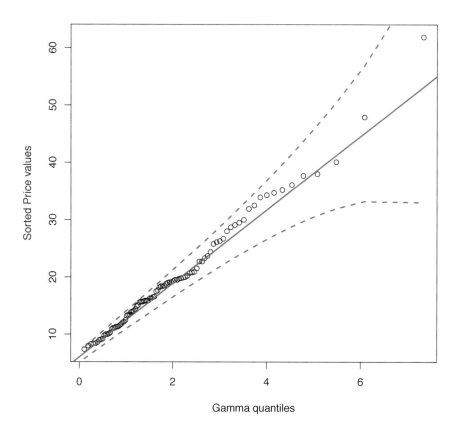

Figure 3.19: Gamma QQ-plot for the Cars93 Price data.

Fig. 3.19 was obtained by specifying the `qqPlot` arguments `dist = ''gamma''`
and `shape = 2`. The shape of the 95% confidence interval is different here than
it is in the normal QQ-plots seen earlier, but all of the data points fall within
this confidence region in Fig. 3.19, suggesting that what appeared to be upper-
tail outliers may just be nominal points in the upper tail of this fairly strongly
asymmetric distribution.

The geyser duration data

Fig. 3.20 presents the same four-plot summary as in the previous two examples,
for the `duration` variable from the `geyser` data frame in the `MASS` package.
These numbers describe the durations in minutes of 299 eruptions of the Old
Faithful geyser in Yellowstone National Park in Wyoming. The scatterplot in the

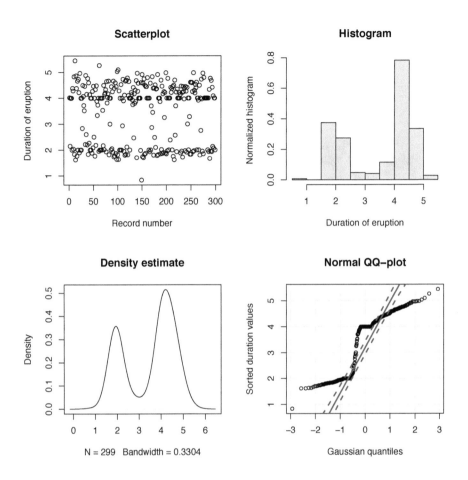

Figure 3.20: Four-plot summary of the MASS geyser duration data.

upper left has an unusual appearance, with most of the data values clustered in two horizontal "bands," reflecting the bimodal distribution of these data values. This characteristic can also be seen in the histogram in the upper right, but even more clearly in the density estimate in the lower left, which suggests that most eruptions last either "about two minutes" or "about four minutes." This bimodal character is also reflected in the pronounced "kink" in the center of the normal QQ-plot shown in the lower right.

The Boston rad data

The plots in Fig. 3.21 summarize the `rad` variable from the `Boston` data frame in the `MASS` package, described as an "index of radial highway accessibility" for the neighborhoods represented in the dataset. The scatterplot in the upper left

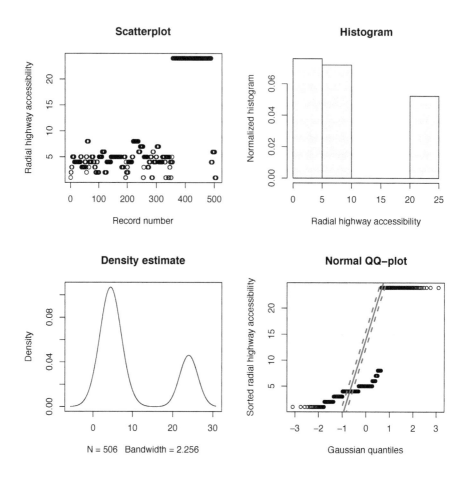

Figure 3.21: Four-plot summary of the MASS Boston rad data.

shows glaring, repeated outliers: the "plateau" in this plot represents repeated data observations that lie well above all of the others. These outliers are also clearly evident in the histogram in the upper right, responsible for the gap in the middle of the plot. The density estimate in the lower left is strongly suggestive of a bimodal distribution, like that seen in the geyser duration data but differing in the complete lack of values in this gap. This gap is more clearly evident in the normal QQ-plot in the lower right, which also shows that *all* of the data values are repeated, reflected in this plot's "stair-step" appearance.

3.4 Visualizing relations between variables

Probably the best-known representation of the relationship between two variables is the scatterplot, used extensively throughout this book. Sometimes, it

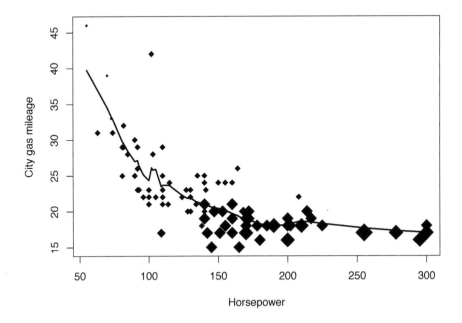

Figure 3.22: Scatterplot of city gas mileage vs. horsepower from the `Cars93` data frame with point sizes defined by cylinders and a `supsmu` reference curve.

is useful to augment this plot with reference lines or curves, or in cases where there are repeated data points, to replace it with the *sunflower plot* introduced in Chapter 2; both of these ideas are discussed in Sec. 3.4.1. Other useful two-variable displays are the boxplot and the *beanplot*, both describing the relationship between a categorical variable and a numerical variable and discussed in Sec. 3.4.2, and the *mosaic plot*, which shows the relationship between two categorical variables and is discussed in Sec. 3.4.3.

3.4.1 Scatterplots between numerical variables

Fig. 3.22 is a scatterplot of the city gas mileage (`MPG.city`) versus horsepower (`Horsepower`) from the `Cars93` data frame in the `MASS` package, with two enhancements. The first is variable point sizes that depend on the value of `Cylinders`, giving the number and/or type of cylinders in the car's design. The six unique values of this variable can be obtained as `table(Cars93$Cylinders)`:

```
##
##      3     4     5     6     8 rotary
##      3    49     2    31     7      1
```

To use these levels to control the size of the points, the scatterplot in Fig. 3.22 was generated with the following plot command:

```
plot(Cars93$Horsepower, Cars93$MPG.city,
     xlab = "Horsepower", ylab = "City gas mileage",
     pch = 18, cex = 0.6 * as.numeric(Cars93$Cylinders))
```

Here, `pch = 18` specifies diamond-shaped points in the scatterplot, and the `cex` specification causes the size to be determined by the levels of the `Cylinders` variable. Note that because this variable includes the level `rotary`, it is necessarily a factor, which is why the conversion `as.numeric` is required. As a result of this conversion, the point size magnification factor `cex` varies from 0.6 for the three-cylinder cars to 3.6 for the single rotary engine (the largest diamond at approximately 250 horsepower). Using this point size scaling makes it possible to see how both horsepower and gas mileage vary with the number of cylinders, similar to the `symbols` plot example presented in Chapter 2.

The second scatterplot enhancement shown in Fig. 3.22 is the addition of a smooth reference curve, generated by the `supsmu` function. This function implements the *supersmoother*, an adaptive scatterplot smoother that is relatively easy to use and typically gives useful reference curves showing the basic trend in the data. (A detailed discussion is beyond the scope of this book; those interested in more details should consult the book by Härdle [36, pp. 181–184].) The code used to generate the smooth curve in Fig. 3.22 was:

```
smooCurve <- supsmu(Cars93$Horsepower, Cars93$MPG.city)
lines(smooCurve, lwd = 2)
```

A useful extension of the scatterplot when some data points are repeated several times is the *sunflower plot* like that shown in Fig. 3.23, constructed using the base graphics function `sunflowerplot`. In a standard scatterplot, these repeated points are indistinguishable from a single data point, but in a sunflower plot, repeated points are represented by "sunflowers" with one "petal" for each repetition. Specifically, data points that are not repeated are represented exactly as in the standard scatterplot. If the same (x, y) pair appears twice in a dataset, the corresponding sunflower plot places a single point at this (x, y) value, but extends two "leaves" from this point, to indicate that the point appears twice. This behavior is illustrated in Fig. 3.23, for the `insulin` values versus the `triceps` values from the Pima Indians diabetes data frame from the `mlbench` package. For example, consider the point at a triceps skin fold thickness of approximately 55 and a serum insulin value of 0: two records exhibit this same pair of values, indicated by the solid point at this (x, y) value and the two vertical lines extending from it, one above and the other below. A more dramatic example is the large dot at the point $(0, 0)$, representing the 227 records that list zero for both of these variables. As emphasized in Sec. 3.3.5, these records represent disguised missing data, since zero values are biologically impossible for either of these variables. The point here is that the sunflower plot gives a clear indication that many records exhibit zero values for both of these variables, something the standard scatterplot cannot do.

```
par(mfrow=c(1,1))
sunflowerplot(PimaIndiansDiabetes$triceps, PimaIndiansDiabetes$insulin,
              xlab = "Triceps skin fold thickness",
              ylab = "Serum insulin value")
```

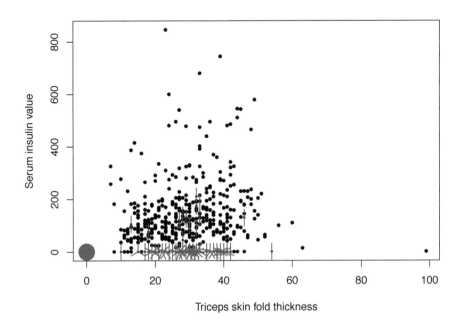

Figure 3.23: Sunflower plot of reported insulin versus triceps values from the Pima Indians diabetes dataset.

3.4.2 Boxplots: numerical vs. categorical variables

Fig. 3.24 is a *variable-width* boxplot summary showing the range of variation of city gas mileage over all records listing each of the six possible Cylinders values in the Cars93 data frame. The option varwidth = TRUE has been used here, which makes the width of each individual boxplot depend on the number of records in each group. Recall that the record counts for these distinct Cylinders values were tabulated earlier. Here, the widest boxplots are those for 4- and 6-cylinder vehicles, which account for 86% of the total, while the narrowest boxplots are those for the rotary engine (one record) and the 3- and 5-cylinder engines, each of which appears three times. It is clear from this plot that gas mileage decreases—quite substantially—as the number of cylinders increases from 3 to 8, and that the single record listing the Cylinders value "rotary" behaves very much like the 8-cylinder records.

```
par(mfrow=c(1,1))
boxplot(MPG.city ~ Cylinders, data = Cars93,
              xlab = "Cylinders", ylab = "MPG.city",
              varwidth = TRUE)
```

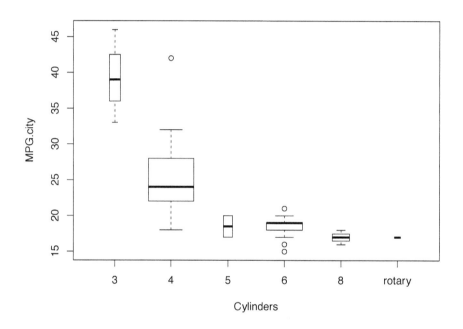

Figure 3.24: Boxplot summary of city gas mileage versus cylinders from the Cars93 data frame.

An extremely useful extension of the boxplot is the *beanplot*, illustrated in Fig. 3.25, showing how the distribution of the statusquo variable in the Chile data frame varies across the four values of the vote variable. Here, the "box" in the boxplot has been replaced by a nonparametric density estimate, giving a more complete view of how the data values are distributed over their feasible range. The variable statusquo is defined as "scale of support for the status-quo" so it is not surprising that those planning to vote "No" in the election (vote value N) exhibit mostly negative statusquo values, which may be seen clearly in Fig. 3.25. Also, the horizontal line across each beanplot represents the mean of the statusquo values in each data subset, which is quite negative here, consistent with the distribution suggested by the beanplot. Similarly, those planning to vote "Yes" (vote value Y) exhibit a distribution of mostly positive statusquo values, with a correspondingly positive mean. In contrast, the undecided group (vote value U) exhibits statusquo values ranging from

```
par(mfrow=c(1,1))
library(beanplot)
beanplot(statusquo ~ vote, data = Chile,
         what = c(1,1,1,0),
         col = c("transparent", "black", "black", "black"))
```

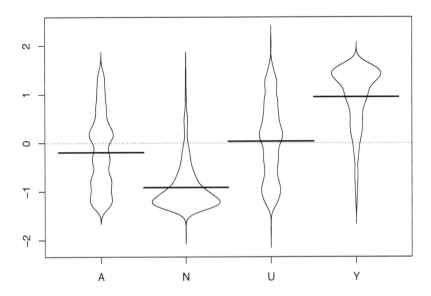

Figure 3.25: Beanplots of `statusquo` vs. `vote` from the `Chile` data frame.

quite negative to quite positive with a mean of zero, while those planning to abstain (`vote` value `A`) also exhibit a wide distribution of values, but more concentrated on negative `statusquo` values, making the average negative.

3.4.3 Mosaic plots: categorical scatterplots

Mosaic plots describe the relationship between two categorical variables, as illustrated in Figs. 3.26 and 3.27. Essentially, these plots are graphical representations of *contingency tables* that tell us how many times the values of two categorical variables occur together in a dataset. Fig. 3.26 illustrates a case where there is no relationship between these variables, resulting in a plot that resembles a checkerboard with roughly even-sized rectangles, suggesting no preferential clustering of values of one variable together with the other. For this plot, two statistically-independent categorical variables were generated:

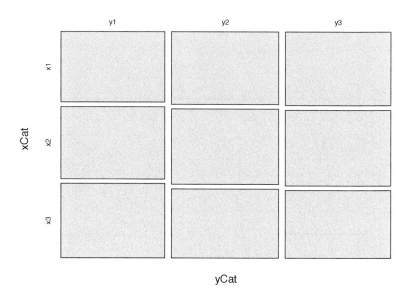

Figure 3.26: Mosaic plot of two unrelated, simulated categorical variables.

```
set.seed(22)
xCat <- sample(c("x1", "x2", "x3"), size = 2000, replace = TRUE)
yCat <- sample(c("y1", "y2", "y3"), size = 2000, replace = TRUE)
```

Since each variable assumes one of three possible values—each equally likely—
and the variables are statistically independent, each value of one variable should
occur about equally often in conjunction with each value of the other. The
mosaic plot shown in Fig. 3.26 reflects this, consisting of nine approximately
square regions, each of approximately the same size. This plot was generated
with the following *R* code:

```
mosaicplot(yCat ~ xCat, main = "")
```

In contrast, Fig. 3.27 shows a case where there is a strong association between
two categorical variables. This plot is based on the six-level `Cylinders` variable and the two-level `Origin` variable from the `Cars93` data frame. Here, the
sections of the mosaic plot vary greatly in size with the two variables, suggesting a strong association between levels. Specifically, note the dashed lines
for the "USA" level of the `Origin` variable when `Cylinders` has the values 3,
5, or "rotary," indicating there are no records in the data frame with these

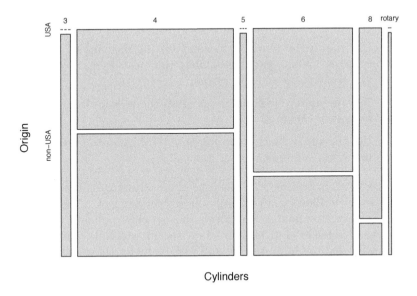

Figure 3.27: Mosaic plot showing the relationship between the Cylinders and Origin variables from the Cars93 data frame.

combinations of values. In practical terms, this reflects the fact that cars with 3 or 5 cylinders or rotary engines were not manufactured in the US in 1993. For the other three, more common values of `Cylinders`, there appears to be a systematic relationship: the majority of 4-cylinder cars were made outside the US, while the majority of 6-cylinder cars and the overwhelming majority of 8-cylinder cars were made in the US. Finally, note that this plot also shows that the most frequent `Cylinders` value in the dataset is 4, followed by 6, then 8, with everything else much less frequent.

3.5 Exercises

1: The results presented for the `anscombe` data frame from the `datasets` package in Section 3.2.3 used the `apply` function to compute the means and standard deviations of the columns of this data frame. This exercise asks you to extend this characterization for the numerical variables in the `fgl` data frame from the `MASS` package:

1a. To see that the restriction to numerical variables is important here, first use the `apply` function to compute the mean of *all* data columns from the `fgl` data frame.

1b. Use the `str` function to see which columns are numerical, and repeat (1a) but restricted only to these columns.

1c. Use the `apply` function to compute the medians of all numerical columns. Which variable exhibits the most dramatic difference between the mean and the median?

1d. Use the `apply` function to compute the standard deviations of all numerical columns.

1e. Use the `apply` function to compute the MAD scale estimator for all numerical columns.

2: It was noted in Section 3.2.4 that one of the advantages of probability theory is that it allows us to quantify our expectations by assuming our data sequence (approximately) conforms to a specific probability distribution like the Gaussian distribution. As a specific example, the `qnorm` function was used to obtain the 5% and 95% limits for the standard Gaussian distribution (i.e., mean zero, standard deviation 1), giving us the range where we expect to see 90% of the values of a sample of Gaussian random variables. Use the `qnorm` function to find the range of values where we expect to see 99% of these values.

3: It was noted in Section 3.2.5 that histograms, while the best known and probably most widely used of the three distributional characterizations presented there, are often not very effective in describing data distributions. This exercise asks you to develop a side-by-side comparison of histograms constructed from: (1) the `HeadWt` variable from the `cabbages` data frame in the `MASS` package; and (2) the `RI` variable from the `fgl` data frame, also in the `MASS` package. Use the `mfrow` and `pty` graphics parameters to set up a side-by-side display of square plots, and use the `truehist` function from the `MASS` package to generate the histograms. Put titles on the histograms: "Cabbage head weight" and "fgl: Refractive Index". Can you confidently say that either variable does or does not appear to have an approximately Gaussian distribution from these plots?

4: In contrast, it was noted that the normal QQ-plot gives a much clearer view of, first, whether the Gaussian assumption represents a reasonable approximation for a particular variable. Repeat Exercise 3, but using the `qqPlot` function from the `car` package instead of the histograms used there. Do these plots give clear indications of whether either variable has an approximately Gaussian distribution?

5: It has been noted repeatedly that, although normal QQ-plots were developed to provide an informal assessment of the reasonableness of the Gaussian distributional assumption, as in Exercise 4, these plots are extremely useful even in cases where we don't necessarily expect the Gaussian distribution to be reasonable. Use the `qqPlot` function to characterize the potassium oxide concentration data (chemical symbol K) from the forensic glass dataset `fgl` in the `MASS` package. Specifically:

5a. Using `mfrow`, set up a side-by-side array of full height plots;

5b. In the left-hand plot, show the estimated potassium oxide density, with the title "Density";

5c. In the right-hand plot, show the normal QQ-plot from the `qqPlot` function, with the title "Normal QQ-plot".

5d. What data features are most prominent from these plots?

6: Section 3.3.2 discussed three ways of detecting univariate outliers. This exercise compares two of them for the magnesium oxide concentration measurements from the `fgl` data frame in the `MASS` package:

6a. Compute the mean and standard deviation from the magnesium oxide data (chemical symbol `Mg`). How many data points are declared outliers by the three-sigma edit rule?

6b. Compute the median and MAD scale estimator from this data. How many data points are declared outliers by the Hampel identifier?

6c. Generate a plot of the magnesium oxide concentration data versus its record number in the dataset and put horizontal lines at the median (dashed line, normal width) and the upper and lower outlier detection limits for the Hampel identifier (dotted line, twice normal width). Based on this plot, which outlier detector seems to be giving the more reasonable results?

7: It was noted in Section 3.3.3 that inliers often arise from disguised missing data, but this is not always the case. Using the approach described there, determine whether there appear to be inliers present in the magnesium oxide concentration data from the `fgl` data frame considered in Exercise 6. Specifically, how many inliers—if any—appear to be present in this data sequence and what are their values?

8: A number of the forensic glass samples in the `fgl` data frame from the `MASS` package list zero concentrations for one or more of the oxides included in the summary. It is possible that these cases are related to the glass sample types, and this is something that can be assessed using the stacked barplots discussed in Chapter 2. Specifically:

8a. Define the logical vector `zeroMg` with the value `TRUE` if `Mg` is zero and `FALSE` otherwise;

8b. Using the `table` function, create `tbl` as the contingency table between `zeroMg` and the `type` values from the `fgl` data frame;

8c. Use the `barplot` function to create a stacked barplot of record counts by `type`, with default colors "black" and "grey" for the nonzero and zero `Mg` records;

8d. Use the `legend` function to put a legend in the upper right corner of the plot, with squares (use `pch = 15`) of the corresponding color and the text "Mg nonzero" and "Mg zero".

9: Exercise 7 in Chapter 2 used the generic `plot` function to create a mosaic plot showing the relationship between the variables `shelf` and `mfr` in the `UScereal` data frame from the `MASS` package. That plot was somewhat crude, however, since the axis labels did not include all of the `mfr` designations. To obtain a more informative mosaic plot, use the `mosaicplot` function with the formula interface for these variables. Specify the `las` argument to make all of the axis labels horizontal. Does this plot suggest a relationship between these variables?

10: Again using the `mosaicplot` function, construct a mosaic plot showing the relationship between the variables `Cult` and `Date` in the `cabbages` data frame from the `MASS` package. As in Exercise 9, use the `las` argument to make the axis labels horizontal. Does this plot suggest a relationship between these variables?

Chapter 4

Working with External Data

One of the great aids to learning to use R is its many built-in data examples, several of which are used extensively in this book. But if we want to go beyond learning how to do things in R with these built-in examples to actually *using* the R programming language to analyze real-world data, it is necessary to somehow get this data into our R session. The purpose of this chapter is to give a reasonably thorough introduction to some useful tools for accomplishing this task—i.e., for getting data into our R session—and those for the closely related task of saving results to external data files that can be used by others.

The rest of this chapter is organized as follows. The brief discussion of computer system organization given in Chapter 1 noted that the data used in an interactive R session is stored in RAM and is *volatile*, meaning that it disappears when the R session ends unless we save it explicitly. In contrast, the external data considered in this chapter is generally contained in *files* that are external to our R session but that can be read from or written to by simple R commands. Thus, Sec. 4.1 begins with a slightly more detailed introduction to how files are organized in a computer system, focusing on some of the useful R commands available for interacting with these files.

For small datasets like examples taken from textbooks or magazines, it is possible to manually type the values into a data frame, as described in Sec. 4.2. *This is generally bad practice and should be avoided, for reasons discussed in Sec. 4.2.2; unfortunately, it is sometimes the easiest and fastest way to obtain a small dataset, as the example discussed in Sec. 4.2.1 illustrates.*

At the other extreme, the Internet represents an increasingly important source of potentially very large data files, and Sec. 4.3 gives a short but useful introduction to some of the ways we can obtain Internet data and get it into an R session. In the most favorable cases, Internet web pages provide us links to data files that can be downloaded and read into our R session using the tools described in later sections of this chapter. One of the main points of these discussions is that data files come in different types, each requiring different R tools to read, and these discussions focus on a few of the most commonly encountered file types. Alternatively, if the data we want to retrieve is not contained in a file but is embedded in the text displayed on a web page, it may be necessary to extract it directly from the web page. Since web pages are rendered in HTML (hypertext markup language), it is necessary in these cases to read and parse HTML files, a topic discussed briefly in Sec. 4.3.2.

One of the most popular and convenient data file formats is the *comma separated value (CSV)* file, discussed in detail in Sec. 4.4. Unfortunately, as discussed in Sec. 4.4.2, these files are easily confused with *spreadsheets*, which are not at all the same, requiring different utilities to access; these utilities are readily available, but it is important to recognize that they are not the same as those required to read and write CSV files. Many other file types exist and, while this book makes no attempt at considering them all, a few other key file types and their main uses are considered in Sec. 4.5, including simple text files, a special format for saving and retrieving arbitrary R objects (RDS files), and a few of the most popular graphics file formats used to save graphical output for use by other programs (e.g., to prepare documents or slide presentations).

An important practical issue that arises frequently is the need to merge data from multiple sources, and Sec. 4.6 introduces the use of the `merge` function in R to combine different data frames. This process is essentially the same as the *database join* used to merge data from two or more tables in a database, and Sec. 4.7 gives a brief introduction to the topic of *relational databases* and some of the R tools available for working with them. Relational databases are particularly useful in storing very large collections of highly structured data and are widely used to store business data. The basic idea is to minimize duplication of information by splitting it up into *tables* and then retrieving information from the relevant tables as needed using *database queries*. These queries are typically written using *structured query language (SQL)* and Sec. 4.7 briefly introduces SQL queries, the `sqldf` package that allows us to execute SQL queries against data frames in R, and the `RSQLite` package that allows us to work with `sqlite3` databases in R. Support for other databases is also covered briefly.

4.1   File management in R

By default, all of the functions used to read external data from files into our interactive R session assume these files exist in our *working directory*, and the `getwd` function can be used to identify this working directory:

```
getwd()
```

```
## [1] "C:/Users/Ron/Documents/IntroRbook/RevisedManuscript"
```

This function returns a text string that identifies the directory where R will look for files read by functions like `read.csv` or `readLines` discussed in subsequent sections of this chapter, and where R will put files created with functions like write.csv or writeLines. It is important to note that directories are arranged in a hierarchical structure, and that the different levels of this hierarchy are specified in the directory name. In this particular case, the directory resides on the built-in hard disk on a PC, designated as the "C-drive," which is why the directory name begins with "C:", followed by a forward slash (/). The next part of the string returned by the getwd function is "Users," indicating that the working directory is located as part of a larger directory named "Users" that is located on the C-drive. This text string is also followed by a forward slash, indicating that the working directory belongs to a subdirectory of "Users" named "Ron". Further text strings separated by additional forward slashes describe the path from this subdirectory down through other subdirectories named "Documents" and "IntroRbook," until we arrive at the local name of the working directory, which is "IntroRSkeleton."

Before going on to a further discussion of file management in *R*, it is important to emphasize two points. First, we can exploit the hierarchical structure of directories to simplify file operations involving directories that are "nearby," either as subdirectories of the working directory, as upper-level directories that contain the working directory, or as "neighboring" directories that are most easily specified as subdirectories of higher-level directories that contain the working directory. This is accomplished using the *relative path specifications* discussed below. The second key point is that the forward slash separation of directory name components is *R*-specific. *In particular, directory specifications given in the command window of a Microsoft Windows machine separate these components using the backslash (\\).* This point is important, for two reasons: first, using the backslash in *R* file path specifications will result in an error message, even on a Windows machine; and second, by *always* using the forward slash file path specification, *R* allows the same code to run on different machines, regardless of how the operating system specifies file paths. In particular, UNIX machines use the forward slash for file specifications in the command window. (It is possible on a Microsoft Windows machine—and sometimes useful—to specify file paths using backslashes, but *R* treats these as special characters that must be "escaped" in the file path, replacing each backslash in the original path specification with a double backslash \\\\.)

The working directory is set when the *R* icon is clicked to start a new *R* session, but it can be changed using the setwd function, as in this example:

```
setwd("C:/NewDirectory")
```

This command changes the current working directory to ``C:/NewDirectory``, *assuming this directory exists;* if not, an error message is returned. In such cases, it is possible to create a directory using the create.dir function; for details, refer to help(create.dir).

R provides a number of functions for learning about what files are contained in the working directory, determining their properties (e.g., sizes, creation times, file protections, etc.), or modifying them (e.g., renaming, moving, or deleting files). The function list.files, when called without any parameters (e.g., as `list.files()`), returns a list of all files contained in the working directory. It is also possible to list files in an alternate directory using the `path` option to specify the desired directory, or to list only those files that contain a specific character string in their name, via the `pattern` option. The following example illustrates how this works:

```
list.files(path = "C:/Users/Ron/Documents/IntroRbook/ExploringDataCode",
           pattern = "Three")

## [1] "ThreeSigma.R"
```

Here, `path` specifies the alternative directory to be searched, while `pattern` specifies that only those files whose names contain the string "Three" are to be returned. This example also provides a useful illustration of the advantages of relative path specification: note that both the working directory, with the local name "IntroRSkeleton," and the alternative directory considered here are sub-directories of one named "IntroRbook." The relative path specification of this alternative directory is "../ExploringDataCode," where the initial two dots are interpreted as "go up one level in the hierarchy," and the forward slash following these dots indicate that the desired directory is a subdirectory of this higher-level directory. Specifically, the following sequence gives exactly the same results as above:

```
list.files("../ExploringDataCode", pattern = "Three")

## [1] "ThreeSigma.R"
```

The `file.info` function allows us to learn details about a file beyond its name, including its size, whether it is a directory or not, whether it is executable or not, its file permissions (e.g., who can read it, write to it, or delete it), and when it was created or modified. As a specific example, applying this function to the file listed in the previous two examples yields this result:

```
file.info("../ExploringDataCode/ThreeSigma.R")

##                                  size isdir mode                  mtime
## ../ExploringDataCode/ThreeSigma.R 350 FALSE  666 2016-02-21 17:15:13
##                                                      ctime              atime
## ../ExploringDataCode/ThreeSigma.R 2016-02-21 17:13:10 2016-02-21 17:13:10
##                                    exe
## ../ExploringDataCode/ThreeSigma.R  no
```

Note that this example has used the relative path specification, but here we must specify the file name and not simply the directory where it resides. For a more detailed discussion of how to interpret these file details, refer to the help file for `file.info`.

We can also create and edit files using the `file.create` and `file.edit` functions; because these functions are so useful for creating and editing *R* program files, they are discussed in Chapter 7 on programming. Finally, files can be renamed, copied, or deleted with the functions `file.rename`, `file.copy`, and `file.remove`, respectively. *Note that the `file.rename` and `file.remove` functions perform the indicated operations without warning messages or asking "are you sure you really want to do this?"* If a file does not exist, it cannot be removed or renamed, and a warning message will result if we attempt to do so, but, otherwise, *R* assumes you know what you are doing when you request a file be renamed or deleted, so be careful.

4.2 Manual data entry

There was a time when manual data entry was fairly common, and the need for it does still arise occasionally, motivating the discussion presented in Sec. 4.2.1. That said, manual data entry is tedious and error-prone, so it should be avoided whenever possible, a point emphasized in Sec. 4.2.2: it is easy to transpose digits or characters, enter correct values in the wrong fields, or skip records entirely.

4.2.1 Entering the data by hand

The following example is based on data required for the data story presented in Chapter 6 that compares two informal economic indices: the *Big Mac index*, based on the price of a MacDonald's Big Mac burger in local currrencies around the world, and the *Grande Latte index*, based on the local price of a Starbucks Grande Latte (approximately 16 oz). In both cases, the prices were converted from the local currency to US dollars at the prevailing exchange rate. Data for the Big Mac index is available as an Excel spreadsheet and this data source forms the basis for the example discussed in Sec. 4.4.2. Data for the Grande Latte index does not appear to be available in either a spreadsheet or a file, however, so it must be extracted from an Internet web page. The values used here for this index were obtained from the following website, accessed on 9/19/2015:

`http://www.coventryleague.com/blogentary/the-starbucks-latte-index`

This blog post reproduces a barchart with numbers appended that was published by the *Wall Street Journal* and *Bloomberg News*, dated Feburary 27, 2013. It is noted that the conversions from local currencies to US dollars were based on the exchange rates for February 20, 2013.

The data for this example consists of 29 city names and prices in US dollars that appear in a barchart included in the blog post described above. As discussed in Sec. 4.3.2, this blog post is rendered in HTML (hypertext markup language), and in favorable cases, the numerical data from this web page could be extracted automatically by downloading this HTML file and parsing it to extract the content we want. Unfortunately, the numerical values of interest here are embedded in a JPEG image, a file format discussed in Sec. 4.5.3, and extracting numerical data from these files is not straightforward. Since the number of data values required here is small—only 29 cities and their associated Grande Latte prices—these values were entered manually.

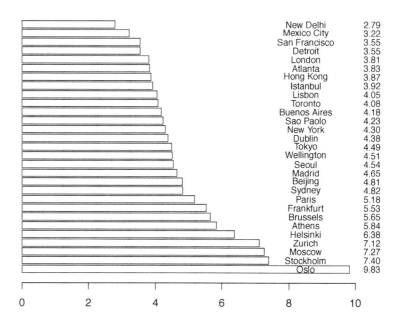

New Delhi	2.79
Mexico City	3.22
San Francisco	3.55
Detroit	3.55
London	3.81
Atlanta	3.83
Hong Kong	3.87
Istanbul	3.92
Lisbon	4.05
Toronto	4.08
Buenos Aires	4.18
Sao Paolo	4.23
New York	4.30
Dublin	4.38
Tokyo	4.49
Wellington	4.51
Seoul	4.54
Madrid	4.65
Beijing	4.81
Sydney	4.82
Paris	5.18
Frankfurt	5.53
Brussels	5.65
Athens	5.84
Helsinki	6.38
Zurich	7.12
Moscow	7.27
Stockholm	7.40
Oslo	9.83

Figure 4.1: City names and Grande Latte prices for the Starbucks Latte index.

Fig. 4.1 is similar to that seen in the blog post, showing the 29 city names and Grande Latte prices and generated with the following R code:

```
Mids <- barplot(LatteIndexFrame$price, col="transparent",
    horiz=TRUE, xlim=c(0,11))
text(8.5,Mids,LatteIndexFrame$city, cex=0.8)
text(10.5,Mids,format(LatteIndexFrame$price,nsmall=2), cex=0.8)
```

The data frame `LatteIndexFrame` has one row for each city in Fig. 4.1, and three columns: `city`, `price`, and `country`. It was constructed by first creating a one-row data frame with the desired column names:

```
LatteIndexFrame <- data.frame(city = "Oslo", price = 9.83, country = "Norway")
```

This initial data frame was then edited using the `fix` command:

```
fix(LatteIndexFrame)
```

This command pops up a spreadsheet-style data entry window with the appropriate column headers, allowing us to either change the initial values we entered or enter additional rows. Here, the final data frame was created by entering additional rows for the 28 other values for `city`, `price`, and `country`.

After manually creating a data frame like this, it is important to save the result in a file. Specifically, the process just described creates an object in your interactive *R* session, which is stored in RAM. As discussed in Chapter 1, RAM is *volatile*, meaning that its contents disappear when the computer is turned off. For this reason, you should make a habit of saving any important data objects you create in an *R* session to a file, something that can be done in several different ways; here, the results were saved to a CSV file using the simple code discussed in Sec. 4.4.1.

4.2.2 Manual data entry is bad but sometimes expedient

Even though manual data entry is sometimes necessary—or at least, the most practical alternative—it is generally bad practice, for two reasons. The first and most obvious is that, if the dataset is very large, the effort required for manual entry may be infeasible: examples include the Australian vehicle insurance dataset `dataCar` from the `insuranceData` package, which has over 67,000 records, and the unclaimed Canadian bank account dataset with over 12,000 records discussed further in later sections of this chapter. In such cases, the only practical way to obtain the data is to obtain or develop whatever computer code may be necessary to acquire and convert the data into a useable format. In practice, this process is sometimes messy and complicated, which is why this chapter is devoted to the task of working with external data.

The second reason manual data entry is bad is that it is error prone, with errors that can take many forms and, in unfortunate cases, be extremely difficult to detect. Specific examples include:

- omitting digits from numerical data (e.g., the year 2006 entered as 206);

- transposing digits in numerical data (e.g., the year 2006 entered as 2060);

- omitting or transposing characters from long character strings (e.g., "unkown" or "uknown" for "unknown");

- omitting one or more fields from a record (e.g., forgetting to enter values for `price` for certain cases in the Grande Latte example);

- entering the correct value in the wrong field (e.g., entering the `city` value in the `country` field in the Grande Latte data example);

- skipping records (e.g., skipping from the "Helsinki" entry to the "Brussels" entry in the Grand Latte example, omitting the "Athens" entry).

For both of these reasons, it is best to avoid manual data entry whenever possible, especially if more than a few records are involved.

Conversely, since much data is initially entered manually by someone before being incorporated into a file or database, it is also extremely important to examine data carefully to detect any data quality errors that may be present. In the hypothetical examples above, it should be possible to at least detect that the years 206 and 2060 represent data entry errors. In other cases, the errors will be harder to detect but may still be possible using some of the exploratory data analysis techniques described in Chapters 3 and 9.

4.3   Interacting with the Internet

At the other end of the data entry spectrum, the Internet is an increasingly important source of large volumes of data. In the simplest cases, the Internet provides links to files that contain the data we want, allowing us to obtain this data in the following three steps:

1. Find the website from which the dataset is available;

2. Click the appropriate links to download the data in the format you want;

3. Read the resulting data file into your *R* session.

Sec. 4.3.1 gives introductory overviews of three Internet data examples, based on three different file formats: a CSV file containing unclaimed Canadian bank account data discussed further in Sec. 4.4.1, an Excel spreadsheet containing the Big Mac economic index data discussed in Sec. 4.4.2, and a text file containing automobile mileage data discussed in Sec. 4.5.1. Following these examples, Sec. 4.3.2 then briefly discusses the more complex situation where the desired data is not available in a file, but must be extracted from the HTML representation of a web page.

   Before proceeding to these examples, it is important to note that the Internet is a potentially volatile data source: web pages can be changed, moved, or taken down. This means that data obtained from the Internet at one date and time can be different or unavailable at a later date and time. For this reason, it is important when working with Internet data sources to indicate when the data values were obtained.

4.3.1   Previews of three Internet data examples

The preceding paragraph listed the three steps required in obtaining data from files that are available via the Internet. The following sections give more complete discussions of these steps for three specific Internet-based data examples.

An automobile gas mileage dataset

Datasets containing automobile gas mileage data are available from the `datasets` package as the `mtcars` data frame, and from the `MASS` package as the `Cars93` data frame. A third automobile gas mileage dataset is available from the University of California at Irvine's Machine Learning Repository, a collection of datasets that have been used extensively as benchmarks for developing and evaluating machine learning algorithms like those discussed in Chapter 10. The website associated with this repository can be viewed from your *R* session with the `browseURL` function:

```
browseURL("http://archive.ics.uci.edu/ml")
```

The only required argument for this function is the path to the website (specifically, its Universal Record Locator or URL), and the effect of this function is to open a new window in your computer's default browser and display the requested web page. From this browser window, you can interact with the web page in exactly the same way as if you had found it using your favorite search engine. Two key points are, first, that for this function to work correctly, it is necessary that your computer be connected to the Internet, and, second, that the URL be specified correctly; if either of these conditions are not satisfied, an error message will be returned when the function is executed. The results presented for this example were obtained on 10-August-2016.

From the web page for the UCI Machine Learning Repository, the easiest way to obtain the desired dataset is to click on the link labeled "view all datasets," which brings up another web page listing all of the datasets available in the archive, in alphabetical order. The dataset of interest for this example is one named "Auto MPG" and scrolling down the page leads to a link to another page giving metadata for this dataset. This web page has two links, one to a text file containing this metadata, and the other to a text file containing the data values themselves. Clicking on this second link—labeled "Data Folder"—leads to another page with further links to several other files. The basis for the example considered here is the file named "auto-mpg.data."

Clicking on this link allows us to view the contents of this file, but the simplest way to obtain the actual data itself is to use the `download.file` function. To use this function, we need the complete URL specification for the file: in some cases, this specification can be difficult to find (e.g., the website may link to the file through another website and the complete path may not show up in our browser), but in this case, the URL appears at the top of the web page as:

```
http://archive.ics.uci.edu/ml/machine-learning-databases/auto-mpg/auto-mpg.data
```

Assigning this value to the variable `URL`, we can download this data file and save it in the file `UCIautoMpg.txt` in the current working directory:

```
download.file(URL, "UCIautoMpg.txt")
```

A detailed discussion of how to read this file using the `readLines` function and how to convert the result into the data frame we want is given in Sec. 4.5.1.

The Canadian unclaimed bank account data

The dataset on which this example is based was obtained from Socrata, another source of freely available data on a variety of different topics. We can connect to this website from within an *R* session using the same basic code as before:

```
SocrataURL <- "https://opendata.socrata.com/"
browseURL(SocrataURL)
```

Many datasets are available from Socrata, but the specific one considered here lists unclaimed Canadian bank account data, described on their website as:

> A list of all the abandoned bank accounts at branches in the Edmonton area and/or registered to addresses in the Edmonton area.

As before, the `browseURL` command brings up a browser window and displays the Socrata web page. Near the top of this page, you should see a text box that displays the phrase "Unclaimed Bank Accounts." Clicking on this link will take you to another page that displays the data in a spreadsheet format, with a number of option buttons, including one labeled "Export." Clicking on this link will bring up a list of options, including one that says "Download a copy of this dataset in a static format," and further lists a number of different *format* options, including both "CSV" and "CSV for Excel." Here, we select the "CSV" option, which downloads the CSV file `Unclaimed_bank_accounts.csv`. This file will be saved in the default download directory on your computer and must therefore be moved into your working directory before it can be used in your R session. The simplest way to do this is to copy the file from the download directory into the working directory; in this example, the following R accomplishes this task:

```
DownloadFile <- "C:/Users/Ron/Downloads/Unclaimed_bank_accounts.csv"
file.copy(DownloadFile, "Unclaimed_bank_accounts.csv")
```

A detailed discussion of how to read the contents of this data file into our R session is given in Sec. 4.4.

The BigMac economic index

According to the website listed below, the Big Mac index was "invented by *The Economist* in 1986 as a light-hearted guide to whether currencies are at their 'correct' level." Beyond this description, on 9-September-2015, the following URL also offered access to a dataset with the values for this index, computed at various months for the years from 2000 to 2015:

http://www.economist.com/content/big-mac-index

Several different variables are included in this dataset, but the one that will be taken as the working definition of the Big Mac index in the discussions presented in this book is the variable `dollar_price`, which gives the price in US dollars of a MacDonald's Big Mac in the specified country. This number is obtained by taking the purchase price of the burger in the local currency and converting it into US dollars using the prevailing exchange rate at the time. A comparison of the BigMac index with the Grande Latte index discussed briefly in Sec. 4.2.1 forms the basis for one of the data stories presented in Chapter 6.

The BigMac index is available as a Microsoft Excel spreadsheet, and it was downloaded from the Internet on 9-September-2015, using the `browseURL` function to access the web page, clicking on the appropriate link to download the file, and copying the file from the download directory into the current working directory. The *R* tool used to read this spreadsheet into an internal *R* data frame was the `xlsx` package, described in Sec. 4.4.2.

4.3.2 A very brief introduction to HTML

It has been noted that Internet web pages are typically rendered in HTML (hypertext markup language). While a detailed introduction to HTML is beyond the scope of this book, the following discussion gives an idea of what it looks like and what is involved in extracting information from it. Paul Murrell gives a very readable introduction to HTML in his book, *Introduction to Data Technologies* [56, Ch. 2], but, essentially, it is a specialized text format that tells a web browser both what to display and how to display it. To illustrate, the `readLines` function discussed in Sec. 4.5.1 reads text files into character vectors, and this function allows the specification of Internet files via their URL. The HTML for the UCI Machine Learning Repository web page can be read as:

```
htmlExample <- readLines("http://archive.ics.uci.edu/ml")
```

Here, the `readLines` function returns a 455 element character vector, each representing one line of the HTML file that creates the UCI Machine Learning Repository web page. The first 10 lines show what HTML looks like:

```
htmlExample[1:10]

##  [1] "<!DOCTYPE html>"
##  [2] "<html>"
##  [3] "<head>"
##  [4] "<title>UCI Machine Learning Repository</title>"
##  [5] ""
##  [6] "<!-- Stylesheet link -->"
##  [7] "<link rel=\"stylesheet\" type=\"text/css\" href=\"assets/ml.css\" />"
##  [8] ""
##  [9] "<script language=\"JavaScript\" type=\"text/javascript\">"
## [10] "<!--"
```

The first line is a DOCTYPE declaration, specifying that what follows is an HTML document, and the fourth line gives the main title of the web page as "UCI Machine Learning Repository." This example illustrates the point made earlier that HTML documents consist of both formatting information that tells a web browser how to render the page, and content that defines what is to be displayed. When using HTML documents as data sources, we typically want to extract this content, discarding the formatting information. The `XML` package provides functions like `htmlTreeParse` and `readHTMLTable` to do this, but using them does require some understanding of how HTML documents are structured. For a general introduction, Murrell's treatment is a good place to start [56, Ch. 2], and much more detailed treatments can be found in the books by Willard [77] or Duckett [23]. For a more detailed discussion of the functions available in the XML package, refer to its associated help file.

4.4   Working with CSV files

One of the most convenient file formats for exchanging data between different software environments is the *comma-separated value or CSV* file, which are organized in rows, with one record per row and the fields in each record separated by commas. The ability of different software environments to read and write CSV files is important because, as discussed in Chapter 1, datasets are often collected, processed, merged, and analyzed by a chain of different people and/or organizations, frequently using different software utilities. In favorable cases, CSV files can serve as a common "data conduit" at every stage in this chain. These files do have their limitations, however, which tend to be software-specific; this important point is discussed further in Sec. 4.4.3.

4.4.1   Reading and writing CSV files

In *R*, the simplest—and most reliable—way to read and write CSV files is to use the functions read.csv and write.csv, respectively. A required parameter for both of these functions is the name of the file to be read from or written to, and this file name must be a character string that corresponds, by default, to a file in the current working directory. Conversely, it is possible to read from or write to files in other directories by including the full path designation in the file name. For the read.csv function, this file name is the only required parameter, although the function supports many optional parameters. The example presented here uses one of these optional parameters—stringsAsFactors—and a detailed discussion of these parameters is given in the help file for this function. The write.csv function requires two arguments: the matrix or data frame containing the data to be written, and a character string specifying the name of the file into which the data is to be written. By default, the data is written into a CSV file in the current working directory, but the data can be written into files in other directories by specifying a complete path, as discussed in Sec. 4.1.

A read.csv example

Because it provides a simple example that will be useful in a number of subsequent discussions, both in this chapter and later ones, the following discussion starts by reading a CSV file that has been downloaded from the Internet as described in Sec. 4.3. Following the simple instructions described there downloads a CSV file named Unclaimed_bank_accounts.csv and saves it in the current working directory. As noted above, this file can be read using the read.csv function, which reads each record from the CSV file into one row of a data frame. The following code reads the file contents into a data frame called unclaimedFrame and uses the head function to examine its first few records:

```
unclaimedFrame <- read.csv("Unclaimed_bank_accounts.csv",
                            stringsAsFactors = FALSE)
head(unclaimedFrame)

##    Last...Business.Name                    First.Name  Balance
## 1             LACASSE        ALPHONSE                   98428.28
## 2             PAYDLI          PETER                     77273.87
## 3             FURTAT  VLADIMIR & VANENTINA              73406.84
## 4             KELM                SELMA                 67510.34
## 5             LIM               SU LENG                 51877.33
## 6             VIG        JOGINDER SINGH                 50254.01
##                   Address              City Last.Transaction
## 1                                                 06/10/1986
## 2        9346 86 AVE NW           Edmonton         08/17/1995
## 3      29 ANDOVER ST  CARLTON NSW 2218 AUS          10/01/1987
## 4        7319 89 ST NW          Edmonton         09/08/1995
## 5    217 FERGUSON PL NW          Edmonton         11/16/1994
## 6 SITE 216 BOX 70 RR 2          St. Albert         01/22/1988
##                                       bank_name
## 1 CANADIAN IMPERIAL BANK OF COMMERCEMMERCE
## 2                          ROYAL BANK OF CANADA
## 3                       TORONTO-DOMINION BANK
## 4                             BANK OF MONTREAL
## 5                       TORONTO-DOMINION BANK
## 6                          ROYAL BANK OF CANADA
```

As noted, the only required argument for the **read.csv** function is the file name, but this example also specifies the optional parameter **stringsAsFactors** as **FALSE**: if we omitted this specification, the data fields containing text strings in the CSV file would be converted to factors in the data frame. As discussed in Chapter 1, factors are a specific data type in *R* used to code and manipulate categorical variables, and they behave differently in a number of important ways from character vectors. In particular, character manipulation functions like **nchar** and **strsplit** used in Sec. 4.4.3 to parse these character strings (see Chapter 8 for further discussion) do not work with factor variables.

A write.csv example

Sec. 4.2.1 described the process of manually creating **LatteIndexFrame**, the data frame that characterizes the Grande Latte index discussed further in Chapter 6. As noted, manual data entry is generally undesirable, but because this dataset is fairly small (29 rows, with 3 columns each) and the data values were embedded in a JPEG image file, manual entry was the most expedient way to obtain it. Since the only thing worse than manual data entry is arguably *repeated* manual data entry, it is important to save the results in a file, and the CSV file is an extremely convenient choice. The following *R* command does this:

```
write.csv(LatteIndexFrame, "LatteIndexFrame.csv", row.names=FALSE)
```

Note that this function is called with three parameters: the first specifies the data frame to be saved, the second is the name of the file to be created, and the third specifies that a separate column of row names—in this case, the numbers 1 through 29—is *not* to be written to the file. The default is to create a column of row names in the CSV file because this is needed by some applications (e.g., Microsoft Excel), but since *R* data frames create these rownames by default when they are read from a CSV file, having an explicit column of row names simply duplicates this information. This is seldom useful and in some cases can cause undesirable side effects; for that reason, whenever data frames are saved to CSV files in the examples presented here, they will be saved with "`row.names = FALSE`" unless there is a specific reason to do otherwise.

4.4.2   Spreadsheets and csv files are *not* the same thing

In the preface to his book, *Introduction to Data Technologies* [56, p. xx], Paul Murrell notes that, on Windows XP machines, the icon for CSV files identifies them as Microsoft Office Excel spreadsheets, and the same is true of newer Windows machines. In fact, this designation is an error: while it is true that the *Excel* software package can *read* and *write* simple CSV files, three distinct things are being confused here. Specifically:

1. a *CSV file* is the simple data file described above, that can be read by Microsoft Excel, along with many other programs, including *R*, *Python*, and Microsoft Access, to name only a few;

2. the *Microsoft Excel spreadsheet program* is a software package that can do many things, including read and write CSV files, perform simple computations and data analyses, and generate plots;

3. a *Microsoft Excel data file* contains the data on which a Microsoft Excel spreadsheet is based, along with much additional information, including display formatting, the code required for internal computations, etc.

In software environments like *R* or *Python*, the main practical difference between a CSV file and a spreadsheet data file is that simple utilities like the `read.csv` function described in Sec. 4.4.1 can work with CSV files, but not with Microsoft Excel data files or other spreadsheet data files, which are organized very differently. For example, spreadsheets often consist of multiple *sheets*, each representing a distinct collection of data elements; further, spreadsheets commonly have embedded computations, where one or more cells in the spreadsheet display results computed from values in other cells of the spreadsheet. An Excel data file contains all of the information necessary to perform these computations, along with all necessary formatting information. As the following example illustrates, these Excel spreadsheet data files can be read into *R* data frames, but this requires the use of additional *R* packages.

The dataset containing the Big Mac index discussed in Chapter 6 is available in the form of the Excel spreadsheet file described above, which can be downloaded and opened with Microsoft Excel, but which can also be read into an *R* data frame in several different ways. The results presented here used the *Java*-based *R* package `xlsx`. Specifically, the file was downloaded under the name

BMfile2000-Jul2015.xls, which contains 22 worksheets, each named with a
month and year. Because it is most consistent with the time period for which
the Grande Latte index data is available, the worksheet of primary interest here
was that for January 2013, which had the name Jan2013. The following R code
converts this worksheet into the data frame BigMacJan2013:

```
library(xlsx)
BigMacJan2013 <- read.xlsx("BMfile2000-Jul2015.xls", sheetName = "Jan2013")
```

The first six records of the resulting file look like this:

```
head(BigMacJan2013)
##      Country local_price  dollar_ex dollar_price  dollar_ppp
## 1 Argentina       19.00   4.9765000     3.817944   4.3504186
## 2 Australia        4.70   0.9590946     4.900455   1.0761562
## 3     Brazil       11.25   1.9933500     5.643766   2.5759057
## 4    Britain        2.69   0.6332120     4.248183   0.6159277
## 5     Canada        5.41   1.0029000     5.394356   1.2387244
## 6      Chile     2050.00 471.7500000     4.345522 469.3872683
##   dollar_valuation dollar_adj_valuation euro_adj_valuation
## 1       -12.580758            31.583765          9.1424392
## 2        12.205424            -3.442778        -19.9104025
## 3        29.224960            89.815124         57.4425667
## 4        -2.729621             6.361586        -11.7781520
## 5        23.514254            21.154868          0.4921682
## 6        -0.500844            43.103425         18.6974462
##   sterling_adj_valuation yen_adj_valuation yuan_adj_valuation
## 1              23.713617          59.93016         37.3100903
## 2              -9.217955          17.35804          0.7592459
## 3              78.462105         130.70600         98.0755902
## 4               0.000000          29.27450         10.9902803
## 5              13.908482          47.25462         26.4273439
## 6              34.544275          73.93144         49.3310684
```

Of primary interest here are the columns Country and dollar_price, giving
the price of a Big Mac purchased in the local currency, converted to US dollars
at the prevailing exchange rate. A detailed discussion of this dataset and its
relationship to the Grande Latte index data described in Sec. 4.2.1 is given in
Chapter 6.

4.4.3   Two potential problems with CSV files

While it has been argued that CSV files represent an extremely convenient and
versatile data exchange format, especially between different software environments like R, *Python*, Microsoft products, or other programming environments
like *SAS*, CSV files are not without their difficulties. The following discussion
considers two issues that are relevant to R applications: first, the possibility
that the read.csv function may change variable names; and second, the problems that can arise when reading CSV files containing character strings with
embedded commas.

The variable name change problem

Valid variable names in *R* can contain periods ("*.*") or underscores ("*_*"), but they cannot obtain other special characters like slashes, blanks, or dollar signs. Other software environments have different naming rules, however, so given a CSV file from an arbitrary source that may have been generated in another—and often unknown—software environment, it is possible that we will encounter CSV files whose headers contain variable names that are not valid in *R*. In such cases, the `read.csv` function automatically converts these variable names to names that are valid in *R*, replacing any invalid characters with periods.

The unclaimed Canadian bank account dataset is a case in point. Downloading the CSV file from the Socrata website and reading it into the data frame `unclaimedFrame` as described in Sec. 4.4.1 yields the following variable names:

```
colnames(unclaimedFrame)

## [1] "Last...Business.Name" "First.Name"          "Balance"
## [4] "Address"              "City"                "Last.Transaction"
## [7] "bank_name"
```

Note that the first variable name here contains four periods (three in succession between "Last" and "Business," and one between "Business" and "Name"), while the second and sixth variables each contain one period, all of which resulted from the replacement of special characters from the original variable names that are not valid in *R* variable names. In fact, these original names can be recovered by using the `readLines` function in *R* instead of `read.csv` to read the data file, which reads the file line-by-line, returning a character vector with one element for each line read from the file. The following *R* code does this and applies the `strsplit` function discussed in detail in Chapter 8 to split the first record into parts separated by commas:

```
altRead <- readLines("Unclaimed_bank_accounts.csv")
strsplit(altRead[1], split=",")

## [[1]]
## [1] "Last / Business Name" "First Name"          "Balance"
## [4] "Address"              "City"                "Last Transaction"
## [7] "bank_name"
```

We can see from this result that the first element of this character vector lists the variable names shown in the browser, and not those shown in the data frame returned by the `read.csv` function. For example, the original name of the first variable ("Last / Business Name") contains three spaces and one forward slash. As noted above, none of these characters are permitted in valid *R* variable names, so the `read.csv` function converts them to "*.*" which is valid in *R* variable names. Similarly, the spaces in "First Name" and "Last Transaction" are also replaced with periods by the `read.csv` function. These differences can cause serious and subtle programming errors if attempts are made to match the original names with the "sanitized" names created by the read.csv function. *One of the important points of this book is that programming is all about automating repetitive tasks.* Thus, programs often access variables by name—indeed, this is generally the preferred way to obtain the variables we need from a file or a data frame—so it is important to be aware of possible changes in variable names like those introduced here by the read.csv function.

Problems of embedded commas

A subtle issue that breaks some CSV file read utilities but not others is the presence of embedded commas in a text field. As a specific example, the standard *R* CSV file read function read.csv handles commas embedded in quoted text strings fine, but the function read.csv.sql in the sqldf package discussed in Sec. 4.7.2 fails in the face of these embedded commas. The following example, based on the unclaimed Canadian bank account dataset, illustrates this point.

In particular, the read.csv function was used in Sec. 4.4.1 to read the CSV file Unclaimed_bank_accounts.csv into a data frame with 12816 records, each with 7 fields. The read.csv.sql function from the sqldf package discussed in Sec. 4.7.2 allows us to use SQL queries to read *selected portions* of a CSV file, something that can be extremely useful if we want only a subset of records from a very large data file, and possibly only a subset of the fields in each record. Unfortunately, this function fails if we attempt to use it to read the unclaimed bank account data CSV file:

```
library(sqldf)
sqldfFrame <- read.csv.sql("Unclaimed_bank_accounts.csv", eol = "\n")
```

This attempt results in the following error message:

```
Error in .local(conn, name, value, ...) :
  RS_sqlite_import: Unclaimed_bank_accounts.csv line 9
    expected 7 columns of data but found 8
```

Since this error message tells us what the problem is, we can use the readLines function to read lines 8 and 9 of the file, and then use the strsplit function discussed in Chapter 8 to parse these text strings into components separated by the comma delimiter. Doing this for line 8 yields:

```
unclaimedLines <- readLines("Unclaimed_bank_accounts.csv")
line8 <- unclaimedLines[8]
strsplit(line8, split = ",")

## [[1]]
## [1] "MYKYTIUK"              "ROSE        "
## [3] "47714.39"              "17315 106 ST NW "
## [5] "Edmonton"              "01/29/1994"
## [7] "TORONTO-DOMINION BANK "
```

while for line 9 we obtain:

```
line9 <- unclaimedLines[9]
strsplit(line9, split = ",")
```

```
## [[1]]
## [1] "\"HIGER EDUCATION GROUP" " UNIVERSITY OF REGINA\""
## [3] ""                        "41615"
## [5] "UNIVERSITY OF REGINA "    "UNKNOWN ALTA "
## [7] "09/12/1995"              "ROYAL BANK OF CANADA "
```

Consistent with the above error message, note that line 8 of the records from the unclaimed bank account CSV file splits into seven components when we parse it using the comma separator, while line 9 splits into eight components. The problem here is that the function read.csv.sql from the sqldf package does not distinguish between commas embedded in a single quoted text string like "HIGER EDUCATION GROUP, UNIVERSITY OF REGINA" and those separating distinct quoted text strings (e.g., "MYKYTIUK" and "ROSE").

The point of this example is to emphasize that subtle features in CSV files can cause standard functions that normally read these files to fail. This is one of the reasons it is important to know something about the read utilities available for other file types. In particular, note that the readLines utility used here assumes only that the file is organized into records terminated by end-of-line characters. Because this function makes no assumptions about the format of these individual records, it provides a useful basis for reading records with complicated internal record formats and examining them to see what those formats are. An example of using this function to extract records with a complicated format and develop a parsing strategy is presented in Sec. 4.5.1.

4.5   Working with other file types

As noted in the introduction to this chapter, external data can be stored in many different file types. While it is not reasonable to attempt an extensive coverage of these different file types and the rationales behind them, the following subsections provide a few representative illustrations. Specifically, Sec. 4.5.1 describes ASCII text files, commonly used to store text data, and the readLines and writeLines functions available in R to read and write these files. Sec. 4.5.2 then describes *RDS files*, special-purpose files that can be used to efficiently save and restore arbitrary R objects (e.g., data frames, complex lists structures, functions, etc.), along with the R functions used to create these files and retrieve the R objects they contain. Finally, Sec. 4.5.3 describes a few of the most popular files for saving graphical objects generated in an R session, making them available for other uses (e.g., inclusion in documents or slide presentations).

4.5.1   Working with text files

Text files represent a data source of growing importance, motivating the inclusion in this book of Chapter 8, which introduces the basics of analyzing text data. The basic R function for reading text data is readLines, a function we have already seen in several different applications. Specifically, this function was used to read raw HTML data in Sec. 4.3.2, and it was used again in Sec. 4.4.3 both to examine the original variable names in a CSV file and to detect the presence of embedded commas in this file. To give a more complete picture of how this function can be useful, the following discussion presents another example, one that uses the `readLines` function to read the automobile gas mileage dataset from the UCI Machine Learning Repository described in Sec. 4.3.1. Here, the format of this dataset is somewhat complicated, requiring us to develop a parsing strategy to extract the fields we want from each record, and the `readLines` function allows us to read these records into a character vector that will serve as the basis for both developing this strategy and implementing it.

To see what the raw data format looks like, the following code reads the dataset described in Sec. 4.3.1 and displays the first record, split into two parts using the functions `nchar` and `strsplit` discussed in Chapter 8. In particular, the `nchar` function tells us how many characters are in the first record, and the `substr` function allows us to split this record into two components, the first consisting of characters 1 through 56, and the second consisting of characters 57 through 84:

```
autoMpgRecords <- readLines("UCIautoMpg.txt")
x <- autoMpgRecords[1]
nchar(x)

## [1] 84

substr(x, 1, 56)

## [1] "18.0   8   307.0      130.0      3504.      12.0   70  1"

substr(x, 57, 84)

## [1] "\t\"chevrolet chevelle malibu\""
```

The symbol "\t" at the beginning of the second part of this record is a tab character, and here it separates two parts of the record. To the left of this tab character is a sequence of eight numbers, separated by varying amounts of whitespace, while to the right is a second character string representing the name of a specific automobile. The names of all of these variables are available in the metadata file associated with the data file, also available from the UCI Machine Learning Repository. To obtain it, we proceed in exactly the same way as we did to obtain the data file, applying the `download.file` function to the following URL to obtain the text file `UCIautoMpgNames.txt`:

http://archive.ics.uci.edu/ml/machine-learning-databases/auto-mpg/auto-mpg.names

Once again, we can read this text file with the `readLines` function. Lines 32 to 44 of this file contain the variable names we want:

```
autoMpgNames <- readLines("UCIautoMpgNames.txt")
autoMpgNames[32:44]

##  [1] "7. Attribute Information:"
##  [2] ""
##  [3] "    1. mpg:            continuous"
##  [4] "    2. cylinders:      multi-valued discrete"
##  [5] "    3. displacement:   continuous"
##  [6] "    4. horsepower:     continuous"
##  [7] "    5. weight:         continuous"
##  [8] "    6. acceleration:   continuous"
##  [9] "    7. model year:     multi-valued discrete"
## [10] "    8. origin:         multi-valued discrete"
## [11] "    9. car name:       string (unique for each instance)"
## [12] ""
## [13] "8. Missing Attribute Values:  horsepower has 6 missing values"
```

Returning to the structure of the data records, this data source appears to be a *fixed width format file*, a file format popular in many languages that can be read using the R function **read.fwf**. This requires that we know the record format—specifically, the width of each field in the record—and provide this information in our call to the **read.fwf** function. Rather than attempt this, the rest of this discussion uses an alternative approach, based on some of the string handling functions discussed further in Chapter 8.

The basic sequence of steps required is the following. First, note that each character vector returned by the **readLines** function shown above includes two internal quotation marks, represented as \", and the first step in the reformatting process is to remove these internal quotation marks from every record. This can be done with the **gsub** function discussed in detail in Chapter 8; specifically, the following code does this:

```
noQuotes <- gsub('\"', '', autoMpgRecords)
```

This function replaces all internal quotes (\") with an empty character, deleting it without leaving any spaces in its place. Further, **gsub** applies this replacement to every element of **autoMpgRecords**, returning the vector **noQuotes**. Given this vector, we apply these steps to each element:

1. Split the element on the tab character (\t) using the **strsplit** function discussed in Chapter 8, giving two character vectors, the first containing the eight numerical variables, and the second containing the car names;

2. Again using the **strsplit** function, split the first of these two character vectors on any occurrence of whitespace, giving a character vector with eight elements, one for each numerical variable in the data record;

3. Convert these eight elements to numeric variables with the **as.numeric** function;

4. Combine the resulting eight numerical variables and the car name character string into one record of the desired data frame.

We could execute these steps manually, but it is both simpler and less prone to error if we incorporate them into the *R* function `ConvertAutoMpgRecords` listed below and discussed in detail in Chapter 8:

```
ConvertAutoMpgRecords

## function(rawRecords){
##    #
##    noQuotes <- gsub('\"', '', rawRecords)
##    #
##    n <- length(noQuotes)
##    outFrame <- NULL
##    for (i in 1:n){
##      x <- noQuotes[i]
##      twoParts <- unlist(strsplit(x, split = "\t"))
##      partOne <- unlist(strsplit(twoParts[1],
##                            split = "[ ]+"))
##      numbers <- as.numeric(partOne)
##      upFrame <- data.frame(mpg = numbers[1],
##                        cylinders = numbers[2],
##                        displacement = numbers[3],
##                        horsepower = numbers[4],
##                        weight = numbers[5],
##                        acceleration = numbers[6],
##                        modelYear = numbers[7],
##                        origin = numbers[8],
##                        carName = twoParts[2])
##      outFrame <- rbind.data.frame(outFrame, upFrame)
##    }
##    return(outFrame)
## }
```

Applying this function to the character vector `autoMpgRecords` read from the original file with the `readLines` function gives the desired data frame, with the following structure:

```
autoMpgFrame <- ConvertAutoMpgRecords(autoMpgRecords)

## Warning in ConvertAutoMpgRecords(autoMpgRecords):  NAs introduced by coercion
## Warning in ConvertAutoMpgRecords(autoMpgRecords):  NAs introduced by coercion
## Warning in ConvertAutoMpgRecords(autoMpgRecords):  NAs introduced by coercion
## Warning in ConvertAutoMpgRecords(autoMpgRecords):  NAs introduced by coercion
## Warning in ConvertAutoMpgRecords(autoMpgRecords):  NAs introduced by coercion
## Warning in ConvertAutoMpgRecords(autoMpgRecords):  NAs introduced by coercion

head(autoMpgFrame)

##   mpg cylinders displacement horsepower weight acceleration modelYear
## 1  18         8          307        130   3504         12.0        70
## 2  15         8          350        165   3693         11.5        70
## 3  18         8          318        150   3436         11.0        70
## 4  16         8          304        150   3433         12.0        70
## 5  17         8          302        140   3449         10.5        70
## 6  15         8          429        198   4341         10.0        70
##   origin                   carName
## 1      1 chevrolet chevelle malibu
```

```
## 2        1              buick skylark 320
## 3        1              plymouth satellite
## 4        1              amc rebel sst
## 5        1              ford torino
## 6        1              ford galaxie 500
```

Note the six warning messages here: these correspond to the fact that six **horsepower** values are missing, as noted in the portion of the metadata for this file shown earlier.

The main point of this discussion has been to illustrate the utility of the **readLines** function in *R*, both to read metadata text files like the one describing the auto-mpg dataset from the UCI Machine Learning Repository, and to help us understand the structure of data records in cases like this one where the format is not a simple CSV file that can be read with the **read.csv** function. It often turns out to be useful to read a few records from a data file and examine them in detail, particularly in cases like the one discussed in Sec. 4.4.3 where tools that *should* work do not, like the **read.csv.sql** example considered there.

4.5.2   Saving and retrieving R objects

The file formats discussed so far—e.g., the CSV files and text files used in the previous examples—are *general-purpose* in the sense that many different software environments can read and write these files. In some cases, however, we wish to save the complete details of an *R* object, to be able to work with it in a future *R* session. That is, recall that whatever *R* objects we create in an interactive session are *volatile*, disappearing when we end the session, unless we do something to save them. In the case of a data frame like that created for the Grande Latte index described in Sec. 4.2.1, this is easily accomplished by saving the contents of the data frame as a CSV file with the **write.csv** function. In the case of more complex *R* objects like the predictive models discussed in Chapters 5 or 10, however, the structure of the object is not well matched to the record structure of the CSV file, so we need an alternative.

The function **saveRDS** allows us to save any *R* object to an external file with the extension **.rds**, and, once we have done this, we can restore the object to a later *R* session with the function **readRDS**. As a simple example, suppose we use the linear modeling function **lm** discussed in Chapter 5 to build a model to predict the gas mileage **mpg** from all other variables in the **mtcars** data frame. The following code fits this model, saving it as **linearModel**:

```
linearModel <- lm(mpg ~ ., data = mtcars)
```

The resulting *R* object is a named list with 12 elements and these names:

```
names(linearModel)
```

```
##  [1] "coefficients"  "residuals"    "effects"    "rank"
##  [5] "fitted.values" "assign"       "qr"         "df.residual"
##  [9] "xlevels"       "call"         "terms"      "model"
```

Further, these list elements have a variety of different structures themselves, including single numbers, numerical vectors, matrices, data frames, and other *R*-specific objects. We can save this object, preserving all of this structure, with the `saveRDS` function:

```
saveRDS(linearModel, "linearModelExample.rds")
```

To recover this object, we use the `readRDS` function, which converts a specified external `.rds` file into an internal *R* object. To see that this function recovers exactly the same object we saved, we can read it into a new *R* object with a different name and compare the results:

```
recoveredLinearModel <- readRDS("linearModelExample.rds")
identical(recoveredLinearModel, linearModel)
```

```
## [1] TRUE
```

The functions `saveRDS` and `readRDS` are especially useful if we are applying tools like those described in Chapters 5 and 10 to large datasets, yielding possibly complicated models that take a long time to fit. It is extremely desirable in such cases to fit these models once and save them, rather than having to re-fit them multiple times and incur the repeated computational cost of doing this.

4.5.3 Graphics files

In an interactive *R* session, graphics functions like `plot` generate results that are displayed to us in a graphics window that is opened the first time we request graphical output. Once we have a result we want to save either for our own records (e.g., an exploratory data visualization to document what we did) or to share with others (e.g., an explanatory data visualization to be used in explaining what we have found), it is desirable to save this result in an external file. Once again, it is important to emphasize that whatever we create in an interactive *R* session is *volatile*, disappearing forever when we terminate our session, unless we do something to save it in our external file system. Graphical results can be saved in several different ways, and the following discussion introduces two of them: PDF files and PNG files. For a more detailed discussion of this topic, refer to Paul Murrell's book on *R* graphics [57, Ch. 9].

As Murrell notes, graphics formats can be divided into two main groups: *vector formats* and *raster formats*. Very roughly, vector formats encode recipes for drawing the graphical display, while raster formats describe the display as a set of points or *pixels*, giving information (e.g., intensity and color values) for each point in the display. In describing the essential differences between these formats, he offers the following summary [57, p. 309]:

> In general, vector formats are superior for images that need to be viewed at a variety of scales, but raster formats will produce much smaller files if the image is very complex.

The PDF file format represents probably the most widely available and most easily shared vector format. Popular raster formats include JPEG files like the one used to display the Grande Latte index data discussed in Sec. 4.2.1, PNG files, and TIFF files, all of which can be used to save graphics files in R. Of these raster formats, consideration is restricted here to PNG files, which Murrell describes as "usually better for statistical plots," and which are compatible with many other software environments like Microsoft Powerpoint, allowing graphical displays saved as PNG files to be incorporated into documents produced in these environments.

The mechanisms for creating these two graphical file types in R are essentially the same, consisting of the following steps:

1. call a function that instructs our R session to divert the output from all upcoming graphics commands to a specified file;

2. execute the R commands required to generate the plot we want;

3. call the function dev.off() to stop saving graphical results to this file and re-direct the output to our interactive graphics window.

Specifically, to save graphics as a PDF file, we first call the function pdf with the name of the PDF file to be created, while to create a PNG file, we call the function png. The following examples illustrate both cases.

Both examples save the plot shown in Fig. 4.2, a variable-width boxplot summary of the variation of gas mileage with number of cylinders in the auto-mpg dataset from the UCI Machine Learning Repository, obtained as described in Secs. 4.3.1 and 4.5.1. For the PDF file, the code is:

```
pdf("AutoMpgBoxplotEx.pdf")
boxplot(mpg ~ cylinders, data = autoMpgFrame,
        xlab = "Cylinders", ylab = "Gas mileage",
        las = 1, varwidth = TRUE)
dev.off()
```

As described above, the first line causes all subsequent graphical output to be directed to the file AutoMpgBoxplotEx.pdf, the next portion of code (i.e., the boxplot function call) creates the desired plot, and the last line reverts the R graphics system to interactive mode.

The code required to save this plot as a PNG file is essentially identical, replacing pdf wherever it appears in the above code with png:

```
png("AutoMpgBoxplotEx.png")
boxplot(mpg ~ cylinders, data = autoMpgFrame,
        xlab = "Cylinders", ylab = "Gas mileage",
        las = 1, varwidth = TRUE)
dev.off()
```

Again, for a more complete discussion of saving graphics results, refer to the book by Paul Murrell [57, Ch. 9].

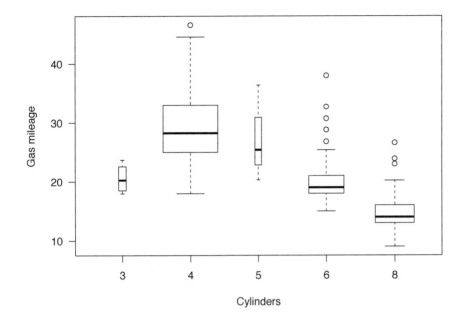

Figure 4.2: Variable width boxplot summary of gas mileage vs. cylinders from the UCI auto-mpg dataset.

4.6 Merging data from different sources

Often, the data we wish to analyze comes from multiple sources. The Big Mac vs. Grande Latte economic index comparison presented in Chapter 6 is a typical example: there, the objective is to assess the general degree of agreement between these indices, and to identify any points of glaring disagreement for possible further investigation. Both datasets are available—the Big Mac index data was obtained from a Microsoft Excel spreadsheet as described in Sec. 4.4.2, while the Grande Latte index data was entered manually as described in Sec. 4.2.1—but they reside in different data frames. For the analysis presented in Chapter 6, it is necessary to merge these two data sources into a common data frame. The following discussion considers the process of performing such data merges.

This process of merging two data sources corresponds to the database *join* operation discussed further in Sec. 4.7.1. Specifically, one of the primary design objectives in a relational database like those discussed in Sec. 4.7 is to minimize duplication of data by storing related values together in (relatively) small tables (e.g., customer-specific details like names and addresses) and joining the results contained in these separate tables when we need a larger, more complete dataset (e.g., a table containing both customer-specific details from one table and transaction-specific details from another).

In R, the `merge` function can be used to merge (or join) two data frames. This function is called with the names of the data frames, and in the simplest cases this is enough: it returns a merged data frame whose rows are defined by the values of the variables (i.e., columns or fields) common to both of the original data frames. For example, if both data frames contain columns labelled `country` but no other common columns, the merged data frame will have one row for each country that appears in both data frames, together with all of the *other* columns from these data frames. If the original data frames have several columns in common, the resulting merged data frame has one row for each combination of these variables that appears in both of the original data frames, again with all of the other columns from both data frames included.

If we want to perform our data frame merge only on a subset of the common columns in the two original data frames, the `merge` function allows us to do this by specifying the optional argument `by`. Thus, if our original data frames have common columns `country` and `price` but we only want to use the `country` column as the basis for our merge, we would specify `by = "country"`. The resulting data frame would then have variables named `country`—containing the common `country` values used in merging the original data frames—along with the separate variables `price.x` containing the `price` values from the first data frame and `price.y` with the corresponding values from the second data frame.

It often happens that the "same variable" in the two original data frames have different names. For example, the labels for the "country" column in these data frames could be `country` and `Nation`. In such cases, we can specify the *join keys*—the variables to be used in merging the data frames—by using the optional arguments `by.x` and `by.y`. That is, if the first data frame has the column `country` and the second data frame has the corresponding column labelled `Nation`, we would specify `by.x = "country"` and `by.y = "Nation"`. The resulting data frame will have one record for each case where the value of `country` in the first data frame matches the value of `Nation` in the second data frame. The label for this column in the merged data frame is taken from the first data frame (i.e., this column would be labelled `country` in this example).

In the economic index example considered in Chapter 6, we have a more serious problem: the two original datasets are not in one-to-one correspondence. Specifically, the Big Mac index data consists of a single value for each of 57 *countries*, while the Grande Latte index data consists of a single value for each of 29 *cities*, some of which are located in the same country. Thus, before we can compare these two indices—the primary focus of the discussion presented in Chapter 6—we must merge them into a data frame based on a common index. Since problems of this type arise frequently—and can be a source of horrendous errors—the rest of this section is devoted to a detailed discussion of how to create this common data frame.

Since cities are uniquely located in countries, the country specification added to the Grande Latte index provides the basis for merging these data frames. That is, the desired result is a data frame with the following four columns:

1. `country`, the country specified for the Big Mac index, and the country where the city is located in the Grande Latte data frame;

2. `city`, the city specified in the Grande Latte data frame;

3. `BigMacIndex`, the value of the Big Mac index for `country`;

4. `GrandeLatteIndex`, the value of the Grande Latte index for `city`.

To obtain this data frame, we first added—alas, manually—the column `country` to the Grande Latte data frame giving the country in which each city resides. We then applied the **merge** function described above, *but it is important to note that our first attempt, described next, is flawed and does not give us quite what we want.* The code for this first attempt looks like this:

```
FlawedMergeFrame <- merge(LatteIndexFrame, BigMacJan2013,
                          by.x="country", by.y="Country")
```

Before discussing what is wrong with this result, it is important to describe what the code is doing. Specifying `by.x = "country"` and `by.y = "Country"` defines the join key for this merge. As noted above, this particular form of specification is necessary because the "country" column has different labels in the source data frames `LatteIndexFrame` and `BigMacJan2013`.

What exactly is wrong with this first attempt? For one thing, the resulting data frame has 13 columns instead of the four we want. This problem is easily corrected by keeping only the desired variables instead of all of them, which also allows us to rename the variables we keep, simplifying the merge. Specifically, the following merge gets us a step closer to what we want:

```
LatteSubset <- data.frame(country = as.character(LatteIndexFrame$country),
                          city = as.character(LatteIndexFrame$city),
                          GrandeLatteIndex = LatteIndexFrame$price,
                          stringsAsFactors = FALSE)
BigMacSubset <- data.frame(country = as.character(BigMacJan2013$Country),
                           BigMacIndex = BigMacJan2013$dollar_price,
                           stringsAsFactors = FALSE)
BetterMerge <- merge(LatteSubset, BigMacSubset)
str(BetterMerge, vec.len = 2)

## 'data.frame': 23 obs. of  4 variables:
##  $ country         : chr  "Argentina" "Australia" ...
##  $ city            : chr  "Buenos Aires" "Sydney" ...
##  $ GrandeLatteIndex: num  4.18 4.82 5.65 4.23 4.08 ...
##  $ BigMacIndex     : num  3.82 4.9 ...
```

This result is better, but it is still not exactly right. Specifically, we want every record from the original Grande Latte index data frame where there is a corresponding country in the Big Mac index data frame. To see that we haven't quite accomplished this, compare the `country` variable from the `BetterMerge` data frame with that from the `LatteSubset` data frame:

```
setdiff(LatteSubset$country, BetterMerge$country)

## [1] "Korea"    "USA"       "England"
```

The `setdiff` function computes the *set difference* between two sets, defined
as those values that are included in the first set (i.e., countries listed in the
`LatteSubset` data frame) but not in the second (i.e., countries absent from
our attempted merge). The first of these countries—Korea—is absent from the
merge because it is not included in the Big Mac index data for January 2013,
but the other two cases arise from different representations for the same country.
Specifically, "USA" in the Grande Latte index data frame appears as "United
States" in the Big Mac index data, while "England" in the Grande Latte data
appears as "Britain" in the Big Mac index data. One solution would be to
manually edit the `LatteSubset` data frame using the `fix` function, but a better
approach is to use the following R code, which can be documented to describe
how the original data was modified:

```
USAindex <- which(LatteSubset$country == "USA")
LatteSubset$country[USAindex] <- "United States"
EnglandIndex <- which(LatteSubset$country == "England")
LatteSubset$country[EnglandIndex] <- "Britain"
FinalMerge <- merge(LatteSubset, BigMacSubset)
str(FinalMerge, vec.len=2)

## 'data.frame': 28 obs. of  4 variables:
##  $ country          : chr  "Argentina" "Australia" ...
##  $ city             : chr  "Buenos Aires" "Sydney" ...
##  $ GrandeLatteIndex : num  4.18 4.82 5.65 4.23 3.81 ...
##  $ BigMacIndex      : num  3.82 4.9 ...
```

Here, it is easily verified that the only country present in the original Grande
Latte index dataset but not in the `FinalMerge` data frame is Korea. This final
merged data frame forms the basis for the results presented in Chapter 6.

4.7   A brief introduction to databases

The final section of this chapter gives a brief introduction to databases, with
emphasis on either interacting with them from R or using them within R via
the `sqldf` and `RSQLite` packages discussed in Secs. 4.7.2 and 4.7.4, respectively. This introduction begins with a brief discussion in Sec. 4.7.1 of the
essential ideas: relational databases and what they are, the key role of queries
in extracting data from relational databases, and SQL or *structured query language*, the specialized language in which these queries are written. Introductory
but more detailed discussions of these topics are given by Paul Murrell in his
book on data technologies [57, Chs. 7-8]. Much more thorough introductions
to databases, their design principles, advantages and disadvantages, and other
details are given in database textbooks like the one by Date [17].

Following this general introduction to relational databases and SQL, the `sqldf` package is introduced in Sec. 4.7.2, which allows us to apply SQL queries to data frames in *R*. This package is built on the `RSQLite` package which provides *R* support for the `sqlite3` database, but the `sqldf` package is introduced first here because it is likely that most readers are more familiar with data frames than with relational databases. Introducing `sqldf` first provides an opportunity to try out SQL queries and appreciate their structure and power before going into the details of working with external databases. These topics are then introduced in Sec. 4.7.3, which gives a general description of the basic mechanics of connecting with external databases and a little information about the main *R* packages for accomplishing this. Sec. 4.7.4 then concludes the chapter with a somewhat more detailed discussion of these issues for the specific case of working with `sqlite3` databases in *R* through the `RSQLite` package.

4.7.1 Relational databases, queries, and SQL

As noted in the introduction to this chapter, databases are used extensively in business to efficiently store very large quantities of data. They are particularly useful in cases where a "flat file" representation (e.g., a CSV file) would involve many records with the same values for certain fields. For example, in a database storing customer transaction histories, if we store the customer's name, address, and billing information along with the details of each individual transaction, customers who have made many purchases will have many such duplicated fields, and if repeat customers are responsible for most of our business, the overall degree of duplication will be substantial. *Relational databases* are designed to overcome this problem by separating the database into individual tables; thus, for example, the name, address, and billing information for each customer is stored in a table with one entry for each customer, which also has a customer ID. Individual purchase information—e.g., item, price, size, color, etc.—can then be stored in a separate table, where one field is the customer ID; while this single field is repeated frequently, the detailed information specific to each customer is not duplicated, potentially saving a great deal of storage space.

Another advantage of this organization—where customer-specific information is stored only once and not many times—is that it greatly simplifies database updates if this information changes. For example, if a customer moves, this database organization allows us to update their address only once, in the customer information table, and not every time it occurs in a larger data table that has both customer information and transaction details.

Many different databases exist, typically consisting of large hardware and software installations, supported by commercial vendors, and the following discussion makes no attempt to give a comprehensive introduction to databases, their design, or their use. Rather, the intent here is to provide a short introduction to the key ideas of relational databases and the support *R* has for interacting with them. At a very high level, we can think of a relational database as a collection of *tables*, each of which contain related subsets of data, together with supporting software that allows us to extract data from related tables, update the data in these tables, delete data or tables, or add new tables. Broadly, then, databases are created and used through a sequence of steps like this:

1. The database is *designed:* this involves deciding what variables are included and how they are organized (standard database texts like that by Date devote considerable attention to database design [17]);

2. The database design is *implemented* in a specific software environment (e.g., Oracle, Teradata, Microsoft SQL Server, Microsoft Access, MySQL, PostgreSQL, or sqlite3), which involves creating and populating the data tables, providing—and typically restricting—access to the data, etc.;

3. Users run SQL queries against the database to extract the specific subsets of data they need.

Generally, databases are in a constant state of flux, with new entries being added to existing tables, new tables being added to handle new types of data, data values being changed to reflect changes in data sources or to correct errors, etc.

Since the most common way R users interact with databases is as consumers of data from databases that have been created by others, the primary focus here is on how queries are executed. This involves three basic steps:

1. Connect to the database from our R session;

2. Execute SQL queries against the database to retrieve the data we want;

3. Close the database connection.

The mechanics of how we connect to a database and disconnect from it depend on the specific database we are using, and these details are considered further in Sec. 4.7.3. Fortunately, the SQL language is (reasonably) well standardized, so the format of the SQL queries we need to run to retrieve the data we want is (largely) independent of the database. As the parenthetical expressions imply, this statement is not *strictly* true, but it is true enough that the example SQL statements discussed here should run on most databases.

A key element of relational databases is the concept of a *table*, which is a rectangular array of data, much like a data frame in R. Specifically, tables are organized by rows and columns, with each row representing a single data record, and each column defining a single variable or field in these records. All data in a relational database is organized in tables, and the purpose of a database query is to extract some or all of the data from one or more tables, or values simply computed from the data in these tables. The SQL language provides a standardized—and very powerful—way of extracting this data, returning another table as the result. The following paragraphs provide a broad overview of what these queries look like, while Sec. 4.7.2 presents a number of specific examples of SQL queries applied to R data frames using the `sqldf` package.

As Murrell notes in his introduction to SQL [57, Sec. 7.2.1], much of what we want to do with a database will be through queries of this general form:

```
SELECT columns (i.e., variables)
   FROM tables
   WHERE row condition
```

In fact, the simplest SQL query has the form "SELECT * FROM table." This query returns a new table with all columns from all records contained in the specified table, effectively making a copy of the original table. In the more general query given above, a subset of columns is selected—a potentially important restriction if we are querying databases with hundreds or thousands of variables per record—and the row condition specifies that we want only those records satisfying some condition on one or more of these variables.

Note also that Murrell's general query includes the phrase "FROM tables," where "tables" is plural. In fact, we can form queries invoving multiple tables, but we must do this in such a way that the resulting table has well-defined values for all fields in all records. In the simplest case, this involves the *inner join* of two tables, corresponding to the data frame merge discussed in Sec. 4.6. Rather than present the general format of queries involving inner joins here, specific examples will be presented and discussed in Sec. 4.7.2.

Another common type of query involves summaries of certain data variables over groups defined by one or more other data variables. In particular, SQL supports functions like AVG to compute averages over groups of records, COUNT to count the number of records in each group, MIN to find the minimum value in each group, or MAX to find the maximum value in each group. These functions are most useful when combined with the GROUP BY syntax to define these groups in terms of variables in a table. Again, examples to illustrate this idea are presented in Sec. 4.7.2.

4.7.2 An introduction to the sqldf package

As noted, the sqldf package provides support for SQL queries like those just discussed against data frames. Since the preceding discussion was a little vague, it is useful to begin this discussion with a few concrete examples to illustrate both specific SQL syntax and the results we can obtain using these queries.

First, recall from Fig. 4.2 that the auto-mpg dataset obtained from the UCI Machine Learning Repository included cars with both 3 and 5 cylinders, which are somewhat unusual. The following query extracts the numbers of cylinders, the model year, and the name for all of these unusual cars:

```
library(sqldf)
strangeCars <- sqldf("SELECT cylinders, modelYear, carName
                              FROM autoMpgFrame
                              WHERE cylinders == 3 OR cylinders == 5")

strangeCars

##    cylinders modelYear          carName
## 1          3        72   mazda rx2 coupe
## 2          3        73        maxda rx3
## 3          3        77        mazda rx-4
```

```
## 4              5        78            audi 5000
## 5              5        79   mercedes benz 300d
## 6              5        80   audi 5000s (diesel)
## 7              3        80        mazda rx-7 gs
```

Here, the first line loads the `sqldf` package to make the `sqldf` function available for our use. Next, the `sqldf` function is called with the SQL query that we want to execute and it returns the table defined by that query. In this particular case, the three variable names following the `SELECT` keyword in the query indicate that we want our resultant table to include the variables `cylinders`, `modelYear`, and `carName`. The `FROM` keyword specifies that we want these variables taken from the data frame `autoMpgFrame`, and the `WHERE` clause selects all records for which either the condition `cylinders == 3` or `cylinders == 5` holds.

The following SQL query uses the `GROUP BY` syntax described in Sec. 4.7.1 to retrieve the average `mpg`, `horsepower`, and `weight` as a function of the number of cylinders, along with the number of records in each group:

```
cylinderSummary <- sqldf("SELECT cylinders, AVG(mpg), AVG(horsepower),
                                 AVG(weight), COUNT(*)
                          FROM autoMpgFrame
                          GROUP BY cylinders")
cylinderSummary

##    cylinders AVG(mpg) AVG(horsepower) AVG(weight) COUNT(*)
## 1          3 20.55000        99.25000    2398.500        4
## 2          4 29.28676        78.28141    2308.127      204
## 3          5 27.36667        82.33333    3103.333        3
## 4          6 19.98571       101.50602    3198.226       84
## 5          8 14.96311       158.30097    4114.718      103
```

Here, the `GROUP BY cylinders` clause groups the data into subsets by their `cylinders` values, and the summary computations requested in the `SELECT` clause by the functions `AVG` and `COUNT` are applied to each group. The use of `COUNT(*)` here is a convenience, requesting counts of all records (i.e., all rows) in each group; we could have obtained the same results with `COUNT(cylinders)`, but the "*" syntax simplifies the query slightly.

As a third and last example, the following SQL query executes a join between two data frames. Specifically, this example creates the SQL query that is equivalent to the final data frame merge presented in Sec. 4.6:

```
query <- "SELECT Lsub.country, Lsub.city, Lsub.GrandeLatteIndex, Msub.BigMacIndex
          FROM LatteSubset AS Lsub INNER JOIN BigMacSubset AS Msub
          ON Lsub.country = Msub.country"
IndexFrame <- sqldf(query)
head(IndexFrame)

##        country      city GrandeLatteIndex BigMacIndex
## 1       Norway      Oslo             9.83    7.842279
## 2       Sweden Stockholm             7.40    6.388096
## 3       Russia    Moscow             7.27    2.425695
## 4  Switzerland    Zurich             7.12    7.124849
## 5      Finland  Helsinki             6.38    5.088563
## 6       Greece    Athens             5.84    4.477935
```

This query is more complicated than the prevous two and it illustrates some details that are worth noting. First, note that we could have used the SELECT * syntax for this query, but since this requests *all* columns from the two data frames after the inner join operation, we would then get two copies of the country column, with identical names. To avoid this, it is necessary to specify exactly which columns we want in our final result, as in this query. The FROM clause in this query illustrates two important points. First, expressions like LatteSubset AS Lsub define *aliases*, allowing us to simplify other terms in the query: e.g., we can specify Lsub.country instead of LatteSubset.country in the SELECT clause. Second, the INNER JOIN specification indicates that we want the *inner join* of these two subsets, a form of data table merge that retains only those records that have values for all of the fields specified in the SELECT clause. Thus, Korea is not included in the final result because, as noted in Sec. 4.6, there is no record for Korea in the Big Mac index data frame BigMacSubset. Finally, the ON keyword in the last line of the query is equivalent to the by argument in the merge function discussed earlier: it specifies the matching conditions for the inner join, determining which records are kept in the final result.

The intent of the discussion presented here has not been to provide a complete introduction to the sqldf package and its capabilities, but rather to do two things. First and foremost was to provide a few specific examples of the syntax of SQL queries and the results that can be obtained with them, since this can be done in the context of data frames, without the need for connecting to external databases. The second objective was to introduce this package as a useful *R* tool for working with data frames. It is true that some of the things that can be done with SQL queries can also be done with other *R* utilities and sometimes with less effort, as illustrated by the inner join SQL syntax from the last example, compared with its equivalent merge implementation in Sec. 4.6. Nevertheless, SQL queries are widely used outside the *R* community and it can be extremely convenient to be able to run a query provided by someone in the database community without having to figure out how to re-write it in more familiar *R* code, particularly in cases where the query is complicated. Indeed, re-writing complicated queries is itself an error-prone process, so there is considerable advantage in being able to use them directly in *R* code, without having to develop new code and verify that it gives equivalent results.

Finally, it is worth noting that the sqldf package also includes the function read.csv.sql, which can be used to run SQL queries against external CSV files. This capability can be extremely useful in cases where we want to extract a small or moderate-sized subset of data from an enormous CSV file. As illustrated in Sec. 4.4.3, this function is more fragile than the standard read.csv function, since it cannot handle character fields with embedded commas, but in cases where our CSV file does not exhibit this problem, using read.csv.sql with the appropriate SQL query can be a much better solution than attempting to read the entire data file into memory and then extracting what we want from it.

4.7.3   An overview of R's database support

The primary use of databases is to provide reliable, permanent storage of large collections of data, organized in an efficient manner that prevents unnecessary duplication, thus reducing total storage requirements and improving the reliability of the results when data values need to be updated. As a corollary, then, databases represent external files, possibly hosted on external servers (i.e., other computers) that we access via some communication network (i.e., either a company-supported intranet of networked computers or an external network accessed via the Internet). To work with a database from an interactive *R* session, then, the following three steps are necessary:

1. Connect to the database;

2. Execute SQL queries to perform the required database operations;

3. Disconnect from the database.

The preceding discussions have given a brief introduction to SQL queries, along with a few examples, to illustrate some of the data access options SQL makes possible, but we can also use SQL queries to create new databases, add or delete tables from them, or modify their contents by adding, deleting, or changing individual data values. Examples of some of these operations are presented in Sec. 4.7.4 for the specific case of `sqlite3` databases.

More generally, *R* provides two basic packages to support connecting to and using external databases, along with a number of database-specific packages that build on these to simplify working with specific databases. The rest of this section gives a very brief introduction to these two packages and some of the databases they support. For more detailed discussions, refer to the short section on databases in *R* in Paul Murrell's book [57, Sec. 9.7.8], the *R Data Import/Export* help manual available through the Help tab in an interactive *R* session, or the `help` files for the database interface packages `DBI` or `RODBC`.

The `DBI` package provides basic support for communicating with a variety of databases, serving as the basis for a number of database-specific *R* packages, including the `RSQLite` package discussed in Sec. 4.7.4. Other databases supported through the `DBI` package are Oracle, through the `ROracle` package, and PostgreSQL, through the `RPostgreSQL` package. The key functions provided by the `DBI` package are the following:

1. `dbDriver` specifies the type of database we want to connect with (e.g., `dbDriver("PostgreSQL")` specifies a `PostgreSQL` database);

2. `dbConnect` connects with a specific database through a device driver created by the `dbDriver` function;

3. `dbGetQuery` allows us to send SQL queries to the database and retrieve the results;

4. `dbDisconnect` terminates our connection with the database.

These functions and simple extensions of them will be illustrated in Sec. 4.7.4 in connection with the `RSQLite` package.

The `RODBC` package provides similar support for the many databases meeting the requirements of the Open Database Connectivity (ODBC) standard. Examples include Microsoft SQL Server, intended for large, commercial applications, Microsoft Access, widely available on personal computers, MySQL, PostgreSQL, and Oracle databases, among many others. The `RODBC` package provides the same basic capabilities as the `DBI` package, but for ODBC-compliant databases, and the three basic functions have different names:

1. `odbcConnect` connects to a specified database;

2. `sqlQuery` sends SQL queries and returns the results;

3. `odbcClose` terminates our connection to the database.

Note that there is no need to specify the type of database here, since we are using the common ODBC connection infrastructure for any compliant database. For a more detailed introduction to this extremely useful package, refer to Brian Ripley's vignette, *ODBC Connectivity*, available through the package's help files.

4.7.4 An introduction to the `RSQLite` package

The *R* package `RSQLite` provides the basis for the following example, illustrating the creation of a new `sqlite3` database, connecting to it, interrogating it to see what it contains, creating new tables in it, and applying SQL queries against these tables. Specifically, this example creates the database file `EconomicIndexDatabase.db` in our working directory, with two tables: one based on the Grande Latte index data frame `LatteSubset` and the other based on the Big Mac index data frame `BigMacSubset`.

The first steps are to load the `RSQLite` package and open a connection to the database `EconomicIndexDatabase.db`. Note that even though this database does not exist initially, establishing the connection will create this file as an empty database and allow us to work with it:

```
library(RSQLite)
conn <- dbConnect(SQLite(), "EconomicIndexDatabase.db")
```

This code uses the `dbConnect` function from the `DBI` package on which `RSQLite` is based to establish a connection to the database `EconomicIndexDabase.db`. As noted, if this database does not exist, establishing this connection creates the file, allowing us to work with it using the `RSQLite` functions described next; conversely, if this database does exist, this call to `dbConnect` establishes the connection required to allow us to work with it. Also, note that the `SQLite()` function appearing as the first argument of the `dbConnect()` call initiates the driver required to communicate with `sqlite3` databases.

To make this new database useful, we need to put something in it, and a particularly easy way to do this is with the function `dbWriteTable`, which creates a new table from a specified data frame in our *R* session. The following two calls to this function create the tables GrandeLatteTable and BigMacTable in our database:

```
dbWriteTable(conn, "GrandeLatteTable", LatteSubset)
dbWriteTable(conn, "BigMacTable", BigMacSubset)
```

A crucial point is that sqlite3 table names are not *case-sensitive, in contrast to R variable names.* Thus, queries using GrandeLatteTable, grandelattetable, or GRANDELATTETABLE will work equally well. The functions dbListTables and dbListFields list the tables in the database and the fields in each table:

```
dbListTables(conn)

## [1] "BigMacTable"      "GrandeLatteTable"

dbListFields(conn, "GRANDELATTETABLE")

## [1] "country"          "city"             "GrandeLatteIndex"

dbListFields(conn, "bigmactable")

## [1] "country"     "BigMacIndex"
```

To see how many records are in each table, we can execute a SQL SELECT COUNT(*) statement on each table:

```
dbGetQuery(conn, "SELECT COUNT(*) AS 'GrandeLatteRowCount' FROM GrandeLatteTable")

##    GrandeLatteRowCount
## 1                   29

dbGetQuery(conn, "SELECT COUNT(*) AS 'BigMacRowCount' FROM BigMacTable")

##    BigMacRowCount
## 1              57
```

Note that the function dbGetQuery issues the SQL statement included in the function call against the database connection conn previously established. The RSQLite package also includes other functions like dbSendQuery that can be used for this purpose: the primary difference is that dbGetQuery sends the query to the database, executes it and returns the result, and clears the connection to make it available for other RSQLite functions. In contrast, the dbSendQuery function only sends the query: to retrieve the results, it is necessary to call the function dbFetch, and then explicitly clear the connection by calling ClearResult. The reason for this difference is that the second approach allows us to retrieve database results in chunks that we can control, while dbGetQuery returns *all* database results. The ability to work with data in chunks can be very important if we are working with a very large database where the results might not all fit in memory (i.e., in the RAM on our computer).

As a less trivial example, suppose we want to compare the distribution of Big Mac index values across those countries where the Grande Latte index values are also available with those where they are not. In particular, note that one reason we might have Big Mac index values but not Grande Latte index values is that Starbucks was not a significant presence in some of these countries. Or, in more economic terms, is there evidence that those countries without Grande Latte index values are either generally poorer or generally richer than those with Grande Latte index values? To explore this question, we need the Big Mac index values for the two groups of countries separately. The following query gives us the Big Mac index values for countries with both indices:

```
query <- "SELECT M.country, M.BigMacIndex FROM BigMacTable AS M
WHERE M.country IN (SELECT country FROM GrandeLatteTable)"
BigMacBoth <- dbGetQuery(conn, query)
```

Here, we are selecting both the country and the Big Mac index values from `BigMacTable`, and the `WHERE` clause here uses the `IN` operator to keep only those records where `country` from `BigMacTable` is in the set of `country` values in `GrandeLatteTable`. Specifically, the `IN` operator allows us to select any value for `M.country` from the list that follows. In this example, that list is obtained using a *subquery:* a query embedded within our main query, designated by enclosing it in parentheses. That is, the subquery "SELECT country FROM GrandeLatteTable" selects all `country` values from `GrandeLatteTable`, and the `IN` operator restricts those records in our main query—requesting all `country` and `BigMacIndex` values from `BigMacTable`—to those matching the subquery.

A simple modification of this query returns the other set of Big Mac index values of interest here, from countries *without* Grande Latte index values:

```
query <- "SELECT M.country, M.BigMacIndex FROM BigMacTable AS M
WHERE M.country NOT IN (SELECT country FROM GrandeLatteTable)"
BigMacOnly <- dbGetQuery(conn, query)
```

The only difference between this query and the previous one is the presence of the negation operator `NOT` in the `WHERE` clause, preceding the `IN` operator. Other than that, both the main query and the subquery are identical to those in the previous example. The results obtained with this query are discussed in the "Big Mac and Grande Latte economic indices" data story in Chapter 6.

As noted in Sec. 4.7.3, when we are finished working with a database, the last thing we do is to disconnect from it. For the `sqlite` databases considered here, this is accomplished with the `DBI` function `dbDisconnect`:

```
dbDisconnect(conn)
```

To conclude this discussion of the `RSQLite` package and `sqlite3` databases, it is important to note a few key points. First, there is much more that can be done with `sqlite3` databases via the `RSQLite` package than has been described here. Specifically, this database supports a full range of SQL queries, the construction and use of *views* or "virtual tables," the creation and use of "triggers" that cause certain database operations to occur in response to other database operations (e.g., deletion of an entry in one table triggering updates of other tables), and the imposition of integrity constraints (e.g., limits on values that can be entered in a database table). A detailed treatment of these features and how they are implemented and used is given in the excellent book by van der Lans [66] devoted to working with `sqlite3` databases. Conversely, it is important to note that `sqlite3` databases are intended for "small" applications. Specifically, `sqlite3` databases do not support the full range of "large database" features seen in commercial products like those by Oracle, Teradata, and others for the creation and management of very large databases across large organizations where different users have different levels of access. For more general discussions of relational databases, their rationale, and structure, refer to database texts like that by Date [17].

4.8   Exercises

1: One of the external data sources discussed in Section 4.3.1 was the UCI Machine Learning Repository, at the following URL:

http://archive.ics.uci.edu/ml

Use the `browseURL` function to examine this website and find information about the Wine Quality Dataset. How many variables (attributes) are included in this dataset, and how many records (instances)?

2: The Wine Quality Dataset considered in Exercise 1 is actually split into three separate files: a text file called `winequality.names`, and two CSV files, one for red wines and the other for white wines. Use the function `download.file` to obtain the text file from the following URL and save it in a file named `UCIwineQualityNames.txt`:

http://archive.ics.uci.edu/ml/machine-learning-databases
  /wine-quality/winequality.names

Once you have obtained this file, read it into your R session with the `readLines` function. What are the 12 variables in the two CSV files?

3: This exercise builds on Exercises 1 and 2 and asks you to download and examine the wine quality file for red wines from the UCI Machine Learning Repository. Specifically:

3a. From the results of Exercise 1, identify the URL for the red wine quality file and use the `download.file` function to put a copy of it in your working directory, under the name `UCIwineQualityRed.csv`.

3b. Using the standard `read.csv` function, read `UCIwineQualityRed.csv` into a data frame. Examine this data frame with the `str` function: is the result what you expect? (E.g., does it have the expected number of columns?)

3c. Repeat (3b) but using the `read.csv2` function, again characterizing the data frame with the `str` function. Is this the result you expect? Looking at the help file for the `read.csv` function, what is responsible for the difference?

4: Once we have successfully extracted the data frame we want from an Internet data source in an unusual format, it may be useful to save it in a more standard format (e.g., a standard CSV file). This exercise asks you to do this, to provide the standard CSV file needed for Exercise 5:

4a. Save the results obtained from Exercise (3c) as a standard CSV file using the `write.csv` function, without row names;

4b. Read this file back into a new data frame with the `read.csv` function. Using the `str` function, compare this new data frame with the original. Are they the same? If not, how do they differ?

5: As discussed in Section 4.7.2, the `sqldf` package allows us to apply SQL queries to *R* data frames, and several examples were presented from the `autoMpgFrame` data frame obtained from the UCI Machine Learning Repository. This exercise considers queries against the `mtcars` data frame from the `datasets` package. Specifically:

5a. Create a `select` query to obtain the subset of the `mtcars` data frame for the four-cylinder cars. Characterize the resulting data frame with the `str` function and tabulate the `gear` variable from this subset.

5b. Repeat (5a) for the eight-cylinder cars. Is the distribution of `gear` variable values similar or very different?

5c. An optional argument for the `sqldf` function is `row.names`, which has the default value `FALSE`. Repeat the data read from (5a) using this argument and use the `rownames` function to obtain the names of the four-cylinder cars.

6: As discussed briefly in Sections 4.3.3 and 4.7.2, the `read.csv.sql` function in the `sqldf` package allows us to use SQL queries to extract a subset of records from CSV files, provided they are "clean" (e.g., without embedded commas in text strings). The only required argument for `read.csv.sql` is the name of the CSV file to be read, and using this default function call returns the complete contents of the file, just as `read.csv` does. This behavior is a consequence of the fact that the default value for the optional argument `sql` that controls the record retrieval is:

```
sql = "select * from file"
```

To obtain a different result, add a `where` clause (see the discussion in Section 4.7.2) to read the CSV file created in Exercise 4, restricting the result to those records where `quality` is greater than 5. Use the `str` function to characterize the resulting data frame. What differences do you notice between this data frame and that obtained in Exercise (4b)?

7: The SQL clause GROUP BY was demonstrated in an example presented in Section 4.7.2 that showed how it can be used to construct summaries of data frames using the sqldf package. This exercise asks you to construct a similar summary based on the mtcars data frame from the datasets package. Specifically, use the GROUP BY clause with the sqldf package to construct a summary listing: (1) the value of gear; (2) the average mpg value over all records listing this gear value; (3) the corresponding average hp value; and (4) the number of records listing each gear value. Display your results.

8: A useful modification to SQL queries like those used in the sqldf package is the ability to assign your own names to variables like AVG(mpg) returned by the query discussed in Section 4.7.2. To assign your own variable name, simply replace the variable designation in the SELECT query—e.g., AVG(mpg)—with a name designation using AS (e.g., AVG(mgp) AS avgMileage). This exercise asks you to repeat Exercise 7, assigning the following names: avgMPG for the average mpg value, avgHP for the average hp value, and N for the number of records. Display your results.

9: Section 4.6 discussed the important topic of merging data frames obtained from different data sources. This exercise asks you to construct two mileage summary data frames similar to that requested in Exercise 8, one from the mtcars data frame and the other from the Cars93 data frame from the MASS package, and then merge them into an overall summary of average horsepower by cylinder from the two sources. Specifically:

   9a. Create the summary data frame sumMT from the mtcars data frame that gives (1) the cyl value named Cylinders; (2) the average hp value named hpMT; and (3) the record count named nMT;

   9b. Create the summary data frame sum93 from the Cars93 data frame that gives (1) the Cylinders value named Cylinders; (2) the average Horsepower value named hp93; and (3) the record count named n93;

   9c. Use the merge function discussed in Section 4.6 to merge these to data frames based on their Cylinders values. Display your results. What differences do you note in the horsepower summaries from the two data sources?

10: If we compare the individual summaries sumMT and sum93 constructed in Exercise 9, we find that they do not include the same range of Cylinders values. By default, the merge function only includes results that are represented in both of the data frames being merged, but the optional logical argument all allows us to retain records that may be absent from one or the other of these data frames. Repeat the merge in Exercise (9c), specifying the all argument to retain all records. Display your results.

Chapter 5

Linear Regression Models

Predictive models are mathematical models—i.e., equations—that allow us to predict some variable of interest from one or more other variables that are believed to be related. The simplest case is that of a "best-fit line" through a scatterplot, which serves as the starting point for this chapter. More specifically, this line-fitting problem is a special case of the more general *linear regression problem*, to which this chapter provides a simple but reasonably thorough introduction. Some of the most important linear regression tools available in *R* are introduced, along with some fundamental ideas that are important in fitting both linear regression models and the more general predictive models discussed in Chapter 10.

5.1 Modeling the whiteside data

The whiteside data frame from the MASS package introduced in Chapter 1 provides a simple, easily interpreted example illustrating various aspects of the problem of fitting linear regression models. Recall that this data frame includes three variables: Temp, a measure of the outside temperature for each week during two different heating seasons, one before insulation was installed and one after, Gas, a measure of the heating gas consumed each week, and Insul, a categorical variable with the two values Before, for those records from the first heating season before insulation was installed, and After, for those records from the second heating season, after insulation was installed.

Fig. 5.1 shows a scatterplot of all of the Gas versus Temp values from this data frame, with each individual observation represented as an open circle. Overlaid on this scatterplot is a straight line representing the best-fit linear model obtained as described in Sec. 5.1.3. Subsequent discussions in this chapter will revisit this example several times to illustrate different aspects of fitting linear models to data and to demonstrate the range of results we can obtain with the linear regression modeling tools available in *R*.

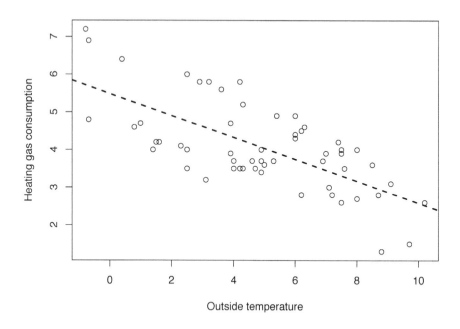

Figure 5.1:   Heating gas consumption vs. outside temperature from the **whiteside** data frame, with its best-fit linear model shown as a dashed line.

5.1.1   Describing lines in the plane

Before describing the line-fitting problem, it is necessary to say something about how lines in the plane are represented mathematically. Several different representations are possible, but this chapter focuses primarily on the *slope-intercept representation*, which describes (almost) any line in the plane in terms of two parameters: the (*y*-axis) *intercept* parameter, and the *slope* parameter. *Vertical lines* cannot be represented in this form, but this representation is consistent with both what the linear model fitting function `lm` returns, and the arguments of the `abline` function that displays lines on plots like Fig. 5.1.

Fig. 5.2 shows the standard parameterization of a straight line in the plane in terms of these two parameters, where the intercept is designated as a and the slope is designated as b. This representation describes the line mathematically by the equation:

$$y = a + bx. \tag{5.1}$$

Thus, when $x = 0$, it follows that $y = a$, meaning that the line *intercepts the y-axis* at the value $y = a$, as indicated by the dotted line labelled "a" in Fig. 5.2. For the b parameter, note that if we move horizontally from some value x to

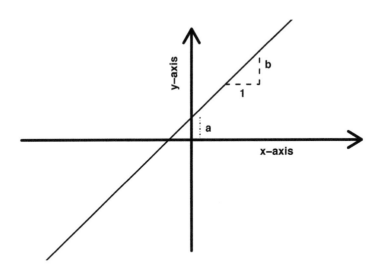

Figure 5.2: The intercept (**a**) and slope (**b**) parameters describing a straight line in the plane.

the value $x + 1$, one unit to the right, the y value increases from $a + bx$ to $a + b(x + 1) = [a + bx] + b$. That is, a one-unit increase in x causes a b-unit increase in y, assuming b is positive; this interpretation is illustrated graphically by the two legs of the triangle drawn in dashed lines in Fig. 5.2. Thus, the larger (i.e., more positive) we make b, the more steeply the line rises to the right in our plot. Conversely, if b is negative, y *decreases* as x increases, corresponding to a line that slopes down and to the right in the plane.

Fig. 5.3 shows four examples of straight lines in the plane, with different values of the intercept parameter a and the slope parameter b. For any finite values of a and b, we can add straight lines to an existing plot with the `abline` function, called with these two values. Thus, the upper left plot shows a line with intercept $a = 1$ and positive slope $b = 1.5$, while the upper right plot shows a line with the same intercept ($a = 1$), but with the negative slope parameter $b = -1.5$. As noted in the above discussion, the positive slope in the left-hand plot corresponds to a line that increases as we move to the right, while the negative slope in the right-hand plot corresponds to a line that decreases as we move to the right. Both lines were generated using the `abline` function, which can be called with the intercept parameter a and the slope parameter

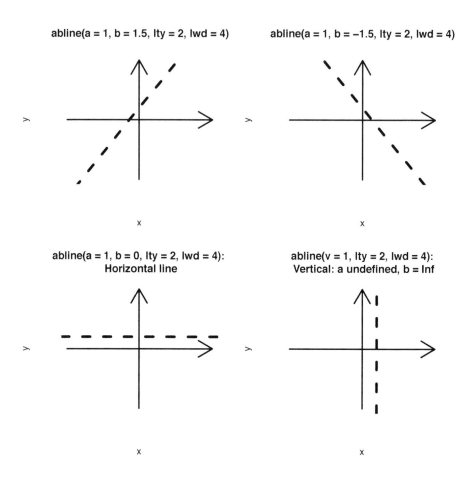

Figure 5.3: Four straight line examples and their `abline()` representations.

b, as indicated in the plot titles. Optional arguments for the `abline` function include the line type, `lty`, and the line width, `lwd`.

The lower two plots in Fig. 5.3 show the special cases of horizontal and vertical lines. The lower-left plot shows the horizontal line $y = a$ for $a = 1$, which can be represented in Eq. (5.1) by setting $b = 0$. Alternatively, the `abline` function can be called with the argument `h = 1` to give us exactly the same result in a slightly simpler fashion. The lower-right plot shows a vertical line which, as noted above, cannot be represented in slope-intercept form: the slope of this line is infinite and the intercept parameter a is undefined; since all vertical lines run parallel to the y-axis, they either intersect this axis everywhere (i.e., for a vertical line that coincides with the axis) or nowhere, as in this example. Vertical lines can be generated with the `abline` function, as in this example, generated by specifying `v = 1`, but they do not correspond to a result that can be obtained from the linear regression modeling procedures described here, in contrast to all of the other lines shown in Fig. 5.3.

5.1.2 Fitting lines to points in the plane

The abline function provides an extremely convenient way of *displaying* straight lines in a plot, once we know the slope and intercept parameters that describe those lines, but given a collection of points like the scatterplot shown in Fig. 5.1, how do we *determine* those parameters? This is the task of linear regression modeling, the primary subject of this chapter.

The basic solution approach is to seek the values $\hat{a}$ and $\hat{b}$ of the intercept and slope parameters that make the points lying on this line best approximate our data points. More specifically, given a collection of N observed x values $\{x_k\}$ and of N observed y values $\{y_k\}$, we want to compute those values $\hat{a}$ and $\hat{b}$ such that this condition holds:

$$y_k \simeq \hat{a} + \hat{b}x_k, \tag{5.2}$$

where the symbol $\simeq$ means "is approximately equal to." By itself, this "equation" gives us a useful statement of our objective, but it does not provide a basis for actually determining the unknown values of $\hat{a}$ and $\hat{b}$. Fortunately, we can re-express Eq. 5.2 in the following equivalent form:

$$y_k = \hat{a} + \hat{b}x_k + e_k, \tag{5.3}$$

where $\{e_k\}$ is a sequence of *prediction errors*, one for each data point in our dataset. Using this representation, we can re-state our objective in fitting a straight line to our data points as "make these prediction errors as close to zero as possible."

Since $\{e_k\}$ is a vector, we can measure its "nearness to zero" in several different ways, generally leading to quite different results. Historically, the most popular approach has been the *method of least squares*, which defines the nearness of $\{e_k\}$ to zero as the sum of its squared values:

$$J_{OLS} = \sum_{k=1}^{N} e_k^2. \tag{5.4}$$

Before describing how this measure can be used to obtain the results we want, two mathematical details are worth noting because they will be important in interpreting our results. First, note that, while the individual prediction errors e_k can be either negative or positive, the squared value e_k^2 can never be negative. This means that, no matter what values are in our prediction error sequence $\{e_k\}$, it is always true that $J_{OLS} \geq 0$. The second important observation is that $J_{OLS} = 0$ can occur if—and only if—all of the prediction errors are zero: $e_k = 0$ for all k. This point is important because if $e_k = 0$ for all k, it follows from Eq. 5.3 that the straight line with intercept $\hat{a}$ and slope $\hat{b}$ coincides *exactly* with all of our data points, corresponding to a *perfect fit* to our data.

The method of *ordinary least squares* is a regression model-fitting procedure that adopts J_{OLS} to measure the size of our prediction error sequence, and chooses those parameters $\hat{a}$ and $\hat{b}$ that make this measure as small as possible. A detailed discussion of the math behind the method of ordinary least squares is beyond the scope of this book, but three points are extremely important here. First, this method leads to a mathematical problem that is easier to solve than those that arise from selecting alternative measures of "nearness to zero." Second, except in cases where the problem is inherently ill-posed (e.g., attempting to fit a linear regression model to a collection of points that lie on a vertical line), the method of ordinary least squares has a unique solution, something that is not true of some of the alternative methods we could consider. Finally, the third key point here is that the method of ordinary least squares is extremely well supported in R, via functions like `lm`, used to fit most of the linear regression models discussed in this chapter. (For those interested in a more detailed discussion of the mathematics behind ordinary least squares and some alternatives, refer to the discussion in *Exploring Data* [58, Ch. 5].)

5.1.3   Fitting the `whiteside` data

The dashed line in Fig. 5.1 represents the following linear model:

```
linearModelA <- lm(Gas ~ Temp, data = whiteside)
```

The first argument of this function—`Gas ~ Temp`—uses R's formula interface to tell the `lm` function to fit a model that predicts `Gas`—the variable named on the left of the `~` symbol—from all variables on the right of this symbol, in this case, `Temp`. The second argument—`data = whiteside`—tells the `lm` function where to find these variables. This function returns a list object, with these elements:

```
names(linearModelA)
```

```
##  [1] "coefficients"  "residuals"     "effects"      "rank"
##  [5] "fitted.values" "assign"        "qr"           "df.residual"
##  [9] "xlevels"       "call"          "terms"        "model"
```

The `coefficients` element contains the intercept and slope parameters for the linear model that best fits the data, and the `abline` function has been designed to look for these parameters when given a model object from the `lm` function:

```
abline(linearModelA, lty = 2, lwd = 2)
```

Here, `lty = 2` specifies a dashed line and `lwd = 2` makes it twice the standard width; the coefficients for this best-fit line are:

```
linearModelA$coefficients
```

```
## (Intercept)        Temp
##   5.4861933  -0.2902082
```

Since the `Temp` parameter is negative, the line slopes downward as seen in Fig. 5.1, and because the `Gas` values are all positive, the intercept parameter must be positive to guarantee positive predictions. The generic `summary` function gives a more complete model description:

```
summary(linearModelA)

##
## Call:
## lm(formula = Gas ~ Temp, data = whiteside)
##
## Residuals:
##     Min      1Q  Median      3Q     Max
## -1.6324 -0.7119 -0.2047  0.8187  1.5327
##
## Coefficients:
##              Estimate Std. Error t value Pr(>|t|)
## (Intercept)   5.4862     0.2357  23.275  < 2e-16 ***
## Temp         -0.2902     0.0422  -6.876 6.55e-09 ***
## ---
## Signif. codes:  0 '***' 0.001 '**' 0.01 '*' 0.05 '.' 0.1 ' ' 1
##
## Residual standard error: 0.8606 on 54 degrees of freedom
## Multiple R-squared:  0.4668,Adjusted R-squared:  0.457
## F-statistic: 47.28 on 1 and 54 DF,  p-value: 6.545e-09
```

Applied to a linear regression model object, the `summary` function displays:

1. the function call used to fit the model, identifying the model type (i.e., the `lm` function), the variables included in the model, and the data frame from which these variables were obtained;

2. Tukey's five-number summary of the *residuals*, corresponding to the prediction errors $\{e_k\}$ in Eq. (5.3) computed from the data and the model parameters;

3. The model coefficients discussed above, along with some related characterizations discussed in the next paragraph;

4. Five *goodness-of-fit* characterizations, discussed briefly below.

The five goodness-of-fit characterizations included in this summary are the *residual standard error*, the *multiple R-squared*, and the *adjusted R-squared*, all described in Sec. 5.2.3, and the *F-statistic* and its associated *p-value*, discussed briefly in the next paragraph and further in Chapter 9.

The portion of the `summary` output labelled "Coefficients" is a data frame giving the estimated model parameters in the `Estimate` column, and three characteristics of these estimates: `Std. Error, t value`, and `Pr(>|t|)`. The first of these columns gives the *standard error* associated with each coefficient, a measure of how accurately it has been estimated. The next column gives the corresponding *t-statistic* derived from the standard error, and the last column gives the *p-value* associated with this *t*-statistic. These characterizations are discussed in detail in Chapter 9, but the essential notion is that small p-values provide supporting evidence that the model parameter is *significant* in the sense that omitting it—equivalent to setting it to zero—would result in a poorer model. Here, both of these p-values are extremely small, suggesting that both of these terms should be included in the model. In this particular example, the p-value associated with the overall model—giving an indication of how well the model fits the data—is equal to the p-value associated with the slope (i.e., `Temp`) parameter, but this is not generally the case. The most important point is that the p-values associated with the individual coefficients are telling us something about the utility of each term in the model, while the p-value given in the last line of the summary is telling us about the overall fit quality of the model.

5.2   Overfitting and data splitting

The example just described adopts the standard practice from classical statistics of using all of the available data both to fit a model and to evaluate its goodness-of-fit. A danger of this approach is *overfitting:* if we increase the complexity of our model by adding more terms, it is guaranteed to fit the data better. If we make the model too complex, we begin to fit *all* of the details in the data, including those reflecting the unavoidable presence of noise, and not just the "general behavior" of interest. This problem is illustrated with a simple example in Sec. 5.2.1 and it is well recognized in the classical statistics community, leading to a number of goodness-of-fit characterizations like the adjusted R-squared measure mentioned in the previous example and discussed further in Sec. 5.2.3. Unfortunately, these characterizations do not extend easily to modeling procedures like the more complex machine learning models introduced in Chapter 10. Consequently, the machine learning community advocates the data splitting approach described in Sec. 5.2.2 to avoid overfitting. The essential idea is to use one part of our dataset to fit predictive models and another part to evaluate model performance. The two model validation tools introduced in Sec. 5.2.3 build on this idea: one is a simple graphical visualization of how well our model fits the data, and the other is a variant of the adjusted R-squared measure that can be applied to both linear regression models and the machine learning models introduced in Chapter 10.

5.2.1   An overfitting example

The following example provides an extreme but simple illustration of the overfitting problem. It is based on the simulated data frame `full` generated by the R code listed below, with 10 rows and two columns, labelled x and y:

```
set.seed(331)
x <- seq(1, 10, 1)
y <- rnorm(10)
full <- data.frame(x = x, y = y)
xT <- sort(sample(x, size = 5, replace = FALSE))
```

```
train <- full[xT, ]
```

Here, x is the sequence of integers from 1 to 10, and y is 10 samples of a simulated zero-mean, unit-variance Gaussian random variable. Looking ahead to the strategy described in Sec. 5.2.2, we randomly select five of these data points and use them to fit a sequence of models of increasing complexity. (The `sort` function imposes an increasing order on `xT`, which is important in using the `lines` function to add model predictions to the data scatterplots.)

It is well known that "two points determine a line," and it is more generally true that n points determine a polynomial of degree $n - 1$. Thus, three points are enough to determine a quadratic polynomial, four points determine a cubic polynomial, and five points determine a fourth-order polynomial of this form:

$$y = a_0 + a_1 x + a_2 x^2 + a_3 x^3 + a_4 x^4. \tag{5.5}$$

There are restrictions on this result—every point must have a distinct x-value, for example—but the key point is that the five data points in the data frame `train` constructed above are sufficient to exactly determine the coefficients a_0 through a_4 in Eq. 5.5. In what follows, we consider a sequence of polynomial models of increasing degree, starting with degree zero—the constant model $y = a_0$—and increasing the degree by 1 until we get to the fourth-order polynomial that fits our five data points exactly. The `lm` function fits all of these models:

```
model0 <- lm(y ~ 1, data = train)
model1 <- lm(y ~ x, data = train)
model2 <- lm(y ~ x + I(x^2), data = train)
model3 <- lm(y ~ x + I(x^2) + I(x^3), data = train)
model4 <- lm(y ~ x + I(x^2) + I(x^3) + I(x^4), data = train)
```

Three points are worth emphasizing here. First, `model0` is the zero-degree constant model that predicts y from the intercept term a_0 only. By default, the `lm` function always includes an intercept term, but the formula y ~ 1 specifies that this is the *only* term in our model. Second, by specifying data = train in all of these calls to the `lm` function, we are using only the randomly selected training subset from our original data frame `full`. Finally, third, in `model2`, `model3`, and `model4`, the `AsIs` function, `I()`, is used to force terms like x^2 to have their usual arithmetic meaning as powers, protecting them from the very different interpretation this symbol has in R's formula interface (this point is discussed further in Sec. 5.6).

To obtain the predictions for all x values in our complete data frame `full`, the generic function `predict` is called with the model name and the optional argument newdata = full:

```
yHat0 <- predict(model0, newdata = full)
yHat1 <- predict(model1, newdata = full)
yHat2 <- predict(model2, newdata = full)
yHat3 <- predict(model3, newdata = full)
yHat4 <- predict(model4, newdata = full)
```

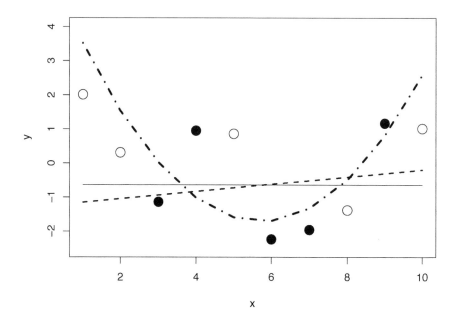

Figure 5.4: Scatterplot of the simulated dataset, overlaid with lines corresponding to the predictions from polynomial models of orders 0, 1, and 2.

Fig. 5.4 summarizes the predictions for the first three models in this sequence. Here, all 10 points from the `full` data frame are represented in the
plot, with the five points in the `train` data frame that were used in fitting
the models represented as solid circles. The solid horizontal line represents the
predictions from `model0`, equal to -0.642 for all x values. The dashed straight
line corresponds to the prediction from `model1`, which has intercept -1.258 and
slope 0.106. The dash-dotted curve in this plot corresponds to the predictions
from `model2`, the quadratic polynomial that best fits the training data. This
model is specified by three parameters: the intercept term a_0, equal to 5.968, a
slope term a_1, equal to -2.685, and a quadratic term a_2, with the value 0.235.
Aside from possibly the points at $x = 4$ and $x = 5$, this quadratic model appears
to fit the overall data much better than the linear and constant models.

Fig. 5.5 shows the corresponding results for the third- and fourth-order poly-
nomials, along with the quadratic polynomial from Fig. 5.4 for comparison. To
show the predictions of these higher-order polynomial models for all of the data
points on the plot, it was necessary to greatly expand the y-axis scale. As we
will see more clearly in the numerical results presented below, the fourth-order
polynomial actually does achieve a perfect fit to the five data points included

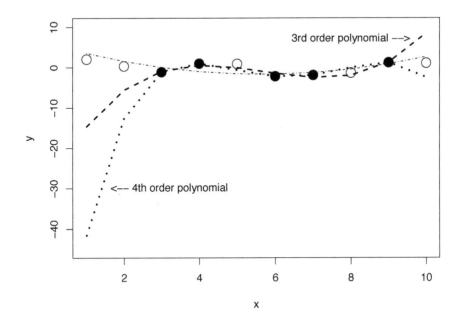

Figure 5.5: Scatterplot of the simulated dataset, overlaid with lines correspond-ing to the predictions from polynomial models of orders 2, 3, and 4.

Order	Training	Validation
0	10.31	13.49
1	10.05	16.77
2	6.03	13.02
3	1.13	369.44
4	0.00	2087.56

Table 5.1: Sum of squared fit errors for the five polynomial models, for the training and validation data subsets.

in the `train` data frame, but the fit to the other five data points is atrocious.

Two numerical characterizations of this fit are given in Table 5.1, which lists the sum of squares fit measure J_{OLS} defined in Eq. 5.4 for each of the five models considered here. The first column of this table gives the values of J_{OLS}

computed from the training data used to fit these models, while the second gives the corresponding values computed from the *validation data*, defined here as all of the data points *not* used in fitting the model parameters. The first thing to note about the numbers in this table is that the training subset fit measures are consistently better—i.e., smaller—for the training data than for the validation data. This is a consequence of the fact that the parameters defining these models were chosen explicitly to minimize this fit measure for the training data subset. In contrast, the corresponding numbers for the validation subset represent a measure of the model accuracy in describing another dataset that is similar but not identical. The second and most important point to note is that the training subset numbers decrease monotonically to zero as the model order increases, while the validation subset numbers generally increase with increasing model order. The last line of this table emphasizes the point of this example: by increasing the model complexity enough—i.e., to a fourth-order polynomial— we have achieved a perfect fit to the training subset, at the expense of a *huge* validation set error measure.

Finally, it important to note that, even though the higher-order polynomial models considered here are *nonlinear functions*, their dependence on the unknown model parameters we are attempting to determine from the data is *linear*. The term "linear regression" refers to problems like this: even if some of the *functions* involved are nonlinear, the model fitting problem remains a linear regression problem if the *parameters* appear linearly. Thus, in Eq. 5.5 the fact that no products, exponentials, or other mathematical functions *of the unknown parameters* a_0 *through* a_4 appear in this equation means that we can use linear modeling tools like the `lm` function to estimate them from data.

5.2.2   The training/validation/holdout split

The previous two plots—especially Fig. 5.5—and the numerical values shown in Table 5.1 provide a dramatic illustration of the problem of overfitting, and they also illustrate the basis for a simple, practical solution to this problem that has emerged from the machine learning community.

Specifically, these results illustrate that we cannot simply judge the quality of a prediction model on the basis of how well it fits the dataset used in building it. By increasing the model complexity enough in this case we were able to achieve a perfect fit to the training data subset, but the prediction errors for the *other* data points in our original dataset were enormous. The basic approach advocated in the machine learning community is a simple extension of what was done here: use one data subset to build our model, and a *different but similar* data subset to evaluate our model. Further, if we make these datasets *mutually exclusive, randomly selected* subsets of the original data, this generally guarantees that these different subsets are similar in character but not identical. Assuming that our original dataset is representative of the phenomenon or situation we are attempting to characterize, this approach gives us exactly what we want: one dataset to build models with and another similar, but distinct, dataset to evaluate our models against. This protects us against overfitting.

More generally, the data splitting approach advocated in the machine-learning community randomly partitions our original dataset into three, mutually exclusive subsets [38, p. 222]:

1. a *training set* used in fitting model parameters;

2. a *validation set* used in making model structure decisions;

3. a *test set* or *holdout set*, used for final model evaluation.

The overfitting example presented in Sec. 5.2.1 used only two of these subsets: a training set consisting of the five points used to fit our polynomial models, and validation set used to show how badly the higher-order polynomial models performed. In the three-way split just described, one way of choosing the best model structure would be to select the polynomial model order giving the smallest validation set value, J_{OLS}. In this case, the quadratic model—seen most clearly in the dash-dotted line in Fig. 5.4—would be selected as the "best" model, consistent with the observations made earlier.

In fact, since the data used in fitting these models is Gaussian random data, the apparent high quality of this quadratic model is an artifact of the small number of data points used in fitting and evaluating this model. This point will be revisited in Chapter 9 in connection with statistical significance and confidence intervals, but the key thing to note here is that our ability to fit and evaluate predictive models depends on our having enough data to do a reasonable job of it. It is for this reason that the three-way split described here has been advocated as the best approach in "data-rich situations" [38, p. 222]. In cases where there is not enough data to reasonably perform this partitioning, the alternative approach of *cross-validation* has been recommended [38, Sec. 7.10]: there, the original dataset is partitioned randomly into K mutually exclusive subsamples (e.g., $K = 5$), a predictive model is fit to a training set consisting of all but one of these subsamples, and evaluation is performed with respect to the other subsample, not used in the fitting. This process is repeated, leaving out each of the K subsamples, fitting the model to the rest, and evaluating performance for the omitted subsample. Averaging these K performance measures then gives an idea of how well the model performs: the approach is not perfect, but it is useful and it has been built into a number of model fitting procedures in R, especially for selecting tuning parameters.

A simple data partitioning procedure

Two key points here are, first, that the partitioning strategy just described is applicable to all predictive model types (e.g., all of the linear regression models discussed in this chapter and all of the other model types discussed in Chapter 10), and, second, that it is very easy to implement in R. To see this second point, note that the following R function, discussed in detail in Chapter 7, implements the three-way random split described here, adopting the 50%/25%/25% partitioning suggested by Hastie, Tibshirani, and Friedman [38, p. 202]:

```
TVHsplit <- function(df, split = c(0.5, 0.25, 0.25),
                     labels = c("T", "V", "H"), iseed = 397){
  #
  set.seed(iseed)
  flags <- sample(labels, size = nrow(df),
                  prob = split, replace = TRUE)
  return(flags)
}
```

As disussed in Chapter 7, the first lines here define a custom *R* function named
`TVHsplit`, called with one required parameter—`df`, a data frame to be partitioned into subsets—and three optional parameters, `split`, `labels`, and `iseed`.
The symbol `#` indicates a skipped (comment) line, and the next line after that
invokes the built-in function `set.seed` to initialize *R's* random number generation system, used by the `sample` function in the next line. This function
draws a random sample with one element for each row of the data frame `df` to
be partitioned, taking the three values defined by the `labels` argument, with
the probabilities specified by the `split` argument. Note that `sample` is called
with `replace = TRUE` so this random sample is drawn *with replacement*. This
means that every element of the vector `flags` has one of the values from `labels`,
regardless of how many rows there are in the data frame `df`. The elements of
the vector `split` determine the relative sizes of the different subsets defined by
these labels. For the default values specified here, the `flags` vector returned by
this function has approximately 50% "T" values, 25% "V" values, and 25% "H"
values. Note that to obtain a 50/50 training/validation split with no holdout
set, we can specify `split = c(0.5, 0.5, 0)`.

The idea of data splitting to avoid overfitting is both important and widely
applicable, so it will reappear throughout this book. Before leaving the current
discussion, however, it is important to emphasize that this partitioning should
be *random*, to avoid the possibility of *systematic differences* between the training, validation, and holdout sets. That is, the best we can reasonably ask of a
predictive model is that it generate accurate predictions for data that is similar to that from which it was built. If we select our training data from one
group—e.g., earlier data records, records from a fundamentally different set of
patients, or characterizations of a different class of materials—we cannot expect
the model to perform well. This point is illustrated with the `whiteside` data
frame in Sec. 5.2.3 where both random partitioning and systematic partitioning
are compared using the tools introduced there.

Application to the `whiteside` data

As noted earlier, the `whiteside` data frame nicely illustrates many of the main
ideas discussed in this chapter. The one disadvantage is that this dataset is
too small to provide a "data rich" example for the three-way data partitioning
strategy just described. It does, however, illustrate some of the disadvantages
of applying random partitioning to small datasets while still clearly illustrating
the basic ideas behind this approach and its typical results.

The following R code uses the `TVHsplit` function just described to create an approximately equal training/validation split for the `whiteside` data frame:

```
TVHflags <- TVHsplit(whiteside, split = c(0.5, 0.5, 0))
trainSub <- whiteside[which(TVHflags == "T"), ]
validSub <- whiteside[which(TVHflags == "V"), ]
```

Here, we have used the vector returned by the `TVHsplit` function to create the training data subset `trainSub` that will be used to fit linear regression models, and the validation subset `validSub` that will be used for model evaluation. Since this partitioning is random, it is important to note that it does not give us an *exact* 50/50 split of the data: the training set has 21 records, while the validation set has 35 records. This difference illustrates one of the disadvantages of applying random partitioning to small datasets: with larger datasets, the sizes of a "random 50/50 split" will be more nearly equal.

With this data partitioning, we can proceed as before and fit a linear regression model using the `lm` function, but this time using only the training subset:

```
linearModelB <- lm(Gas ~ Temp, data = trainSub)
```

Given this model, we can generate the predicted heating gas consumption for both the training and validation datasets by calling the `predict` function with the appropriate `newdata` argument:

```
GasHatBT <- predict(linearModelB, newdata = trainSub)
GasHatBV <- predict(linearModelB, newdata = validSub)
```

In cases like this one where the training and validation sets are not exactly the same size, we cannot directly compare the values for the fit measure J_{OLS} defined in Eq. (5.4) since larger datasets are likely to give larger J_{OLS} values. In fitting our model, however, minimizing the sum of squared errors J_{OLS} is equivalent to minimizing the *mean squared error (MSE)*, defined as:

$$MSE = \frac{1}{N} \sum_{k=1}^{N} e_k^2. \tag{5.6}$$

When we fit a model to a training dataset via the method of ordinary least squares, we are choosing our model parameters to minimize the training set J_{OLS} value or, equivalently, its MSE value, so we generally expect the training set MSE value to be smaller than the validation set MSE value. For the `whiteside` data frame, we have the following results for the training data:

```
ek <- GasHatBT - trainSub$Gas
mseBT <- mean(ek^2)
mseBT

## [1] 0.5179938
```

and the following results for the validation data:

```
ek <- GasHatBV - validSub$Gas
mseBV <- mean(ek^2)
mseBV
```

```
## [1] 0.9776627
```

The large difference seen here is suggestive of overfitting, but as subsequent discussions will show, the real reason for this difference is a consequence of both the small size of the dataset and its heterogeneous underlying structure. Specifically, recall that this dataset contains "Before" and "After" data, and these two subsets behave differently. We can begin to understand both of these issues with the simple graphical tool described next.

5.2.3   Two useful model validation tools

Given a model that has been fit to one dataset—e.g., the training subset in the data partitioning just described—the two tools described next can be used to evaluate its performance against another dataset (e.g., the validation subset). Both of these tools have the advantage that they can be used with *any* model that predicts a numerical response, including all of the models discussed in this chapter, along with many of those discussed in Chapter 10. The first tool is a simple graphical characterization that allows us to see systematic deviations of various types, as well as individual data points that are predicted poorly by the model. The second tool is a numerical measure of how well the model predicts the validation data, based on ideas from classical statistics.

The predicted vs. observed plot

Given a model that generates a set $\{\hat{y}_k\}$ of predictions of some variable y and its observed values $\{y_k\}$, an extremely useful graphical diagnostic is the plot of $\{\hat{y}_k\}$ against $\{y_k\}$. If the model predicts perfectly, then $\hat{y}_k = y_k$ for all k and all points will lie on an equality reference line with unit slope and zero intercept. For more realistic models, the points $(y_k, \hat{y}_k)$ will not lie exactly on this reference line, but for a "good" model, they should lie near it, while for a "poor" model, many or most of these points will lie well away from this line, exhibiting either systematic deviations from it or substantial "scatter" around it.

Fig. 5.6 provides two illustrations of this idea, based on `linearModelB` fit to the training dataset `trainSub` constructed from the `whiteside` data frame. The plot on the left shows the model predictions `GasHatBT` for the training dataset, plotted against the observed `Gas` values from this dataset, while the plot on the right shows the predicted versus observed values from the validation dataset `validSub`. In both cases, the open circles correspond to records with `Insul` value "After," and the solid circles correspond to the "Before" data. Note that points falling *below* the reference line in this plot represent observations that are *under-predicted* by the model, while points lying above the line represent observations that are *over-predicted*. An extremely important observation here is that in

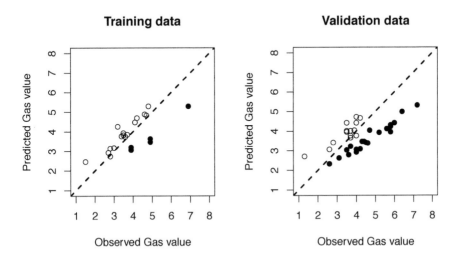

Figure 5.6: Predicted vs. observed plots for `linearModelB`: training data (left), and validation data (right); solid circles in both plots are "Before" records.

both plots, the "Before" observations are almost always under-predicted while the "After" observations are almost always over-predicted.

Another important point here is that the split of "Before" and "After" data points appears very different between the training and validation subsets, evident from the pronounced difference in the numbers of solid circles in the left-hand and right-hand plots. In fact, the Before/After split for the training dataset is 5/16, while that for the validation dataset is 21/14. This is a consequence of randomly sampling a small dataset, particularly when subsets of the partitions are themselves important. In particular, with only 56 data records, if we want training and validation subsets, each with approximately the same number of "Before" and "After" records, each of these cases will have approximately 14 records. This number is small enough that random sampling variations can be quite substantial, and one way of seeing this is to change the random seed argument `iseed` for the `TVHsplit` procedure. Here, the default value is `iseed = 397`, and changing this value to 398 gives 14 training records with `Insul = ''Before''` and 14 training records with `Insul = ''After''`.

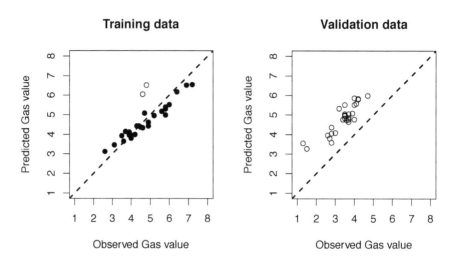

Figure 5.7: Predicted vs. observed plots for `badModel`: training data (left), and validation data (right); solid circles in both plots are "Before" records.

Predicted vs. observed plots can also be used to illustrate the importance of random training/validation splits rather than systematic splits. Since the `whiteside` data frame has 56 records, one obvious way of obtaining an *exact* 50/50 split is to take the first half of the data for the training set and the second half of the data for the validation set. As the following example illustrates, this is an extremely bad idea here. The following R code performs this data split, fits a linear model as before, and generates predictions for both data subsets:

```
badTrain <- whiteside[1:28, ]
badValid <- whiteside[29:56, ]
badModel <- lm(Gas ~ Temp, data = badTrain)
badGasHatT <- predict(badModel, newdata = badTrain)
badGasHatV <- predict(badModel, newdata = badValid)
```

Fig. 5.7 shows the predicted versus observed plots constructed from this model for both subsets, again with the "Before" points marked as solid circles. An obvious feature of these plots is the extreme imbalance of "Before" and "After" points: the training subset has only two "After" points, while the validation subset has *only* "After" points. Also, the "Before" points in the training data plot lie very close to the equality reference line, indicating that they are being fit extremely well, while the two "After" points are badly over-predicted. For the validation data, *all* of the observations are badly over-predicted. These plots provide a clear illustration of the dangers of systematic data partitioning: since the "training" dataset here consists almost entirely of "Before" observations, the linear regression model predicts these points quite well. Unfortunately, the "After" data behaves very differently, and since the "validation" subset consists only of these "After" points, this subset is predicted very badly.

The validation R-squared measure

The preceeding two examples provided nice illustrations of the advantages of looking at plots of predicted versus observed responses, but sometimes we want simple numerical characterizations. These cannot be as informative as the predicted versus observed plots, but they can be extremely useful in summarizing differences in the performance of a group of models, especially if this group includes many different models.

In classical statistics, the *R-squared statistic* describes the extent to which a linear regression model with an intercept term explains the variation seen in the data. Specifically, the (unadjusted) R^2 measure is defined by [22, p. 91]:

$$R^2 = 1 - \frac{\sum_{k=1}^{N} (\hat{y}_k - y_k)^2}{\sum_{k=1}^{N} (y_k - \bar{y})^2}, \tag{5.7}$$

where $\hat{y}_k$ is the predicted value of the kth data observation y_k, and $\bar{y}$ is the average of the observed data values. As this statistic is traditionally defined, no data splitting has been done, so the linear regression model has been fit to the complete set of N data observations. Also, note that $\bar{y}$ represents the best-fit intercept only model, and no linear regression model that includes an intercept term can give a poorer fit than this: worst case, we could always set all of the other coefficients to zero and recover this intercept-only model. Further, note that the predictions for this poorest-quality regression model are $\hat{y}_k = \bar{y}$ for all k, from which it follows that $R^2 = 0$ for this case. In general, we can expect a linear regression model to do better than this, achieving a smaller residual sum of squares (the numerator in the ratio of sums in Eq. (5.7)), implying $R^2 > 0$. The smallest possible value for this numerator term is zero, achieved for regression models that fit the data perfectly so that $\hat{y}_k = y_k$ for all k. In this case, R^2 achieves its maximum possible value of 1.

Unfortunately, using the R-squared measure just defined to evaluate linear regression models ignores the issue of overfitting discussed at length in Sec. 5.2.1. To address this problem, the *adjusted R-squared statistic* was introduced, effectively adding a complexity penalty to models that provide a better fit by adding more terms, as in the example described in Sec. 5.2.1. This measure is defined as [22, p. 92]:

$$R_a^2 = 1 - \frac{[\sum_{k=1}^{N} (\hat{y}_k - y_k)^2]/(N - p)}{[\sum_{k=1}^{N} (y_k - \bar{y})^2]/(N - 1)}, \tag{5.8}$$

where p is the number of parameters in the linear regression model, counting the intercept term. Note that the effect of this modification is to penalize models with more parameters (i.e., larger values of p). Recall that both the R-squared measure defined in Eq. (5.7) and the adjusted R-squared measure defined in Eq. (5.8) are included in the summary that R returns for linear regression models fit by the `lm` function.

The primary practical limitation of the adjusted R-squared measure defined in Eq. (5.8) is that it does not extend readily to the more complex—and increasingly more popular—machine learning models introduced in Chapter 10. In particular, structures like random forest and boosted tree models cannot be simply characterized by a small set of model parameters like linear regression model parameters can, so it is not obvious what value of p—if any—is appropriate for these much more complex models. If, however, we adopt the data partitioning strategy described in Sec. 5.2.2 and use the training subset to fit our models, we can evaluate their performance using the unadjusted R-squared measure defined in Eq. (5.7) computed from the validation or holdout subsets. Since the holdout set should be saved for final model evaluations, a useful approach to deciding between several candidate models is to compare their performance with respect to the *validation R-squared measure*, defined as:

$$R_V^2 = 1 - \frac{\sum_{k \in V} (\hat{y}_k - y_k)^2}{\sum_{k \in V} (y_k - \bar{y})^2}, \tag{5.9}$$

where the notation $k \in V$ here means that the numerator and denominator sums are evaluated only using data from the validation dataset.

It is important to note that $\bar{y}$ here represents the average of the response variable y computed from the *validation set only*. This point is important because, since the regression model considered here has been fit to the training set, its intercept term corresponds to the *training set mean*, which is generally *not* equal to the validation set mean. As a consequence, it is possible for validation R-squared values to be *negative* for really poor models. In particular, this is almost always the case for intercept-only models fit to the training data set (see Exercise 4). Conversely, since the smallest possible value for the numerator sum in Eq. (5.9) is zero—achieved if and only if the model predicts the validation data perfectly—the maximum possible value for R_V^2 is still 1.

A simple R function to compute the validation R-squared measure is listed here and discussed in detail in Chapter 7:

```
ValidationRsquared

## function(validObs, validHat){
##   #
##   resids <- validHat - validObs
```

```
##     yBar <- mean(validObs)
##     offset <- validObs - yBar
##     num <- sum(resids^2)
##     denom <- sum(offset^2)
##     Rsq <- 1 - num/denom
##     return(Rsq)
## }
## <bytecode: 0x00000000053158f8>
```

This function is called with the vectors `validObs` and `validHat` of observed and predicted validation set responses. The names of the function and its arguments include "validation" or "valid" to emphasize that this (unadjusted) R-squared measure should be computed from validation data after the model has been fit to a distinct but similar training set. It can, however, be used with the training set observations and predictions to compute the original R-squared statistic defined in Eq. (5.7). Since this measure does nothing to correct for overfitting, it is generally optimistic relative to the validation set value, suggesting the model is better than it actually is.

This point can be seen for `ModelB`, the linear regression model discussed in Sec. 5.2.2 that predicts `Gas` from `Temp` in the `whiteside` data frame. The training set value is 0.5579813, while the validation set value is only 0.2955658. These values will serve as useful references for some of the more complex models fit to this dataset described in the following sections.

5.3   Regression with multiple predictors

The linear regression models considered so far predict one numerical response variable from a single numerical covariate or prediction variable. Often, we have several variables that are each capable of providing partial predictions and, in these cases, we can usually achieve better predictions by incorporating two or more of these variables. The discussion in Sec. 5.3.1 provides a simple example, based on the `Cars93` data frame from the `MASS` package. Conversely, if we do include more than one prediction variable, it is important that they each contribute sufficient *independent* information to the prediction; cases where this requirement is not met lead to the problem of *collinearity*, discussed in Sec. 5.3.2.

Before going into these more detailed discussions, it is important to say something about how we set up and solve linear regression modeling problems involving several numerical predictors. Mathematically, the formulation is a simple extension of the problem of fitting lines in the plane discussed in Sec. 5.1.2. To see this point, first re-formulate the line fitting problem slightly by introducing different notation:

$$y_k = a_0 + a_1 x_k + e_k. \tag{5.10}$$

The only difference between this expression and Eq. (5.3) is that the intercept parameter, originally denoted $\hat{a}$, is now denoted a_0, and the slope parameter, originally denoted $\hat{b}$, is now denoted a_1. Next, suppose we want to fit a model that involves several prediction variables, which we denote by x_1 through x_p. To fit a linear regression model that includes these predictors, we expand Eq. (5.10) to:

$$
\begin{aligned}
y_k &= a_0 + a_1 x_{1,k} + a_2 x_{2,k} + \cdots + a_p x_{p,k} + e_k \\
&= a_0 + \sum_{i=1}^{p} a_i x_{i,k} + e_k,
\end{aligned}
\qquad (5.11)
$$

where $x_{i,k}$ denotes the kth observation of the variable x_i in the dataset. As before, to obtain the best-fit values for the model parameters a_0 through a_p, we minimize the value of the ordinary least squares fit measure defined in Eq. (5.4). As we will see in the following examples, the values of the model parameters that minimize this fit measure can be obtained using the `lm` function in *R*.

5.3.1   The `Cars93` example

To provide a concrete example of a linear regression model with multiple numerical predictors, the following paragraphs construct three models to predict the `Horsepower` variable from the `Cars93` data frame. The first of these models uses only the `Weight` variable as a predictor and the second uses only the `Price` variable as a predictor. The third model uses both of these variables as predictors, illustrating how a linear regression problem with more than one numerical predictor is set up and how the results are evaluated.

Because it is generally good practice and because `Cars93` is large enough to support it, this example applies the data partitioning strategy described in Sec. 5.2.2. Specifically, this example performs a (roughly) 50/50 split of the data into training and validation subsets, using the following *R* code:

```
Cars93Flag <- TVHsplit(Cars93, split = c(0.5, 0.5, 0.0))
Cars93T <- Cars93[which(Cars93Flag == "T"), ]
Cars93V <- Cars93[which(Cars93Flag == "V"), ]
```

As in the `whiteside` example considered earlier, note that the random split here is not exactly a 50/50 partitioning: the training set contains 39 records, while the validation set contains 54. Still, the sizes of these data subsets are comparable and they allow us to fit our data models to the training set while retaining a statistically similar validation set to evaluate model performance.

As noted, the first two models considered here predict `Horsepower` from either `Weight` (for `Cars93Model1`) or `Price` (for `Cars93Model2`). The motivation for fitting these models is the roughly linear appearance of the scatterplots in Fig. 5.8, which show the dependence of `Horsepower` on `Weight` (left-hand plot) and `Price` (right-hand plot) for the training data. The dashed lines in these plots are reference lines generated by `Cars93Model1` and `Cars93Model2`, respectively. The *R* code to generate these models is:

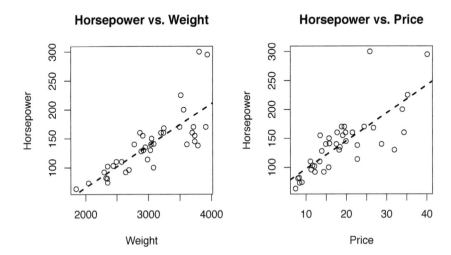

Figure 5.8: Scatterplots of **Horsepower** vs. two prediction variables: **Price** (left-hand plot) and **Weight** (right-hand plot).

```
Cars93Model1 <- lm(Horsepower ~ Weight, data = Cars93T)
Cars93Model2 <- lm(Horsepower ~ Price, data = Cars93T)
```

Note that while there is considerable scatter around both of the reference lines in Fig. 5.8, they do both capture the overall trend seen in most of the data, with the exception of a few points like the two high-horsepower vehicles in the training dataset. An indication of the relative quality of these model fits can be obtained by comparing the **summary** results for these models:

```
summary(Cars93Model1)

##
## Call:
```

```
## lm(formula = Horsepower ~ Weight, data = Cars93T)
##
## Residuals:
##     Min      1Q  Median      3Q     Max
## -56.714 -19.477  -0.695   9.070 103.835
##
## Coefficients:
##              Estimate Std. Error t value Pr(>|t|)
## (Intercept) -79.734564  28.869408  -2.762  0.00889 **
## Weight        0.072510   0.009375   7.734 3.05e-09 ***
## ---
## Signif. codes:  0 '***' 0.001 '**' 0.01 '*' 0.05 '.' 0.1 ' ' 1
##
## Residual standard error: 32.38 on 37 degrees of freedom
## Multiple R-squared:  0.6179,Adjusted R-squared:  0.6075
## F-statistic: 59.82 on 1 and 37 DF,  p-value: 3.048e-09

summary(Cars93Model2)

##
## Call:
## lm(formula = Horsepower ~ Price, data = Cars93T)
##
## Residuals:
##     Min      1Q  Median      3Q     Max
## -73.091 -16.395  -2.969  14.466 126.335
##
## Coefficients:
##              Estimate Std. Error t value Pr(>|t|)
## (Intercept)  49.2095    13.3371   3.690 0.000719 ***
## Price         4.8239     0.6504   7.417 7.94e-09 ***
## ---
## Signif. codes:  0 '***' 0.001 '**' 0.01 '*' 0.05 '.' 0.1 ' ' 1
##
## Residual standard error: 33.22 on 37 degrees of freedom
## Multiple R-squared:  0.5979,Adjusted R-squared:  0.587
## F-statistic: 55.01 on 1 and 37 DF,  p-value: 7.943e-09
```

The adjusted R-squared values for these models are 0.6075 for the model based on `Weight` and 0.5870 for the model based on `Price`. We can also compare the validation R-squared values for these models:

```
HorsepowerHat1V <- predict(Cars93Model1, newdata = Cars93V)
ValidationRsquared1 <- ValidationRsquared(Cars93V$Horsepower, HorsepowerHat1V)
HorsepowerHat2V <- predict(Cars93Model2, newdata = Cars93V)
ValidationRsquared2 <- ValidationRsquared(Cars93V$Horsepower, HorsepowerHat2V)
```

These validation R-squared values are listed in the titles of the predicted vs. observed plots for these models shown in Fig. 5.9. The left-hand plot gives the results for `Cars93Model1` based on `Weight`, while the right-hand plot gives the results for `Cars93Model2` based on `Price`. Note that while the validation R-squared numbers given in the title of each plot are in very rough agreement with the adjusted R-squared values returned by the **summary** function for these models, the results are different enough that they disagree on which model fits

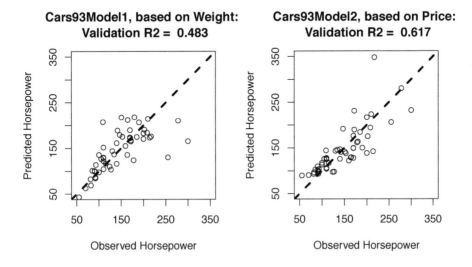

Figure 5.9: Validation set predictions vs. observed Horsepower for Cars93Model1 (left-hand plot) and Cars93Model2 (right-hand plot).

better: the adjusted R-squared values prefer Cars93Model1 based on Weight, while the validation R-squared values prefer Cars93Model2 based on Price. While it should not be taken as a rigid rule, if we must choose, the validation R-squared value should generally be the better guide, since it is based on an independent validation dataset not used in fitting the model. Also, note that the validation R-squared measure can be used to compare linear regression models like this one with other model types (e.g., the machine learning models discussed in Chapter 10), where adjusted R-squared values are generally not defined.

To construct a linear regression model that uses both Weight and Price as predictors, we again use the lm function, listing both prediction variables to the right of the tilde symbol in the formula expression defining the model:

```
Cars93Model3 <- lm(Horsepower ~ Weight + Price, data = Cars93T)
summary(Cars93Model3)

##
## Call:
## lm(formula = Horsepower ~ Weight + Price, data = Cars93T)
##
## Residuals:
##     Min      1Q  Median      3Q     Max
## -51.566 -12.071   0.776   8.713 106.716
##
## Coefficients:
##               Estimate Std. Error t value Pr(>|t|)
## (Intercept) -45.29651   27.15468  -1.668 0.103974
## Weight        0.04424    0.01154   3.834 0.000488 ***
## Price         2.72336    0.78028   3.490 0.001294 **
## ---
## Signif. codes:  0 '***' 0.001 '**' 0.01 '*' 0.05 '.' 0.1 ' ' 1
##
## Residual standard error: 28.38 on 36 degrees of freedom
## Multiple R-squared:  0.7145,Adjusted R-squared:  0.6986
## F-statistic: 45.04 on 2 and 36 DF,  p-value: 1.591e-10
```

Note that since the adjusted R-squared value is larger for this model than for either of the models based on one of these covariates alone, this result suggests that using both covariates together is beneficial in this case. As before, we can construct the predicted vs. observed plot for this model and compute the validation R-squared value from the validation set predictions:

```
HorsepowerHat3V <- predict(Cars93Model3, newdata = Cars93V)
ValidationRsquared3 <- ValidationRsquared(Cars93V$Horsepower, HorsepowerHat3V)
```

The predicted vs. observed plot constructed from these values is shown in Fig. 5.10, which also gives the validation R-squared value for this model. Comparing the three validation R-squared values, it is clear that the model incorporating both covariates performs better than the models based on either one alone. Here, this conclusion agrees with that based on the adjusted R-squared, but for the reasons noted earlier, the validation R-squared values are regarded as the better basis for this decision.

Finally, this example illustrates the advantage of predicted vs. observed plots like those shown in Figs. 5.9 and 5.10 over the scatterplots shown in Fig. 5.8. Specifically, while these scatterplots are extremely useful in characterizing relationships between a numerical response variable and a single predictor, they do not extend to the case of multiple predictors, as in Cars93Model3. We could develop three-dimensional scatterplots to show the relationship between one numerical response variable and two numerical predictors, but this approach raises two difficulties. First, these three-dimensional plots are harder to construct and use effectively than two-dimensional scatterplots are, and, second, this approach does not extend to models with more than two predictors. Conversely, the predicted vs. observed plot is applicable to *any* model that predicts

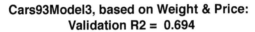

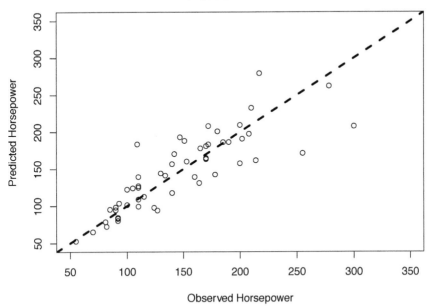

Figure 5.10: Validation set predictions vs. observed Horsepower for Cars93Model3, based on both Weight and Price.

a numerical response variable from an arbitrary number of predictors, even if those predictors are not numerical, as in the models considered in Sec. 5.4.

5.3.2 The problem of collinearity

It was noted that if multiple predictors are to be included in a linear regression model, they should each contribute independent information. Unfortunately, this condition is not always met, as the following example illustrates. There, we attempt to improve the prediction of Horsepower from the Cars93 data frame from the Price variable by also including the related variables Min.Price and Max.Price. As we will see in Chapter 9, the (product-moment) correlation coefficient measures the degree of linear association between numerical variables, quantifying the tendency of their scatterplot to approximate a straight line. The maximum possible value of this correlation coefficient is 1, indicating an exact linear relationship. Computing these correlations between these three price variables, we see that they all exhibit strong linear associations:

```
cor(Cars93T$Min.Price, Cars93T$Price)

## [1] 0.9702696
```

```
cor(Cars93T$Price, Cars93T$Max.Price)

## [1] 0.9773524

cor(Cars93T$Min.Price, Cars93T$Max.Price)

## [1] 0.8970989
```

Since these variables do not satisfy the "independent information" condition noted above, they should not be included together in a linear regression model. To see this point, consider the following model:

```
Cars93Model4 <- lm(Horsepower ~ Min.Price + Price + Max.Price, data = Cars93T)
summary(Cars93Model4)

##
## Call:
## lm(formula = Horsepower ~ Min.Price + Price + Max.Price, data = Cars93T)
##
## Residuals:
##     Min      1Q  Median      3Q     Max
## -64.365 -15.084  -5.233  16.982 115.886
##
## Coefficients:
##             Estimate Std. Error t value Pr(>|t|)
## (Intercept)    45.86      13.95   3.289   0.0023 **
## Min.Price     -82.86      99.98  -0.829   0.4128
## Price         168.07     199.33   0.843   0.4049
## Max.Price     -80.52      99.36  -0.810   0.4232
## ---
## Signif. codes:  0 '***' 0.001 '**' 0.01 '*' 0.05 '.' 0.1 ' ' 1
##
## Residual standard error: 33.63 on 35 degrees of freedom
## Multiple R-squared:  0.6101,Adjusted R-squared:  0.5766
## F-statistic: 18.25 on 3 and 35 DF,  p-value: 2.665e-07
```

Several points are worth noting here. First, comparing the (unadjusted) R-squared values for `Cars93Model2` based on `Price` alone (0.59788) with that for this model (0.6100535), we see that adding the two covariates `Min.Price` and `Max.Price` improves the training set R-squared value slightly, as it must. Comparing the adjusted R-squared values, however, we see that adding these variables does not appear to improve the model fit: the single variable model gives 0.5870119, while the three-variable model gives 0.5766295. These overall fit differences are fairly minor, but if we examine the model coefficients and their standard errors we see that they are dramatically different. In particular, recall that these results for the model based on `Price` alone were:

```
##             Estimate Std. Error  t value     Pr(>|t|)
## (Intercept) 49.209549 13.3371011 3.689674 7.185044e-04
## Price        4.823872  0.6503778 7.417030 7.942763e-09
```

The key details to note here are, first, the positive coefficient for the `Price` covariate, with a value of about 5, and, second, the very small p-values associated with both this coefficient and the intercept coefficient, indicating that both of these parameters are highly significant (see Chapter 9 for a discussion of statistical significance). In contrast, the corresponding results shown above for the three-variable model are very different. First, the `Price` coefficient in this model is almost 35 times larger in this model than in the single-variable model, while the other two variables exhibit very large *negative* values. Second, note that the magnitudes of the standard errors are actually larger than the coefficient values for all three of these variables. This observation is closely related to the third point, which is that *none* of the three variables in this model are statistically significant (i.e., none of their p-values are small enough to be deemed significant; see Chapter 9).

These results illustrate the statistical problem of *collinearity:* if we attempt to build a linear regression model that includes several covariates obeying an *exact* linear relationship—i.e., if one or more of the variables can be written exactly as a linear combination of the others—it can be shown that the linear regression model coefficients cannot be determined by ordinary least squares [58, Sec. 11.2]. Similarly, if the covariates exhibit a "near linear-dependence" as in this case (e.g., very large correlation coefficients), it is not impossible but still difficult to estimate linear model coefficients using the method of ordinary least squares. Commonly, this difficulty manifests itself in exactly the ways seen here: estimated coefficients that are very different from what we expect (e.g., the factor of 35 between the `Price` coefficient values in the single-variable and three-variable models), with large standard errors, leading to increased p-values and lowered assessments of statistical significance.

Another common consequence of collinearity is an extreme sensitivity to minor changes in the data values. That is, if we modify our training dataset and re-fit the model, normally we expect that (relatively) small changes in the data should yield (relatively) small changes in the model we obtain. In the face of collinearity, however, the resulting model changes may be extremely large. To see this point, consider the following alternative data partitioning, obtained by specifying a different random seed:

```
Cars93Flag2 <- TVHsplit(Cars93, split = c(0.5, 0.5, 0.0), iseed = 101)
Cars93T2 <- Cars93[which(Cars93Flag2 == "T"), ]
Cars93V2 <- Cars93[which(Cars93Flag2 == "V"), ]
```

If we re-fit the single variable model based on `Price` alone, we obtain results that are somewhat different, but the main features are consistent: the `Price` coefficient has a moderate, positive value with very small associated standard errors and p-values, indicating strong statistical significance:

```
Cars93Model5 <- lm(Horsepower ~ Price, data = Cars93T2)
summary(Cars93Model5)

##
## Call:
## lm(formula = Horsepower ~ Price, data = Cars93T2)
##
```

```
## Residuals:
##      Min       1Q  Median       3Q      Max
## -87.523 -19.508  -1.424  21.626  74.147
##
## Coefficients:
##              Estimate Std. Error t value Pr(>|t|)
## (Intercept)  66.9483     9.2279   7.255 2.15e-09 ***
## Price         3.8380     0.4105   9.349 1.22e-12 ***
## ---
## Signif. codes:  0 '***' 0.001 '**' 0.01 '*' 0.05 '.' 0.1 ' ' 1
##
## Residual standard error: 31.22 on 51 degrees of freedom
## Multiple R-squared:  0.6315,Adjusted R-squared:  0.6243
## F-statistic: 87.41 on 1 and 51 DF,  p-value: 1.216e-12
```

In contrast, we obtain very different results if we re-fit the three-variable model to this alternative training dataset:

```
Cars93Model6 <- lm(Horsepower ~ Min.Price + Price + Max.Price, data = Cars93T2)
summary(Cars93Model6)

##
## Call:
## lm(formula = Horsepower ~ Min.Price + Price + Max.Price, data = Cars93T2)
##
## Residuals:
##      Min       1Q  Median       3Q      Max
## -75.605 -17.163  -0.646  14.967  61.472
##
## Coefficients:
##              Estimate Std. Error t value Pr(>|t|)
## (Intercept)   62.679      8.338   7.517 1.05e-09 ***
## Min.Price     99.366     70.588   1.408    0.166
## Price       -187.496    140.931  -1.330    0.190
## Max.Price     92.906     70.410   1.320    0.193
## ---
## Signif. codes:  0 '***' 0.001 '**' 0.01 '*' 0.05 '.' 0.1 ' ' 1
##
## Residual standard error: 27.88 on 49 degrees of freedom
## Multiple R-squared:  0.7177,Adjusted R-squared:  0.7004
## F-statistic: 41.52 on 3 and 49 DF,  p-value: 1.694e-13
```

Here, the `Price` coefficient is *negative* and, as before, none of the model parameters exhibit *p*-values small enough to be deemed statistically significant.

In practical terms, the primary consequence of collinearity in linear regression models is that it effectively prevents interpretation of the model coefficients. That is, the coefficients in a linear regression model *should* give us an idea of both the magnitude and the direction of a predictor variable's influence on the response: a positive coefficient on a variable like `Price` in the models considered here means that increases in `Price` are associated with increases in `Horsepower`, while a negative coefficient indicates the opposite dependence. Models that attempt to include highly similar covariates—like `Price`, `Min.Price`, and `Max.Price`—that vary together can seriously interfere with model interpretation, as in this example. The issue of model interpretation is an important and sometimes complicated one, especially for more complex machine learning models, and it is considered in detail in Chapter 10.

5.4 Using categorical predictors

The linear regression modeling problems discussed up to now attempt to predict a numerical response variable from one or more numerical predictor variables. As noted several times in previous discussions, however, not all data variables of interest are numerical. In particular, one extremely important class is categorical variables like the `Insul` variable in the `whiteside` data frame, which takes two distinct values: `Before`, meaning that the associated data observations were made before home insulation was installed, and `After`, meaning the data observations were made after the insulation was installed.

A key characteristic of categorical variables is that we cannot perform arithmetic operations like addition or multiplication with them. Still, as the following example illustrates, categorical variables can be incorporated into linear regression models as predictors, and these variables can improve prediction quality significantly. Building models whose *response* variable is categorical requires fundamentally different methods, and some of these are introduced in Chapter 10 (e.g., logistic regression models and decision trees).

To incorporate a binary predictor like `Insul` into a linear regression model, we define an associated binary *numerical* variable that takes the values 0 or 1, depending on the value of the categorical variable. The `lm` function in R does this automatically: if we specify `Insul` as a predictor, the binary variable `InsulAfter` is created with the value 1 if `Insul` = 'After' and 0 if `Insul` = 'Before'. We can then construct a linear regression model of the form:

$$\text{Gas}_k = a_0 + a_1\text{Temp}_k + a_2\text{InsulAfter}_k + e_k. \tag{5.12}$$

As a specific example, consider the following linear regression model:

```
linearModelC <- lm(Gas ~ Temp + Insul, data = trainSub)
summary(linearModelC)

##
## Call:
## lm(formula = Gas ~ Temp + Insul, data = trainSub)
##
## Residuals:
##      Min      1Q   Median      3Q     Max
## -0.71455 -0.17706  0.04795  0.21904  0.50421
##
## Coefficients:
##              Estimate Std. Error t value Pr(>|t|)
## (Intercept)   6.35689    0.18495  34.370  < 2e-16 ***
## Temp         -0.28907    0.02345 -12.329 3.26e-10 ***
## InsulAfter   -1.54623    0.16348  -9.458 2.09e-08 ***
## ---
```

```
## Signif. codes:  0 '***' 0.001 '**' 0.01 '*' 0.05 '.' 0.1 ' ' 1
##
## Residual standard error: 0.3182 on 18 degrees of freedom
## Multiple R-squared:  0.926,Adjusted R-squared:  0.9177
## F-statistic: 112.6 on 2 and 18 DF,  p-value: 6.686e-11
```

Consistent with Eq. (5.12), this model has three parameters: an intercept, a coefficient for the Temp variable, and a coefficient for the binary InsulAfter variable created automatically by the lm function. Further, note that all three of these model parameters are highly significant (i.e., their associated p-values are very small; see Chapter 9 for a discussion of p-values), and the adjusted R-square measure for this model (0.9177346) is *much* larger for this model than for the linear regression model based on Temp alone (0.5347171).

As before, we can also characterize this model in terms of its validation set performance, generating validation set predictions using the predict function:

```
GasHatCV <- predict(linearModelC, newdata = validSub)
```

These predictions can then be used to generate both the predicted vs. observed plot for this model shown in the right-hand plot in Fig. 5.11, and the validation R-squared value shown in the title. The corresponding results for linearModelB, built from the same training dataset but including only the Temp covariate are shown in the left-hand plot, along with its validation R-squared value. It is clear from these results that including the categorical predictor Insul greatly improves the performance of this predictive model.

Before concluding this discussion, it is important to note three points that are closely related to topics that will be discussed later. First, we can re-write Eq. (5.12) as the following *conditional* model:

$$
\text{Gas}_k = \begin{cases} a_0 + a_1 \text{Temp}_k + e_k & \text{if Insul} = \text{Before} \\ (a_0 + a_2) + a_1 \text{Temp}_k + e_k & \text{if Insul} = \text{After.} \end{cases} \tag{5.13}
$$

In other words, linearModelC can be viewed as a linear regression model like linearModelB that predicts Gas from Temp, but with an intercept term that depends on the value of the Insul variable: this intercept has the value a_0 if Insul is Before but it has the different value $a_0 + a_2$ if Insul is After. Note, however, that the *slope* parameter a_1 is the same in both of these "component models." If the slope parameter were allowed to vary with the Insul variable, we would say there is an *interaction* between the variables Temp and Insul. This topic is discussed in detail in Sec. 5.5, where this specific example is considered.

Second, although the approach just described has only been demonstrated for a two-level categorical variable, it is easily extended to categorical variables with an arbitrary number of levels. The key difference is that, instead of replacing the original categorical variable with a single binary variable, we replace it with a collection of closely related binary variables. Specifically, if a categorical predictor x can assume any one of L distinct levels, we can incorporate it into linear regression models by replacing it with a collection of $L-1$ binary variables defined as follows:

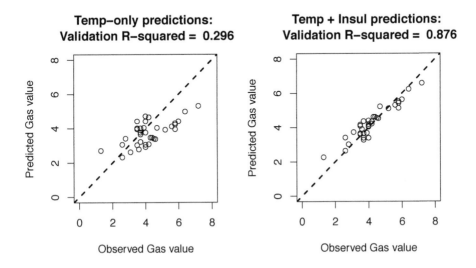

Figure 5.11: Predicted vs. observed plots for `linearModelB` (left) and `linearModelC` (right).

0. Select a *base level* that does not explicitly appear in the model;

1. Define binary variable b_1 to have the value 1 if x assumes the first non-base level, and 0 otherwise;

2. Define binary variable b_2 to have the value 1 if x assumes the second non-base level, and 0 otherwise;

... Define binary variables b_3 through b_{L-2} analogously;

$L-1$. Define binary variable b_{L-1} to have the value 1 if x assumes the $L-1^{th}$ non-base level, and 0 otherwise.

Note that if x assumes the base level, *all* of the binary variables b_1 through b_{L-1} defined here will be zero. Given these $L-1$ binary variables, we incorporate them into a prediction model in the same way as for a binary categorical variable. Specifically, for a linear regression model incorporating a numerical predictor v and an L-level categorical predictor x, our model would be written as:

$$y_k = a_0 + a_1 b_{1,k} + a_2 b_{2,k} + \cdots + a_{L-1} b_{L-1,k} + a_L v_k + e_k. \tag{5.14}$$

Note that this model has $L + 1$ coefficients: one for the intercept term (a_0), $L-1$ for the categorical variable x (a_1 through a_{L-1}), and one for the numerical variable (a_L). As in the case of a binary categorical variable like `Insul`, the `lm` function in R automatically takes care of constructing these binary variables, allowing us to use a syntax like the following to build models involving arbitrary categorical predictors:

```
linearModel <- lm(y ~ x + v, data = trainingSet)
```

Finally, the third key point is that some categorical variables exhibit "thin" levels, defined as levels that only rarely occur in the data. A specific example is the `rotary` value for the categorical variable `Cylinders` in the `Cars93` data frame, which only appears once. In such cases, if we adopt the recommended practice of randomly partitioning of the data into training, validation, and possibly holdout subsets, one or more of these thin levels may not be represented in all of the subsets. If a level is present in either the validation or holdout subsets that is not present in the training subset, no binary variable is constructed for this level and no model parameter is estimated for it. Thus, it is not possible to generate predictions from the validation and/or holdout records listing this thin level. This point is important and is considered in Chapter 10 as part of a second look at data partitioning.

5.5   Interactions in linear regression models

We saw that in Eq. (5.13) that the linear regression model including both the real variable `Temp` and the categorical variable `Insul` could be written as two standard linear models, each predicting `Gas` from `Temp` with the same slopes, but different intercept terms. In practical terms, this means that the dependence of `Gas` on `Temp` in this model does not involve the value of the other variable `Insul`. If this `Temp` dependence *did* vary with the value of `Insul`, we would say that there is an *interaction* between the variables `Temp` and `Insul`. More specifically, for a binary categorical variable like `Insul`, such an interaction is equivalent to a model that has this form:

$$\text{Gas}_k = \begin{cases} a_0 + a_1 \text{Temp}_k + e_k & \text{if Insul} = \text{Before} \\ (a_0 + a_2) + (a_1 + b_{12})\text{Temp}_k + e_k & \text{if Insul} = \text{After.} \end{cases} \tag{5.15}$$

The key point to note here is that both the intercept term and the slope associated with the `Temp` variable depend on the value of the `Insul` variable. Defining the indicator variable `InsulAfter` as before, Eq. (5.15) can be rewritten as:

$$\text{Gas}_k = a_0 + a_1 \text{Temp}_k + a_2 \text{InsulAfter}_k + b_{12} \text{Temp}_k \times \text{InsulAfter}_k + e_k. \tag{5.16}$$

The main difference between this equation and Eq. (5.12) is the presence of the product term Temp × InsulAfter with its associated coefficient b_{12}.

As before, models of this form are easily fit in R using the linear modeling function lm. In this case, we include both the individual Temp and Insul terms in our model—often called the *main effects*—along with the interaction between these variables using the * notation in the formula interface:

```
linearModelD <- lm(Gas ~ Temp * Insul, data = trainSub)
summary(linearModelD)

##
## Call:
## lm(formula = Gas ~ Temp * Insul, data = trainSub)
##
## Residuals:
##       Min       1Q   Median       3Q      Max
## -0.68220 -0.09362  0.00906  0.18244  0.40830
##
## Coefficients:
##                  Estimate Std. Error t value Pr(>|t|)
## (Intercept)       6.72575    0.26330  25.544 5.31e-15 ***
## Temp             -0.36225    0.04504  -8.044 3.39e-07 ***
## InsulAfter       -2.01845    0.29633  -6.811 3.03e-06 ***
## Temp:InsulAfter   0.09609    0.05160   1.862     0.08 .
## ---
## Signif. codes:  0 '***' 0.001 '**' 0.01 '*' 0.05 '.' 0.1 ' ' 1
##
## Residual standard error: 0.2984 on 17 degrees of freedom
## Multiple R-squared:  0.9385, Adjusted R-squared:  0.9277
## F-statistic: 86.48 on 3 and 17 DF,  p-value: 1.692e-10
```

Note that by specifying the predictors as Temp * Insul here, we do not have to explicitly indicate both of the main effect terms and the product term, but they all appear in the model summary, with the product term represented as Temp:InsulAfter. Also, note that as in the case of categorical main effects, the binary variable InsulAfter is automatically created by the lm function. Comparing the adjusted R-squared values for this model with that for linearModelC, which includes both main effects but not their interaction term, it appears that including the interaction term does improve the model slightly.

Once again, this interpretation agrees with what we obtain from the validation set characterization, constructed exactly as before. First, we generate the validation set predictions using R's generic predict function:

```
GasHatDV <- predict(linearModelD, newdata = validSub)
```

Given these predictions, we can generate the predicted vs. observed response plot for this model and compute the associated validation R-squared value. An example is shown in Fig. 5.12, which compares the results for linearModelC, without the interaction term, with linearModelD, that includes the interaction term. We can see that the predictions from the model with the interaction term generally fall closer to the equality reference line than those from the model

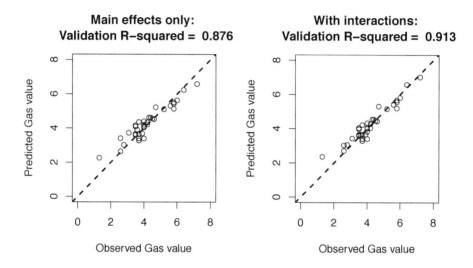

Figure 5.12:   Predicted vs. observed plots for `linearModelC` (left) and `linearModelD` (right).

without the interaction term, consistent with the larger validation R-squared value seen when the interaction term is included.

This example included the interaction between a numerical variable (`Temp`) and a two-level categorical variable (`Insul`), but the basic approach extends to linear regression models that include interactions between terms of arbitrary type: both numerical, both categorical, or mixed. This example was chosen for its simplicity and because including the interaction term gave a clearly better model. More generally, including an interaction between a K-level and an L-level categorical variable involves products between the $K-1$ binary variables used to represent one variable and the $L-1$ binary variables used to represent the other, for a total of $(K-1) \times (L-1)$ terms, which can quickly lead to a model with a lot of parameters. In the case of two numerical variables, their interaction is typically modeled by including the product of these variables in the linear regression model. That is, if our model includes the numerical main effects u and v, the model with interactions would take the form:

$$y_k = a_0 + a_1 u_k + a_2 v_k + b_{12} u_k v_k + e_k. \qquad (5.17)$$

The key feature of a linear regression model with interactions between predictors is that the influence of one of the predictors on the response variable depends on the value of the other predictor. We can emphasize that the impact of v_k on y_k in the model defined in Eq. (5.17) by re-writing it in the following form, where the effective coefficient of the v_k variable varies linearly with u_k:

$$y_k = a_0 + a_1 u_k + (a_2 + b_{12} u_k) v_k + e_k. \qquad (5.18)$$

The intent of this discussion has been to introduce, first, the idea of including interactions in predictive models—i.e., the idea that the influence of one predictor variable can depend on the value of others—and, second, the simple syntax that R uses with the linear modeling function `lm` to fit models with interactions. Further treatment of topics like how to interpret interactions or decide which ones might be worth considering is beyond the scope of this book. For more complete discussions of these questions, refer to a linear regression text like the one by Draper and Smith [22].

5.6 Variable transformations in linear regression

It has been noted earlier that the term "linear regression" refers to the problem of fitting predictive models that depend linearly on the unknown parameters. This means it is possible, and, as the following example illustrates, sometimes extremely useful to build linear regression models that involve nonlinear transformations of one or more of the prediction variables. As a specific example, consider the scatterplot in Fig. 5.13, which shows the relationship between the `MPG.city` variable in the training subset of `Cars93` data frame from the `MASS` package and the `Horsepower` variable. In addition, a curved trend line generated by the `supsmu` function has been added, emphasizing that while there appears to be a systematic relationship between horsepower and city gas mileage—i.e., gas mileage decreases as horsepower increases—this trend is not linear.

One mathematical function that exhibits a similar shape to the nonlinear trend seen here is the *reciprocal function:*

$$f(x) = 1/x, \qquad (5.19)$$

and this observation provides the motivation to construct and compare the following two models. The first is the linear regression model that attempts to predict city gas mileage from horsepower:

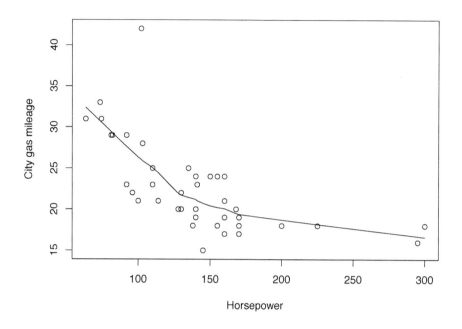

Figure 5.13: City gas mileage vs. horsepower from the training subset of the Cars93 data frame.

```
Cars93Model7 <- lm(MPG.city ~ Horsepower, data = Cars93T)
summary(Cars93Model7)

##
## Call:
## lm(formula = MPG.city ~ Horsepower, data = Cars93T)
##
## Residuals:
##     Min      1Q  Median      3Q     Max
## -7.3180 -3.3977 -0.5402  2.3619 16.7286
##
## Coefficients:
##              Estimate Std. Error t value Pr(>|t|)
## (Intercept) 32.27740    2.00913  16.065  < 2e-16 ***
## Horsepower  -0.06869    0.01349  -5.092 1.06e-05 ***
## ---
## Signif. codes:  0 '***' 0.001 '**' 0.01 '*' 0.05 '.' 0.1 ' ' 1
##
## Residual standard error: 4.298 on 37 degrees of freedom
## Multiple R-squared:  0.412,Adjusted R-squared:  0.3961
## F-statistic: 25.92 on 1 and 37 DF,  p-value: 1.062e-05
```

Note that although the coefficient associated with the Horsepower variable is highly significant, the adjusted R-squared value is fairly low, suggesting that this predictive model is not a great one. We can attempt to obtain a better model by considering one of the following general form:

$$\text{MPG.city}_k = a_0 + a_1(1/\text{Horsepower}_k) + e_k, \tag{5.20}$$

which can be fit using the following call to the `lm` function:

```
Cars93Model8 <- lm(MPG.city ~ I(1/Horsepower), data = Cars93T)
summary(Cars93Model8)


##
## Call:
## lm(formula = MPG.city ~ I(1/Horsepower), data = Cars93T)
##
## Residuals:
##      Min      1Q  Median      3Q     Max
## -6.0006 -2.3569 -0.1525  1.9348 16.7910
##
## Coefficients:
##                  Estimate Std. Error t value Pr(>|t|)
## (Intercept)        11.018      1.846   5.967 6.95e-07 ***
## I(1/Horsepower)  1447.496    216.785   6.677 7.66e-08 ***
## ---
## Signif. codes:  0 '***' 0.001 '**' 0.01 '*' 0.05 '.' 0.1 ' ' 1
##
## Residual standard error: 3.775 on 37 degrees of freedom
## Multiple R-squared:  0.5465,Adjusted R-squared:  0.5342
## F-statistic: 44.58 on 1 and 37 DF,  p-value: 7.663e-08
```

Several important points should be noted here. First and foremost, comparing the adjusted R-squared values for these two models suggests that the model that includes the reciprocal transformation applied to the `Horsepower` variable fits the data much better than the model where this predictor enters linearly. The second crucial point is the use of the `AsIs` function `I()` in the `lm` call. As noted in the overfitting example presented in Sec. 5.2.1, this function is used to enforce the usual arithmetic interpretation of terms like "1/Horsepower," protecting it from possible special interpretation by the formula interface in the `lm` function call. *This is good practice: the way the formula interface sometimes interprets mathematical terms can be mysterious, so protecting transformations like logs, reciprocals, and powers with the `AsIs` function can save a certain amount of frustration, debugging, and/or embarrassing errors.*

Once again, we obtain further support for including the reciprocal transformation in our model by comparing the validation set results with and without this transformation. This summary is presented in Fig. 5.14, which presents predicted vs. observed city gas mileage plots for the model where `Horsepower` enters linearly (left-hand plot), and where the reciprocal transformation is applied to `Horsepower` (right-hand plot). It is clear both from the general shape of these plots and the associated validation R-squared values that the model with the transformation gives the better predictions.

This short introduction to transformations in linear regression models was included to emphasize two points. First, as this example illustrates, for a given

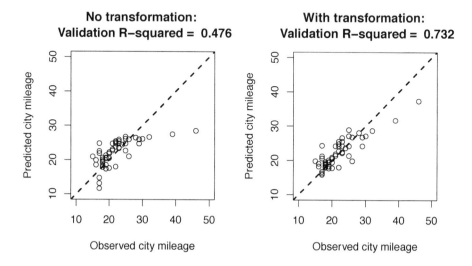

Figure 5.14: Predicted vs. observed MPG.city from un-transformed Horsepower (left), and with the reciprocal transformation (right).

prediction task (e.g., predicting city gas mileage), the variables we have available do not always come to us in their most useful form. Sometimes, as this example again illustrates, much better predictions can be obtained by applying a simple transformation to one or more of our prediction covariates. The second point of this discussion has been to illustrate that the lm function in R makes it easy to try these transformations and see whether they do improve our models, and if so, how much. A detailed discussion of transformations and how to select them is beyond the scope of this book, but a few transformations that are often useful with numerical variables include the following:

1. the log transformation, especially when the range of the variable covers many orders of magnitude;

2. the square root transformation, sometimes a useful alternative to the log transformations when zeros are present in the data;

3. the reciprocal transformation used here;

4. simple power transformations like x^2 or x^3.

Note that the log transformation is only useful with strictly positive variables, and the reciprocal transformation is most useful for these variables (in particular, the variable must never be zero). For a further treatment of transformations, refer to the discussion in *Exploring Data* [58, ch. 12].

5.7   Robust regression: a very brief introduction

The notion of outliers or anomalous data points was introduced in Chapter 1, and their influence and detection were examined in Chapter 3. The purpose of the following discussion is to illustrate, first, the potentially severe sensitivity of standard linear regression procedures to the presence of outliers in the data, and, second, to give a very brief introduction to *robust* or *outlier-resistant* regression methods. The specific method used here is `lmrob`, a function from the `robustbase` package that was demonstrated in connection with the `UScereal` data frame in Chapter 2 and is discussed further in Chapter 9. This procedure represents an outlier-resistant alternative to the standard linear regression procedure `lm` used in all of the previous examples presented in this chapter.

The basis for the example considered here is the `whiteside` data frame from the `MASS` package, with the `Temp` and `Gas` values both modified for a single observation. This particular modification was motivated by the popular if unfortunate practice of representing missing numerical values with a single, absurdly large numerical code like 9999. This convention probably stems from a time when datasets were relatively small, primarily consisting of numerical data spanning a relatively limited range of values, and shared with a relatively small group of co-workers. In such a setting, analysts would know to look for these large numbers and leave them out of their analyses or apply some simple imputation strategy, but the practice has persisted to some extent even to today, where datasets are often extremely large and the data collection and data analysis communities are completely separate.

Here, suppose the `Temp` and `Gas` values for observation 40 from the `whiteside` data frame were not observed, but were recorded using the missing value code 9999. In what follows, we compare the model coefficients obtained under the following four scenarios:

1. an ordinary least squares model fit using the `lm` procedure, applied to the original (unmodified) `whiteside` data;

2. an ordinary least squares model fit using the `lm` procedure, applied to the modified `whiteside` data;

Scenario	Procedure	Dataset	Intercept	Slope
1	lm	whiteside	5.4862	−0.2902
2	lm	whitesideMod	−0.8048	1.0001
3	lmrob	whiteside	5.3722	−0.2713
4	lmrob	whitesideMod	5.4157	−0.2760

Table 5.2: Model coefficients under the four scenarios described in the text.

3. a robust regression model fit using the lmrob procedure, applied to the original whiteside data;

4. a robust regression model fit using the lmrob procedure, applied to the modified whiteside data.

The modified dataset used in Scenarios (2) and (4) was constructed here as:

```
whitesideMod <- whiteside
whitesideMod$Temp[40] <- 9999
whitesideMod$Gas[40] <- 9999
```

The models in these four scenarios were obtained using the following R code:

```
model1 <- lm(Gas ~ Temp, data = whiteside)
model2 <- lm(Gas ~ Temp, data = whitesideMod)
model3 <- lmrob(Gas ~ Temp, data = whiteside)
model4 <- lmrob(Gas ~ Temp, data = whitesideMod)
```

Essentially, the lmrob procedure detects outlying data observations and downweights them, fitting a linear regression model that reflects the behavior of the nominal or "non-outlying" portion of the data. Here, this nominal portion of the data represents the original whiteside data frame, so what we hope to see from the robust regression procedure is approximately the same result in Scenarios (3) and (4) that we obtain using standard linear regression applied to the uncontaminated dataset in Scenario (1). As the following results demonstrate, the results obtained in Scenario (2) are dramatically different, being totally dominated by the outlying data values in the modified data frame.

Specifically, Table 5.2 compares the intercept and slope values estimated under these four modeling scenarios. Comparing Scenarios (1), (3), and (4), we see that we do indeed obtain approximately the same results: the intercept parameter is roughly +5.4 and the slope is between −0.27 and −0.29. In contrast, for Scenario (2)—standard linear regression based on the contaminated data—the intercept and slope both have the opposite sign from these other three results: the intercept is approximately −0.8 and the slope is essentially +1. *One of the important points of this example is the dramatic difference in outlier sensitivity of the linear regression procedures* lm *and* lmrob. In particular, while the lm procedure returns results for the outlier-contaminated dataset that are nothing like those obtained from the original dataset, the lmrob procedure returns results that differ only by $\sim$ 1% to 2% between these two cases.

Further, if we examine the detailed summary information for the linear regression model returned by the lm function with the outlier-contaminated data, we find an R^2 value consistent with a perfect fit to the data:

```
summary(model2)

##
## Call:
## lm(formula = Gas ~ Temp, data = whitesideMod)
##
## Residuals:
##     Min      1Q  Median      3Q     Max
## -7.3959 -2.6206 -0.1956  2.4546  8.8049
##
## Coefficients:
##               Estimate Std. Error t value Pr(>|t|)
## (Intercept) -0.8048458  0.4967767   -1.62    0.111
## Temp         1.0000751  0.0003718 2689.91   <2e-16 ***
## ---
## Signif. codes:  0 '***' 0.001 '**' 0.01 '*' 0.05 '.' 0.1 ' ' 1
##
## Residual standard error: 3.682 on 54 degrees of freedom
## Multiple R-squared:      1,Adjusted R-squared:      1
## F-statistic: 7.236e+06 on 1 and 54 DF,  p-value: < 2.2e-16
```

In addition to providing a further illustration of the potential impact of outliers on ordinary least squares linear regression procedures like lm, this result also emphasizes the point made earlier that perfect fits to the data are rarely useful. In this particular case, the introduction of the outlying pair of data values at (9999, 9999) into a set of numerical data spanning a total range from −0.8 to 10.2 effectively creates two clusters of points: the outlying pair constitutes one cluster, while *all* of the nominal data points constitute the other, essentially concentrated at zero. This view is illustrated in Fig. 5.15, which shows a scatterplot of the modified **whiteside** data values, overlaid with the standard linear regression line from the lm model from Scenario (2). We can see that this line does give an essentially perfect fit to these two "point clusters."

While this example may appear extreme, it illustrates a problem that does arise in practice, and it emphasizes three points that should always be kept in mind when analyzing data. First, even a single outlier can cause "standard" procedures like linear regression analysis, correlation analysis, or simple descriptive statistics to fail catastrophically. (For example, a naive interpretation of the Scenario (2) results would say that "installing insulation caused heating gas consumption to increase *enormously*," clearly not a reasonable conclusion.) Second, this example illustrates that standard, if unfortunate, practices like numerical missing data codes can introduce glaring outliers. In a way, the severity of these outliers is an advantage, since it makes them easy to detect if we look

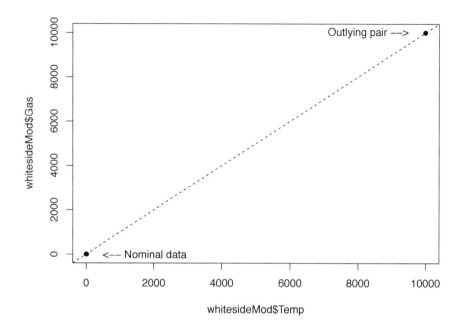

Figure 5.15: Modified `whiteside` data and `lm` regression line.

at the data in any way at all, in contrast to cases like the zero-coded missing data seen in the Pima Indians diabetes dataset discussed in Chapter 3 where this coding artifact is much less obvious. Finally, this example again emphasizes the importance of somehow *looking* at our data before building models or basing decisions on it. Even if plotting the data is inconvenient, a preliminary look at simple descriptive statistics like means and standard deviations can reveal problems like those seen here. For example, the mean `Temp` value from the modified dataset is 183.35 Celsius: an average winter temperature well above the boiling point of water should raise suspicions.

A detailed discussion of robust regression is beyond the scope of this book, but these ideas are discussed further in Chapter 9, and some of the references cited there provide very useful—and very detailed—treatments.

5.8   Exercises

1: Section 5.1.3 considered a linear regression model that predicts heating gas consumption from outside temperature, based on the `whiteside` data frame in the `MASS` package. This problem asks you to consider a similar analysis, exploring the relationship between `calories` and two possible predictors, `fat` and `potassium`, from the `UScereal` data frame in the `MASS` package. Specifically:

1a. Using the `mfrow` and `pty` graphics parameters discussed in Chapter 2, set up a side-by-side array of two square plots;

1b. In the left-hand part of the array, create a scatterplot of `calories` versus `fat`;

1c. Fit a standard linear regression model that predicts `calories` from `fat` and use the `summary` function to characterize this model;

1d. Add a dashed reference line to this plot, of twice standard width, based on the linear model predictions from (1c), and add a title giving the adjusted R-square value from the summary;

1e. Repeat steps (1b), (1c), and (1d) to model the relationship between `calories` and `potassium`. Based on the adjusted R-square values, which model appears to give better predictions?

2: As discussed briefly in Section 5.7, standard linear regression procedures are extremely sensitive to the presence of outliers, motivating the development of robust procedures like `lmrob` in the `robustbase` package that have better outlier resistance. This exercise asks you to repeat Exercise 1, but using the `lmrob` procedure to fit linear regression models instead of the standard linear regression models used there. Specifically:

2a. Using the `mfrow` and `pty` graphics parameters discussed in Chapter 2, set up a side-by-side array of two square plots;

2b. In the left-hand part of the array, create a scatterplot of `calories` versus `fat`;

2c. Fit a regression model using `lmrob` that predicts `calories` from `fat` and use the `summary` function to characterize this model;

2d. Add a dashed reference line to this plot, of twice standard width, based on the linear model predictions from (1c), and add a title giving the adjusted R-square value from the summary;

2e. Repeat steps (1b), (1c), and (1d) to model the relationship between `calories` and `potassium`. Based on the adjusted R-square values, which model appears to give better predictions?

3: The predicted vs. observed plot was introduced in Section 5.2.3 as a useful way of characterizing model performance, even in cases involving more than one prediction variable where scatterplots overlaid with prediction lines like those considered in Exercises 1 and 2 are not appropriate. These plots were advocated based on the Training/Validation split discussed in Section 5.2.2, but they are even useful in cases where the dataset is too small to split conveniently. This exercise asks you to fit prediction models using both of the variables considered in Exercises 1 and 2, by both methods, and compare them using predicted vs. observed plots based on the complete dataset. Specifically:

3a. Using the `mfrow` and `pty` graphics parameters discussed in Chapter 2, set up a side-by-side array of two square plots;

3b. Fit a linear regression model that predicts `calories` from both `fat` and `potassium`, and use the `summary` function to characterize this model;

3c. Use the `predict` function to generate the predicted `calories` values from this model;

3d. In the left-hand position of the plot array, create a plot of the predicted `calories` versus the observed `calories` for this model, with dashed equality reference line of twice normal width. Include a title giving the adjusted R-squared value for this model;

3e. Fit a robust regression model using the `lmrob` function that predicts `calories` from both `fat` and `potassium`, and use the `summary` function to characterize this model;

3f. Use the `predict` function to generate the predicted `calories` values from this model;

3g. In the right-hand position of the plot array, create a plot of predicted `calories` versus the observed `calories` for this model, with dashed equality reference line of twice normal width. Include a title giving the adjusted R-squared value for this model. Which of the models in Exercises 1, 2, and 3 seem to give the best predictions?

4: The validation R-squared measure was introduced in Section 5.2.3 as a useful alternative to the adjusted R-squared measure for characterizing the goodness-of-fit for arbitrary prediction models. Consider a dataset $\mathcal{D}$ that has been randomly split into two mutually exclusive subsets, a training set $\mathcal{T}$ and a validation set $\mathcal{V}$, and suppose that their means $\bar{y}_T$ and $\bar{y}_V$ are, while similar, not equal. Show that the validation R-squared measure for the intercept-only model $\hat{y}_k = \bar{y}_T$ is negative. (Hint: start by expressing the numerator in the expression for validation R-squared in terms of both $(\bar{y}_T - \bar{y}_V)^2$ and $(y_k - \bar{y}_V)^2$.)

5: The `Chile` data frame in the `car` package is large enough for the data partitioning described in Section 5.2.2, and it also illustrates a number of other points, some of which will be examined in subsequent exercises.

5a. The working code for the `TVHsplit` function listed in Section 5.2.2 essentially consists of two lines: one to initialize the random number generator with the `set.seed` function, and the other to create the vector `flags` by calling the `sample` function. Using the default `iseed` value from the `TVHsplit` function, create and display a vector named `flags` that assigns each record from the `Chile` data frame randomly to the value `T` or `V`, each with probability 50%.

5b. Using the `flags` vector constructed in (5a), build two subsets of the `Chile` data frame, one named `ChileTrain` for every record with

flags value T, and the other named ChileValid for every record
with flags value V. Display the dimensions of these data frames.

5c. The Chile data frame exhibits missing data. Compute the means of
the non-missing statusquo values from the data frames ChileTrain
and ChileValid. Do these results support the assertions made about
$\bar{y}_T$ and $\bar{y}_V$ in Exercise 4?

6: The summaries generated for standard linear regression models fit using
the lm function provide standard characterizations like the adjusted R-
squared value that are widely used in characterizing model quality, but
the graphical predicted vs. observed plot introduced in Section 5.2.3 can
give us a lot more insight into model performance.

6a. Using the ChileTrain data frame constructed in Exercise 5, fit a linear regression model that predicts statusquo from all other variables
in the dataset. Show the summary characterization for this model and
note the adjusted R-squared value;

6b. Using the predict function, generate statusquo predictions from
this model for both the ChileTrain and ChileValid data frames
from Exercise 5. Construct a side-by-side array of square plots showing predicted vs. observed statusquo values for both datasets, giving
them the titles "Training set" and "Validation set", respectively. Include dashed equality reference lines to help judge model quality. Do
these plots suggest this prediction model is a good one?

7: This exercise and the next three provide a simple example of step-by-step
model building, illustrating a number of extremely damaging errors and
showing how they can be corrected. The basis for these exercises is the
cpus data frame from the MASS package, based on a published comparison
of computer performance from 1987. One of these variables is the measured performance perf and the objective of these exercises is to build
a linear regression model to predict this variable from the others in the
dataset. This exercise illustrates the importance of looking at the data to
avoid naive mistakes in selecting prediction variables. Specifically:

7a. First, use lm to build a naive prediction model firstModel that
predict perf from all other variables in the dataset. Look at the
summary response for this model: what indications does this result
give that the result is a uselessly poor model?

7b. Apply the str function to the cpus data frame: which variable exhibits one unique level for every record in the dataset?

7c. Repeat (7a) but removing the variable identified in (7b) to generate secondModel, and compute the corresponding summary results.
Which variable appears to be the best predictor of perf?

8: For the linear regression model constructed in Exercise (7c), construct
a side-by-side plot array showing, in the left-hand plot, the predicted

response from `secondModel` versus the observed `perf` value, and in the right-hand plot, the predicted response versus the variable identified in Exercise (7c) as most predictive. Include dashed reference lines in both plots. Do these plots support the conclusion that this variable is most important?

9: For different reasons, the variables identified in Exercises 7 and 8 are not appropriate covariates to use in a useful prediction model. Build a third linear regression model, named `thirdModel`, that excludes both of these predictors and generate the `summary` results for this model. Comparing the significance results for these two models, which variables are important contributors to the third model that were not important contributors to the second model?

10: Linear regression models are best suited to situations where the response variable is approximately normally distributed. In cases where this is not true, variable transformations like those described in Section 5.6 are sometimes used to obtain better models. This last exercise explores this idea for the `cpus` dataset:

10a. Using the `qqPlot` function from the `car` package, construct side-by-side normal QQ-plots for the `perf` response variable (left-hand plot) and its logarithm (right-hand plot). Which of these variables is better approximated by a normal distribution?

10b. Using the `AsIs` function `I()` to protect the transformation, build a linear regression model `fourthModel` that predicts the log of the response variable and display its summary.

10c. Generate predictions from this model and construct side-by-side plots of the predicted versus observed responses for `thirdModel` on the left and for `fourthModel` on the right, with equality reference lines in both plots.

Chapter 6

Crafting Data Stories

Most of this book is concerned with the art of exploratory data analysis, but the ultimate objective of most data analyses is a presentation of the results to others. Here, the term *data stories* refers to summaries that support this objective; that is, these data stories are intended for an audience other than the data analyst. This means that both the content and the format must be matched to the needs of that audience.

6.1 Crafting good data stories

In general terms, an effective data story may be viewed as a text-based or text-augmented version of the explanatory data visualizations described in Chapter 2. In fact, the comments from Iliinsky and Steele quoted in Chapter 2 are worth repeating here since they actually represent a useful working definition of a data story [44, p. 8]:

> Whoever your audience is, the story you are trying to tell (or the answer you are trying to share) is *known to you at the outset*, and therefore you can design to specifically accommodate and highlight that story. In other words, you'll need to make certain *editorial decisions* about what information stays in, and which is distracting or irrelevant and should come out. This is a process of selecting focused data that will support the story you are trying to tell.

The only modification required here would be to change the phrase "selecting focused data" in the last sentence to "selecting focused data and summary text." In some cases—e.g., the executive summaries discussed in Sec. 6.2.1—the entire data story is too short to actually include an explanatory data visualization, but a typical data story does include one or more of these visualizations, along with sufficient explanatory text that the reader can understand clearly what the author is attempting to say about the data, and why.

6.1.1   The importance of clarity

Much has been written about the disastrous launch of the Space Shuttle Challenger on January 28, 1986, which exploded 73 seconds later, killing the entire seven-member crew. The cause of this explosion was quickly determined to be the failure of two rubber O-rings, a concern raised by the engineering staff of the rocket manufacturer, Morton Thiokol, the night before the launch. Their concerns were based on the history of partial failures in previous launches under low temperatures, together with the very cold temperatures predicted for the day of this January launch. In his book, *Visual Explanations* [65, pp. 38–53], Edward Tufte devotes 16 pages to the story of the decision to launch and the way the Thiokol engineers presented the data on which their recommendation not to launch was based. Tufte's account is based on an extensive examination of the many publications that came out in the wake of the Challenger disaster, and his primary focus is on the ineffectiveness of the data story used to convey the concern—ultimately dismissed by those in charge of the launch decision— that cool temperatures were related to prior O-ring problems. In particular, he argues that [65, p. 40]:

> Regardless of the indirect cultural causes of the accident, there was a clear proximate cause: an inability to assess the link between cool temperature and O-ring damage on earlier flights.

To make his case, Tufte shows several of the 13 charts prepared by the Thiokol engineers in support of their argument that the launch should be delayed. His point is that these charts do not clearly convey any link between low external temperatures and O-ring damage, for several reasons. For example, one of the charts Tufte shows provides many details about the nature and extent of O-ring damage on previous launches, but two crucial details are missing: first, there is no overall damage measure to convey how severe the damage was on the prior launches when it did occur, and, second, the launch temperatures were not shown. Further confusing the situation, these charts referred to each case by a NASA launch number, a different Thiokol solid rocket motor number, and in some, but not all, cases, a launch date. Also, subsequent charts narrowed the focus to one specific damage mechanism which only occurred twice.

As a more effective alternative, Tufte presents a plot similar to Fig. 6.1 showing an overall damage index that he constructed from the available data, versus the temperature of each launch. Further, this plot is extended to include the forecast launch temperature range for January 28, 1986, emphasizing how much colder the Challenger launch was than any prior launch. As Tufte notes, this plot is highly suggestive of more severe O-ring damage at lower temperatures; in fact, he points out that *every* launch at temperatures colder than 66 degrees exhibited O-ring damage, and the predicted temperature for the Challenger launch was at least 37 degrees colder than this. Fig. 6.1 is a sunflower plot constructed from the data frame `SpaceShuttle` in the `vcd` package, which includes both the temperatures for each shuttle launch prior to the Challenger disaster, and Tufte's damage index. Although this plot shows that some O-ring

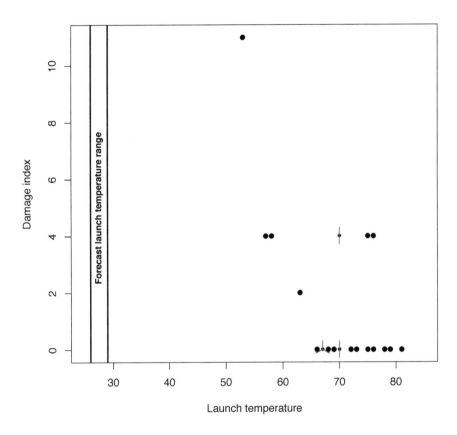

Figure 6.1: Space Shuttle O-ring damage index vs. launch temperature.

damage did occur at higher temperatures, both the dramatic increase in the damage index with decreasing temperature and the *much* lower temperature of the January 28, 1986 launch are clearly evident in this plot.

6.1.2 The basic elements of an effective data story

Essentially, an effective data story briefly describes one or more datasets, clearly explains why we are looking at them, and succinctly summarizes what we have learned from analyzing them. In his book on writing, W. Ross Winterowd notes that the six questions, "Who? What? Where? When? Why? How?," are often used by newspaper reporters in formulating accounts of human events [80, pp. 66–67]. The following variations on these questions represent a good starting point in crafting an effective data story:

1. *Who is our intended audience?* The answer to this question will influence both the length and content of our data story, an important point discussed further in Sec. 6.2;

2. *What data source are we examining, and what do we conclude from it?* Since we are describing data analysis results, it is important to be succinct but unambiguous about both what our conclusions are and the data on which they are based;

3. *Where did this data come from?* Since "the same" data can come from several different sources, it is important to at least provide a reference to the source of the data on which our story is based;

4. *When was this data generated?* Even in cases where time does not explicitly enter our analysis, knowing when the data values were collected can be important (e.g., were these national sales figures collected before or after the Great Recession?);

5. *Why are we analyzing this dataset?* Together with the questions of what data source we are analyzing and what we conclude from it, this is probably the most important question to be addressed in any data story;

6. *How did we arrive at our conclusions?* Data stories for non-technical audiences should not go into great technical detail, but enough detail should be given—or references to sources of these details cited—to leave the reader comfortable that a systematic analysis was conducted.

As noted, the first question—Who is our intended audience?—determines much, and this point is discussed in detail in Sec. 6.2. Clear partial answers to questions (2) and (5) about what data source are we examining and why should be given at the beginning of the data story to make it clear what the document is about. In some cases, the answers to questions (3), (4), and/or (6) are central enough to our data story that they too should be discussed right at the beginning, at least briefly. For example, if our data story extends an earlier one, either updating or replacing the original data with newer data, or applying a different analysis to the same data, these details represent an important part of the motivation for our analysis and should be stated clearly from the outset.

6.2   Different audiences have different needs

As noted, different audiences may have very different needs, based on their level of prior exposure to the problem being discussed, and the level of detail they want. The following subsections describe three very different styles of data stories, intended for very different audiences. The first is the *executive summary* or *abstract* described in Sec. 6.2.1, a very short document that is intended to summarize highlights for one of two audiences: first, those (e.g., executives) who need to be informed of key details but have neither the time nor the desire to read a longer, more complete account; and second, those who read the summary in a longer document first as part of their decision of whether to read any further. The second type of data story is the *extended summary* described in Sec. 6.2.2, and it is the main focus of this chapter, illustrated with three specific examples in Sec. 6.3. This type of document gives a much more complete description than an abstract or executive summary, but it does not attempt to present the "complete details" that may be required by a select few. This third level of documentation is described briefly in Sec. 6.2.3.

6.2.1   The executive summary or abstract

As noted, the executive summary or abstract represents a very short, highly distilled data story, typically intended for one of two audiences. As the name "executive summary" implies, the first of these audiences consists of executives in large organizations, who are typically making decisions about many different things, often on the basis of limited information. In such cases, the purpose of the executive summary is to convey two things: the first is the essential information they need to make their decision, and the second is confidence that the author of the executive summary knows what they are talking about and their conclusions and/or recommendations are correct. The second audience is the *potential reader* of a longer document, such as a technical journal publication or conference paper, who is reading the abstract to decide whether the complete paper is relevant and/or interesting enough for them to take the time to read.

To illustrate the content of a typical executive summary, consider the following hypothetical example, with the title "Recommendations on the Replacement of Xfer in Our PolyGoodies Line":

> Xvendor, the only manufacturer of the chain transfer agent Xfer used in the production of our PolyGoodies premium polymer product line, has announced that it will stop offering Xfer for sale in January. Candidate alternatives to Xfer include GoodX, sold by Yvendor, and BetterX, sold by Zvendor. In our semi-works, we have prepared 20 small batches of PolyGoodies with each of these three chain transfer agents and evaluated them with respect to elongation-at-break, bulk modulus, and color fastness, the three primary release properties for the product. Results for bulk modulus and color fastness were comparable for all three chain transfer agents, while elongation-at-break was substantially poorer for GoodX than with Xfer and slightly better for BetterX. Based on these results, we recommend an aggressive program to transition from Xfer to BetterX for the PolyGoodies line before the end of the year.

Several points are worth noting here. First and foremost, considerable background is assumed on the part of the reader: for example, it is clear that this summary is intended for a manager or an executive with a company that manufactures polymer products, but this document does not explain what terms like "chain transfer agent" or "elongation-at-break" mean. In fact, the intended readers may not fully understand what these terms mean either, but it is assumed that these readers appreciate their importance. Also, note that while this document summarizes the results of a data analysis (i.e., an analysis of the physical properties data obtained from 60 batches of polymeric material, manufactured using three different chain transfer agents), the only numerical detail given is the number of batches manufactured for each case. Also, this format is obviously too short to include or explain any type of plot.

Similarly, the abstract of a technical paper is typically part of either a publication in a specialized journal or the proceedings of a conference on a specialized topic. In either case, considerable background knowledge is again assumed: there is no need, for example, to explain that polymers are long-chain molecules whose physical properties depend on both their composition and their structure to the readers of a journal like *Journal of Polymer Science* or to the attendees of a polymer science technical meeting.

6.2.2   Extended summaries

Extended summaries provide longer, more detailed treatments of data analysis results than the paragraph-length executive summaries or abstracts just described. In fact, these shorter summaries can form the basis for a useful "lead paragraph" that helps orient the reader at the beginning of the document, but extended summaries are generally intended for a larger, more varied audience. As a consequence, these documents usually assume a less knowledgeable reader, so this lead paragraph should at least make forward references to discussions of basic details included in the body of the text. In particular, this paragraph should clearly state what dataset we are analyzing and why, at least in general terms, and indicate any other key points that will be addressed. The rest of the document then expands on this first paragraph, briefly noting key details and elaborating on our most important findings.

Since explanatory data visualizations can convey a lot of information more simply than words can, typical data stories make extensive use of these visualizations, but two points are important here. First, be judicious: carefully select both the main points you wish to convey, and the visual details that will help to convey them. Second, be sure to include sufficient explanatory text to make it clear what the reader is seeing in these plots through the use of clear axis labels, legends, and text annotations in the plot where appropriate. Also, explanatory text in the body of the document that is associated with a plot should be placed near it as possible. The more you can do to help the reader *easily* see and understand the key details in your plots, the more likely they are to understand the message you are attempting to convey in the data story.

The data story examples presented in Sec. 6.3 are all extended summaries, from 3 to 5 pages long with between 1 and 4 figures, depending on the nature of the material presented. Note that while these stories relate to economics (the Big Mac and Grande Latte indices), sociology (demographic and environmental influences on housing prices), and insurance (the Australian vehicle loss data), none of these summaries assume specialized knowledge in any of these areas.

6.2.3 Longer documents

In general, we can expect an inverse relationship between the length of a document and the number of people who will read it. Unfortunately, it is also true that there is a direct relationship between the content that can be included in a document and its length. The focus of this chapter on the extended summary described in Sec. 6.2.2 is motivated by the fact that, in many cases, a document of a few pages with a few figures represents an optimum in the trade-off between making a document detailed enough to be useful, and short enough for people to read. Still, there are occasions where longer documents are necessary.

One reason for creating longer documents is the need to capture key data acquisition and analysis details. For example, the discussion of the Big Mac and Grande Latte economic indices presented in Sec. 6.3.1 briefly describes what these indices are and how they appear to relate, but nothing is said about the process of obtaining the data. These details are not important for the general reader, but they are crucial to anyone wishing to extend or update the analysis; fortunately, they have been described in reasonable detail in Chapter 4. Specifically, the Big Mac index was obtained by reading an Excel spreadsheet from the URL listed here into a data frame with the `xlsx` package, and the Grande Latte index values were obtained by manual data entry from the blog post cited here because they were not available in a format that could be easily read into an *R* data object. Similarly, another important detail discussed in Chapter 4 was the task of merging these data sources into a common data frame that could be used to create the scatterplot shown in Fig. 6.4. This merge was complicated by the fact that the Big Mac index is country-based, with one value for each country, while the Grande Latte index is city-based. Since cities can be placed uniquely within countries but not *vice versa*, it was necessary to re-base the Grande Latte index values on countries before this data merge could be accomplished. Finally, the task of constructing the separate Big Mac index subsets for those countries with and without Grande Latte indices was accomplished using SQL queries, also described in Chapter 4.

The key point here is that summary information like that just described needs to be recorded *somewhere* if there is any likelihood of our ever wanting to share our analysis procedures with others, extend our results, or simply re-create them. This need provides the main motivation for *reproducible research*, a topic discussed briefly in Chapter 11.

6.3 Three example data stories

This final section provides three examples of data stories, each in the extended summary form described in Sec. 6.2.2. Each of the following subsections presents one of these examples, giving the data story as a stand-alone summary intended for a general reader who is interested in the questions and conclusions associated with the example, but not the details of the analysis or the data preprocessing.

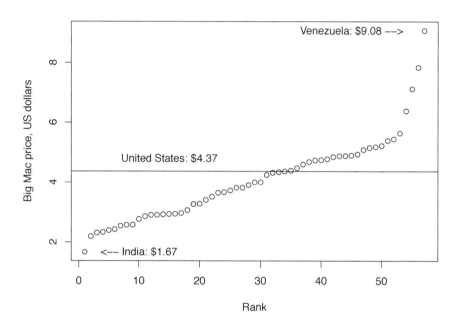

Figure 6.2: Scatterplot of the rank-ordered January 2013 Big Mac index.

6.3.1   The Big Mac and Grande Latte economic indices

In 1986, an article in *The Economist* proposed the *Big Mac index*, based on the local price of a McDonalds Big Mac in different countries, as "a light-hearted guide to whether currencies are at their 'correct' level." The following URL provides access to a dataset with the values of this index and several related variables for selected months from 2000 to 2015:

http://www.economist.com/content/big-mac-index

Fig. 6.2 shows a plot of the January 2013 Big Mac index based on this data for 57 countries, ranked from the smallest value to the largest. The horizontal line in the plot represents the Big Mac index value for the United States ($4.37). The smallest index value is that for India, at $1.67, while the largest is that for Venezuela, at $9.08. Both of these extremes are indicated on the plot.

The website also gives a somewhat more serious description of the index:

> It is based on the theory of purchasing-power parity (PPP), the notion that in the long run exchange rates should move towards the rate that would equalise the prices of an identical basket of goods and services (in this case, a burger) in any two countries. For example,

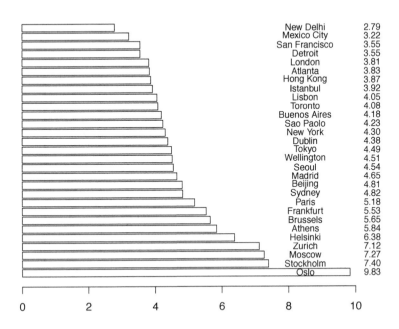

New Delhi	2.79
Mexico City	3.22
San Francisco	3.55
Detroit	3.55
London	3.81
Atlanta	3.83
Hong Kong	3.87
Istanbul	3.92
Lisbon	4.05
Toronto	4.08
Buenos Aires	4.18
Sao Paolo	4.23
New York	4.30
Dublin	4.38
Tokyo	4.49
Wellington	4.51
Seoul	4.54
Madrid	4.65
Beijing	4.81
Sydney	4.82
Paris	5.18
Frankfurt	5.53
Brussels	5.65
Athens	5.84
Helsinki	6.38
Zurich	7.12
Moscow	7.27
Stockholm	7.40
Oslo	9.83

Figure 6.3: City names and prices for the Starbucks Grande Latte index.

the average price of a Big Mac in America in July 2015 was \$4.79; in China it was only \$2.74 at market exchange rates. So the "raw" Big Mac index says that the yuan was undervalued by 43% at that time.

Sometime after the Big Mac index was defined, the *Grande Latte index* shown in Fig. 6.3 was proposed, based on the local price of a Starbucks Grande Latte (approximately 16 oz), converted from the local currency to US dollars at the prevailing exchange rate. The following blog post, accessed on 9/19/2016, reproduces a barchart with numbers appended that was published by the *Wall Street Journal* and *Bloomberg News*, dated Feburary 27, 2013 and based on exchange rates for February 20, 2013:

http://www.coventryleague.com/blogentary/the-starbucks-latte-index

Fig. 6.3 is an approximate reconstruction of the barchart from this blog post.

Fig. 6.4 is a scatterplot of the Grande Latte index against the Big Mac index, to give an idea of how the two indices compare in the 24 countries where they are both available. Since they are both attempting to measure about the same thing

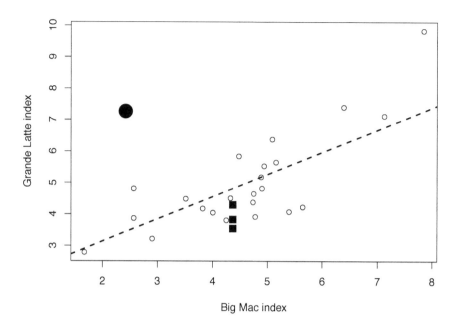

Figure 6.4: Scatterplot of the Grande Latte index against the Big Mac index.

in about the same way, we might expect them to show about the same results. Conversely, if we believe that the demographic characteristics of most of those people buying Big Macs differ a lot from those of most people buying Grande Lattes, we might expect to see this difference reflected in a disparity between the two indices, reflecting differences in the prevalence of those demographic characteristics in the different countries. This plot suggests that these indices do tend to vary together—in countries where Big Macs are more expensive, so are Grande Lattes, in general—but the agreement is not perfect. The dashed line in the plot represents an ordinary least squares fit to the data, which generally goes through the center of the data points, with (mostly) comparable scatter on either side of the line. One of the most interesting exceptions is the larger, solid circle at a Big Mac index value of approximately 2.5 and a Grande Latte index value a little greater than 7. This point corresponds to Russia, where—at least in early 2013—the Starbucks Grande Latte was much more expensive in local currency than the MacDonalds Big Mac. Also, the cluster of three solid squares below the dashed line represent four US cities, which share the country-based Big Mac index value of 4.5 but exhibit three different Grande Latte index values, giving an indication of how much this index can vary within one country where currency differences are not an issue.

```
## log="y" selected
```

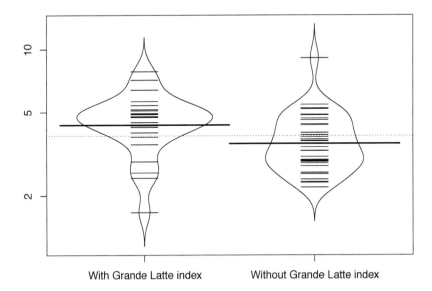

Figure 6.5: Beanplot comparison of Big Mac index values for countries with and without Grande Latte index values.

Finally, it is worth noting that the Big Mac index values are available for many more places than the Grande Latte index values are. This observation raises the possibility that the countries where Grande Latte index is available (i.e., where Starbucks has stores) may be systematically different than those where the Big Mac index is available. For example, maybe the Starbucks countries tend to be richer than the McDonalds countries. We can gain some insight into this possibility by comparing the distribution of Big Mac index values for those countries with and without Grande Latte index values. Fig. 6.5 gives a beanplot summary comparing the distributions of these two sets of Big Mac index values, with the beanplot on the left summarizing those countries with Grande Latte index values reported, and the beanplot on the right summarizing those countries without Grande Latte index values. The wide lines across each beanplot give the average of the values in the beanplot, and it is clear that the average Big Mac index value for countries reporting Grande Latte index values is larger than those without Grande Latte index values. Looking at the individual values, it is clear that there are a couple of outliers—e.g., the largest Big

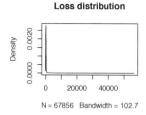

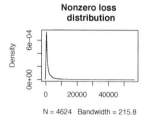

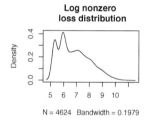

Figure 6.6: Three density estimates from the Australian vehicle insurance data.

Mac index value is reported for Venezuela, which does not have a Grande Latte index reported, while the smallest value is reported for India, which does—but overall, it does appear that the Big Mac index values are generally larger for countries that list Grande Latte index values than for those that do not.

6.3.2 Small losses in the Australian vehicle insurance data

In their book, *Generalized linear models for insurance data* [19], one of the example datasets de Jong and Heller consider summarizes Australian vehicle insurance claims and losses for 67,856 single-driver, single-vehicle policies. In addition to claim and loss data, this dataset also includes information about the value, age, and type of car, the gender and age of the driver, and the region where the car is based. This dataset is available in *R* as the data frame `dataCar` in the `insuranceData` package.

Fig. 6.6 gives three views of the loss data from this dataset. The top left plot was constructed from all of the loss data in the dataset, and it shows the behavior typical of insurance loss data. Specifically, the only detail visible in

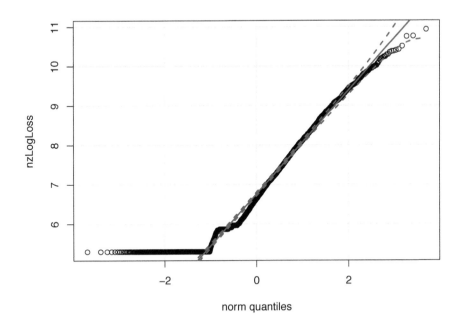

Figure 6.7: Normal QQ-plot of the nonzero log loss data from the Australian vehicle insurance dataset.

this plot is the extremely large peak at zero, consistent with the fact that only 6.8% of the policies characterized in this dataset experienced losses. The middle plot in Fig. 6.6 shows the density estimate constructed from only the nonzero losses, and this plot also exhibits typical behavior for data of this type: the distribution appears highly skewed, with a high concentration of small losses, and a rapidly decaying probability for large losses. The wide range of these nonzero loss values, from 200 to 55,922, motivates the use of a log transformation to obtain a more detailed view of their behavior. The lower right plot shows the density estimate obtained from these log-transformed nonzero losses, and here we see pronounced multimodal character, with two very prominent peaks, a third that is less well defined, and a gradually decaying right tail. Multimodal distributions are known to occur in certain kinds of insurance loss data, such as comprehensive insurance policies that cover both small claims like broken windshields and very large claims like car theft.

Fig. 6.7 shows a normal quantile-quantile plot constructed from the log-transformed nonzero loss data, motivated in part by the fact that insurance losses are sometimes assumed to exhibit a lognormal distribution. This plot includes 95% confidence limits around the Gaussian reference line, and the fact

log="y" selected

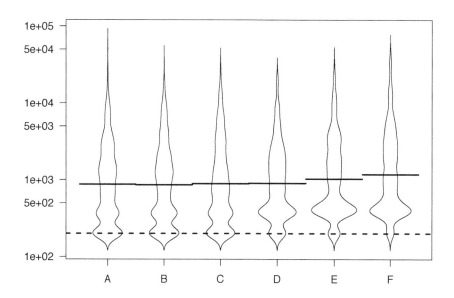

Figure 6.8: Log-scale beanplot summary of nonzero losses by `area`. The dashed line represents the minimum nonzero loss of 200.

that the upper portion of this plot mostly falls within these limits suggests that this approximation may be fairly reasonable. The most dramatic feature of this plot, however, is the flat lower tail, which corresponds to the smallest nonzero loss value in the dataset and which indicates that this single value is repeated quite frequently. In fact, this value is *exactly* 200, occurring 695 times and representing 15% of the nonzero losses seen in the dataset.

The meaning of this minimum nonzero loss is unclear, but it is interesting to note that its prevalence seems to vary significantly with region (the `area` variable in the dataset). Fig. 6.8 gives a log-scale beanplot summary of the nonzero losses for the different values of this variable. The dashed line across the bottom of the plot represents the minimum nonzero loss value, and the width of the "bump" in each plot reflects the prevalence of these minimum loss records in each region. These minimum losses appear most prevalent in areas A, B, and C, less so in area D, still less in area E, and fairly rare in area F.

6.3.3 Unexpected heterogeneity: the Boston housing data

The Boston data frame in the MASS package is based on data described in a paper by Harrison and Rubinfeld published in 1978 [37] that examined the relationship between air quality—as quantified by nitrogen oxide concentration—and median house price in 506 census tracts in and around the city of Boston. The authors applied standard linear regression analysis to this dataset, and in their book on regression diagnostics, Belsley, Kuh, and Welsch revisit this analysis to look for individual data points that may be unusually influential [6, pp. 229–261]. The analysis presented here is motivated by the observation that there appears to be extreme heterogeneity in the behavior of some of the other explanatory variables included in this dataset. To address this heterogeneity, a *model-based recursive partitioning (MOB)* model [84] is built that effectively constructs separate— and very different—linear models for census tracts within the city of Boston and those constituting the surrounding suburbs. This model suggests that the dependence of median house price on nitrogen oxide concentration was *much* greater in Boston than it was in the suburbs.

As in the models described by Harrison and Rubinfeld [37] and Belsley, Kuh, and Welsch [6], the response variable considered here is the median housing price (medv), and this response is predicted from all other variables included in the data frame. A linear regression model fit to this dataset gives what appears to be a reasonable fit, based on the following summary statistics:

```
BostonLinearModel <- lm(medv ~ ., data = Boston)
summary(BostonLinearModel)

##
## Call:
## lm(formula = medv ~ ., data = Boston)
##
## Residuals:
##      Min      1Q  Median      3Q     Max
## -15.595  -2.730  -0.518   1.777  26.199
##
## Coefficients:
##                Estimate Std. Error t value Pr(>|t|)
## (Intercept)  3.646e+01  5.103e+00   7.144 3.28e-12 ***
## crim        -1.080e-01  3.286e-02  -3.287 0.001087 **
## zn           4.642e-02  1.373e-02   3.382 0.000778 ***
## indus        2.056e-02  6.150e-02   0.334 0.738288
## chas         2.687e+00  8.616e-01   3.118 0.001925 **
## nox         -1.777e+01  3.820e+00  -4.651 4.25e-06 ***
## rm           3.810e+00  4.179e-01   9.116  < 2e-16 ***
## age          6.922e-04  1.321e-02   0.052 0.958229
## dis         -1.476e+00  1.995e-01  -7.398 6.01e-13 ***
## rad          3.060e-01  6.635e-02   4.613 5.07e-06 ***
## tax         -1.233e-02  3.760e-03  -3.280 0.001112 **
## ptratio     -9.527e-01  1.308e-01  -7.283 1.31e-12 ***
## black        9.312e-03  2.686e-03   3.467 0.000573 ***
## lstat       -5.248e-01  5.072e-02 -10.347  < 2e-16 ***
## ---
## Signif. codes:  0 '***' 0.001 '**' 0.01 '*' 0.05 '.' 0.1 ' ' 1
```

```
##
## Residual standard error: 4.745 on 492 degrees of freedom
## Multiple R-squared:  0.7406,Adjusted R-squared:  0.7338
## F-statistic: 108.1 on 13 and 492 DF,  p-value: < 2.2e-16
```

Given the very small p-values seen here, all of the variables in this dataset appear to exhibit substantial predictive power except for `indus`, the proportion of non-retail business acres per town, and `age`, the proportion of owner-occupied units built prior to 1940. In particular, note that the coefficient of the `nox` variable, which was of primary interest in the original paper, is negative and highly significant: the higher the nitrogen oxide concentration, the lower the median property value.

The examination of the Boston housing dataset presented by Belsley, Kuh, and Welsch made use of *single-row diagnostics* that measure the influence of individual observations by effectively removing them from the dataset, refitting the model without them, and looking at how much this deletion influences the modeling results. They note that while census tracts from the city of Boston only account for about 25% of the total records (131 of 506), these neighborhoods account for the majority of the points found to be influential using these deletion diagnostics (40 of 67, or approximately 60%). They go on to say [6, p. 239]:

> While we did not explore the point further, one might speculate that this central-city behavior differs systematically from that of the surrounding towns.

Fig. 6.9 provides supporting evidence for this conjecture, consisting of two sunflower plots constructed from the Boston housing dataset. The plot on the left shows the property tax rate versus the *index of accessibility to radial highways*, while the plot on the right shows the *zoning index*, defined as "the proportion of a town's residential land zoned for lots greater than 25,000 square feet," again plotted against the index of accessibility to radial highways. In the sunflower plot on the left, it is clear that there is a large collection of neighborhoods that all exhibit the same very high index of accessibility to radial highways and also exhibit the second-highest tax rate seen. Similarly, the sunflower plot on the right shows that these same neighborhoods all exhibit the smallest possible zoning index value (i.e., zero). While many other neighborhoods also exhibit this same zoning index value, the clear picture that emerges from these sunflower plots is that this group of neighborhoods differs systematically in several ways from the others in the dataset. In fact, based on the identifications given by Belsley, Kuh, and Welsch [6, p. 230], this cluster corresponds to the census tracts in the city of Boston, motivating the model considered next.

The class of MOB models introduced by Zeileis *et al.* [84] combines the useful features of linear regression models with those of decision tree models, to obtain models that can be interpreted as *conditional regression models:* the dataset is partitioned into subsets based on certain variables, and separate linear regression models are obtained for each subset. That is, the model structure consists of a decision tree that, on the basis of the values of specified *partitioning*

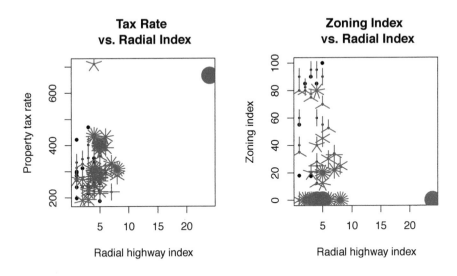

Figure 6.9: Two sunflower plots: tax rate vs. radial highway index (left) and zoning index vs. radial highway index (right).

variables, assigns each record in the dataset to a terminal node. There, rather than generating a single prediction as in a standard decision tree model, the MOB model includes a full linear regression model at each node, predicting all data records that have been mapped into that node on the basis of a set of *regression covariates*. The `lmtree` function in the `partykit` package fits MOB models of the type just described, given a response variable, a set of regression covariates, and a set of partitioning variables:

```
library(partykit)
mobModel <- lmtree(medv ~ . - rad | rad, data = Boston)
print(mobModel)

## Linear model tree
##
## Model formula:
## medv ~ . - rad | rad
##
## Fitted party:
## [1] root
## |   [2] rad <= 8: n = 374
## |       (Intercept)          crim            zn         indus          chas
## |     -15.478326943   1.157199959   0.013122209   0.019253225   0.982844966
## |               nox            rm           age           dis           tax
## |      -8.190276027   9.286125701  -0.052662410  -0.904354138  -0.009209309
```

```
## |            ptratio           black          lstat
## |         -0.632631215   0.016617227  -0.061425213
## |    [3] rad > 8: n = 132
## |            (Intercept)          crim              zn         indus            chas
## |          71.853737983  -0.146771540              NA            NA     8.904654942
## |                   nox            rm             age           dis             tax
## |         -37.768541844  -1.733560847    0.030781767  -3.403274178              NA
## |                ptratio         black           lstat
## |                     NA   0.003648496  -0.810801998
##
## Number of inner nodes:    1
## Number of terminal nodes: 2
## Number of parameters per node: 13
## Objective function (residual sum of squares): 6589.884
```

Here, the response variable to be predicted is medv as before, all variables except rad have been taken as potential regression covariates, and rad has been specified as the partitioning variable. The rationale for this choice of partitioning variable was its apparent effectiveness—as seen in Fig. 6.9—in separating the inner city Boston census tracts from the suburban tracts. The print function applied to this model shows its general structure: a two-node tree has been fit to the data, one corresponding to all of the suburban rad values (i.e., all census tracts with rad value between 1 and 8), and the other corresponding to the inner city neighborhoods (i.e., the only rad value greater than 8 is rad = 24, which corresponds to the inner city neighborhoods). The print summary also gives us the coefficients of the best fit linear regression model for each of these nodes, based on the specified regression covariates.

Note that these models are very different. In particular, note that the coefficient of the nox variable in the inner city model is *much* larger than that in the suburban model. In fact, comparing these coefficients with those from the linear regression model presented earlier, note that the coefficient here for the suburban census tracts is about half of what it was before, while that for the inner city census tracts is more than twice what it was before. Also, note that the coefficient for the rm variable—defined as the average number of rooms per dwelling—actually has opposite signs for the inner city and suburban data subsets: this suggests that more rooms brought higher prices in suburbia in 1978, but lower prices in the city of Boston.

It is not clear how far to push the analysis and interpretation of the models presented here, but the comparison of the linear regression model and the regression tree-based MOB model does support the conjecture of Belsley, Kuh, and Welsh that the inner city and suburban neighborhoods may behave differently with respect to the influence of various factors on housing prices.

Chapter 7

Programming in R

The previous chapters have focused on using R in an interactive session, assuming you have started up an interactive R session by clicking on a desktop icon, which opened a command window. Working inside this command window, we have discussed in some detail how to generate various types of plots, how to enter data either manually or from various external sources, how to perform various data characterization and analysis tasks, and how to save our results so they don't disappear forever when we end our interactive R session. Other than a few illustrative examples like the custom functions TVHsplit and ValidationRsquared discussed in Chapter 5, everything has been done through a sequence of manual operations inside this interactive R session window. This chapter introduces the fundamentals of *programming* in R, focused on creating our own custom R functions. This chapter assumes little or no prior programming experience, with R or any other programming language, and may therefore be largely unnecessary for those who do have prior programming experience. That said, the R-specific details presented here should be useful to those with prior programming experience in other languages but not in R.

7.1 Interactive use versus programming

Given all we can do using R's built-in functions in an interactive session, some may wonder why it is worth the effort to learn how to create your own functions, but there are two good reasons to do this. The one most people encounter first is that creating our own functions can save us a lot of typing in interactive R sessions when we find ourselves doing the same or highly similar things repeatedly. By encapsulating these tasks in an R function, we can greatly reduce the amount of typing we need to do, while at the same time reducing the likelihood of mistakes. The second reason for developing our own custom R functions is that the result clearly documents what we have done, providing a basis for others to understand and build on it. Indeed, this aspect of program development is closely related to the concept of *reproducibility* introduced in Chapter 11.

7.1.1   A simple example: computing Fibonnacci numbers

The *Fibonacci numbers* are perhaps the best known of all numerical sequences, featured in popular books and movies and defined by the recursion relation:

$$F_i = F_{i-1} + F_{i-2}, \tag{7.1}$$

for $i = 3, \ldots, n$, and initialized by $F_1 = F_2 = 1$. The first 10 elements are:

1, 1, 2, 3, 5, 8, 13, 21, 34, 55

The name "Fibonacci numbers" refers to the fact that the sequence defined in Eq. (7.1) appeared in the book *Liber abaci (Book of the abacus)* published by Leonardo of Pisa—also known as Leonardo Fibonacci—in 1202. There, the recursion relation defining these numbers appeared as the solution to a problem about the growth rate of a population of rabbits. It is ironic that most people know the name Fibonacci from this sequence, when the most significant impact of his book was its introduction of Arabic numerals to European merchants, making commercial transactions *much* easier than with Roman numerals [54].

The Fibonacci sequence is useful here because it provides a simple motivating example for the development of computer programs. Specifically, while the sequence of the first 10 Fibonacci numbers listed above is easy enough to generate by hand, suppose we wanted to know the value of the 30th Fibonacci number, or the 100th or the 1000th? It is easy enough to generate programs that will answer these questions, but attempting to do so manually quickly becomes tedious, error prone, and ultimately infeasible. The following sequence walks through the process of developing procedures to compute Fibonacci numbers, starting with a brute force approach that takes no advantage of computer programming at all and concluding with a simple custom R program that can be used to compute as many Fibonacci numbers as we want.

Crude brute force

Since R provides basic arithmetic capabilities, the simplest—and least useful—approach to computing the Fibonacci numbers would be to use R to implement the brute force computations. For example, the following sequence of steps computes the first five Fibonacci numbers:

```
Fib1 <- 1
Fib2 <- 1
Fib3 <- Fib2 + Fib1
Fib4 <- Fib3 + Fib2
Fib5 <- Fib4 + Fib3
```

It is important to emphasize that this is *not* programming: we are using R to do the computations, but this sequence is a fully manual one, with all of the attendant disadvantages of manual operations. It is tedious and error prone and does not scale at all well: generating the 100th Fibonacci number this way would be as punitive as having to write "I will not talk in class" 100 times on the blackboard (a form of punishment popular with elementary school teachers in the mid- to late twentieth century).

Brute force with vectors

A critical limitation of the previous example that seriously limits our ability to improve it is the fact that every element of the sequence has its own name: the first element is `Fib1` and the last one is `Fib5`. We can take a useful step towards a better solution by defining our Fibonacci sequence as a single vector with multiple elements instead of a collection of distinct variables, each with its own name. The next "program" in this sequence would be:

```
Fib <- vector("numeric", length = 5)
Fib[1] <- 1
Fib[2] <- 1
Fib[3] <- Fib[2] + Fib[1]
Fib[4] <- Fib[3] + Fib[2]
Fib[5] <- Fib[4] + Fib[3]
```

It is worth emphasizing that this example is no more an actual program than the last one, although the introduction of the vector `Fib` is a step in the right direction, forming the basis for the next example.

Using loops: a first actual program

Control structures represent an important component of any programming language, and these will be discussed further in Sec. 7.2.3. One of the simplest of these control structures is the *for loop* used here:

```
n <- 30
Fib <- vector("numeric", length = n)
Fib[1] <- 1
Fib[2] <- 1
for (i in 3:n){
  Fib[i] <- Fib[i-1] + Fib[i-2]
}
```

This example represents an extremely simple program, but an actual program nonetheless. Specifically, this code sequence automates the "grunt work," allowing us to answer one of the questions posed earlier—i.e., the 30th Fibonacci number is 832,040—with *a lot* less typing than would be required using either of the previous two procedures. The key to this automation is the use of the `for` loop at the end of the code listing, which causes the indented lines of code to be repeated for each value of the index `i` specified in the `for` statement. Note that the first two Fibonacci numbers are still specified explicitly, as in the previous two examples, but this is necessary to set up the `for` loop: specifically, the vector indices in the loop start at 3 rather than 1 so that the terms appearing on the right-hand side of the assignment statement in the loop are well defined for all values of the loop counter `i`.

A shorter but poorer alternative

In the interest of saving key strokes, we could modify the previous example, "simplifying" it as follows:

```
Fib <- vector("numeric", length = 30)
Fib[1] <- 1
Fib[2] <- 1
for (i in 3:30){
  Fib[i] <- Fib[i-1] + Fib[i-2]
}
```

Unfortunately, this "simplification" comes at the expense of *hard-coding* the Fibonacci sequence length. This practice is strongly discouraged because it makes programs hard to change as our needs change. For example, changing this key program parameter from 30 to 100 requires changing both the `length` argument in the first line of the program and the upper limit of the `for` loop at the end of the program. If we make only one of these changes, we will either get a different result than we want, or our program will terminate with an error. Specifically, if we change the `length` = 30 argument to `length` = 100 but we do not change the upper limit of the `for` loop, we will correctly generate the first 30 Fibonacci numbers, but the last 70 elements of the `Fib` vector will have their initial value of zero. Conversely, if we increase the upper limit of the `for` loop from 30 to 100 but do not increase the length of the `Fib` vector in the first line, the program will halt with an error when we attempt to create the 31st Fibonacci number since the element `Fib[31]` does not exist.

An initialization trick

The fact that the elements of the vector `Fib` are given initial values when the vector is created suggests the following initialization trick. Since the first two elements of the Fibonacci sequence have the fixed value 1, we can initialize the vector so *all* of its elements have this value, and then modify elements 3 through n using the same for loop as before. The easiest way to do this is with the `rep` function, which returns a vector with n elements, each containing the value specified by its first argument. This trick yields the following program:

```
n <- 30
Fib <- rep(1, n)
for (i in 3:n){
  Fib[i] <- Fib[i-1] + Fib[i-2]
}
```

The `Fibonacci` custom function

The final example converts the previous one into a custom *R* function:

```
Fibonacci <- function(n){
  Fib <- rep(1, n)
  for (i in 3:n){
    Fib[i] <- Fib[i-1] + Fib[i-2]
  }
  return(Fib)
}
```

This example illustrates the typical structure of an R function:

```
FunctionName <- function(Argument1, ..., ArgumentN){
  (Body: executable code goes here)
  return(Result)
}
```

The first line of a function definition gives the name of the function, to the left of the assignment arrow, with the keyword `function` to the right of this arrow, followed by parentheses and an opening bracket. If the function operates on arguments—and most functions do—the names of these arguments go between the parentheses, and the opening bracket at the end of the line indicates that the body of the function follows. The body can include any valid R code, and if the function returns a result—and again, most functions do—this is best done explicitly by calling the `return` function. The function definition concludes with a closing bracket that matches the opening bracket at the end of the first line.

In the case of the function `Fibonacci` listed above, the only argument is n, which defines the length of the Fibonacci sequence to be generated. The body of this function consists of the R code from the previous example, except for the first line that defined n: this code is not needed here since the argument n in the function call supplies this value. To use this function, we call it with the desired Fibonacci sequence length n and it returns the vector containing the first n Fibonacci numbers:

```
Fibonacci(10)

## [1]  1  1  2  3  5  8 13 21 34 55
```

Note that this sequence is identical to the one shown at the beginning of this example, providing a (partial) verification that the function has been written correctly (i.e., it accurately generates the desired results). This idea—testing a function by applying it to a case where we know the correct answer and verifying that this is the result we get—is an extremely important idea discussed further in Sec. 7.3.4. Finally, note that this function can be used to answer the two "extreme Fibonacci" questions posed earlier: what are the values of the 100th and the 1000th Fibonacci numbers? To do this, we simply call the `Fibonacci` function with n = 1000 to obtain the vector of the first 1000 Fibonacci numbers and then look at the 100th and 1000th elements of this vector:

```
FibThousand <- Fibonacci(1000)
FibThousand[100]

## [1] 3.542248e+20

FibThousand[1000]

## [1] 4.346656e+208
```

7.1.2   Creating your own functions

At a high level, the task of writing your own function consists of these steps:

1. Decide what you want your function to do:

   a. Decide on the name for your function;

   b. Decide what *arguments* the program should be called with;

   c. Decide what computations should occur in the body of the function;

   d. Decide what data object the function should return.

2. Create an external file that will contain your function, e.g.:

   a. Use the `file.create` function to create the file;

   b. Use the `file.edit` function to allow you to edit it;

   c. Create the text of the function and save the result.

3. Use `source` to bring your function into your interactive *R* session;

4. Test your program on some examples with *known* results;

5. Decide whether your program is working:

   a. If so, celebrate and use your program;

   b. If not, groan and begin the debugging process.

Designing your function

If you use *R* very much, once you start creating functions, you will probably create many of them. For this reason, it is a good idea to give your function a name that allows you—and anyone else you might choose to share it with—to see as clearly as possible what the function does from its name alone. This name will become the keyword in the first line of code that defines the function, so it must be one word, although "CamelCase" concatenations of words are acceptable, often allowing for more informative names (e.g., `FindOutliers`). Also, it is good practice to use this function name as the name of the file you create to hold the function, the step discussed next; for example, if you are creating the function named `FindOutliers`, the file name should be `FindOutliers.R`, which tells you immediately when looking in the directory, first, that this file contains *R* source code, and, second, what function it contains.

In general, *R* works with *data objects*, created either by simple assignments or reading the contents of external files. Functions in *R* then *typically* operate on one or more *arguments*—i.e., data objects passed to the function when it is called—to create and return a new data object, although there are exceptions. Specifically, some functions are called for their *side effects*, such as the generation of a plot or the creation of an external file. With the important exception of those that generate plots, most custom functions *should not* exhibit side effects.

In particular, an extremely useful feature of R *functions is that variables used in the body of the function are only defined locally.* This means that if we work with a variable in the body of a function, even if it has the same name as one that exists in our R session, nothing we do inside the body of the function will modify this external variable. Conversely, R does support a global assignment operator that allows us to "reach outside" the body of the function and modify external variables, but this is generally very bad practice and should be avoided since it can cause extremely subtle problems that may be nearly impossible to diagnose and correct.

A detailed discussion of function arguments is given in Sec. 7.2.1, but the key point in initially developing a function is that these arguments should define any externally supplied data objects we need to work with in the body of the function. The body then includes the executable R code that turns these input data objects into the output data object we wish to return. A useful approach to designing a program is to construct an "IPO table," that defines the Inputs, Processing, and Output for our function:

I: List the inputs, i.e., the arguments that will appear on the first line of our function definition;

P: Outline the sequence of processing steps in the body of the function that will be used to turn the inputs into an output;

O: Describe the output data object that the function will return.

The more explicitly we can define these elements of our program initially, the easier the program development task will be.

Creating your function file

Since objects in an interactive R session are volatile and disappear when our session ends, it is important to save any custom function we create in an external file. As noted, this file should have the same name as the function it contains, with the file extension ".R" so we know the file contains R code.

To create this file, we need to use a text editor. While it is not the best possible choice for creating programs, the default text editor invoked by the file.edit function is an acceptable choice and, for that reason, the program development procedure described here uses it. For example, the following sequence was used to create and edit the file ComputeIQD.R for the example discussed at the end of this section:

```
file.create("ComputeIQD.R")
file.edit("ComputeIQD.R")
```

The first step in this sequence creates a new file named ComputeIQD.R, and the second step opens a new window that allows you to edit this file. You enter the program text into this file and, when you are done, *save the result*. The program can then be accessed by calling the source function discussed below.

As Murrell notes in his book, *Introduction to Data Technologies* [56, Sec. 2.4], better editors for program development are available with features like automatic indentation, parenthesis matching, syntax highlighting, and line numbering. *If you have such an editor available and want to use it, the only change in the program development process described here would be to create and save your program using this editor; after that, you would load the program into your interactive* R *session using the* source *function described below.*

One of the other points Murrell emphasizes in his discussion of writing code is that word processing programs like Microsoft Word or Open Office Writer are *not* acceptable text editors for creating program source files. *Specifically, these programs include a lot of "invisible help features" that are intended to improve the appearance of written documents and simplify certain formatting tasks, but these features can have disastrous consequences for computer programs.* In particular, these word processing programs typically include invisible formatting characters and auto-capitalization features that will introduce fatal syntax errors in what was entered as a syntactically correct program.

Using source

The built-in function source is called with the name of a file containing R code, which is read, parsed, and, assuming there are no syntax errors, executed. If we call source with the name of a file containing an R function, this function will be checked for syntax errors and, if none are found, it will be loaded into our interactive R session, making it available for use until we terminate our R session. In the case of the function ComputeIQD discussed at the end of this section, the call would be source("ComputeIQD.R").

Testing your code

The final step in developing a custom R function is testing it. An unfortunate reality is that, when we create a new function, it seldom does what we want it to the first time we attempt to use it. Often, we have *syntax errors*, essentially "grammatical errors" in the R programming language, so that the R code interpreter cannot figure out what we are asking it to do. Common syntax errors include things like missing keywords (e.g., leaving function out of the function definition line), missing or mismatched parentheses or curly braces, or omitted assignment operators. As noted, the source function will check for these and let us know if they are present, and these must be corrected before we can even attempt to use our function. More subtle problems are logic errors in the body of our function, and these will often result in test cases giving us the wrong answers. It is for this reason that we should always test our function before we start to use it and depend on it, running it on test cases where we know what result we *should* get and verifying that this is the result we actually do get. This important topic is discussed further in Sec. 7.3.4.

The ComputeIQD example

To illustrate the sequence just described, consider the creation of a function to compute the interquartile distance discussed in Chapter 3 from a numerical vector x. The name of this program will be ComputeIQD, which will be saved in the file ComputeIQD.R. The IPO table for this example looks like this:

I: The only input argument is numerical vector x;

P: Processing done by the body of the function:

1. Compute the upper and lower quartiles of the x vector;
2. Compute the IQD as the difference between these quartiles.

O: Output: return the value of IQD.

As noted earlier, we create and edit the source file with these commands:

```
file.create("ComputeIQD.R")
file.edit("ComputeIQD.R")
```

The code we enter into this file with the text editor will look like this:

```
ComputeIQD <- function(x){
    #
    quartiles <- quantile(x, probs = c(0.25, 0.75))
    IQD <- as.numeric(quartiles[2] - quartiles[1])
    return(IQD)
}
```

The first line includes the function name (ComputeIQD), the assignment arrow and the keyword function, the input argument x in parentheses, and the opening curly bracket tells R that the body of the function follows. The next line contains only the comment character "#", included here to provide space and make the function easier to read; more generally, informative comments are good practice, a point discussed in Sec. 7.3.2. The next two lines perform the two computation steps outlined in the IPO table above, and the last indented line returns the IQD value computed from x. The final line is the closing curly brace that tells the R interpreter the functon definition is complete.

Loading our function with source, we can test it:

```
source("ComputeIQD.R")
set.seed(9)
x <- rnorm(100)
upQ <- quantile(x, probs = 0.75)
loQ <- quantile(x, probs = 0.25)
upQ - loQ

##      75%
## 1.167729

ComputeIQD(x)

## [1] 1.167729
```

The fact that these two numerical values—the one we computed manually and the one returned by our function—are equal gives us at least partial verification that the function has been implemented correctly.

7.2   Key elements of the R language

The following discussions give a brief introduction to the key elements of the *R* programming language used in examples like those discussed in Sec. 7.4. For a very readable and more extensive introduction, refer to *An Introduction to R*, one of the PDF help files included with all *R* installations. This note is periodically updated by the *R* Core Team as part of their general *R* maintenance effort, and it touches on almost all of the topics covered in this chapter, frequently in greater detail, and it covers a number of other topics not addressed here.

The primary focus of this chapter is on writing custom functions in *R* so the following discussions begin with functions and their argument types. The computations in the body of a function make use of various data types and control structures, but the most important of these have already been introduced. One exception is the extremely flexible list data type discussed in Sec. 7.2.2, widely used in data objects returned by functions; specifically, *R* allows a function to return a single object, but returning a multi-element list allows us to return essentially anything. Because control structures are not widely used interactively in *R*, Sec. 7.2.3 describes a few of these structures that are most useful in developing custom functions.

7.2.1   Functions and their arguments

We have seen the basic structure of *R* functions in a number of examples, but this structure is crucial to the following discussion, so it bears repeating here:

```
FunctionName <- function(Argument1, ..., ArgumentN){
  Body: sequence of R statements that compute something
  return(Result)
}
```

The body of the function can contain any valid *R* code necessary to compute the output value `Result` that the function returns. The `return` statement in the last line is not strictly necessary—if it is omitted, the function will return the last value computed in the body code—but it is good practice because including this line makes it clear exactly what value the function is returning. Much of the code appearing in the body of the function consists of computations that we would do manually, with the probable exception of control structures like `for` loops or `if`/`else` sequences. For this reason, the rest of this section focuses on how to effectively specify and use the arguments passed to the function, and Sec. 7.2.3 discusses these control structures. Also, because list structures can be used to return complicated, multiple-component results from a function, these are discussed in Sec. 7.2.2.

What data values does a function use?

The primary focus of this discussion of R functions is on the arguments with which a function is called. This is the most obvious and most important source of the data values a function uses in computing the result returns, or in the case of a plot function, in forming the graphical display it plots. There are, however, two other sources of data available in the body of a function. The first is any collection of variables defined inside the function body in the form of assignments that create new, intermediate variables required in the computations. Note that since these computations are *local*, within the body of the function, these intermediate variables vanish when the function returns its result.

The other source of data objects available in the body of an R function is *free variables*, which are variables that are neither passed as arguments in the function call nor created within the body of the function. *One of the key features of R is that free variables are obtained from the* enclosing environment *for the function*. In the simplest case, this enclosing environment is the R session from which the function was called. This feature can be useful when R finds "the right value" for these free variables, but in cases where the function finds something else, this mechanism can lead to some extremely subtle program bugs. For this reason, it is a good idea to check the code in your function body carefully to make sure all variables either appear as arguments in the function declaration or are created locally within the body of the function.

Required vs. optional arguments

As the following discussions illustrate, function arguments can be classified in several different ways. Probably the most important distinction is between *required arguments*, which must be present in a function call, and *optional arguments*, which can be omitted. If we omit a required argument in a function call, R responds with an error, something like the following:

```
rnorm()

## Error in rnorm():  argument "n" is missing, with no default
```

The `rnorm` function returns a sequence of n simulated Gaussian random numbers, but to do this, we must specify how many numbers we want. That is, n is a required parameter, and failing to specify it causes execution to halt and return the error message shown above.

Optional arguments can be omitted from a function call because they are included in the function definition with *default values*, which are used if the function is called without specifying these arguments. Continuing the above example, the built-in function `rnorm` has two optional arguments, **mean** with a default value of 0, and **sd**, with a default value of 1. One way of seeing what these optional arguments and their defaults are is with the **head** function:

```
head(rnorm)
```

```
##
## 1 function (n, mean = 0, sd = 1)
## 2 .Call(C_rnorm, n, mean, sd)
```

Like `plot` and `summary`, the function `head` is generic, returning results that depend on the class of object it is called with. When called with a function name, as in this example, the first line it returns includes the `function` keyword, followed by the argument list in parentheses, separated by commas. Required arguments are listed without default values (like `n` in this example), and optional arguments are listed with the argument name, followed by an equal sign, followed by the default value. Thus, we see that the default value for the optional argument `mean` is 0 and for `sd` is 1, as noted above.

Named vs. positional arguments

When we define a function in *R*, all arguments—both required and optional—must be listed in the parentheses following the keyword `function` in the first line of the function definition. This means that all of these arguments have a natural order, set when the function is defined, and *R* allows us to specify their values *by this position* if we want to. Returning to the `rnorm` function example, note that the argument `n` appears first, followed by the argument `mean`, with the `sd` argument occurring last. Thus, we obtain the same three random numbers—assuming we have set the seed to initialize the random number generator—whether we specify these arguments explicitly by name or implicitly by position:

```
set.seed(9)
rnorm(n = 3, mean = 1, sd = 0.1)
```

```
## [1] 0.9233204 0.9183542 0.9858465
```

```
set.seed(9)
rnorm(3, 1, 0.1)
```

```
## [1] 0.9233204 0.9183542 0.9858465
```

In practice, positional arguments are often used because they are convenient, but it is generally better practice to explicitly specify all arguments by name, with the possible exception of functions with only one required argument. The reason for this recommendation—especially for custom functions—is that if arguments are added or removed as a function is modified, code using function calls by position that worked before may suddenly stop working.

Approximate name matching

If we adopt the good practice of giving meaningful names to function arguments, these names can become inconveniently long. One of the convenient features that the *R* function interface provides is *partial name matching*, which helps overcome this difficulty. Specifically, whenever we call a function and specify arguments by name, *R* will accept *unambiguous partial matches* for these names. For the `rnorm` function, note that the names `m`, `me`, `mea`, or `mean` all unambiguously match the `mean` argument. Thus, we can specify the mean and standard deviation arguments to this function with the shortened names `m` and `s`:

```
set.seed(9)
rnorm(n = 3, mean = 1, sd = 0.1)

## [1] 0.9233204 0.9183542 0.9858465

set.seed(9)
rnorm(n = 3, m = 1, s = 0.1)

## [1] 0.9233204 0.9183542 0.9858465
```

As with positional arguments, the use of approximate name matching can be extremely convenient, but it should not encourage the habit of using cryptic abbreviations for arguments because this makes the code harder to interpret.

The "..." argument

A very special optional argument that appears in some functions is "...", which allows us to pass arbitrary named arguments to other functions that are called inside our function's body. This ability is particularly useful in custom plotting functions where it allows us to pass any of the very long list of plot parameters and local arguments to a base graphics function like `plot` or `barplot` without having to specify their names in advance. This technique is illustrated in the function `PredictedVsObservedPlot` listed and described in Sec. 7.4.3.

The NULL trick

The "NULL trick" is an extremely useful idea in cases where we have a relatively complicated default behavior that we want to allow users to override. The basic idea is to include an optional argument in our function definition and give it the default value `NULL`. In the body of the function, we then test to see whether this parameter is `NULL`, and if it is, we execute the code necessary to implement our complicated default behavior; otherwise, we allow the user to specify a non-default value for this parameter.

This idea is also illustrated in the function `PredictedVsObservedPlot` listed and described in Sec. 7.4.3, where it is used to specify default values for the common x- and y-axis limits in a plot of model predictions versus observed response values. There, the optional argument specified this way is `xyLimits`, which is a two-element numerical vector giving common minimum and maximum axis limits for both the x- and y-axes. The default value for this argument is `NULL` and the body of the function tests for this default with the `is.null` function, which returns `TRUE` if the argument is `NULL` and `FALSE` otherwise. If this argument does have the `NULL` value, the common minimum axis limit is computed in the body of the function as the minimum of the observed and predicted response vectors being plotted, and the common maximum axis limit is computed as the maximum of these values.

Argument validation with `stopifnot`

Although it represents a specific control structure to be used in the body of a function, the built-in R function `stopifnot` is extremely useful in validating function arguments. This is something that is typically not done when we are first developing functions, but it is a good idea, especially if our functions are to be shared with others. The `stopifnot` function is called with a collection of logical statements and it causes function execution to abort, returning an error message if any of these logical statements do not evaluate to `TRUE`. Otherwise, the function returns `NULL` and execution continues. The following examples show how this function works:

```
n <- 3
sd <- 1
stopifnot(n > 0, sd > 0)
sd <- -1
stopifnot(n > 0, sd > 0)

## Error:  sd > 0 is not TRUE
```

Advice commonly heard in programming circles is "fail hard and fail early," meaning that it is better to catch an error and abort (i.e., fail hard) early in a computational sequence, before we have either wasted a lot of time before ultimately failing or, worse yet, obtained and returned an incorrect result without offering any clue that anything is wrong.

7.2.2   The `list` data type

The essential data structure that made it possible in Sec. 7.1.1 to move from brute force, manual computation to a simple loop to compute Fibonacci numbers was the vector. In R, vectors can be numeric, character, or logical, and they may be of essentially arbitrary length. Their one key feature is that every element must be of the same type, a restriction that makes many vector-based computations possible. For example, adding numeric vectors is possible because addition is well defined between their elements. There are circumstances, however, where we need more flexibility.

The *list* data type in R is a vector-like structure whose elements do not have to be the same type and can contain *any* valid R object, making them extremely flexible. To see this flexibility, consider the following example:

```
xList <- vector("list", 4)
xList[[1]] <- 3
set.seed(xList[[i]])
xList[[2]] <- rnorm
xList[[3]] <- xList[[2]](n = 10)
```

```
xList[[4]] <- mean(xList[[3]])
xList

## [[1]]
## [1] 3
##
## [[2]]
## function (n, mean = 0, sd = 1)
## .Call(C_rnorm, n, mean, sd)
## <bytecode: 0x00000000287526d8>
## <environment: namespace:stats>
##
## [[3]]
##  [1] -0.96193342 -0.29252572  0.25878822 -1.15213189  0.19578283
##  [6]  0.03012394  0.08541773  1.11661021 -1.21885742  1.26736872
##
## [[4]]
## [1] -0.06713568
```

Much is happening here. The first line creates a four-element list (i.e., a vector of type "list" with length 4), and the second line assigns the first element of this list the integer value 3. Note that list elements are accessed with two square brackets ("[[" and "]]") instead of one, as in the case of vectors. The next line accesses this list element and uses its value as the argument for the set.seed function to initialize the random number generator, and the following line assigns the built-in Gaussian random number generator function rnorm to the second list element. The next line retrieves this function and applies it, with the argument n = 10 to obtain the third list element, and the following line computes the mean of this sequence of random numbers and assigns it to the fourth list element. The last line displays the four elements of the list, showing that, indeed, the different list elements can contain essentially anything.

This extreme flexibility makes lists a popular data type for function return values in R. Specifically, the return statement at the end of a function body allows us to return a single data object. If we make this data object a list, we can return as many things—and as many different types of things—as we want. Further, we can assign names to these list elements and use these names with the "gets" operator ("$") to access these list elements:

```
names(xList) <- c("RandomSeed", "GeneratorFunction", "RandomSequence",
                  "SequenceMean")
xList$RandomSeed

## [1] 3

xList$GeneratorFunction

## function (n, mean = 0, sd = 1)
## .Call(C_rnorm, n, mean, sd)
## <bytecode: 0x00000000287526d8>
## <environment: namespace:stats>

xList$RandomSequence
```

```
##  [1] -0.96193342 -0.29252572  0.25878822 -1.15213189  0.19578283
##  [6]  0.03012394  0.08541773  1.11661021 -1.21885742  1.26736872

xList$SequenceMean

## [1] -0.06713568
```

Unfortunately, this flexibility comes at a price, actually two prices. First, there are many computations we can do with vectors that we cannot do with lists, even if they contain the same data values. For example:

```
vectorSequence <- xList$RandomSequence
sd(vectorSequence)

## [1] 0.8657293

listSequence <- as.list(vectorSequence)
sd(listSequence)

## Error:  is.atomic(x) is not TRUE
```

Here, we have extracted the `RandomSequence` element from the named list `xList`, which is a vector, so we can apply the standard deviation function `sd` and obtain a meaningful result. Next, we use the `as.list` function to convert this vector to a list containing exactly the same values and attempt to repeat the computation, which fails.

Finally, the second price paid for the added flexibility of the list structure is that when we can do computations on lists, these are slower than the corresponding computations on vectors. Consequently, if our function can return results using a simpler data structure like a vector or a data frame, this is probably preferable since we can do more computations more efficiently with these structures than we can with lists. Conversely, if we need to return a wide variety of different kinds of results, a list may be the only reasonable option. In particular, many modeling functions return lists, allowing them to return individual components of many different types.

7.2.3   Control structures

Like all programming languages, R supports several control structures, but some are used more frequently than others. The following discussions focus on what are probably the two most common control structures—`for` loops and `if-else` structures—and then introduces the vector `ifelse` function that can be extremely useful and is probably not as well known as it should be.

Loop structures

While R generally discourages the use of loops (see Sec. 7.2.4 for a discussion of this point), the language does support several types of loops. The only one of these discussed here is the `for` loop that we saw in the Fibonacci sequence programs discussed in Sec. 7.1.1. The basic structure of this loop is the following:

```
for (Element in Set){
  Computations for each Element value
}
```

Probably the most common `for` loop implementation is like the ones seen in Sec. 7.1.1, where `Set` is a sequence of integers (e.g., the sequence `3:n` of integers from 3 through `n`), and `Element` is a specific integer in this sequence, often denoted by a single letter like `i` or `j`. As in the body of a function, the indented code in the `for` loop can consist of any valid sequence of executable *R* statements. The basic purpose of the `for` loop is to perform a set of computations for each element of `Set`, and these computations typically depend on this element value and are often stored in the corresponding element of a vector or list.

Because it is a simple, representative example, the first Fibonacci sequence program discussed in Sec. 7.1.1 is repeated here to show the detailed structure of a typical `for` loop:

```
n <- 30
Fib <- vector("numeric", length = n)
Fib[1] <- 1
Fib[2] <- 1
for (i in 3:n){
  Fib[i] <- Fib[i-1] + Fib[i-2]
}
```

If-else sequences

Besides loops, the other control structures that arise most frequently in *R* programs are those based on single `if` statements, `if/else` pairs, or chains of `if/else` pairs. The general structure of the single `if` statement is:

```
if (Condition){
  Computations to perform when Condition holds
}
```

Here, `Condition` is any logical statement that evaluates to `TRUE` or `FALSE`: if `TRUE`, the indented block of code is executed; otherwise, this code is skipped. This construction is used twice in the function `PredictedVsObservedPlot` described in Sec. 7.4.3, once to implement the "NULL trick" described in Sec. 7.2.1, and once to specify whether to add a dashed line to the plot.

The `if/else` pair is a more flexible control structure that allows one set of code to be executed if `Condition` is `TRUE` and another set to be executed if it is `FALSE`. The format of this structure is:

```
if (Condition){
  First set of computations, when Condition holds
} else {
  Second set of computations, when Condition does not hold
}
```

In fact, this structure can be extended to make it more flexible by nesting multiple if/else pairs, along the following lines:

```
if (Condition1){
  The Condition1 computations
} else {
  if (Condition2){
    The Condition2 computations
  } else {
    if (Condition3){
      The Condition 3 computations
    } else {
      The "None of the Above" computations
    }
  }
}
```

The ifelse function

The ifelse function is called with a logical vector test and two vectors of possible return values, yes and no, each of the same length as test. For every element of test that is TRUE, ifelse returns the corresponding element of yes, and for every element of test that is FALSE, it returns the corresponding element of no. This function is useful for tasks like replacing missing values with zero:

```
set.seed(3)
x <- rnorm(10)
x
```

```
## [1] -0.96193342 -0.29252572  0.25878822 -1.15213189  0.19578283
## [6]  0.03012394  0.08541773  1.11661021 -1.21885742  1.26736872
```

```
y <- sqrt(x)
```

```
## Warning in sqrt(x):  NaNs produced
```

```
y
```

```
## [1]       NaN       NaN 0.5087123       NaN 0.4424735 0.1735625 0.2922631
## [8] 1.0566978       NaN 1.1257747
```

```
z <- ifelse(is.na(y), 0, y)
z
```

```
## [1] 0.0000000 0.0000000 0.5087123 0.0000000 0.4424735 0.1735625 0.2922631
## [8] 1.0566978 0.0000000 1.1257747
```

Note that since the square root function only returns numerical results for positive values of its argument, the vector y generated here contains missing values (i.e., "NaN" values, meaning "Not a Number"). The ifelse function is then used to replace these missing values ("NaN" values are detected by the is.na function, along with "NA" values, which mean "Not Available"). That is, if the test passes—i.e., if y is missing—the corresponding values are set to zero; otherwise, they are left alone.

Missing values and if-else logic

In his book on databases, Date discusses the problems that arise when special values are used to represent missing data, leading to three-valued logic [17, Ch. 18]. Specifically, in the if and ifelse functions just described, it has been assumed that a logical expression can return only two values, either TRUE or FALSE. In three-valued logic, the possibility that one or more of the variables involved in the logical expression may have a missing or unknown value means that the result of the logical expression may be missing or unknown. If we adopt R's default missing value notation NA, simple logical expressions like x > 0 can return the three possible values TRUE, FALSE, or NA, as in the following example:

```
x <- c(-1, 0, 1, 2, 3, NA, 4, 5, -6)
x > 0
```

```
## [1] FALSE FALSE  TRUE  TRUE  TRUE    NA  TRUE  TRUE FALSE
```

As Date points out in his discussion, the use of three-valued logic can lead to a number of practical complications. The following examples illustrate two of these complications in R's use of three-valued logic. First, consider the consequences of using the logical expression x > 0 from the above example in the following ifelse function call:

```
y <- ifelse(x >= 0, paste(x, "non-negative"), paste(x, "negative"))
y
```

```
## [1] "-1 negative"    "0 non-negative" "1 non-negative" "2 non-negative"
## [5] "3 non-negative" NA               "4 non-negative" "5 non-negative"
## [9] "-6 negative"
```

In the description of the ifelse function given above, it was assumed that the logical expression—here x > 0—would return either TRUE, causing the first of the two following expressions to be returned as the result, or FALSE, causing the second expression to be returned. In this example, when the expression x > 0 returns the missing value NA, this value is also returned by the ifelse function. While it is difficult to argue in favor of any other result as being more reasonable here, the result is unexpected if we have not considered the possibility that x may have missing values, and in some cases this can lead to program errors.

Alternatively, in the if-else control structure discussed earlier, missing values in the logical expression in the if statement will cause program execution to terminate with an error. As a specific example, if there were no missing x values, the following if-else sequence would yield a similar result to the ifelse example just described, but missing x values cause the loop to halt when the if statement encounters a missing value:

```
for (xEl in x){
  if (xEl >= 0){
    print(paste(xEl, "non-negative"))
  } else {
    print(paste(xEl, "negative"))
  }
}
```

```
## [1] "-1 negative"
## [1] "0 non-negative"
## [1] "1 non-negative"
## [1] "2 non-negative"
## [1] "3 non-negative"
```

```
## Error in if (xEl >= 0) {:  missing value where TRUE/FALSE needed
```

The main points of this discussion are, first, that missing values do have a significant impact on the logical expressions that drive if-else logic, and, second, that different R constructs based on this logic behave differently.

NULL is *not* the same as NA

An extremely important but subtle point is that R's two special values NA and NULL are *not* the same. Specifically, the special value NA is R's standard indicator for missing data and it influences if-else logic in the way just described. In contrast, the special value NULL that forms the basis for the "NULL trick" described in Sec. 7.2.1 behaves very differently, as this example illustrates:

```
y <- c(-1, 0, 1, 2, 3, NULL, 4, 5, -6)
y
```

```
## [1] -1  0  1  2  3  4  5 -6
```

Unlike the previous example, where we can assign the value NA to the sixth element of the nine-element vector x, using NULL in the same way causes the sixth element to be *omitted*, giving us an eight-element vector. The following examples further illustrate the difference between the NA and NULL values in R:

```
a <- NA
length(a)
```

```
## [1] 1
```

```
is.na(a)
```

```
## [1] TRUE
```

```
is.null(a)
```

```
## [1] FALSE
```

```
b <- NULL
is.null(b)
```

```
## [1] TRUE

length(b)

## [1] 0

is.na(b)

## Warning in is.na(b): is.na() applied to non-(list or vector) of type 'NULL'

## logical(0)

if (b > 0){
  print("b is positive")
}

## Error in if (b > 0) {: argument is of length zero

c <- seq(0, 9, 1)
c

##  [1] 0 1 2 3 4 5 6 7 8 9

c[5] <- NA
c

##  [1]  0  1  2  3 NA  5  6  7  8  9

c[6] <- NULL

## Error in c[6] <- NULL: replacement has length zero

c

##  [1]  0  1  2  3 NA  5  6  7  8  9
```

The first example, a, shows that NA is a value that can be assigned to a variable, giving an object of length 1 that can be tested for with the is.na function, and for which the is.null function returns the value FALSE, emphasizing that NA and NULL are different. The second example, b, shows that NULL can be assigned to a variable and this value is detected by the is.null function; further, this value has zero length (like the character string blank, '') and it causes an error in logical expressions like is.na(b) or b > 0. The third example, c, shows that we can assign the value NA to any element of an existing vector, but that attempting to assign the value NULL fails, with no change to the original vector.

Finally, note that the value NULL arises frequently and is particularly useful in dealing with named list elements, as the following example illustrates:

```
zList <- list(a = "a", b = "B", c = NA, d = NULL, e = "eee")
zList

## $a
## [1] "a"
##
```

```
## $b
## [1] "B"
##
## $c
## [1] NA
##
## $d
## NULL
##
## $e
## [1] "eee"
```

Note that here we have constructed a list whose d element has the value NULL, like the variable b in the previous example. If we attempt to re-assign the element e to have this value, however, it is removed from the list:

```
zList$e <- NULL
zList
```

```
## $a
## [1] "a"
##
## $b
## [1] "B"
##
## $c
## [1] NA
##
## $d
## NULL
```

This behavior is closely related to the fact that if we apply the "gets" operator $ to extract a non-existent list element, it returns NULL, as in this example:

```
zList$f
```

```
## NULL
```

The key point is that, while they may sound similar, the special values NA and NULL behave very differently in R and are useful in different circumstances.

7.2.4   Replacing loops with apply functions

Loops are an extremely important control structure, present in all computer languages, including R as we have just seen. In R, however, explicit loops are often slower than other program constructs that allow us to do the same thing. A specific example is simple numerical operations on vectors. That is, given vectors x and y, we can obtain the vector whose elements contain the sums of each element of x and y directly as x + y. In languages that do not support vector arithmetic operations, we would have to compute the vector sum by looping over the individual elements, e.g.:

```
#
#  Don't do this!
#
n <- length(x)
z <- vector("numeric", length = n)
for (i in 1:n){
  z[i] <- x[i] + y[i]
}
```

We should never do this in *R* because it is much slower than computing the vector sum directly.

In more complex cases, simple vector operations like sums or differences may not be available, but even in those cases, there are sometimes alternatives in *R* to using loops. In particular, the apply family of functions—apply, lapply, and sapply—can often be used to greatly speed computations that would seem to be natural candidates for slower loop structures. Conceptually, the simplest of these functions is probably lapply, which is called with a list and a function, and returns another list, each element of which is the result of applying the function to the corresponding element of the original list. For example, the following code applies the **head** function to each element of the four-element list xList discussed in Sec. 7.2.2:

```
lapply(xList, head)

## $RandomSeed
## [1] 3
##
## $GeneratorFunction
##
## 1 function (n, mean = 0, sd = 1)
## 2 .Call(C_rnorm, n, mean, sd)
##
## $RandomSequence
## [1] -0.96193342 -0.29252572  0.25878822 -1.15213189  0.19578283  0.03012394
##
## $SequenceMean
## [1] -0.06713568
```

Since applying **head** to single numbers simply returns the number, the first and last element of the list returned by lapply here are simply the corresponding elements of the original list, but the other two elements show the expected results when **head** is applied to a function (i.e., we see the argument list and default values) or a vector (i.e., we see the first six elements).

The sapply function behaves similarly, except that is called with a *vector* and a function instead of a list and a function, and it returns a *list*, each of whose elements is the result of the function applied to the corresponding element of the vector. This arrangement provides for the possibility that the results of each function call will have a different size or a different structure so the result would not represent a vector of the same length as the original. The following example illustrates this point:

```
set.seed(3)
sapply(c(1, 2, 4, 8), rnorm)

## [[1]]
## [1] -0.9619334
##
## [[2]]
## [1] -0.2925257  0.2587882
##
## [[3]]
## [1] -1.15213189  0.19578283  0.03012394  0.08541773
##
## [[4]]
## [1]  1.1166102 -1.2188574  1.2673687 -0.7447816 -1.1312186 -0.7163585
## [7]  0.2526524  0.1520457
```

In particular, note that each element of the list returned by `sapply` is a vector
whose length is equal to the corresponding element of the vector with which it
was called.

Finally, the `apply` function is called with a matrix or data frame, a `MARGIN`
parameter, and a function to be applied to the rows (if `MARGIN = 1`) or columns
(if `MARGIN = 2`) of this input data object. Here, the output is a vector, as in
the following example:

```
apply(mtcars, MARGIN = 2, FUN = ComputeIQD)

##      mpg       cyl      disp        hp      drat        wt      qsec
##  7.37500   4.00000 205.17500  83.50000   0.84000   1.02875   2.00750
##       vs        am      gear      carb
##  1.00000   1.00000   1.00000   2.00000
```

In this case, by specifying `MARGIN = 2` and `FUN = ComputeIQD`, we have applied
the interquartile distance function `ComputeIQD` described at the end of Sec. 7.1.2
to each column of the `mtcars` data frame from the `dataset` package. The result
is a named vector of IQD values, with the names corresponding to the column
names in the `mtcars` data frame.

The key point of this discussion is that the `apply` family of functions is
both easy to use and much faster than the equivalent `for` loop implementation.
Replacement of loops with these functions is not always possible, but it often
is, and, in those cases, it should be considered.

7.2.5 Generic functions revisited

The concept of *generic functions* was introduced in Chapter 2 where it was illustrated with the `plot` function, whose behavior depends on the class of the object
with which it is called. In the brief discussion of object-oriented programming
in *R* given there, it was noted that *R* supports three different object-oriented
systems, based on *S3 objects*, *S4 objects*, and *reference classes*, and it was also
noted that the S3 object system is both the simplest and the one we encounter
most frequently. Associated with S3 objects are *methods* for generic functions

like `plot`, so that when this function is called with an object belonging to a specific S3 class, the function performs the actions appropriate to that class. The following discussion introduces the process of creating your own S3 classes and their associated methods, which consists of the following steps:

1. define a new class name for your S3 objects;

2. use the `class` function to assign objects to this class (but carefully: see the following discussion);

3. create a new method for any existing generic function for this class, or create a new generic function with a method for this class.

This process can be extremely useful if you find yourself frequently creating similarly structured R objects (e.g., named lists with the same names) and need to do the same things over and over again to get the plot you want or compute and display the summary details you need. The rest of this discussion covers the following topics: first, defining new S3 classes and creating objects; then, adding new methods to an existing generic function; and, finally, creating a new generic function. For more detailed discussions of these topics, refer to Hadley Wickham's book, *Advanced R* [74, Sec. 7.2].

Defining and using new S3 classes

The `class` function in base R can be used both to inquire what class any R object has, and to assign it a new class. As Hadley Wickham notes in his discussion of object-oriented programming in R, assigning an object a new class can be dangerous, since if we make inappropriate assignments, we can cause useful generic functions to fail [74, p. 106]. As a specific illustration, he considers an object returned by the linear regression modeling function `lm` discussed in Chapter 5, which returns an object of class "lm," e.g.:

```
modelEx <- lm(Gas ~ Temp, data = whiteside)
class(modelEx)

## [1] "lm"
```

An extremely useful generic function in R is `summary`, which has many associated methods, including those for objects of class "lm" and "data.frame." In his discussion, Wickham shows that if we re-assign the class of `modelEx` from "lm" to "data.frame," the `summary` function won't fail outright (i.e., it won't return an error message), but it no longer returns a useful result.

To avoid this problem, Wickham advocates using *constructors*, functions that typically check the consistency of the original object with the new class before making the new class assignment. A useful variation on this theme is to create custom functions that make the new class assignment within the body of the function. That way, the object returned by the function belongs to the new S3 class you want to create, so it can be used by the methods you create for this object class. As a simple example, consider the following function:

```
AnnotatedGaussianSample <- function(n, mean = 0, sd = 1, iseed = 33){
  #
  set.seed(iseed)
  x <- rnorm(n, mean, sd)
  ags <- list(n = n, mean = mean, sd = sd, seed = iseed, x = x)
  class(ags) <- "AnnotatedGaussianSample"
  return(ags)
}
y <- AnnotatedGaussianSample(100)
class(y)

## [1] "AnnotatedGaussianSample"
```

The first eight lines here define the function `AnnotatedGaussianSample`, which is called with one required argument and three optional arguments, and it returns a five-element list giving the values of these arguments and a vector of Gaussian random variables generated from them. Before returning this object, the function assigns it the class "AnnotatedGaussianSample," as seen in the final line of code here: the previous line calls the function with `n = 100`, which returns the object `y` that has this class. We can obtain a more complete view of the structure of this object—including its class—with the `str` function:

```
str(y)

## List of 5
##  $ n   : num 100
##  $ mean: num 0
##  $ sd  : num 1
##  $ seed: num 33
##  $ x   : num [1:100] -0.1359 -0.0408 1.0105 -0.1583 -2.1566 ...
##  - attr(*, "class")= chr "AnnotatedGaussianSample"
```

The only reason for defining a new object class like this is that it allows us to define and use methods for existing generic functions, or to define new generic functions with methods that do useful things with objects of this new class. These topics are considered next.

Adding methods to a generic function

Adding a method to an existing generic function is extremely easy. Specifically, if `foo` is an existing generic function and you have defined an object class "new-Class," creating a new method for `foo` only requires writing a function named `foo.newClass` that does what you want `foo` to do when presented with an object of class "newClass." As a specific example, creating the following function adds a `summary` method for objects of class "AnnotatedGaussianSample" created by the function described above:

```
summary.AnnotatedGaussianSample <- function(x){
  #
  print(paste("Sample size:", x$n))
  print(paste("Mean:", x$mean))
```

```
print(paste("Standard deviation:", x$sd))
print(paste("Random seed:", x$seed))
print("Sample quantiles:")
quantile(x$x)
}
```

Given an object of class "AnnotatedGaussianSample," this method extracts the values of the arguments used to set up and call the rnorm function, displays them, and then computes and displays Tukey's five-number summary for the resulting random Gaussian sequence. To invoke this method, we simply apply the generic function to the object:

```
summary(y)

## [1] "Sample size: 100"
## [1] "Mean: 0"
## [1] "Standard deviation: 1"
## [1] "Random seed: 33"
## [1] "Sample quantiles:"
##         0%          25%          50%          75%         100%
## -2.17699470 -0.58222699  0.01886743  0.70752861  2.73116619
```

For comparison, suppose we created exactly the same Gaussian random number sequence, but directly, without calling the function AnnotatedGaussianSample. This would give different results:

```
set.seed(33)
x <- rnorm(n = 100, mean = 0, sd = 1)
class(x)

## [1] "numeric"

summary(x)

##      Min.   1st Qu.   Median      Mean   3rd Qu.      Max.
## -2.17699  -0.58223  0.01887   0.05949   0.70753   2.73117
```

More generally, we can create methods for any existing generic function to make it work with our new class using exactly the same approach illustrated here for the generic summary function. To see what methods have been defined for a given class, use the methods function:

```
methods(class = "AnnotatedGaussianSample")

## [1] summary
## see '?methods' for accessing help and source code
```

Creating new generic functions

It is also extremely easy to create new generic functions. For example, to create a new generic function named foo, all we need is the following code:

```
foo <- function(x){UseMethod("foo")}
```

As it stands, this new generic function is useless because it doesn't have any associated methods. These are created in exactly the same way as before. For example, we can define a method for S3 objects of class "AnnotatedGaussianSample"
by creating a function whose name is that of the generic function (i.e., `foo`),
followed by a period and the name of our S3 class:

```
foo.AnnotatedGaussianSample <- function(x){
  #
  print("This is the foo method for AnnotatedGaussianSample S3 objects")
  summary(x)
}
```

To use this function for the S3 object y of this class that we created above, we
simply call the parent function `foo`:

```
foo(y)
```

```
## [1] "This is the foo method for AnnotatedGaussianSample S3 objects"
## [1] "Sample size: 100"
## [1] "Mean: 0"
## [1] "Standard deviation: 1"
## [1] "Random seed: 33"
## [1] "Sample quantiles:"
##          0%          25%          50%          75%         100%
## -2.17699470 -0.58222699  0.01886743  0.70752861  2.73116619
```

There is, however, an extremely important additional detail when defining our
own generic functions: these functions need a *default method* that is called for
objects that do not have special methods pre-defined for them. To see this point,
consider what happens if we call our `foo` function with the numeric object x
defined in the previous example:

```
foo(x)
```

```
## Error in UseMethod("foo"):  no applicable method for 'foo' applied to an object
of class "c('double', 'numeric')"
```

To overcome this problem, define the method `foo.default`, e.g.:

```
foo.default <- function(x){
  #
  print("This is the default foo method")
}
```

Now, if we call `foo` with the numeric object x, we no longer get an error message:

```
foo(x)
```

```
## [1] "This is the default foo method"
```

7.3 Good programming practices

The next four sections offer some advice on good programming practices, intended to make program development easier and less painful, and to make the end results more useful both to yourself (especially some months after you have written them and forgotten exactly what they do and how) and to others.

7.3.1 Modularity and the DRY principle

In his book, *Introduction to Data Technologies*, Paul Murrell makes repeated reference to the *DRY principle*, an acronym for "Don't Repeat Yourself" [56, p. 35]. He credits this principle to a programming book by Hunt and Thomas [43] and notes that it means there should only be *one copy* of any important piece of information in a program. We have already encountered this idea in the discussion of why hard-coded parameters are a bad idea in Sec. 7.1.1. In practical terms, this means that key program parameters should be defined *once* by assigning them to a variable, which is then used in the program every time that parameter is required. As noted in Sec. 7.1.1, this practice makes it easier to change programs as they evolve and makes it less likely to "partially change them," causing program bugs that we have to spend time finding and fixing.

Another manifestation of the DRY principle is the notion of *modularity:* very often, we find ourselves doing the same thing, or almost the same thing, repeatedly. In particular, we often find that new programs require us to implement code that looks disturbingly familiar. As with avoiding hard-coded parameters, we can save ourselves a lot of work and improve reliability by identifying these repetitive subtasks and splitting them out as their own separate functions. If we make these functions flexible enough through the careful specification of calling arguments, we can re-use them in many different applications, greatly reducing the amount of code we have to write, test, and debug.

7.3.2 Comments

It was noted in the discussion of the `ComputeIQD` program example presented at the end of Sec. 7.1.2 that the symbol "#" indicates a comment in program code. Specifically, anything following this symbol on a line is ignored by the *R* interpreter, which gives us a mechanism for including explanatory comments in their code. Probably no one comments their program code as well as they should, but to the extent that *informative* comments can be included in programs and functions, this practice can be extremely helpful, both in making it clear to others what we have done and in reminding ourselves what we have done some months after we have done it.

Perhaps the best commented one of the examples included in Sec. 7.4 is the function `PredictedVsObservedPlot` discussed in Sec. 7.4.3. This example includes a header block that explains briefly what the program does and what the parameters are, and this is followed by more specific comments explaining how the function interprets the default `NULL` value for the `xyLimits` argument.

General guidelines are that the comments should *briefly* explain key details like what the program does and what the arguments mean, and also clarify any subtle, non-obvious details in the code. In general, program comments may be viewed as short and highly specialized forms of program documentation, a much larger topic discussed briefly in Chapter 11.

7.3.3   Style guidelines

Google is one of a number of large companies that uses R extensively, enough so that they have developed a programming style guide for R and made it publicly available from the following URL:

`https://google.github.io/styleguide/Rguide.xml`

These guidelines are reasonably short but they offer advice on a variety of different programming practices, including naming program files (they should end with the ".R" extension), indentation (two spaces, no tabs), and assignment statements (use "`<-`" and not "`=`"), with simple code examples that meet and violate their guidelines. The note concludes with a "parting words" section that offers this advice:

> The point of having style guidelines is to have a common vocabulary of coding so people can concentrate on what you are saying, rather than on how you are saying it. We present global style rules here so people know the vocabulary. But local style is also important. If code you add to a file looks drastically different from the existing code around it, the discontinuity will throw readers out of their rhythm when they go to read it. Try to avoid this.

7.3.4   Testing and debugging

It is easy enough to make mistakes in developing programs that the results should *always* be tested, at the very least by applying them to a few test cases where you know what to expect, verifying that this is what you get. For simple graphics functions like `PredictedVsObservedPlot` described in Sec. 7.4.3, testing may be as simple as using it to generate a few plots to make sure they look right. In the more typical case of a function that returns a data object, specific test cases should be developed where the result is known and direct comparisons can be made. The `testthat` R package has been developed to facilitate this type of testing, including functions like `expect_equal` that verify the result a function returns is what we expect. This package can be used to develop test suites that run multiple test cases, something that becomes increasingly important as our functions become more complex and flexible.

If our function does *not* give the expected results, we are faced with the task of debugging it: looking into the code and trying to find out where we have made mistakes. As noted in Sec. 7.1.2, if we are using the `source` function to load our programs from files, it checks for syntax errors and provides some information about what these errors are. As a result, syntax errors are generally the easiest to correct, and this must be done before we can use our program, anyway. The more difficult—and more frustrating—errors are program logic errors, and these are typically the ones that we find on testing. What we usually discover, sometimes after considerable effort, is that the program is doing exactly what we asked it to, but this wasn't really what we wanted it to do.

This is one area where modularity can be extremely helpful. If we build our programs up from small, specialized modules, each designed to do one specific thing, thoroughly testing each module as we go, our final program is more likely to work correctly. Also, in cases where we discover that it isn't working correctly, we should be able to determine whether one of these modules is failing or our top-level program is combining the results from these modules incorrectly.

7.4 Five programming examples

Since many people learn best by example, five simple R program examples are included here to illustrate key details in developing useful custom functions in R. The first two examples were described in Chapter 5 in connection with building and evaluating linear regression models. The third example generates the predicted vs. observed plots discussed but not listed there. The last two examples support exploratory data analysis and were discussed briefly in Chapter 3.

7.4.1 The function `ValidationRsquared`

The function `ValidationRsquared` was described in detail in Chapter 5 where the validation R-squared measure was proposed as a useful measure of prediction quality for predictive models of arbitrary type. This function is relatively simple and relatively standard in that it is called with two arguments, both required, and it returns a numerical result. The code listing follows:

```
ValidationRsquared

## function(validObs, validHat){
##   #
##   resids <- validHat - validObs
##   yBar <- mean(validObs)
##   offset <- validObs - yBar
##   num <- sum(resids^2)
##   denom <- sum(offset^2)
##   Rsq <- 1 - num/denom
##   return(Rsq)
## }
## <bytecode: 0x00000000053158f8>
```

The representation shown here—and in all subsequent examples—is what we get if we type the function name in an interactive R session: unlike the function definition, which includes the function name and the assignment arrow, here the first line only shows the **function** keyword and the arguments. Since neither of these arguments have default values specified, both are required. The main body of the function consists of the six lines of R code that compute the return value, Rsq, from the two required arguments. The final line of code returns this numerical value.

7.4.2   The function TVHsplit

The function TVHsplit was also described in detail in Chapter 5 as part of the discussion of data partitioning to avoid overfitting. This function is called with four arguments, one required and the other three optional with default values specified in the function definition. Specifically, these arguments are:

1. the required argument df, a data frame with N rows;

2. the optional argument split, a three-component numerical vector;

3. the optional argument labels, a three-component character vector;

4. the optional integer argument iseed.

This function initializes R's random sampling system with the iseed parameter and then calls the sample function to sample with replacement from the values defined by labels with the probabilities defined by split. The number of samples drawn corresponds to the number of rows of the data frame df, and the result returned is a character vector that assigns each row of the data frame to one of the subsets specified by the labels variable. The code listing follows:

```
TVHsplit

## function(df, split = c(0.5, 0.25, 0.25),
##                        labels = c("T", "V", "H"), iseed = 397){
##     #
##     set.seed(iseed)
##     flags <- sample(labels, size = nrow(df), prob = split,
##                     replace = TRUE)
##     return(flags)
## }
## <bytecode: 0x0000000006dc5e00>
```

7.4.3   The function PredictedVsObservedPlot

The function PredictedVsObservedPlot constructs the predicted vs. observed response plot advocated in Chapter 5 for graphically assessing the performance of predictive models. This function uses all of the argument types described in Sec. 7.2.1, including both optional and required arguments, an optional argument specified using the "NULL trick," and optional named arguments passed to R's basic plot function via the "..." argument. More specifically, this function accepts the following arguments:

1. Observed is a required numeric vector of observed values to be plotted;

2. `Predicted` is a required numeric vector of model predictions of `Observed`;

3. Optional argument `xyLimits` is either `NULL` or a two-component vector specifying common values for the `xlim` and `ylim` arguments of the `plot` function. The default value is `NULL`;

4. `refLine` is an optional logical argument specifying whether a dashed reference line is to be added to the plot. The default value is `TRUE`;

5. Optional named parameters may be passed to the `plot` function via the "..." argument.

If `xyLimits` has its default value `NULL`, the common x- and y-axis limits are determined by the minimum and maximum values of the required arguments `Observed` and `Predicted`. If optional argument `refLine` is `FALSE`, the dashed equality reference line is not added to the plot. This function is called for its side effects and does not return a data value. A program listing follows:

```
PredictedVsObservedPlot

## function(Observed, Predicted,
##                              xyLimits = NULL,
##                              refLine = TRUE, ...){
##     #
##     ############################################################
##     #
##     #  Plot predicted vs. observed responses with common
##     #  axis limits, passing additional arguments to plot().
##     #  If refLine = TRUE, a dashed equality reference line
##     #  is added to the plot.
##     #
##     ############################################################
##     #
##     #  If xyLimits is NULL, compute from extremes of
##     #  Observed and Predicted
##     #
##     if (is.null(xyLimits)){
##        xyLimits <- c(min(Observed, Predicted),
##                      max(Observed, Predicted))
##     }
##     plot(Observed, Predicted, xlim = xyLimits,
##          ylim = xyLimits, xlab = "Observed response",
##          ylab = "Predicted response", ...)
##     if (refLine){
##        abline(a = 0, b = 1, lty = 2, lwd = 2)
##     }
## }
```

7.4.4 The function `BasicSummary`

The function `BasicSummary` was described in Chapter 3 to generate a preliminary data summary for a data frame specified by the required argument `df`. The only other argument is `dgts`, an integer specifying the precision of the fractional results returned, with a default value of 3. This function returns a data frame with one row for each column of df and the following columns:

1. variable, the name of the corresponding column of df;

2. type, the class of the variable, as defined by the class function;

3. levels, the number of distinct values exhibited by the variable;

4. topLevel, the most frequently occurring value;

5. topCount, the number of times the most frequent value occurs;

6. topFrac, the fraction of records represented by topCount;

7. missFreq, the number of missing values exhibited by the variable;

8. missFrac, the fraction of records represented by missFreq.

The code listing follows:

```
BasicSummary

## function(df, dgts = 3){
##   #
##   ################################################################
##   #
##   #   Create a basic summary of variables in the data frame df,
##   #   a data frame with one row for each column of df giving the
##   #   variable name, type, number of unique levels, the most
##   #   frequent level, its frequency and corresponding fraction of
##   #   records, the number of missing values and its corresponding
##   #   fraction of records
##   #
##   ################################################################
##   #
##   m <- ncol(df)
##   varNames <- colnames(df)
##   varType <- vector("character",m)
##   topLevel <- vector("character",m)
##   topCount <- vector("numeric",m)
##   missCount <- vector("numeric",m)
##   levels <- vector("numeric", m)
##   for (i in 1:m){
##     x <- df[,i]
##     varType[i] <- class(x)
##     xtab <- table(x, useNA = "ifany")
##     levels[i] <- length(xtab)
##     nums <- as.numeric(xtab)
##     maxnum <- max(nums)
##     topCount[i] <- maxnum
##     maxIndex <- which.max(nums)
##     lvls <- names(xtab)
##     topLevel[i] <- lvls[maxIndex]
##     missIndex <- which((is.na(x)) | (x == "") | (x == " "))
##     missCount[i] <- length(missIndex)
```

```
##    }
##    n <- nrow(df)
##    topFrac <- round(topCount/n, digits = dgts)
##    missFrac <- round(missCount/n, digits = dgts)
##    #
##    summaryFrame <- data.frame(variable = varNames, type = varType,
##                               levels = levels, topLevel = topLevel,
##                               topCount = topCount, topFrac = topFrac,
##                               missFreq = missCount, missFrac = missFrac)
##    return(summaryFrame)
## }
## <bytecode: 0x000000003d38cee0>
```

Note that in counting the distinct levels, the standard missing value designation NA is counted as a level. Also, in counting missing values for missFreq, NA, blanks, and spaces are all counted as missing values.

7.4.5 The function FindOutliers

Although it is quite simple compared to many of the built-in functions in R, this last example is the most complex one considered here, calling four other, external functions to perform key computations. Specifically, the top-level function FindOutliers is called with the required argument x and three optional threshold arguments, passed to the functions ThreeSigma, Hampel, and BoxplotRule. These functions each return two-component lists with elements up and down, representing upper and lower outlier detection limits for the x values computed by a different method. FindOutliers then calls the function ExtractDetails with x and these three detection limit vectors to identify the individual points in x that are deemed outliers by each method, and the results are combined into the following four data frames:

1. sumFrame has one row for each method, giving the method name, the total number of data observations in x, the number of missing values in x, the number of outliers detected, the upper and lower outlier detection limits, and the minimum and maximum non-outlying values;

2. threeFrame gives the indices, values, and class (upper or lower outlier) of each outlier detected in x under the three-sigma edit rule;

3. HampelFrame gives the corresponding results for the Hampel identifier;

4. boxFrame gives the corresponding results for the boxplot outlier rule.

These four data frames are returned as the components of a four-element list; a listing of the program code follows:

```
FindOutliers

## function(x, t3 = 3, tH = 3, tb = 1.5){
##    #
##    threeLims <- ThreeSigma(x, t = t3)
```

```
##    HampLims <- Hampel(x, t = tH)
##    boxLims <- BoxplotRule(x, t = tb)
##    #
##    n <- length(x)
##    nMiss <- length(which(is.na(x)))
##    #
##    threeList <- ExtractDetails(x, threeLims$down, threeLims$up)
##    HampList <- ExtractDetails(x, HampLims$down, HampLims$up)
##    boxList <- ExtractDetails(x, boxLims$down, boxLims$up)
##    #
##    sumFrame <- data.frame(method = "ThreeSigma", n = n,
##                           nMiss = nMiss, nOut = threeList$nOut,
##                           lowLim = threeList$lowLim,
##                           upLim = threeList$upLim,
##                           minNom = threeList$minNom,
##                           maxNom = threeList$maxNom)
##    upFrame <- data.frame(method = "Hampel", n = n,
##                          nMiss = nMiss, nOut = HampList$nOut,
##                          lowLim = HampList$lowLim,
##                          upLim = HampList$upLim,
##                          minNom = HampList$minNom,
##                          maxNom = HampList$maxNom)
##    sumFrame <- rbind.data.frame(sumFrame, upFrame)
##    upFrame <- data.frame(method = "BoxplotRule", n = n,
##                          nMiss = nMiss, nOut = boxList$nOut,
##                          lowLim = boxList$lowLim,
##                          upLim = boxList$upLim,
##                          minNom = boxList$minNom,
##                          maxNom = boxList$maxNom)
##    sumFrame <- rbind.data.frame(sumFrame, upFrame)
##    #
##    threeFrame <- data.frame(index = threeList$index,
##                             values = threeList$values,
##                             type = threeList$outClass)
##    HampFrame <- data.frame(index = HampList$index,
##                            values = HampList$values,
##                            type = HampList$outClass)
##    boxFrame <- data.frame(index = boxList$index,
##                           values = boxList$values,
##                           type = boxList$outClass)
##    outList <- list(summary = sumFrame, threeSigma = threeFrame,
##                    Hampel = HampFrame, boxplotRule = boxFrame)
##    return(outList)
## }
## <bytecode: 0x0000000028e7c700>
```

The function **ThreeSigma** is called with the numerical vector **x** and the threshold parameter **t3** obtained from the calling arguments of the **FindOutliers** function, returning a two-element list with the upper and lower outlier detection limits. The code for this function is:

```
ThreeSigma

## function(x, t = 3){
##    #
##    mu <- mean(x, na.rm = TRUE)
```

```
##   sig <- sd(x, na.rm = TRUE)
##   if (sig == 0){
##     message("All non-missing x-values are identical")
##   }
##   up <- mu + t * sig
##   down <- mu - t * sig
##   out <- list(up = up, down = down)
##   return(out)
## }
## <bytecode: 0x0000000009275820>
```

Hampel is called with the numerical vector x and the threshold parameter tH from FindOutliers, returning the Hampel identifier outlier detection limits:

```
Hampel

## function(x, t = 3){
##   #
##   mu <- median(x, na.rm = TRUE)
##   sig <- mad(x, na.rm = TRUE)
##   if (sig == 0){
##     message("Hampel identifer implosion: MAD scale estimate is zero")
##   }
##   up <- mu + t * sig
##   down <- mu - t * sig
##   out <- list(up = up, down = down)
##   return(out)
## }
## <bytecode: 0x0000000005fc7d00>
```

BoxplotRule is called with the numerical vector x and the threshold parameter tb from FindOutliers, returning the boxplot outlier limits:

```
BoxplotRule

## function(x, t = 1.5){
##   #
##   xL <- quantile(x, na.rm = TRUE, probs = 0.25, names = FALSE)
##   xU <- quantile(x, na.rm = TRUE, probs = 0.75, names = FALSE)
##   Q <- xU - xL
##   if (Q == 0){
##     message("Boxplot rule implosion: interquartile distance is zero")
##   }
##   up <- xU + t * Q
##   down <- xU - t * Q
##   out <- list(up = up, down = down)
##   return(out)
## }
## <bytecode: 0x0000000010175150>
```

ExtractDetails is called with the numerical vector x and upper and lower outlier detection limits up and down, returning a list with these elements:

1. nOut, the number of outliers detected;

2. `lowLim`, the lower outlier detection limit;

3. `upLim`, the upper outlier detection limit;

4. `minNom`, the smallest non-outlying value;

5. `maxNom`, the largest non-outlying value;

6. `index`, an index vector pointing to all outliers;

7. `values`, the vector of values for all identified outliers;

8. `outClass`, the vector of outlier types (upper or lower).

A code listing for this function follows:

```
ExtractDetails

## function(x, down, up){
##    #
##    outClass <- rep("N", length(x))
##    indexLo <- which(x < down)
##    indexHi <- which(x > up)
##    outClass[indexLo] <- "L"
##    outClass[indexHi] <- "U"
##    index <- union(indexLo, indexHi)
##    values <- x[index]
##    outClass <- outClass[index]
##    nOut <- length(index)
##    maxNom <- max(x[which(x <= up)])
##    minNom <- min(x[which(x >= down)])
##    outList <- list(nOut = nOut, lowLim = down,
##                    upLim = up, minNom = minNom,
##                    maxNom = maxNom, index = index,
##                    values = values,
##                    outClass = outClass)
##    return(outList)
## }
## <bytecode: 0x000000002792e050>
```

7.5   R scripts

As noted at the beginning of this chapter, *R* programs can be run from the operating system command prompt rather than from within an interactive *R* session. These programs are sometimes called either *scripts* or *batch programs* and, while they are *almost* identical to program sequences that we would run interactively, there are a few subtle differences. A discussion of developing *R* scripts is beyond the scope of this book, but for a useful introduction to this topic, refer to the appendix on "Invoking R" in the document *An Introduction to R*, maintained by the *R* Core Team and included as part of the *R* help distribution.

7.6 Exercises

1: The function `ComputeIQD` developed in Section 7.1.2 computed the upper and lower quartiles required to compute the interquartile distance using *R*'s built-in `quantile`. By default, this function returns a named vector of quantiles, and the `ComputeIQD` function removes these names with the `as.numeric` function. An alternative is to use the optional logical argument `names` for the `quantile` function to suppress these names. Revise the `ComputeIQD` function listed in Section 7.1.2 to obtain a new function `ComputeIQD2` that does not use the `as.numeric` function but which gives the same results as `ComputeIQD`. Verify this by applying `ComputeIQD2` to the Gaussian random sample example presented in the text.

2: Chapter 3 introduced the notion of outliers and their influence on *location estimators* like the mean (highly sensitive to outliers) and the median (largely insensitive to outliers). Another, much less well-known location estimator is *Gastwirth's estimator*, defined as:

$$x_{Gast} = 0.3x_{(1/3)} + 0.4x_{(1/2)} + 0.3x_{(2/3)}, \qquad (7.2)$$

where $x_{(1/3)}$ represents the $1/3$ quantile of x, $x_{(1/2)}$ represents the $1/2$ quantile (i.e., the median), and $x_{(2/3)}$ represents the $2/3$ quantile. Using the `quantile` function to compute these quantiles, create a function named `Gastwirth` that is called with x and returns x_{Gast}. Apply this function to the `fibre` variable from the `UScereal` data frame in the `MASS` package: how does this location estimator compare with the mean? With the median?

3: The `Gastwirth` function developed in Exercise 2 does not handle missing data:

   3a. To see this point, apply the `Gastwirth` function to the `statusquo` variable from the `Chile` data frame in the `car` package;

   3b. To address this issue, create a new function, `Gastwirth2`, that accepts an optional logical argument `dropMissing` with default value `TRUE`. Apply this function to the `statusquo` variable.

4: A useful diagnostic plot for predictive models is that of prediction error versus the observed response. This exercise asks you to create a simple function to generate these plots, using the ... mechanism discussed in Section 7.2.1 to pass arguments to the generic plot function. Specifically, create a function called `ErrVsObserved` with two required arguments named `obs` and `pred`, that uses the `dots` mechanism to pass any named optional argument to the `plot` function. The `ErrVsObserved` function should do the following two things:

   1. Call `plot` to create a scatterplot of the prediction error (`pred - obs`) versus the observed response;

2. Add a dashed horizontal line at zero, with twice the standard width.

To verify that your function works, use it to generate a side-by-side array of prediction error versus observed Gas values from the whiteside data frame in the MASS package for these two models: (1) a basic model that predicts Gas from Temp alone, and (2) a model that predicts Gas from Temp and Insul with their interaction term included.

5: The NULL trick was described in Section 7.2.1 as a simple way of providing complicated default values for optional parameters, but it can also be useful in simpler cases. This exercise asks you to use this trick to provide a default title to the plot generated by the function ErrVsObserved from Exercise 4. Specifically:

5a. Create the function ErrVsObserved2 that is identical to the function in Exercise 4 except for the optional argument titleString. Use the NULL trick to make this default title "Error vs. observed".

5b. Use ErrVsObserved2 to re-generate the side-by-side plots, with the left-hand plot using the default title and the right-hand plot using the alternative title "Second model".

6: It was noted in the discussion of transformations given at the end of Chapter 5 that the square root transformation is only applicable to non-negative numerical values. Also, the ifelse function introduced in Section 7.2.3 was used to implement a square root variation that replaced the missing values for $\sqrt{x}$ when x was negative with the value zero. This exercise asks you to use the ifelse function to implement the following modified square root transformation that can be useful for arbitrary numerical data vectors. Specifically, use the ifelse function as the basis for a simple custom function named ModSqrt that computes the following transformation:

$$T(x) = \begin{cases} \sqrt{x} & x \geq 0 \\ -\sqrt{|x|} & x < 0. \end{cases} \qquad (7.3)$$

To show the nature of this transformation, create a random sample of 100 zero-mean, unit-variance Gaussian random variables (use the set.seed function to specify a random seed of 73) and generate a plot of $T(x)$ versus x for this sequence. To better see the character of this transformation, add a dotted reference line corresponding to no transformation (i.e., $y = x$).

7: The apply function introduced in Section 7.2.4 provides a very useful way of characterizing the columns in a data frame of numerical variables. This exercise asks you to build and demonstrate a simple function to characterize *numerical* data frames using the apply function. Specifically:

7a. Create a function called DescribeDataFrame that is called with the one required argument numericDataFrame and no optional arguments. This function uses the apply function to compute and return a data frame with one row for each column in `numericDataFrame` and four columns, giving the mean, median, standard deviation, and MAD scale estimator for each column of the original data frame. The row names of the output data frame should be the column names of the input data frame.

7b. Demonstrate your function by applying it to the `mtcars` data frame.

8: It was noted in Section 7.2.1 that the `stopifnot` function can be very useful in flagging errors. This exercise asks you to do the following:

8a. Modify the function `DescribeDataFrame` from Exercise 7, creating the new function `DescribeDataFrame2` that does two things. First, use the `apply` function to apply the `is.numeric` function to all columns of the input data frame, returning a logical vector named `VariablesAreNumeric`. Second, use the `stopifnot` function to terminate execution if all columns of the input data frame are not numeric;

8b. Apply this new function to the `mtcars` data frame: do you get the same results?

8c. Apply both the original function from Exercise 7 and the new function to the `Cars93` data frame from the `MASS` package: which one gives you a clearer indication of what is wrong?

9: The list data type in *R* was discussed in Section 7.2.2, where it was noted that its primary advantage is its flexibility: the elements of a list can be anything we want them to be, and they don't all have to be the same type. The primary disadvantage is that lists are somewhat harder to work with. For example, if `index` is a vector of integers that point to elements of the vector `xVector`, the following assignment returns the elements of `xVector` that are identified by the elements of `index`:

```
subVector <- xVector[index]
```

If we convert `xVector` to a list `xList` with the `as.list` function and then attempt to use `index` in the same way to extract the corresponding elements of `xList`, the operation fails. This exercise asks you to create a simple function called `ExtractSublist` with three requirements. First, it is called with two required arguments, `xList` and `index`, and no optional arguments. Second, it creates the logical vector `elementsInList` with one element for each element of `index` that is `TRUE` if that element points to an element of `xList` and `FALSE` otherwise; the function uses this vector with the `stopifnot` function used in Exercise 8 to halt execution if `index` points to non-existent elements of `xList`. Third, if `index` does point only to elements of `xList`, `ExtractSublist` extracts these elements and returns them as a new list. Test your function with these test cases:

```
set.seed(13)
xVector <- rnorm(20)
xList <- as.list(xVector)
indexA <- c(1, 2, 4, 8, 16)
indexB <- c(1, 4, 16, 24)
```

That is, call **ExtractSublist** with the arguments **xList** and **indexA** for the first case and the arguments **xList** and **indexB** for the second. Do these test cases give the correct results?

10: In Section 7.2.2, the following constructor for a class of annotated Gaussian random samples was presented:

```
AnnotatedGaussianSample <- function(n, mean = 0, sd = 1, iseed = 33){
    #
    set.seed(iseed)
    x <- rnorm(n, mean, sd)
    ags <- list(n = n, mean = mean, sd = sd, seed = iseed, x = x)
    class(ags) <- "AnnotatedGaussianSample"
    return(ags)
}
```

To demonstrate the utility of defining this class, a **summary** method was implemented for these objects. This exercise asks you to create the corresponding **plot** method for these objects that does the following:

1. Create a scatterplot of the random samples $x(k)$ versus their sample number k, with labels "Sample number, k" and "Sample value, x(k)" using the default **plot** method;

2. Use the ... mechanism to allow named optional arguments to be passed to the default **plot** method;

3. Add a solid horizontal line at the mean argument used to generate the random sample;

4. Add dashed horizontal lines at the mean plus or minus one standard deviation, again based on the arguments used to generate the random sample;

5. Add dotted horizontal lines at the mean plus or minus two standard deviations;

6. Add a two-line title: the top line should read "Annotated Gaussian sample:" and the second line should list "N = " followed by the size of the random sample, followed by a comma and "seed = " followed by the random seed used to generate the sample.

Demonstrate this plot method by generating an annotated Gaussian sample **y** of length 100 and the default arguments in the constructor function listed above. Use the ... mechanism to specify solid circles for the data points and y-axis limits from -3 to 3.

Chapter 8

Working with Text Data

Text data has always been an important *potential* source of useful information, provided we can extract key details from it. With the increasing availability of *digitized* text data from the Internet and other sources (e.g., internal company repositories), there is growing interest in the automated analysis of text data, motivated by problems like the following:

1. You have a large collection of free-form documents from which you would like to extract keyword summaries and see how they depend on time or other potentially related variables (e.g., what are customers complaining about, and how does it vary by geographic region or product type?);

2. You have short text strings that contain location, manufacturer, or other key information embedded within them, and you want to extract this information and turn it into a potentially useful categorical variable;

3. You have what *should* be a categorical variable (e.g., "city") with only a few valid values, but because of misspellings and other anomalies, each valid value appears in many different, invalid representations: can we build an automated procedure to "clean" this variable?

A key challenge of working with text data is that it is, by nature, *unstructured*, coming in a wide and growing variety of formats (e.g., PDF files, text embedded in HTML web pages, raw text files, or hard copy that must be scanned or manually entered to make it available in digital form), lengths (everything from 140-character twitter tweets to multivolume reports by government agencies that run to thousands of pages), languages, and styles (e.g., from completely free-form texts to more structured formats like surveys with fixed questions and mostly multiple-choice answers). Specialized text analysis methods have been developed to deal with problems like these, and this chapter provides an introduction to some of the simplest of these methods and their implementation in *R*. A key theme of this discussion is that certain forms of text preprocessing are critically important if we are to obtain useful results from automated text analysis, regardless of the nature of our objectives.

8.1 The fundamentals of text data analysis

Like categorical variables, text strings are not amenable to the arithmetic operations that form the basis for the analysis of numerical data variables. Unlike the discrete levels of a categorical variable, however, it is generally useful to decompose text strings into smaller components, which can then be converted to numbers (e.g., counts of all words, counts of specific words, counts of specific word combinations, etc.) that are amenable to a wide range of numerical characterizations. Sec. 8.1.1 gives a high-level overview of the steps typically involved in working with text data and some of the most useful R packages for implementing these steps. To make these ideas more concrete, they are illustrated in Sec. 8.1.2 with a simple example.

8.1.1 The basic steps in analyzing text data

At a very high level, the analysis of text data consists of the following steps:

1. Get the text to analyze;

2. *Normalize* the text, eliminating irrelevant details, e.g.:

    2a. Remove case effects: convert all text to lower or upper case;

    2b. Remove nuisance characters: punctuation, symbols, and numbers;

    2c. Parse the text into words;

    2d. Remove non-informative *stopwords*;

    2e. *Stem* the text, removing word endings like "ing," "s," or "ed";

3. Convert the normalized text to numbers;

4. Apply mathematical data analysis tools to these numbers;

5. Interpret these analysis results in terms of the original text data.

The problem of obtaining the text to analyze was discussed in some detail in Chapter 4, although some aspects of this problem are discussed further here. The task of normalizing or preprocessing the text to eliminate irrelevant differences is critically important and unique to text data analysis, so it is discussed briefly in the following paragraphs and illustrated in the example presented in Sec. 8.1.2, demonstrating why these steps are important by showing what happens when you omit them. Given normalized text data, a variety of specialized techniques have been developed to convert it into numbers that are amenable to many different forms of mathematical analysis; Sec. 8.5 introduces a few of these techniques and some of their R implementations. Sec. 8.1.2 presents one example—the construction and use of *document-term matrices* or *document-feature matrices* to find frequently occurring terms in text documents—and the examples in Sec. 8.6 provide other, more detailed illustrations of these methods, along with the interpretation of their results.

At a high level, R supports two types of text analysis tools. The first is the class of *character functions* built into base R that implement simple operations on character vectors. Some of the most important of these functions are discussed in detail in Sec. 8.2, including **nchar** that counts characters, **grep** that finds elements of a character vector that contain a specified substring, and **gsub** that replaces one substring with another. Another extremely useful character function is **tolower**, which converts all characters in the elements of a character vector to lower case: this function is discussed further in the following paragraphs and in the example presented in Sec. 8.1.2. The other class of text analysis tools consists of specialized packages like **tm** (the basic *text mining* package in R) or **quanteda**, both of which will be used in examples presented throughout this chapter. In addition, other text analysis packages are also available in R, like the java-based **qdap** package, developed for "quantitative discourse analysis," and the **topicmodels** package, developed to support the classification of documents by topic. These packages are not covered in this book, but the basic introduction to text data analysis provided here should provide useful background for those interested in exploring them on their own.

An important difference between these two types of text analysis tools is the range of data types they support. Specifically, the character functions work with *character vectors* like the following example:

```
cv <- c("This vector is a character vector.",
         "It has three elements.",
         "Each element is a text string.")
```

Functions like **nchar**, **grep**, and **gsub** discussed in Sec. 8.2 accept character vector arguments, returning numerical characterizations (e.g., **nchar** returns the number of characters in each element of the character vector), new character vectors derived from the original (e.g., **gsub** performs character substitutions), or lists of components (e.g., the **strsplit** function splits each element of a character vector into smaller pieces, returning a list of these pieces).

Higher-level text analysis packages like **tm** and **quanteda** support operations on other text-based data structures. For example, the **tokens** function from the **quanteda** package decomposes the elements of a character vector into *tokens*, typically corresponding to *words:*

```
library(quanteda)
tokenVector <- tokens(cv)
tokenVector

## tokens from 3 documents.
## text1 :
## [1] "This"      "vector"   "is"       "a"        "character" "vector"
## [7] "."
##
## text2 :
## [1] "It"       "has"      "three"    "elements" "."
##
## text3 :
## [1] "Each"     "element"  "is"       "a"        "text"     "string"   "."
```

Note that this function returns a named list object, where each element is a character vector containing the individual "words" extracted from the original character vector. The term "words" is included in quotes here because the basic `tokens` function includes punctuation marks like periods as tokens in these character vectors. This behavior is typically not what we want, but it is easily corrected using the `remove_punct` argument discussed in Sec. 8.1.2. The results returned by the `tokens` function correspond to a *bag-of-words* representation of the text, since each element of this list corresponds to a *bag* or *multiset*, a generalization of the mathematical notion of a *set*. Specifically, a set is a collection of *unique* elements, each of which appears only once, while a multiset generalizes this idea by allowing multiple copies of the elements [63, p. 44]. Thus, the collections of tokens in `Component` 2 and `Component` 3 in the above example correspond to sets, while the collection of tokens in `Component` 1 is a multiset or bag, since the token "vector" appears twice. It is important to allow for multiple copies of tokens in analyzing text documents since the number of times each term appears forms a useful basis for text characterization, as subsequent examples will demonstrate.

Another important text data structure supported by higher-level text analysis packages is the *corpus*, defined as a collection of text documents. In the `quanteda` package, the `corpus` function converts the character vector `cv` defined above into a corpus of three documents, one for each element of the character vector:

```
cvCorpus <- corpus(cv)
summary(cvCorpus)

## Corpus consisting of 3 documents:
##
##    Text Types Tokens Sentences
##   text1     6     7         1
##   text2     5     5         1
##   text3     7     7         1
##
## Source:  C:/Users/Ron/Documents/IntroRbook/RevisedManuscript/* on x86-64 by Ron
## Created: Thu Mar 01 09:31:52 2018
## Notes:
```

Here, the generic function `summary` returns useful information about the corpus, including the number of documents it contains, the number of "types" or *unique* terms in each document, the total number of tokens in each document, and the number of sentences in each document. To see the texts themselves, we can extract them with the `texts` function:

```
corpusTexts <- texts(cvCorpus)
head(corpusTexts)

##                                 text1                                text2
## "This vector is a character vector."            "It has three elements."
##                                 text3
##     "Each element is a text string."
```

Note that here, because `cvCorpus` is an extremely small corpus, we could have simply applied the `texts` function directly to obtain the same result, but a corpus will often contain a large number of documents, so it is usually best to assign the vector of texts to an *R* object like `corpusTexts` and then look at the results either by element or with useful utility functions like **head** to see the first few texts or `tail` to see the last few. Also, as we will see, much of what we want to do with text data using the **quanteda** package is based on *document-feature matrices*, created with the `dfm` function: the key point here is that this function will accept character vectors, tokens, or corpus objects.

The `tm` package also provides extensive support for working with corpus objects. For example, the `tm_map` function provides special machinery to apply transformations to every document in a corpus, returning another corpus. This function supports a number of transformations, whose names are returned by the `getTransformations` function:

```
library(tm)
getTransformations()

## [1] "removeNumbers"      "removePunctuation" "removeWords"
## [4] "stemDocument"       "stripWhitespace"
```

Note that these functions perform many of the important preprocessing operations listed earlier, but this list does *not* include a case normalization operation. It is easy to construct these functions, however, using the `tm` functions `tm_map`, `content_transformer`, and the basic *R* string handling functions like `tolower` and `toupper`. For example, the following sequence converts all documents in a corpus `X` to lower case:

```
Xlower <- tm_map(X, content_transformer(tolower))
```

In fact, we can perform most or all of the text preprocessing operations listed above using only the lower-level character vector functions provided in base *R*, but the support included in text analysis packages like `tm` and `quanteda` generally makes this much easier. That said, it is important to read the documentation for these procedures and carefully examine the results they return, because these results are not always what we expect.

8.1.2 An illustrative example

The following example illustrates the text analysis steps just described, based on the metadata text file `UCIautoMpgNames.txt` for the UCI auto-mpg dataset discussed in Sec. 8.2.6. As noted, the first step in this process is getting the data: here, this is a simple matter of using the `readLines` function, which reads each record from the external text file into one element of a character vector. We can then see what we have with the `head` function; specifying `n = 4` lets us look at the first four lines of text:

```
UCImetadata <- readLines("UCIautoMpgNames.txt")
head(UCImetadata, n = 4)
```

```
## [1] "1. Title: Auto-Mpg Data"
## [2] ""
## [3] "2. Sources:"
## [4] "    (a) Origin:   This dataset was taken from the StatLib library which is"
```

Altogether, this file contains 45 records, and we can see from this sample that some of these lines are blank, others are short, and others are much longer. Applying the **tokens** function from the **quanteda** package converts each element of this character vector into the bag-of-words format discussed in Sec. 8.1.1, represented as a list with one element for each element of the original character vector. Invoking the **head** function without specifying the optional argument **n** shows the first six elements of this list:

```
library(quanteda)
firstTokens <- tokens(UCImetadata)
head(firstTokens)
```

```
## tokens from 6 documents.
## text1 :
## [1] "1"         "."        "Title"    ":"        "Auto-Mpg" "Data"
##
## text2 :
## character(0)
##
## text3 :
## [1] "2"         "."        "Sources"  ";"
##
## text4 :
##  [1] "("        "a"        ")"        "Origin"   ":"        "This"     "dataset"
##  [8] "was"      "taken"    "from"     "the"      "StatLib"  "library"  "which"
## [15] "is"
##
## text5 :
## [1] "maintained" "at"        "Carnegie"  "Mellon"    "University"
## [6] "."          "The"       "dataset"   "was"
##
## text6 :
## [1] "used"       "in"         "the"       "1983"      "American"
## [6] "Statistical" "Association" "Exposition" "."
```

Note that since the second element of the character vector **UCImetadata** is the empty string (""), the corresponding element of the bag-of-words list is also empty ("**character(0)**").

For text data that can be viewed as a collection of "documents," a very useful characterization tool is the *document-feature matrix*, discussed in detail in Sec. 8.5.1. This matrix has one row for each document in the collection and one column for each *feature* that appears in one or more documents, with each element telling us how often a feature appears in a document. As the following example illustrates, these matrices can be used to identify the most frequently occurring features in a document collection. Here, each element of the character vector UCImetadata is treated as a document, and the features are the individual "words" extracted by the tokens function. The following *R* code constructs the document-feature matrix from the firstTokens bag-of-words shown above and lists the 20 most frequently occurring features:

```
UCIdfm1 <- dfm(firstTokens, tolower = FALSE, verbose = FALSE)
topfeatures(UCIdfm1, n = 20)

##            .          :        the          ,          "          (
##           28         19         11          8          8          7
##            )         of         in continuous    dataset       1993
##            7          7          6          6          5          4
##     discrete         is        The          7          3    Quinlan
##            4          3          3          3          3          3
##    attribute        mpg
##            3          3
```

These results illustrate the importance of the text preprocessing tasks listed in Steps (2a) and (2b) in the procedure outlined in Sec. 8.1.1. In particular, note that the two most frequently appearing terms in this document are the punctuation marks period (".") and colon (":"), and the fourth through eighth most frequent terms are also punctuation marks. Other terms on this list of frequent features are the numbers 1993, 7, and 3. Further, note that the words "the" and "The" are counted separately, occurring 11 and 3 times, respectively, because we have specified tolower = FALSE in the call to the dfm function. Generally, we do not want our results to depend on whether a word is capitalized or not, so the default value for the optional argument tolower for the dfm function is TRUE; this default was over-ridden here to show the consequences of not performing case normalization. Similarly, we can eliminate the spurious influence of symbols by setting the remove_symbols argument to TRUE in the tokens call, eliminate punctuation by setting remove_punct to TRUE, and setting remove_numbers to TRUE (the default values for all of these arguments is FALSE). These specifications give the following top feature list:

```
secondTokens <- tokens(UCImetadata,
                       remove_symbols = TRUE,
                       remove_punct = TRUE, remove_numbers = TRUE)
UCIdfm2 <- dfm(secondTokens, tolower = TRUE, verbose = FALSE)
topfeatures(UCIdfm2, n = 20)

##          the         in         of continuous    dataset
##           14          8          7          6          5
##    attribute   discrete         is    quinlan        mpg
##            5          4          3          3          3
##       values multi-valued       data          a     origin
##            3          3          2          2          2
##         this        was    statlib    library university
##            2          2          2          2          2
```

These preprocessing steps have helped a lot, but now the most frequently occurring terms are the non-informative *stopwords* "the," "in," and "of." We

can remove these terms by applying the `tokens_remove` function, specifying
`stopwords("english")` as the features to be removed:

```
thirdTokens <- tokens_remove(secondTokens, stopwords("english"))
UCIdfm3 <- dfm(thirdTokens, verbose = FALSE)
topfeatures(UCIdfm3, n = 20)
```

##	continuous	dataset	attribute	discrete	quinlan
##	6	5	5	4	3
##	mpg	values	multi-valued	data	origin
##	3	3	3	2	2
##	statlib	library	university	learning	information
##	2	2	2	2	2
##	original	instances	attributes	number	horsepower
##	2	2	2	2	2

Note that stopword removal requires a list of the words to be removed. One of
the advantages of packages like `quanteda` and `tm` is that they provide stopword
lists, like that returned by the `quanteda` function `stopwords` used here. That
is, the construction of stopword lists is a non-trivial task, and while the use
of application-specific stopwords can substantially improve the performance of
some text analysis procedures [61], the 174-element English-language stopword
list provided by the `quanteda` package is an excellent first step.

Removing these stopwords, the remaining features give us a better idea of
what the original document is about, but some undesirable distinctions persist. For example, the word "attribute" appears five times, while its plural
"attributes" counts as a separate word that appears twice. Often, we want to
consider words that differ only in endings like "s," "ed," or "ing" as variants of
a single parent word, all to be counted together. This can be accomplished by
stemming the text, removing these endings and retaining only the word "stems."
In `quanteda`, this task is supported by the `tokens_wordstem` function:

```
fourthTokens <- tokens_wordstem(thirdTokens)
UCIdfm4 <- dfm(fourthTokens, verbose = FALSE)
topfeatures(UCIdfm4, n = 20)
```

##	attribut	continu	dataset	origin	discret	quinlan
##	7	6	5	4	4	3
##	mpg	instanc	valu	multi-valu	data	statlib
##	3	3	3	3	2	2
##	librari	univers	use	learn	informat	predict
##	2	2	2	2	2	2
##	number	horsepow				
##	2	2				

Note that now the common stem "attribut" of both "attribute" and "attributes"
is counted together, appearing seven times and becoming the most frequent word
in the document. This example also illustrates that the results returned by automated stemming procedures like `tokens_wordstem` or the `stemDocument` transformation in the `tm` package are often not "words" as we usually think of them.
In particular, note that the final "e" has been removed from both "attribute"
and "discrete," while the "continuous" has been shortened to "continu."

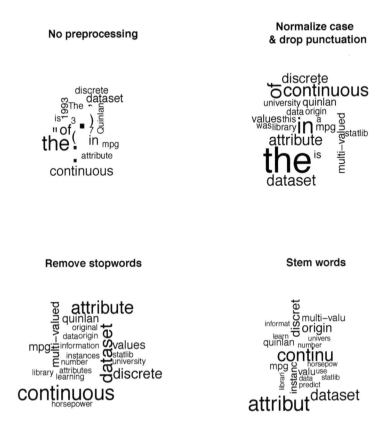

Figure 8.1: Four wordclouds showing the effects of text preprocessing.

An increasingly popular way to represent frequent terms in text data is with *wordclouds*, which display words in different sizes, depending on how frequently they occur in a document. The *R* package `wordcloud` provides a simple way of generating wordclouds, given a table of word frequencies. Fig. 8.1 shows four examples, each constructed from the results returned by the `topfeatures` function for one of the examples just discussed. The top-left wordcloud represents the original summary, constructed from the bag-of-words `firstTokens` with no preprocessing. The predominance of punctuation marks and numbers is immediately apparent here, and the distinction between the capitalized word "The" and its lower-case form "the" is apparent on closer inspection.

The top-right wordcloud in Fig. 8.1 represents the results obtained after case conversion and the removal of punctuation marks, numbers, and special symbols. Here, the largest words are "the," "in," and "of," reflecting the fact that these stopwords occur most frequently in the cleaned token set `secondTokens`. Removing these stopwords yields the token set `thirdTokens`, used to construct the lower-left wordcloud which is seen to be dominated by the most frequent remaining terms "continuous," "attribute," and "dataset." Finally, the lower right wordcloud was constructed from the stemmed token set `fourthTokens`, and again the influence of this preprocessing step is clearly apparent, since the stems "attribut" and "continu" are most prominent in this wordcloud.

8.2   Basic character functions in R

As noted in Sec. 8.1.1, aside from specialized text analysis packages like `tm` and `quanteda`, much text characterization and manipulation is possible using *R's* basic character functions. The following discussions introduce several of these, including `nchar` for counting characters in a text string, `grep` to search for specific substrings in a character vector, the character substitution functions `sub` and `gsub`, the `strsplit` function for splitting text strings into components around a specified substring, and the `paste` function for combining shorter strings into longer ones.

8.2.1   The `nchar` function

The `nchar` function counts the number of characters in a character string or a vector of character strings. As a specific illustration, consider the `First.Name` field from the unclaimed bank account data frame `unclaimedFrame` discussed in Chapter 4. The first 20 elements of this sequence are:

```
firstNames <- as.character(unclaimedFrame$First.Name)
firstNames[1:20]

##  [1] "ALPHONSE          "
##  [2] "PETER          "
##  [3] "VLADIMIR & VANENTINA          "
##  [4] "SELMA          "
##  [5] "SU LENG          "
##  [6] "JOGINDER SINGH          "
##  [7] "ROSE          "
##  [8] ""
##  [9] "LOUISE ANNE          "
## [10] "FLORENCE          "
## [11] "ROBERTA          "
## [12] "EUROPEAN CONSULTANTS LTD          "
## [13] ""
## [14] "EVA          "
## [15] "STEVE          "
## [16] ""
## [17] "KERRY          "
## [18] "LOTHAR OR EHLSCHEID  HELEN          "
## [19] "NEVILLE D          "
## [20] "TAHANY          "
```

An important point to note here is the number of trailing blanks appearing in these names, indicated by the relatively large distance between the end of most names and its closing quotation mark. This observation is important in part because the **nchar** function, which counts characters, includes blanks (and other whitespace like tabs or newlines) in this count. To see this point, consider the results when we apply **nchar** to the first name "ALPHONSE" by itself, and then again to the first element of the **firstNames** sequence:

```
nchar("ALPHONSE")

## [1] 8

nchar(firstNames[1])

## [1] 19
```

The difference in these results reflects the 11 spaces after the name "ALPHONSE" in the first element of **firstNames**. The presence or absence of whitespace like this in character strings is important because it influences many text analysis results. For example, if we ask, "Is the first name in the first record 'ALPHONSE'?," simple logic will not give us the answer we want:

```
firstNames[1] == "ALPHONSE"

## [1] FALSE
```

To obtain the result we want, the safest way is to remove the trailing whitespace using the substitution functions discussed in Sec. 8.2.4, but we can use the function **grep** described in the next section to ask the closely related question, "Does the first name in the first record *contain* 'ALPHONSE'?."

Note that if the **nchar** function is called with a character vector instead of a single string, it returns a numerical vector giving the lengths of each element of the character vector. For example, applying **nchar** to the first 20 elements of **firstNames** yields these results, shown graphically in Fig. 8.2:

```
nchar(firstNames[1:20])

##  [1] 19 13 29 12 14 21 11  0 21 15 14 33  0 14 12  0 16 33 19 17
```

More specifically, Fig. 8.2 is a horizontal barplot with the **First.Name** values printed across each bar and the length of each string appearing at the right. Note that three of these string lengths are zero, corresponding to the blanks appearing in the names shown above. The possibility of missing entries in text data is a particularly important one, discussed further in Sec. 8.2.3, since missing text data can be represented in many different ways.

Another extremely important aspect of the **nchar** function is that it does not work with factor variables. This point is important because, as noted several times, character vectors are included, by default, as factors in data frames. While this default can be overridden when we create the data frame, usually it

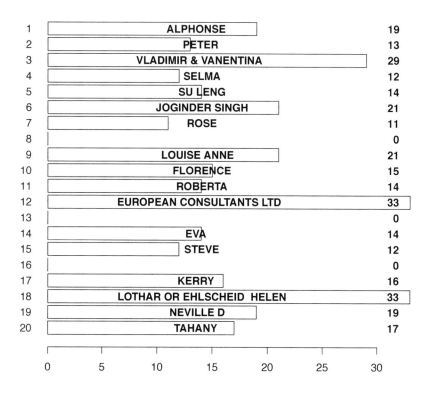

Figure 8.2: The first 20 `First.Name` strings and their lengths.

is not. As an illustration of this point, consider the `cpus` data frame from the MASS package that forms the basis for the simple text analysis example discussed in Sec. 8.6.2. Applying the `str` function to this data frame, we see that the text variable `name` appears in the data frame as a factor:

```
str(cpus)

## 'data.frame': 209 obs. of  9 variables:
##  $ name  : Factor w/ 209 levels "ADVISOR 32/60",..: 1 3 2 4 5 6 8 9 10 7 ...
##  $ syct  : int  125 29 29 29 29 26 23 23 23 23 ...
##  $ mmin  : int  256 8000 8000 8000 8000 8000 16000 16000 16000 32000 ...
##  $ mmax  : int  6000 32000 32000 32000 16000 32000 32000 32000 64000 64000 ...
##  $ cach  : int  256 32 32 32 32 64 64 64 64 128 ...
##  $ chmin : int  16 8 8 8 8 8 16 16 16 32 ...
##  $ chmax : int  128 32 32 32 16 32 32 32 32 64 ...
##  $ perf  : int  198 269 220 172 132 318 367 489 636 1144 ...
```

```
## $ estperf: int  199 253 253 253 132 290 381 381 749 1238 ...
```

To see that the **nchar** function gives useful results when this variable is converted to a text string but not when it is represented as a factor, consider the first 10 values of this variable:

```
namesAsFactor <- cpus$name
nchar(namesAsFactor[1:10])

## Error in nchar(namesAsFactor[1:10]):  'nchar()' requires a character vector

namesAsCharacter <- as.character(cpus$name)
nchar(namesAsCharacter[1:10])

## [1] 13 13 13 14 14 13 15 15 15 15
```

*To emphasize the importance of working with character vectors rather than factors, the examples in this chapter generally include explicit **as.character** conversions.*

8.2.2 The grep function

The function **grep** allows us to search for the presence of a substring in a specified character vector. Specifically, given a substring **subStr** and a character vector **charVec**, the call **grep(subStr, charVec)** returns a list of the elements of **charVec** that contain **subStr** one or more times. As a simple example, consider the results obtained when we search for the letter "O" in the first 20 elements of the text vector **firstNames**:

```
index <- grep("O", firstNames[1:10])
index

## [1]  1  6  7  9 10

firstNames[index]

## [1] "ALPHONSE       "  "JOGINDER SINGH     "  "ROSE         "
## [4] "LOUISE ANNE    "  "FLORENCE        "
```

In many of the applications of **grep** considered in this book, the function is used as in this example, to return an index into a text vector that points to all elements containing the specified substring. In cases where we want the elements themselves, however, it is possible to use the optional logical argument **value**. This argument has the default value **FALSE**, but setting it to **TRUE** returns the elements themselves, as in this example:

```
strings <- grep("O", firstNames[1:10], value = TRUE)
strings

## [1] "ALPHONSE       "  "JOGINDER SINGH     "  "ROSE         "
## [4] "LOUISE ANNE    "  "FLORENCE        "
```

The `grep` function also has a number of other optional arguments, probably the most important of which are the logical arguments `fixed`, discussed in detail in Sec. 8.3, `ignore.case`, and `invert`, all of which have default values `FALSE`. By setting `ignore.case` to `TRUE`, we can achieve case-insensitive string matching; to see the effect of this parameter, consider the following two examples:

```
grep("o", firstNames[1:10], value = TRUE)

## character(0)

grep("o", firstNames[1:10], value = TRUE, ignore.case = TRUE)

## [1] "ALPHONSE          "  "JOGINDER SINGH       " "ROSE        "
## [4] "LOUISE ANNE       "  "FLORENCE      "
```

Specifying `invert = TRUE` inverts the search criterion, returning all elements of the vector that do *not* contain the indicated substring. For example, we can find all of the first 20 `First.Name` values that do not contain the letter "E":

```
grep("E", firstNames[1:20], value = TRUE, invert = TRUE)

## [1] ""                    ""                "          "
## [4] "TAHANY"             "
```

Comparing this result with those shown in Fig. 8.2, we can see that the only ones of the first 20 `First.Name` strings that do not contain the letter "E" are the three empty strings and the name "TAHANY."

8.2.3 Application to missing data and alternative spellings

As emphasized in Chapter 3, missing data is an extremely important problem in practice, in part because there is no universal representation for it. This problem is particularly acute for text data since there are essentially no limits on the range of possible representations for missing text values, as the following example illustrates. This code uses the `grep` function discussed in Sec. 8.2.2 to search for elements of the `City` character vector containing some variation of "unknown," using the stem "unk":

```
City <- as.character(unclaimedFrame$City)
missingCity <- grep("unk", City, ignore.case = TRUE, value = TRUE)
table(missingCity)

## missingCity
##      UNKNOW ALTA M1M 1M1              UNKNOWN ????              UNKNOWN AB
##                        1                       121                      46
##          UNKNOWN  AB               UNKNOWN ALTA              UNKNOWN BC
##                       87                        40                       6
##          UNKNOWN BC               UNKNOWN DEU              UNKNOWN GBR
##                        4                         1                       1
##      UNKNOWN UNKNOWN  UNKNOWN UNKNOWN UNKNOWN              UNKNOWN USA
##                       97                         1                       1
##      UNKNOWN,UNKNOWN           UNKNOWN,UNKNWON              UNKNWON AB
##                     1070                         1                       1
```

The first line extracts the variable `City` from the `unclaimedFrame` data frame and converts it to a character string, while the second uses the `grep` function with `ignore.case = TRUE` to eliminate case effects and `value = TRUE` to return text strings instead of integer indices. The third line uses the `table` function to count each unique `City` value whose lower-case representation contains the substring "unk." The key point here is that there are 15 distinct matches, each representing a *different* coding for "city missing" and altogether accounting for 1478 records from the dataset. In addition, missing `City` values are also frequently encoded as blanks, as the following result demonstrates:

```
blankCity <- which(City == " ")
length(blankCity)

## [1] 3510
```

Finally, it is interesting to note, first, how frequently some of these different missing value representations occur relative to the most common non-missing city values, and, second, the presence of different representations even among these non-missing values. The following example illustrates this point:

```
cityList <- sort(table(City), decreasing = TRUE)
cityList[1:10]

## City
##           Edmonton              UNKNOWN,UNKNOWN            EDMONTON
##               3569                        3510                 859
##       EDMONTON,AB              EDMONTON CAN            EDMONTON AB        St. Albert
##                732                         290                 219               172
##     Sherwood Park              UNKNOWN ????
##                137                         121
```

Here, note that "Edmonton" with a capital "E" and no further elaboration is the most frequently ocurring `City` value, followed in second and third place by the missing city representations blank (" ") and "UNKNOWN, UNKNOWN." The next four entries in this list are alternative designations for Edmonton, all in upper case and with or without abbreviations designating either the province ("AB" for "Alberta") or the country. Further, note that "EDMONTON,AB" and "EDMONTON AB" differ only by a comma, but they appear as distinct values. *This is one of the critical aspects of analyzing text data: "small" changes that humans recognize as "the same thing" are distinct character strings.* In fact, if we consider all of the different character strings that include some form of "edmonton," we find that there are many of them:

```
edmontonIndex <- grep("edmonton", City, ignore.case = TRUE)
edmontonRecords <- City[edmontonIndex]
edmontonTable <- table(edmontonRecords)
sort(edmontonTable, decreasing = TRUE)

## edmontonRecords
##              Edmonton                      EDMONTON
##                  3569                           859
```

```
##              EDMONTON,AB              EDMONTON CAN
##                     732                     290
##              EDMONTON AB              EDMONTON,ALTA
##                     219                      96
##             EDMONTON, AB         EDMONTON ALBERTA
##                      19                       6
##              EDMONTON BC         EDMONTON, ALBERTA
##                       5                       3
## DMONTON EDMONTON EDMONTON    ONTON EDMONTON EDMONTON
##                       2                       2
##               ,EDMONTON                EDMONTON
##                       1                       1
##          EDMONTON ALBER   EDMONTON EDMONTON EDMON
##                       1                       1
##   EDMONTON EDMONTON EDMONT  EDMONTON EDMONTON EDMONTO
##                       1                       1
## EDMONTON EDMONTON EDMONTON              EDMONTON ON
##                       1                       1
##          EDMONTON T5X 3A5 PQ     EDMONTON, ALBERTA T5
##                       1                       1
##           EDMONTON, ATLA.           EDMONTON, BC
##                       1                       1
##             EDMONTON,AN           EDMONTON,ATLA
##                       1                       1
##   MONTON EDMONTON EDMONTON   N EDMONTON EDMONTON EDMO
##                       1                       1
## NTON AB EDMONTON AB EDMONT   NTON EDMONTON EDMONTON E
##                       1                       1
##        OF AB EDMONTON ALTA      SOUTH EDMONTON ALTA
##                       1                       1
```

Here, "edmonton" appears in 32 different forms, accounting for 5822 records out of 12,816 (45.4%), while "Edmonton" appears in only 3569 records (27.8%).

Why is this important? Often, we want to characterize data by groups (e.g., "what is the balance in unclaimed bank accounts in Edmonton?"). In such cases, it is important to be certain that the group of interest is defined by the records we are considering. If we answer this question based on the "Edmonton" records, the result is 2.31e+06, while if we answer based on all of these various "edmonton" records, the result is 3.11e+06, larger by approximately 35%.

8.2.4   The sub and gsub functions

The functions sub and gsub perform string substitutions, differing in that sub replaces *the first occurrence only* of the target substring, while gsub replaces *all occurrances*. Both functions have the same three required arguments, appearing in the same order:

1. pattern is the substring to be replaced;

2. replacement is the replacement text for pattern;

3. x is the character vector where pattern is to be replaced.

In addition, both of these functions support four optional logical arguments, the most useful of which are `ignore.case` and `fixed`, both with default values `FALSE`. Both of these arguments have the same effect here as they do in the `grep` function discussed in Sec. 8.2.2. The effect of the `fixed` argument is particularly important: by default, `pattern` is treated as a regular expression, meaning that certain special characters—i.e., regular expression metacharacters—can trigger much more extensive substitutions than you might intend. This point is discussed in detail in Sec. 8.3 where examples of both straight text substitutions (i.e., with `fixed = TRUE`) and regular expression substitutions are presented.

Both the basic behavior of the functions `sub` and `gsub` and their differences are easily seen in the following example:

```
x <- "google search on oolong and oobleck associates"
sub(pattern = "o", replacement = "X", x)

## [1] "gXogle search on oolong and oobleck associates"

gsub(pattern = "o", replacement = "X", x)

## [1] "gXXgle search Xn XXlXng and XXbleck assXciates"
```

In both cases, the substring to be replaced is "o" and the replacement value is "X," which contains no regular expression metacharacters so it is treated as straight text. The `sub` function finds the first occurrence of "o" and replaces it with "X" but makes no other changes in the text string. In contrast, the `gsub` function replaces all occurrences of "o" in the text string with "X."

The `gsub` function is particularly useful in simple text preprocessing where we wish to remove nuisance characters like numbers, punctuation marks, or other special symbols and replace them with blanks, spaces, or other alternative strings. Since some punctuation marks are important regular expression metacharacters, a detailed discussion of this application is deferred until Sec. 8.3, but the following examples illustrate the basic idea:

```
x <- "eddie coyle, a/k/a 'fingers eddie'"
gsub("/", "", x)

## [1] "eddie coyle, aka 'fingers eddie'"

gsub("'", "", x)

## [1] "eddie coyle, a/k/a fingers eddie"

gsub("a/k/a", "also known as", x)

## [1] "eddie coyle, also known as 'fingers eddie'"
```

The first substitution replaces the slashes in the term "a/k/a" with blanks, reducing it to "aka," while the second removes the single quotes around "fingers eddie." The final substitution replaces the abbreviation "a/k/a" with its full equivalent "also known as."

8.2.5   The `strsplit` function

The function `strsplit` splits each element of a character vector around a specified substring. The required arguments for this function are:

1. `x`, the character vector to be split;

2. `split`, the substring on which the elements of `x` are to be split.

Since `x` is a character vector and each element of this vector is split into substrings separated by `split`, the number of these substrings in each split cannot be determined in advance, so this function returns a list object rather than a vector, with each list element containing all of the split components of the corresponding element of `x`. As with the functions `grep`, `sub`, and `gsub`, the argument `split` is treated as a regular expression unless the optional argument `fixed` is set to `TRUE`.

The following examples illustrate how the `strsplit` function works:

```
x <- c("a comma b comma c", "a, b, c", "a comma (,) b comma (,) c")
strsplit(x, split = ",")

## [[1]]
## [1] "a comma b comma c"
##
## [[2]]
## [1] "a"   " b" " c"
##
## [[3]]
## [1] "a comma ("   ") b comma (" ") c"
```

In this example, `x` is a three element vector and `split` is the punctuation mark ",", so the above function call asks `strsplit` to split each element of `x` on these commas. As noted, the result is a list, each element of which contains the results obtained by splitting the corresponding element of `x`. Since the first element contains no commas (they have been replaced by the word "comma"), no splitting occurs and the result returned is the complete character string. The second element does contain commas, and here the split results in three components; note that since the split is on the comma, no spaces are removed, leaving leading spaces in front of the letters "b" and "c" here. Finally, the third element contains commas, but enclosed in parentheses: here again, the string is split into three components, obtained by omitting the commas and retaining everything else.

The `strsplit` function is particularly useful in parsing text strings into words separated by spaces, although it is important to note that if we specify a single space as our `split` argument, text strings with multiple spaces occurring together will be parsed into a sequence containing "blank" or "empty" words. The following example illustrates this point:

```
x <- "This is a sequence with  some   extra    spaces"
strsplit(x, " ")
```

```
## [[1]]
## [1] "This"      "is"        "a"        "sequence" "with"     ""
## [7] "some"      ""          ""         "extra"    ""         ""
## [13] ""          "spaces"

strsplit(x, " ")

## [[1]]
## [1] "This is a sequence with" "some"
## [3] " extra"                  ""
## [5] "spaces"

strsplit(x, "  ")

## [[1]]
## [1] "This is a sequence with" "some" "extra"
## [3] " spaces"

strsplit(x, "   ")

## [[1]]
## [1] "This is a sequence with  some   extra"
## [2] "spaces"
```

In the first split here, based on the single space, we obtain 14 "words," of which 6 are empty, reflecting the fact that a word followed by two spaces is treated as a first word, separated by one space from a second space; this second space gets parsed into an empty word. If we change our split character to two spaces instead of one, we split our sequence into "words" separated by two spaces, which fails to parse the first five words here since they are separated by single spaces. Increasing our split character to three or four spaces results in the successively fewer number of "words" shown in the later examples. *As will be shown in Sec. 8.3, it is easy to obtain the result we want—i.e., a parsing of the original string into words separated by one or more spaces—using regular expressions.*

8.2.6 Another application: `ConvertAutoMpgRecords`

Chapter 4 discussed the automobile mileage dataset `autoMpg` from the UCI Machine Learning Repository, downloaded as a text file with fixed-width records containing numerical fields separated by spaces and a text field with embedded quotation marks, separated by a tab character. The custom *R* function `ConvertAutoMpgRecords` was developed to extract the data from these fields and save it to a data frame with the following format:

```
head(autoMpgFrame, n = 3)

##   mpg cylinders displacement horsepower weight acceleration modelYear
## 1  18         8          307        130   3504         12.0        70
## 2  15         8          350        165   3693         11.5        70
## 3  18         8          318        150   3436         11.0        70
```

```
##    origin                         carName
## 1       1 chevrolet chevelle malibu
## 2       1           buick skylark 320
## 3       1           plymouth satellite
```

This function is listed here and discussed below:

```
ConvertAutoMpgRecords

## function(rawRecords){
##    #
##    noQuotes <- gsub('\"', '', rawRecords)
##    #
##    n <- length(noQuotes)
##    outFrame <- NULL
##    for (i in 1:n){
##      x <- noQuotes[i]
##      twoParts <- unlist(strsplit(x, split = "\t"))
##      partOne <- unlist(strsplit(twoParts[1],
##                                 split = "[ ]+"))
##      numbers <- as.numeric(partOne)
##      upFrame <- data.frame(mpg = numbers[1],
##                            cylinders = numbers[2],
##                            displacement = numbers[3],
##                            horsepower = numbers[4],
##                            weight = numbers[5],
##                            acceleration = numbers[6],
##                            modelYear = numbers[7],
##                            origin = numbers[8],
##                            carName = twoParts[2])
##      outFrame <- rbind.data.frame(outFrame, upFrame)
##    }
##    return(outFrame)
## }
## <bytecode: 0x0000000009ddfe20>
```

This function has only one argument, and it is required: the character vector read from the external text file containing the autoMpg data. The first statement in the body of the function invokes gsub to remove every escaped double quote (i.e., the string "\"") from this character vector. Considering the first element shows the effect of this substitution (here, the character string has been split into two parts with the substr function for clarity):

```
x <- autoMpgRecords[1]
xA <- substr(x, 1, 50)
xA

## [1] "18.0   8   307.0      130.0      3504.      12.0   "

xB <- substr(x, 51, 84)
xB

## [1] " 70  1\t\"chevrolet chevelle malibu\""

noQuotesB <- gsub('\"', '', xB)
noQuotesB

## [1] " 70  1\tchevrolet chevelle malibu"
```

Since the `gsub` function works on vectors, we can apply it once to remove all escaped double quotes from the character vector `rawRecords`, which corresponds
to `autoMpgRecords`. The `for` loop in the program then operates on each element of this cleaned text vector, first splitting it on the tab character ("\t")
that separates all of the numerical fields from the text field, and then parsing
the numerical portion into individual numerical fields. This is accomplished by
the two calls to the `strsplit` function inside the `for` loop. In both cases, the
function `unlist` is used to convert the list output of `strsplit` into a vector:
since `strsplit` is being called here with a character vector of length 1, it returns a list of length 1, whose only element is a character vector of components
before and after the split character; `unlist` gives us this vector. The first time
`strsplit` is called, this vector has two components, the first being the text
string containing all of the numerical fields before the tab character, and the
second being the car name that follows the tab character:

```
noQuotes <- gsub('\"', '', x)
twoParts <- unlist(strsplit(noQuotes, "\t"))
twoParts

## [1] "18.0   8   307.0    130.0    3504.    12.0   70  1"
## [2] "chevrolet chevelle malibu"
```

This second component of this vector (`twoParts[2]`) becomes the `carName` field
in the data frame built at the end of the loop, but the first component is further split by the second call to the `strsplit` function, which deserves careful
discussion. Specifically, as in the example discussed at the end of Sec. 8.2.5,
the numerical fields here are separated by varying amounts of whitespace. To
effectively separate these fields, we invoke `strsplit` with `split` specified as the
following regular expression, discussed in detail in Sec. 8.3:

```
unlist(strsplit(twoParts[1], "[ ]+"))

## [1] "18.0"  "8"    "307.0" "130.0" "3504." "12.0"  "70"    "1"
```

This vector is converted from character to numeric with the `as.numeric` function, and its elements define columns of the data frame built at the end of the
`for` loop, based on knowledge of what these fields are. The basis for these assignments were the field names obtained from the metadata associated with the
data file, as discussed in Chapter 4.

8.2.7 The `paste` function

The `paste` function combines shorter text strings into longer ones, and it has
been used in examples presented in earlier chapters. In typical applications,
this function is called with a collection of *R* objects, each of which is either a
character string or converted into one, and these components are concatenated
into a longer text string. The following example illustrates this basic behavior:

```
n <- 4
textString <- paste("The value of pi to", n, "digits is", round(pi, digits = n))
textString
```

```
## [1] "The value of pi to 4 digits is 3.1416"
```

Here, **paste** is called with four components, two text strings, the integer **n**, and the R expression **round(pi, digits = n)**, which returns the numerical value of π, rounded to four digits. The function returns a text string composed of these four components, converted to character strings if they are not already in this form, separated by single spaces. This result is obtained because the optional argument **sep** that defines the separator between components has the default value "space" (i.e., " "). If we had specified **sep = ''; ''**, we would have obtained this result instead:

```
n <- 4
textString <- paste("The value of pi to", n, "digits is", round(pi, digits = n),
                    sep = ";")
textString
```

```
## [1] "The value of pi to;4;digits is;3.1416"
```

If one or more of the R objects passed to **paste** are vectors, they are concatenated term-by-term, returning a vector of the same length as the longest component, as in the following example:

```
paste(c("A", "B"), seq(1, 10, 1), c("x", "y", "z"), sep = ":")
```

```
##  [1] "A:1:x"  "B:2:y"  "A:3:z"  "B:4:x"  "A:5:y"  "B:6:z"  "A:7:x"
##  [8] "B:8:y"  "A:9:z"  "B:10:x"
```

Here, the longest of the three vectors passed to **paste** is the sequence of numbers from 1 to 10, so the result is a vector of length 10. Each element of this vector consists of three characters separated by colons (":"), with the first element from the two-component character vector c(''A'', ''B''), the second from the numerical sequence, and the third from the three-component character vector c(''x'', ''y'', ''z''). Elements of the shorter vectors are *recycled*, used in order from the first element to the last, repeating the cycle from the first again until all elements of the output vector are created.

The **paste** function can also be used to concatenate the elements of a character vector into a single string by specifying a non-**NULL** value for the optional argument **collapse**. In this case, the elements of the character vector are separated by the value of **collapse** in forming the output text string. The following pair of examples illustrates how this works. The first generates the sequence of integers from 0 to **n**, separated by commas and spaces:

```
n <- 7
paste(seq(0, n, 1), collapse = ", ")
```

```
## [1] "0, 1, 2, 3, 4, 5, 6, 7"
```

The second example uses the **paste** function in both of the ways just described, first constructing a "header string," **str1**, to appear at the beginning of the final character string. Next, **str2** uses **paste** with **collapse** to convert the vector of **mtcars** column names into a single character string, separated by commas and spaces as in the previous example. Finally, **paste** is used again to concatenate the two character strings **str1** and **str2**, using the default single space separator between these components:

```
str1 <- "mtcars variables:"
str2 <- paste(colnames(mtcars), collapse = ", ")
paste(str1, str2)

## [1] "mtcars variables: mpg, cyl, disp, hp, drat, wt, qsec, vs, am, gear, carb"
```

8.3 A brief introduction to regular expressions

Regular expressions are text strings that consist of both *literal characters*, which have their usual meaning, and *metacharacters*, which have special interpretations that are useful in searching for and/or replacing portions of text in a longer text string. All of the string search and replacement functions described in Sec. 8.2 work with regular expressions by default: this provides tremendous power in working with text strings, but it can also cause horrendous errors if you fail to recognize a metacharacter for what it is, or if you use one incorrectly. The following discussion gives an introduction to regular expressions and their use, but for a more detailed introduction, refer to Paul Murrell's book [56], especially Secs. 9.9.2 and 9.9.3, and his brief reference guide in Chapter 11.

8.3.1 Regular expression basics

To illustrate clearly what regular expressions are, it is useful to begin with a simple example, and this one shows how widely regular expressions are used in *R*. Recall from Chapter 4 that the **list.files** function returns a list of all files in our current working directory. This function accepts several optional arguments to restrict our search, which can be useful if we have a lot of files. One of these arguments is **pattern**, which specifies partial matches in the file names and is interpreted as a regular expression. The following specifications retrieve the indicated files, if any, from our working directory:

1. **list.files(pattern = "R")** lists file names that contain the letter R;

2. **list.files(pattern = "R$")** lists file names that *end* with the letter R;

3. **list.files(pattern = ".R")** lists file names that contain the letter R, *preceded by anything else* (i.e., *not* at the beginning of the file name);

4. **list.files(pattern = "R.")** lists file names that contain the letter R, *followed by anything else* (i.e., *not* at the end of the file name).

Note that since R is case-sensitive, all of these searches look for files that contain *upper case "R"* in the indicated arrangement within the string. In all of these examples, the letter "R" is a literal character in the regular expression, while the characters "$" and "." are regular expression metacharacters. The effect of these metacharacters is to modify the string matching criterion in the ways indicated in the above examples.

In cases where we want to match a special character like "$" in a text string, we can sometimes turn off the regular expression matching machinery and treat our pattern as all literal text. *In particular, this is possible for all of the string search and replacement functions discussed in Sec. 8.2, by specifying the optional argument* `fixed = TRUE`. Conversely, note that this option is not available for the `pattern` argument of the `list.files` function, but this is less of a limitation than it seems since most of these special characters are not valid in file names, anyway. The major exception is "." which is a metacharacter in regular expressions meaning "match anything." In particular, note that file search (3) listed above does not return only those files that end with ".R", indicating that they contain R code, but rather *all* files whose names contain the letter "R" anywhere in the string except at the beginning.

Since the "." metacharacter is so powerful, it is worth a second illustration. Recall from Sec. 8.2.2 that the `grep` function searches for matches in text strings and, if we specify the optional argument `value = TRUE`, it returns the matched text rather than an index pointing to that text. Further, as with `list.files`, the `pattern` argument for `grep` is, by default, a regular expression. To see the effect of the "." metacharacter in regular expressions, consider these examples:

```
x <- c("This is a vector of text strings",
       "The first contains no period, but the second one does.",
       "The third contains ...", "The fourth is like the first")
grep(".", x, value = TRUE)

## [1] "This is a vector of text strings"
## [2] "The first contains no period, but the second one does."
## [3] "The third contains ..."
## [4] "The fourth is like the first"

grep(".", x, value = TRUE, fixed = TRUE)

## [1] "The first contains no period, but the second one does."
## [2] "The third contains ..."
```

In the first example, the period functions as a regular expression metacharacter, so the search instruction is, effectively, "return all character strings that contain any character," and the consequence is that the search returns all elements of the character vector `x`. In the second example, specifying `fixed = TRUE` turns the regular expression machinery off, so the period is treated as a literal, and the search returns only those elements of `x` that contain at least one period.

Recognizing the power of the "." metacharacter in regular expressions is particularly important when replacing text, since its unintentional use can lead to some extremely serious errors. For example, suppose we want to remove periods from a text string. We can use the `gsub` function discussed in Sec. 8.2.4 to replace periods with blanks, but this first attempt fails catastrophically:

```
x <- "One string ... with both embedded ellipses (...) ... and a final period."
gsub(".", "", x)

## [1] ""
```

*Note that because "." is interpreted as the "any character" metacharacter here, this application of **gsub** replaces all characters in the text string x with blanks, reducing the entire string to a blank.* To obtain what we want—i.e., removal of all periods from this text string—we need to specify `fixed = TRUE`:

```
x <- "One string ... with both embedded ellipses (...) ... and a final period."
gsub(".", "", x, fixed = TRUE)

## [1] "One string  with both embedded ellipses ()  and a final period"
```

Because regular expressions are both extremely powerful and somewhat counterintuitive (at least until you have substantial experience with them), Paul Murrell offers the following advice on working with them [56, p. 334]:

> ... it is important to remember that regular expressions are a small computer language of their own and should be developed with just as much discipline and care as we apply to writing any computer code. In particular, a complex regular expression should be built up in smaller pieces in order to understand how each component of the regular expression works before adding further components.

Because examples are extremely helpful in learning exactly what different regular expressions do, the rest of this discussion consists of examples that illustrate specific regular expression metacharacters and common constructions. For much more complete discussions of regular expressions, refer to books like that by Friedl [26] or the on-line help available at the following URL:

http://www.regular-expressions.info

8.3.2 Some useful regular expression examples

The rest of this section consists of a short list of some particularly useful regular expression constructs and examples. It is important to emphasize that these constructs can be combined in many creative ways to generate powerful tools for finding and replacing character sequences in text strings. The following examples cannot begin to convey the range of possibilities: the best way to learn this is to try a lot of different regular expression constructs yourself. For more details, see the references cited at the end of Sec. 8.3.1.

Character sets

Square brackets in a regular expression—i.e., an opening bracket ("["), followed by a bunch of text, followed by a closing bracket ("]")—define a *character set* that can be used to match any character in the set. As a simple example, the following code replaces all vowels in a text string with an underscore ("_"):

```
x <- "This text string is a typical example, not very interesting."
gsub("[aeiouy]", "_", x)

## [1] "Th_s t_xt str_ng _s _ t_p_c_l _x_mpl_, n_t v_r_ _nt_r_st_ng."
```

Another common use for square brackets is to indicate character ranges, like "all lowercase letters" (denoted "[a-z]") or "all numbers" (denoted "[0-9]"), or restricted ranges like "all uppercase letters from A through L" (denoted "[A-L]"). The following example illustrates these applications:

```
x <- "A note from Benny Thornton about Eddie Coyle, 32."
gsub("[a-z]", "_", x)

## [1] "A ____ ____ B____ T_____ _____ E____ C____, 32."

gsub("[0-9]", "_", x)

## [1] "A note from Benny Thornton about Eddie Coyle, __."

gsub("[A-L]", "_", x)

## [1] "_ note from _enny Thornton about _ddie _oyle, 32."
```

As Murrell notes in his discussion of regular expressions, most characters that would normally function as metacharacters in a regular expression function as literals when enclosed between square brackets, but there are exceptions. For example, the period (".") functions as the literal "period" when it appears between squre brackets, as in the following example:

```
x <- "This string is another string.  It is a three-part string. With periods."
gsub("g.", "g;", x)

## [1] "This string;is another string;  It is a three-part string; With periods."

gsub("g[.]", "g;", x)

## [1] "This string is another string;  It is a three-part string; With periods."
```

Note that the first example here replaces the regular expression "g." with "g;" and, here, the period functions as a metacharacter meaning "any character," so all occurrences of the letter "g" followed by anything else are replaced with "g;". In the second case, including the period between square brackets causes it to be interpreted as a literal, so only the strings "g." are replaced with "g;".

An important exception is the metacharacter caret ("^"): when this character appears in a standard regular expression, it is interpreted to mean "strings that *start* with the indicated text," but if this metacharacter appears at the beginning of a sequence between square brackets, it means "anything *except* the following characters." The following examples illustrate this behavior:

```
x <- "A note from Benny Thornton about Eddie Coyle, 32."
gsub("[^a-z]", "_", x)

## [1] "__note_from__enny__hornton_about__ddie__oyle_____"

gsub("[^0-9]", "_", x)

## [1] "_____32_"

gsub("[^A-L]", "_", x)

## [1] "A_____B_____E____C_____"
```

In the first example, the regular expression "`[^a-z]`" in the `gsub` function means "replace any character *except* lower-case letters with an underscore." Thus, capital letters, spaces, punctuation marks, and numbers are replaced with underscores, but none of the lower-case letters are. In the second example, everything except the digits 0 through 9 is replaced with an underscore, and in the final example, everything except the uppercase letters A through L is replaced with an underscore.

Repetition metacharacters

The metacharacters "?", "*", and "+" in a regular expression cause different degrees of repetition, and they can be particularly useful in conjunction with character sets defined by opening and closing square brackets. These metacharacters cause the following subpattern matching behavior:

- the metacharacter "?" matches either zero or exactly one time;
- the metacharacter "*" matches *zero or more* times;
- the metacharacter "+" matches *one or more* times.

The following substitution examples illustrate this behavior:

```
x <- "This is a text string; it is...it is."
gsub("is[.]", "_", x)

## [1] "This is a text string; it _...it _"

gsub("is[.]?", "_", x)

## [1] "Th_ _ a text string; it _..it _"

gsub("is[.]*", "_", x)

## [1] "Th_ _ a text string; it _it _"

gsub("is[.]+", "_", x)

## [1] "This is a text string; it _it _"
```

Note that in these substitutions, since "."—normally the "replace everything" metacharacter—is enclosed in square brackets, it is treated as a literal. Thus, the first substitution replaces "is." wherever it is found with an underscore, and this match occurs twice: once in "is..." and again at the end of the sentence with "is." The second substitution is more aggressive, replacing either "is" or "is." with an underscore, but it does not replace the entire sequence "is..." with an underscore since the substitution rule is "replace 'is' followed by either no period or exactly one period." The third substitution uses the still more aggressive substitution rule "replace 'is' followed by zero or more periods," causing "is..." to be replaced with an underscore. Finally, the fourth substitution rule is "replace 'is' followed by one or more periods," so the places where "is" occurs in the string without any trailing periods are not replaced.

A particularly useful application of these repetition operators is in parsing text strings into words. This was one of the fundamental text analysis preprocessing tasks introduced in Sec. 8.1.1, and it is one of the tasks performed by the `tokens` function in the `quanteda` package. As demonstrated in Sec. 8.1.2, this function splits every text string in a character vector into a "bag of words," represented by a list with one element for each element of the character vector. Each list element contains a character vector whose components are the individual words in the corresponding element of the original character vector. The following example illustrates this conversion:

```
library(quanteda)
x <- c("This string is a string with three elements.",
          "This element is the second element,",
          "and this is the third.")
xTokens <- tokens(x)
xTokens

## tokens from 3 documents.
## text1 :
## [1] "This"     "string"   "is"        "a"         "string"   "with"
## [7] "three"    "elements" "."
##
## text2 :
## [1] "This"     "element"  "is"        "the"       "second"   "element"  ","
##
## text3 :
## [1] "and"      "this"     "is"        "the"       "third"    "."
```

The `tokens` function supports three different methods for splitting text strings into words specified by the `what` argument, including the default method `word` and two alternatives: a "faster but dumber" method, `fasterword`, and the "dumbest but fastest method," `fastestword`. This last method performs a string split on spaces, similar to the following example:

```
x <- c("This string is a string with three elements.",
          "This element is the second element,",
          "and this is the third.")
strsplit(x, split = " ")
```

```
## [[1]]
## [1] "This"     "string"     "is"      "a"      "string"    "with"
## [7] "three"    "elements."
##
## [[2]]
## [1] "This"     "element"  "is"       "the"     "second"   "element,"
##
## [[3]]
## [1] "and"    "this"    "is"    "the"    "third."
```

Note that while this result is almost identical to the previous one, there are some subtle differences that can be extremely important. In particular, note that the first example declares the punctuation marks "." and "," to be words, while the `strsplit` result does not, instead declaring words to be anything separated by a single space. As shown in Sec. 8.1.2, the handling of punctuation marks is an extremely important topic, but it is useful to conclude this discussion with an example illustrating another potentially important subtlety, one where the repetition operator "+" can prove extremely useful.

Specifically, in the above example, words were defined as anything separated by single spaces and, in this case, this working definition of a word was satisfactory. If, however, a text string contains *multiple* spaces, this working definition is no longer adequate but needs to be replaced by "words are substrings separated by one or more spaces." This example, which contains embedded spaces, illustrates the difference:

```
x <- "This is a sequence with  some   extra     spaces"
strsplit(x, " ")

## [[1]]
## [1] "This"     "is"      "a"      "sequence" "with"      " "
## [7] "some"     " "       " "      "extra"    " "         " "
## [13] " "       "spaces"

strsplit(x, "[ ]+")

## [[1]]
## [1] "This"     "is"      "a"      "sequence" "with"     "some"
## [7] "extra"    "spaces"
```

Note that splitting this string on the single space character decomposes it into 14 "words," six of which are blank: since words are defined as being separated by a single space, any extra spaces appear as "blank words." In the second case, by splitting on one or more spaces, the result is the eight recognizable words we expect to see.

Alternatives and groups

Sometimes, we want to find, replace, or split on one of several different patterns, depending on what is found in the text string. For example, some short substrings may appear either as stand-alone words or as part of a longer word, as in this case:

```
x <- "this is an example, that's what this is"
gsub("is", "__", x)
```

```
## [1] "th__ __ an example, that's what th__ __"
```

Here, we have replaced every ocurrence of "is" with a double underscore, including both the word "is" and the final two letters of the word "this." If we are searching for specific words—especially short words—this may not be the result we want. We can get closer to what we want by putting spaces before and after the two letters in our word:

```
x <- "this is an example, that's what this is"
gsub(" is ", "__", x)
```

```
## [1] "this__an example, that's what this is"
```

While this modification no longer changes the letters "is" in the word "this," it also does not pick up the final "is" because this word appears at the end of the text string and is not followed by a space. What we really want is to search for the two letters "is," preceded by either a space or the beginning of the text string and followed by either a space or the end of the text string.

The grouping metacharacters "(" and ")", together with the "|" alternation metacharacter, allow us to make more flexible specifications like this in regular expressions. For example, the search pattern "(a|A)" means "either lower case 'a' or upper case 'A'. " For the example considered here, the following specification accomplishes what we want:

```
x <- "this is an example, that's what this is"
gsub("(^| )is( |$)", "__", x)
```

```
## [1] "this__an example, that's what this__"
```

The beginning of this search string ("(^|)") indicates that we are searching for patterns that either occur at the beginning of the character string or begin with a leading space, while the end of the search string ("(|$)") indicates that this pattern must end either with a space or occur at the end of the text string. Here, we have achieved the replacement we want.

As a final example, the following regular expression appears on the cover of Paul Murrell's *Introduction to Data Technologies* book [56]:

^((HT|X)M|SQ)L|R$

The following example shows what this regular expression matches:

```
x <- c("Book about HTML, XML, SQL, and R.", "HTML", "XML", "SQL", "R", "etc.")
gsub("^((HT|X)M|SQ)L|R$", "keyTopic", x)
```

```
## [1] "Book about HTML, XML, SQL, and R." "keyTopic"
## [3] "keyTopic"                          "keyTopic"
## [5] "keyTopic"                          "etc."
```

In particular, note that Murrell's cover expression matches "HTML," "XML," "SQL," or "R," but only as stand-alone words that are neither preceded nor followed by anything else.

To conclude, it is worth emphasizing the point that Murrell makes about treating the development of regular expressions as a programming exercise, approached carefully and systematically. It is extremely easy to make "small" mistakes that have enormous consequences. For example, an extra space at the beginning of Murrell's clever example causes it to fail, matching only one of the four intended terms:

```
x <- c("Book about HTML, XML, SQL, and R.", "HTML", "XML", "SQL", "R", "etc.")
gsub(" ^((HT|X)M|SQ)L|R$", "keyTopic", x)

## [1] "Book about HTML, XML, SQL, and R." "HTML"
## [3] "XML"                               "SQL"
## [5] "keyTopic"                          "etc."
```

8.4 An aside: ASCII vs. UNICODE

By default, the *R* functions considered in this book use ASCII encoding for text data, which stores each character as a single byte in memory. The term ASCII is an abbreviation for American Standard Code for Information Interchange, and this encoding is adequate to represent all of the keys on the standard American English keyboard. If we need to work with text in other languages, however, the 256 characters that can be represented as a single byte may not be enough: Japanese Kanji ideographs, for example, represent several thousand different characters. To address issues like this one, the UNICODE encoding has been developed to allow computers to work with any character in any language. The most common versions of UNICODE encodings are the one-byte UTF-8 encoding and the two-byte UTF-16 encoding.

The key point of this discussion is to be aware that different encodings exist, and this can cause difficulties if one encoding is assumed for text data that has actually been created under a different encoding. A useful function available in the `tools` package is `showNonASCII`, which can be applied to a character vector to find any non-ASCII characters. Applying this function to the unclaimed bank account character vector `unclaimedLines` reveals that the first 60 characters of element 4528 of this vector contains non-ASCII text:

```
library(tools)
x <- unclaimedLines[4528]
y <- substr(x, 1, 60)
showNonASCII(y)

## 1:  WILLINGDON <e2><80><93> 1980 HOMECOMING        ,,217.83,C 0  K EWON
```

The terms "<e2><80><93>" represent UNICODE characters in this text string.

A detailed discussion of encodings is beyond the scope of this book, but for those needing to know more, Paul Murrell gives a more detailed discussion in his book, *Introduction to Data Technologies* [56, Sec. 5.2.7]. For a still more detailed discussion, refer to the book by Friedl [26].

8.5   Quantitative text analysis

It was noted at the beginning of this chapter that, for analysis, text data must ultimately be converted into numbers that are amenable to mathematical characterizations. The following discussions introduce several tools for converting text data into numbers and then performing various types of text-specific analysis using these numbers. All of the techniques described here are illustrated further in the examples presented in Sec. 8.6.

8.5.1   Document-term and document-feature matrices

One of the simplest and most useful numerical characterizations of a collection of text data is the *document-term matrix*, along with minor variations on this idea like the *term-document matrix* and the *document-feature matrix*. Essentially, a document-term matrix is a rectangular array of numbers, where each row represents a document, each column represents a word or term that may be present in the document, and the numbers in the matrix tell us how many times each term appears in each document. The term-document matrix is the transpose of this structure, where each row represents a term and each column represents a document. Finally, a document-feature matrix is a generalization of the document-term matrix where rows again represent documents, but the columns can be more general *features* like word pairs or *bigrams*, triples or *trigrams*, or higher-order *n-gram sequences* of n successive words.

The text mining package `tm` includes the functions `DocumentTermMatrix` and `TermDocumentMatrix`, which each take a corpus of documents and generate the desired matrix from it. The `tm` package does not support document-feature matrices, but the `quanteda` does via the `dfm` function, which can incorporate n-grams of arbitrary order as features. In addition, the `dfm` function accepts either a corpus or a character vector as its one required argument, and it supports a wide range of logical arguments that can be used to turn on any of the preprocessing steps discussed in Sec. 8.1.1. Because of this greater flexibility, the following examples use the `dfm` function from the `quanteda` package to demonstrate both document-term matrices, where each feature corresponds to a word or *unigram*, and more general document-feature matrices where the features can include n-grams of arbitrary order.

These examples are based on the following simple text vector, which has three components, each of which will be treated as a document in constructing document-term and document-feature matrices:

```
textVector <- c("this document has three elements",
                "each element of this document is made of text strings",
                "each text string in this document is made of words")
```

Applying the `dfm` function to this vector yields the corresponding document-term matrix:

```
exampleDTM <- dfm(textVector, verbose = TRUE)

## Creating a dfm from a character input...
##    ...   lowercasing
##    ...   found 3 documents, 15 features
##    ...   created a 3 x 15 sparse dfm
##    ...   complete.
## Elapsed time:  0.03 seconds.
```

Setting `verbose = TRUE` causes `dfm` to display information about how the matrix is constructed. In particular, note that lower case-conversion is applied by default, but other preprocessing steps like stopword removal are not. Because this document-feature matrix is small, it is useful to examine it here to see the typical structure of these matrices:

```
exampleDTM

## Document-feature matrix of: 3 documents, 15 features (46.7% sparse).
## 3 x 15 sparse Matrix of class "dfmSparse"
##         features
## docs    this document has three elements each element of is made text
##    text1   1       1   1    1        1    0      0     0  0  0    0
##    text2   1       1   0    0        0    1      1     2  1  1    1
##    text3   1       1   0    0        0    1      0     1  1  1    1
##         features
## docs    strings string in words
##    text1    0       0   0    0
##    text2    1       0   0    0
##    text3    0       1   1    1
```

Each row of this matrix corresponds to a document, and the columns correspond to the words that occur in any document, in their order of appearance when going from the first document to the last. Thus, each position in the matrix corresponds to a specific document/term combination that contains the number of times the term occurs in the document. In a large document collection, many terms appear in only a few documents, so many of these entries—in fact, often the majority of these entries—are zero, motivating the use of special *sparse matrix techniques* as noted in the `dfm` listing shown here.

As we saw in Sec. 8.1.2, the `topfeatures` function returns a list of the most frequently occurring features in a document-feature matrix, an extremely useful tool in dealing with larger matrices like those considered in Sec. 8.6. By default, this function returns the top 10 features, in decreasing order of frequency, although both this number and the sort order may be changed. For the matrix `exampleDTM`, the top 10 features are:

```
topfeatures(exampleDTM)
```

```
##      this document        of     each       is     made     text      has
##         3        3         3        2        2        2        2        1
##     three elements
##         1        1
```

The most frequent terms in this document collection are "this," "document," and "of," each appearing three times, followed by "each," "is," "made," and "text," all occurring twice, with all other terms appearing only once.

To illustrate the difference between a document-term matrix and a document-feature matrix, consider the `dfm` function with the optional `ngrams` specification, which allows us to specify that we want to consider n-grams as features, consisting of sequences of n words appearing in order. In this particular example, we will specify n-grams of orders 1 and 2, corresponding to simple words and *bigrams*. Specifically, consider the following R code:

```
exampleDFM <- dfm(textVector, ngrams = 1:2, verbose = FALSE)
topfeatures(exampleDFM, n = 32)
```

```
##             this        document  this_document                of           each
##                3               3              3                 3              2
##               is            made           text       document_is        is_made
##                2               2              2                 2              2
##          made_of             has          three          elements   document_has
##                2               1              1                 1              1
##       has_three three_elements        element           strings   each_element
##                1               1              1                 1              1
##       element_of         of_this        of_text      text_strings         string
##                1               1              1                 1              1
##               in           words      each_text       text_string      string_in
##                1               1              1                 1              1
##          in_this        of_words
##                1               1
```

All of the frequent features we saw in the document-term matrix constructed earlier are still present, with the same frequencies as before, but now we have additional features. These are bigrams like "this document," which appears three times, three bigrams that each appear twice, and 13 other bigrams that each appear once.

We saw in the example discussed in Sec. 8.1.2 that the feature frequencies returned by the `topfeatures` function applied to a document-feature matrix can be used to generate wordclouds that help us visualize important features in text. An application of these matrices to characterizing a much longer document is presented in Sec. 8.6.1.

8.5.2   String distances and approximate matching

The text analysis packages `tm` and `quanteda` characterize words or features in documents. There are applications, however, like the detection and correction of misspellings that involve *approximate matches*, and there the focus is on individual characters in different text strings. These applications can be approached using the **stringdist** package, which implements several distance measures between text strings, allowing us to quantify their similarity and construct approximate matches between strings that are "not too different."

Approximate string matching with amatch

Approximate string matching is supported in the **stringdist** package by the function amatch, which has two required arguments and several optional arguments. The required arguments are x, a character vector of strings we wish to find approximate matches for, and table, a character vector representing the lookup table for matching. Two very useful optional arguments are method, discussed in detail below, and maxDist, discussed next.

To see how the amatch function works, consider the following simple example. First, define the character vector x of strings for which we want approximate matches and table containing the possible matches:

```
x <- c("app", "application", "applied", "apple", "approximate", "basic", "module")
table <- c("application", "base", "center", "mode")
```

Note that the second element of x matches the first element of table exactly, but all of the other elements of x are at best approximate matches to elements of this table. The amatch function returns an index vector with one element for each element of x that points to the closest approximate match in table. Applying this function with its default options gives these results:

```
index <- amatch(x, table)
index

## [1] NA  1 NA NA NA NA NA

table[index]

## [1] NA             "application" NA            NA            NA
## [6] NA             NA

x

## [1] "app"          "application" "applied"     "apple"       "approximate"
## [6] "basic"        "module"
```

The second element of this return vector points to the first element of table, reflecting the exact match just described, but all other elements of this vector are NA, indicating that amatch was unable to find a sufficiently good approximate match in table. This result illustrates the importance of the optional arguments maxDist and method noted above.

The amatch function is based on the **stringdist** function discussed in some detail below, first calling this function to compute the distances between each element of x and all elements of table. If one or more elements of table exhibit **stringdist** values that are less than maxDist, the element with the

smallest of these values is selected as the best match and a pointer to this
element of `table` is returned. Otherwise, `NA` is returned, indicating that no
acceptable match was found. Thus, the results we obtain—and the practical
utility of these results—depend on both the method used in computing string
distances and the maximum distance threshold. The default method is "osa"
described below, and the default maximum distance threshold is 0.1; increasing
this argument to 2 finds three matches, which can be seen most clearly by using
the return value from `amatch` directly as an index into `table`:

```
table[amatch(x, table, maxDist = 2)]

## [1] NA                 "application" NA               NA             NA
## [6] "base"             "mode"

x

## [1] "app"              "application" "applied"        "apple"        "approximate"
## [6] "basic"            "module"
```

This increase in `maxDist` yields the same exact match as before, but now the
function also returns approximate matches between the term "basic" in x and
"base" in `table`, and between "module" in x and "mode" in `table`. Increasing
this threshold further to 4 expands the set of approximate matches, but the new
matches are probably not what we expect:

```
table[amatch(x, table, maxDist = 4)]

## [1] "base"             "application" NA               "base"         NA
## [6] "base"             "mode"
```

In particular, note that the elements "app" and "apple" are both matched to
"base," even though "application" seems the closer match. This result is a
consequence of the way the osa (*optimal string alignment*) method works. If
we change from this default distance to the *Jaro-Winkler distance*, developed
by the US Bureau of Census for approximate name and address matching, we
obtain very different results, at least for appropriately chosen `maxDist` values.
In particular, specifying `method = 'jw'` with the default `maxDist` value gives
the same results as the default osa method, but if we increase this maximum
distance parameter to 0.4, we obtain much better approximate matches:

```
table[amatch(x, table, method = "jw", maxDist = 0.4)]

## [1] "application" "application" "application" "application" "application"
## [6] "base"             "mode"

x

## [1] "app"              "application" "applied"        "apple"        "approximate"
## [6] "basic"            "module"
```

The main points of this example are, first, that the approximate matching results we obtain with the `amatch` function depend on both the string distance

measure and the maximum distance threshold we use, and, second, that these two selections must be coordinated. For example, increasing `maxDist` from its default value of 0.1 to 0.4 with the default osa method has no impact at all, in marked contrast to the jw method. For this reason, if we are to obtain useful approximate string matching results with the `amatch` function, it is necessary to have some intuition about the different string distance options that can be specified with the `method` argument.

String distances and similarities

The string distance function `stringdist` in the R package of the same name implements 10 different string distance measures, specified by the optional `method` argument, and it returns non-normalized distance values. A closely related function is `stringsim`, which uses these same 10 string distance measures as the basis for computing string similarities, which vary between 0 for strings that are not similar at all, to 1 for strings that are identical. The following discussion gives a brief overview of both of these functions, first examining the unexpected results obtained from the approximate string matching example just described. Since the `amatch` function calls the `stringdist` function to compute string distances, consider the distances between "app" and all of the candidate matches in `table` under the default osa method:

```
table

## [1] "application" "base"        "center"      "mode"

stringdist("app", table)

## [1] 8 3 6 4
```

The surprising result seen here is that "application"—the term we probably view as the best match to "app"—actually has the largest osa string distance of any of the elements of `table`. This surprising behavior arises from the fact that the osa distance between text strings of very different lengths is inherently large, a consequence of the way this distance is defined (see the second example for a discussion). Thus, the best osa match for "app" is "base" because the two strings are about the same length and they share one letter, whereas "application" is much longer and thus deemed much farther away. Changing the method to "jw" emphasizes matches in similarly positioned characters in the two strings, with much less emphasis on differences in string length:

```
table

## [1] "application" "base"        "center"      "mode"

stringdist("app", table, method = "jw")

## [1] 0.2424242 0.4722222 1.0000000 1.0000000
```

Here, the closest match to "app" is the expected result "application" and all of the other elements of `table` are substantially farther away.

As noted, one advantage of the string similarity function `stringsim` over the string distance function `stringdist` is that similarities are easier to interpret because they are normalized. In particular, recall that similarites range between 0 and 1, with small similarities indicating very dissimilar strings and values near 1 indicating nearly identical strings. These similarities are based on string distances, but there are some subtle and possibly surprising differences between these two characterizations. For example, if we performed our approximate matching on string similarities instead of string distances, we would match "app" to "application" instead of "base" as `amatch` does:

```
table

## [1] "application" "base"        "center"      "mode"

stringsim("app", table)

## [1] 0.2727273 0.2500000 0.0000000 0.0000000
```

In particular, the normalization used in computing string similarities depends on the string lengths. To see this point, consider the following examples. First, compute the string distances between "app" and the following three terms:

```
stringdist("app", c("bas", "base", "basement"))

## [1] 3 3 7
```

Next, consider the corresponding distances between "app" and the following terms, each of the same length as before, but with no letters in common:

```
stringdist("app", c("xxx", "xxxx", "xxxxxxxx"))

## [1] 3 4 8
```

Now, consider the string similarities between "app" and the original three terms:

```
stringsim("app", c("bas", "base", "basement"))

## [1] 0.000 0.250 0.125
```

Here, the similarity between "app" and "bas" is zero because the distance between these strings is the maximum possible for three-character match strings. Adding one more character, the distance between "app" and "base" is still only 3, while the maximum possible distance is now 4, giving a larger string similarity even though both strings still have only the character "a" in common, in a different position. Expanding the match string to "basement," the string similarity is smaller, but still not zero, since the distance between these strings is 7, compared with a maximum possible distance of 8.

The reason for examining this case in detail is that these results are counter-intuitive: how can "app" exhibit a greater similarity to "application" than to "base"—as we expect—when the *distance* between "app" and "application" is *larger*? As this example shows, the reason lies in how the distances are normalized in computing similarities. As a consequence, we may obtain different "best match" results using string similarities—where we must implement the matching ourselves—than with the `amatch` function, based on string distances.

A second example

As a second example, consider the unique `City` values containing the substring "UNK," and examine their distances and similarities to to the string "UN-KNOWN." These `City` values are:

```
uniqueMissing

##  [1] "UNKNOWN ALTA "            "UNKNOWN ???? "
##  [3] "UNKNOWN,UNKNOWN "          "UNKNOWN BC "
##  [5] "UNKNOWN UNKNOWN "          "UNKNOWN BC"
##  [7] "UNKNOWN AB "               "UNKNOWN AB"
##  [9] "UNKNOWN USA "              "UNKNOWN DEU "
## [11] "UNKNOWN UNKNOWN UNKNOWN "  "UNKNWON AB "
## [13] "UNKNOW ALTA M1M 1M1 "      "UNKNOWN GBR "
## [15] "UNKNOWN,UNKNWON "
```

As noted, the default method for `stringdist` is "osa" or *optimal string align-ment*, which counts the numbers of *deletions, insertions, substitutions,* and *adjacent character transpositions* required to turn one string into the other. This measure is closely related to the *edit distance* or *Levenshtein distance* [63, Sec. 10.8], which does not allow transpositions; for this example, both methods give identical results. The osa string distances and similarities are:

```
osaDists <- stringdist("UNKNOWN", uniqueMissing)
osaDists

## [1]  6  6  9  4  9  3  4  3  5  5 17  5 14  5  9

osaSims <- stringsim("UNKNOWN", uniqueMissing)
osaSims

##  [1] 0.5384615 0.5384615 0.4375000 0.6363636 0.4375000 0.7000000 0.6363636
##  [8] 0.7000000 0.5833333 0.5833333 0.2916667 0.5454545 0.3000000 0.5833333
## [15] 0.4375000
```

Note that these string distances vary from 3 to 17, while the string similarities are bounded between 0 and 1, as noted above. In contrast, the Jaro-Winkler distances span a much narrower range, as noted earlier:

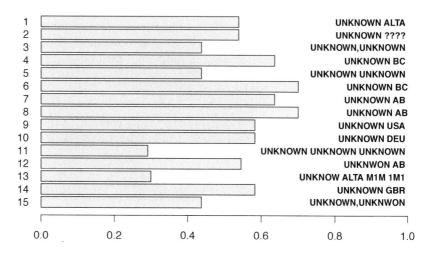

Figure 8.3: OSA similarities between "UNKNOWN" and the indicated text.

```
stringdist("UNKNOWN", uniqueMissing, method = "jw")

##  [1] 0.1538462 0.1538462 0.1875000 0.1212121 0.1875000 0.1000000 0.1212121
##  [8] 0.1000000 0.1388889 0.1388889 0.2361111 0.1688312 0.2809524 0.1388889
## [15] 0.1875000
```

Because the normalization of string similarities allows comparisons across different methods, the rest of this discussion focuses on string similarities to highlight
these differences.

A horizontal barplot of the osa string similarities computed above is shown
in Fig. 8.3, with the elements of uniqueMissing listed at the right and the
corresponding element numbers at the left. These similarities vary from 0.292 to
0.7, with the two most similar text strings being numbers 6 ("UNKNOWN BC")
and 8 ("UNKNOWN AB"). Also, note that "UNKNOWN AB " (string number
7), with an extra space at the end, has a slightly lower similarity. The least
similar string under this measure is number 11, which repeats "UNKNOWN"
three times, separated by spaces.

For comparison, Fig. 8.4 shows the results obtained by four other similarity
measures available in the stringsim function. Because labels like those included
in Fig. 8.3 would be too small to read, the different elements of uniqueMissing
are referred to by number in the four plots in Fig. 8.4, in the same order as
they appear in Fig. 8.3. The upper left plot shows the similarities computed for

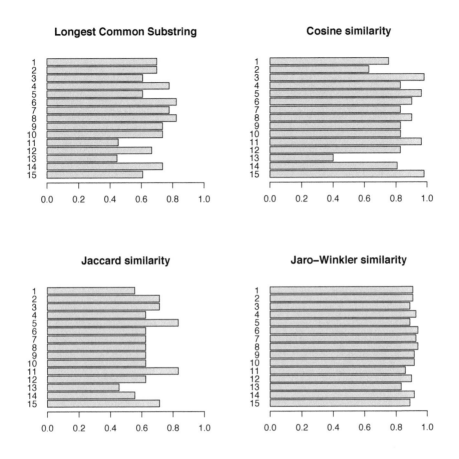

Figure 8.4: String similarities as in Fig. 8.3, by four different methods.

method = 'lcs', corresponding to the *longest common subsequence* method, which may be viewed as a variation of the edit distance described above that only allows deletions and insertions, but no substitutions. Not surprisingly, these results are quite similar to those obtained for the default osa method in Fig. 8.3, although they are not identical. The upper right plot shows the results obtained by the *cosine similarity*, based on comparisons of n-gram counts in the two text strings. Here, the results are very different, as may be seen by considering those strings whose computed similarity to the reference "UNKNOWN" exceed 0.95:

```
## [1] "UNKNOWN,UNKNOWN "        "UNKNOWN UNKNOWN "
## [3] "UNKNOWN UNKNOWN UNKNOWN " "UNKNOWN,UNKNWON "
```

Note that this set of "very highly similar sequences" includes the least similar sequence (number 11, "UNKNOWN UNKNOWN UNKNOWN ") under the osa method. The lower left plot shows the Jaccard string similarities, which measures the fraction of n-grams in common between the two strings. These similarities are generally larger than the osa values (ranging from 0.45 to 0.83, versus 0.29 to 0.7), and the best matches to "UNKNOWN" under this similarity measure are strings number 5 ("UNKNOWN UNKNOWN ") and 11 ("UNKNOWN UNKNOWN UNKNOWN "). Finally, the lower right plot shows the Jaro-Winkler string similarities, similar to those discussed earlier, but with a slight difference. In particular, the Jaro-Winkler measure is an extension of the Jaro measure that gives extra weight to matches in the first four characters in both strings. By default, the `jw` method for the `stringsim` function actually computes the Jaro measure; to get the full Jaro-Winkler measure, it is necessary to specify a positive value for the optional argument `p`. The results shown in the lower right in Fig. 8.4 use the full Jaro-Winkler measure with `p = 0.1`, a commonly recommended choice [67]. Since the first four characters of all 15 of the text strings considered here match the first four characters of "UNKNOWN" exactly, the Jaro-Winkler string similarities are all quite high here, ranging from 0.83 to 0.94.

A more detailed discussion of string distances and similarities is beyond the scope of this book, but for an excellent overview, refer to the paper by van der Loo [67]. The key point here is that the choice of method can make a large difference in practice, a point further emphasized in Sec. 8.6.3, where the problem of cleaning up the `City` values from the unclaimed bank account dataset is examined in more detail.

8.6   Three detailed examples

This chapter began by citing three motivating applications for text data analysis: the automated characterization of large document collections, the extraction of useful categorical covariates from text strings, and the cleaning of text data that *should* represent a useful categorical covariate, but exhibits misspellings or other text-based data anomalies that render it much less useful. The following three examples provide simple illustrations of each of these applications. The first, described in Sec. 8.6.1, applies text analysis tools from the `quanteda` package to characterize the text in an obscure book published in 1904, illustrating the first of these three applications. Next, Sec. 8.6.2 considers the second application, extracting the manufacturer's name from a longer text string that uniquely defines each of 209 computers. The objective of this example is to construct a mathematical model that predicts computer performance from a variety of characteristics, and the key question of interest here is whether the computer manufacturer represents a covariate that can improve our model's performance. The last example applies several of the tools described in this chapter to clean up the extremely messy `City` variable from the Canadian unclaimed bank account dataset introduced in Chapter 4 and examined further in Sec. 8.6.3.

8.6.1 Characterizing a book

A practical problem of growing importance is that of characterizing a large collection of documents without reading them all. Typical applications include classifying a group of documents or searching for documents relating to a certain topic from a collection that is too large to be read through completely. For example, suppose we have tens of thousands of call logs from customer service representatives and we want to answer questions like "what are the top 10 customer concerns this quarter, and how do they compare with those for the previous two quarters?" The example presented here is in a somewhat different vein, but it addresses the same basic question: how can we use the text analysis techniques described in this chapter to obtain a useful characterization of a collection of documents without having to read them all? More specifically, this example considers the 34 chapter-length sections of the 1904 novel *Born Again* by Alfred Lawson as individual documents in a corpus, applying the tools from the `quanteda` package.

The text analyzed here was obtained from *Project Gutenberg*, which "offers over 53,000 free ebooks" according to their website (`www.gutenberg.org`, accessed 2/26/2017). Like the Socrata website discussed in Chapter 4, Project Gutenberg offers its materials in a variety of formats, and one of the documents available as an ASCII text file is Lawson's novel. Lawson was an extremely colorful character whose life is described in the biography by L.D. Henry [40]: his accomplishments included a stint as a minor-league baseball pitcher in the late nineteenth century, the founding of two aeronautical magazines in the very early days of aircraft (he appears to have introduced the term "aircraft" into widespread use [40, p. 61]), the founding of an organization called "The Direct Credits Society" during the Great Depression of the 1930's, and the development of his own version of science called "Lawsonomy," defined as "the knowledge of Life and everything pertaining thereto." Lawson was the subject of a chapter in Martin Gardner's book, *Fads and Fallacies in the Name of Science*, who described *Born Again* as "the worst work of fiction ever published" [29]. This example applies the text analysis tools described earlier in this chapter to gain some insight into the contents of this novel without having to read every word.

We can use the `download.file` function discussed in Chapter 4 to download the ASCII text file from Project Gutenberg and save it as `LawsonBornAgain.txt` in our working directory. Reading this file with the `readLines` function and looking at the first six records gives us a preliminary idea of its contents:

```
BornAgain <- readLines("LawsonBornAgain.txt")
head(BornAgain)

## [1] "The Project Gutenberg EBook of Born Again, by Alfred Lawson"
## [2] ""
## [3] "This eBook is for the use of anyone anywhere at no cost and with"
## [4] "almost no restrictions whatsoever.  You may copy it, give it away or"
## [5] "re-use it under the terms of the Project Gutenberg License included"
## [6] "with this eBook or online at www.gutenberg.org"
```

The result is a character vector with 6195 elements, and further examination shows that the complete text may be divided into these parts:

1. A Project Gutenberg preamble, identifying the source and the document;

2. A short section by Lawson labeled "DEDICATION";

3. The main body of the text consists of 33 sections, each labelled with the word "CHAPTER," followed by a final section labelled "EPILOGUE";

4. A section of short comments by Lawson labelled "STRAY SHOTS";

5. The Project Gutenberg license agreement.

The analysis that follows is based entirely on the text included in (3), represented as a corpus of 34 individual documents. The first step in this analysis is then to extract this portion of the text, consisting of that from the element of the BornAgain vector that contains "CHAPTER I" to the element that contains "(THE END.)". The following R code extracts this subset:

```
firstLine <- grep("CHAPTER I$", BornAgain)
lastLine <- grep("(THE END.)", BornAgain, fixed = TRUE)
LawsonText <- BornAgain[firstLine:lastLine]
```

Note that the first line of this code searches for the regular expression "CHAPTER I$", requiring that the line of text *end* with "CHAPTER I" to avoid obtaining a sequence of indices into lines of text listing "CHAPTER II," "CHAPTER III," or "CHAPTER IV," among other possibilities. Also, note that the second line of code specifies `fixed = TRUE` so that the period in "(THE END.)" is *not* treated as a regular expression. The result consists of 5611 lines of text, the first six and last six of which are:

```
head(LawsonText)

## [1] "CHAPTER I"
## [2] ""
## [3] "Judging from my own experience it is my opinion that many strange and"
## [4] "wonderful events have happened during the past in which man took part,"
## [5] "that have never been recorded."
## [6] ""

tail(LawsonText)

## [1] "gamblers, sneaks, loafers, spongers, and all other kinds of human"
## [2] "parasites to grow fat off the labors of those who toil. Say that I shall"
## [3] "take up the work where John Convert left off, and devote the remainder"
## [4] "of my life and all of my wealth towards the cause he advocated.'\""
## [5] ""
## [6] "(THE END.)"
```

The next step in our analysis is to parse this text into vectors representing the individual chapters of the novel. We do this by using the `grep` function to construct indices pointing to the markers separating the sections, e.g.:

```
CHindex <- grep("CHAPTER", LawsonText)
EPindex <- grep("EPILOGUE", LawsonText)
ENDindex <- grep("(THE END.)", LawsonText, fixed = TRUE)
```

From these indices, we construct pointers to the beginning and end of each chapter:

```
firstIndices <- c(CHindex + 1, EPindex + 1)
nFinal <- length(CHindex)
lastIndices <- c(CHindex[2:nFinal] - 1, EPindex - 1, ENDindex - 1)
```

Given these pointers, we can construct a corpus where each chapter-length segment is represented initially as a single character vector, and then tokenized to create a bag-of-words corpus that will serve as the basis for subsequent analysis. These tasks are best accomplished by defining a function, e.g.:

```
BuildCorpus <- function(textVector, first, last){
  #
  n <- length(first)
  cVector <- vector("character", n)
  for (i in 1:n){
    cVector[i] <- paste(textVector[first[i]:last[i]], collapse = " ")
  }
  outCorpus <- corpus(cVector)
  return(outCorpus)
}
```

Using this function, we can create a corpus of chapters, tokenize it, and remove English-language stopwords:

```
LawsonCorpus <- BuildCorpus(LawsonText, firstIndices, lastIndices)
LawsonTokens <- tokens(LawsonCorpus, what = "word",
                                     remove_numbers = TRUE,
                                     remove_punct = TRUE,
                                     remove_symbols = TRUE)
LawsonClean <- tokens_remove(LawsonTokens, stopwords("english"))
```

Figs. 8.5 and 8.6 present a set of wordclouds characterizing the most frequent terms appearing in selected chapters. The basis for this selection was a preliminary visual examination of the wordclouds generated for all 34 of the book's chapters, retaining only eight to give a broad indication of some of the themes from the book. These wordclouds were generated using the following custom R function, which extracts the ith chapter from the text defined by **tokenList**, constructs a document-feature matrix from these tokens (by default, this process converts the text to lower case), and uses the results to construct a wordcloud:

```
ChapterWordcloud <- function(tokenList, i, topK = 20, iseed = 11, ...){
  #
  DFM <- dfm(tokenList[[i]])
  tf <- topfeatures(DFM, n = topK)
  set.seed(iseed)
  wordcloud(names(tf), tf, ...)
}
```

Figure 8.5: Wordcloud summaries of four selected chapters from *Born Again*.

The upper left plot in Fig. 8.5 characterizes Chapter 2, featuring the words "ship" and "captain," suggesting an ocean voyage. The upper right wordcloud summarizes Chapter 3, where the words "life" and "god" are most prominent, accompanied by "alone," "courage," and "strength," suggesting a contemplation of mortality (if you read the text, you learn that our first-person protagonist has been thrown overboard by the crew of the ship he was sailing on and left to drown). The lower left wordcloud summarizes Chapter 7, and it introduces a number of interesting new terms, including "arletta," "apeman," and "noah." The first few lines of text from this chapter provide a partial explanation:

```
index7 <- firstIndices[7]
LawsonText[index7:(index7 + 5)]
```

```
## [1] ""
```

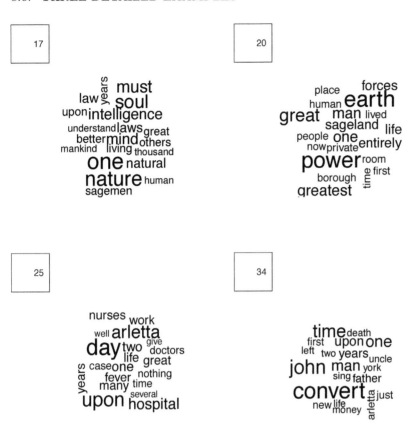

Figure 8.6: Wordclouds for *Born Again* chapters 17, 20, 25, and 34 (Epilogue).

```
## [2] "\"And so you inform me that there is nothing left of beautiful Sageland"
## [3] "but a heap of ruins surrounded by the sea,\" mused the lovely--the idea"
## [4] "struck me to name her Arletta--\"tell me what happened to the rest of my"
## [5] "people.\""
## [6] ""
```

If we read further, our protagonist explains to Arletta—who has been asleep for over 4000 years, until she was awakened by his kiss—that her world ("Sageland") was washed away by the Great Flood, leading to the story of Noah's ark, and thus the appearance of "noah" in the lower left wordcloud. The lower right wordcloud for Chapter 8 features words like "gold," "selfishness," and "swine."

Fig. 8.6 presents wordcloud summaries of four more chapters from *Born Again*. The upper left plot characterizes Chapter 17, which is the first one in which the word "sagemen" appears as a frequent term. Searching the text, we find that the word first appears in Chapter 14, but only once, so it does not qualify as a frequent term until Chapter 17, where it appears five times. Similarly, even though "sageland" appears for the first time in the text quoted above from Chapter 7, it appears as a frequent term for the first time in Chapter 20, where it appears five times; this may be seen in the wordcloud in the upper right in Fig. 8.6 summarizing Chapter 20. The wordcloud in the lower left summarizes Chapter 25, which features the words "arletta," "hospital," and "doctors." Again, a brief look at the text of this chapter reveals that it is set in a hospital where the protagonist is being treated, and it features *two* Arletta's: memories of the one from Sageland, and another, younger one who visits him regularly. Finally, the lower right plot is a wordcloud summarizing the Epilogue, where the words "john" and "convert"—the name of the central character of the novel—appears for the first time, along with the word "death," which features prominently in this final chapter of the book.

The key point of this example has been that the use of text analysis, especially wordcloud summaries based on document-feature matrices constructed from text data, can provide useful pointers into documents or parts of documents in a large collection that may be of interest, without having to read the entire collection. In cases where the collection is extremely large—e.g., thousands or tens of thousands of documents—it may not be feasible to read the entire collection even if we wanted to. In such cases, analysis techniques like those illustrated here can be used to identify portions of the collection that are worth reading or otherwise characterizing more carefully.

Finally, although wordclouds are often rendered in black-and-white, as in Figs. 8.5 and 8.6, colored wordclouds can make it easier to distinguish between terms with different frequencies. This point is illustrated in Fig. 9.27 in the group of color figures at the end of Chapter 9, which shows the bottom right plot from Fig. 8.6 characterizing the words that appear most frequently in chapter 34 (the Epilogue). The code used to generate this plot was:

```
ChapterWordcloud(LawsonClean, 34, scale = c(4, 0.5),
                            colors = rainbow(20))
```

Here, the optional argument `colors` for the `wordcloud` function is passed via the ... mechanism discussed in Chapter 7 from the `ChapterWordcloud` call. The function `rainbow` has been used to generate a collection of 20 colors, which are assigned to the words in the wordcloud based on their frequencies (i.e., words like "uncle" and "money" with the same frequency appear in the same color).

8.6.2   The `cpus` data frame

The `cpus` data frame in the `MASS` package characterizes the performance of 209 computers, based on a 1987 publication [24]. The first six records are:

```
head(cpus)
```

```
##              name syct mmin  mmax cach chmin chmax perf estperf
## 1  ADVISOR 32/60  125  256  6000  256    16   128  198     199
## 2  AMDAHL 470V/7   29 8000 32000   32     8    32  269     253
## 3  AMDAHL 470/7A   29 8000 32000   32     8    32  220     253
## 4  AMDAHL 470V/7B  29 8000 32000   32     8    32  172     253
## 5  AMDAHL 470V/7C  29 8000 16000   32     8    16  132     132
## 6  AMDAHL 470V/8   26 8000 32000   64     8    32  318     290
```

This data frame includes the text variable `name`, a numerical performance measure (`perf`), the predicted performance from a model developed by the authors (`estperf`), and six numerical computer characteristics. The `name` variable includes both the computer manufacturer and a specific model number, with one unique value for each record, so it is not a useful prediction covariate. This example proceeds in three steps to show how the basic character manipulation functions in R can be used to extract a categorical variable from `name` that can be incorporated in linear regression models. First, we build two baseline linear regression models that predict performance from the numerical computer characteristics. Next, we construct a categorical variable `mfgr` from the `name` text field, and finally, third, we include `mfgr` in our regression models to see how much it improves their performance.

Since linear regression model parameters are closely related to correlation coefficients, we can make preliminary variable selections by examining the correlation matrix of all of the numerical variables from the `cpus` data frame. This matrix is shown in Fig. 8.7 using the `corrplot` function from the `corrplot` package discussed in Chapter 9, where the variable names are listed on the diagonal. Ellipses showing the direction and magnitude of the correlation between each pair of variables appear above the diagonal, and the numerical correlation values appear below the diagonal. Examining the correlations with the performance variable `perf`, the most strongly associated other variables seem to be `mmax` (the maximum main memory), `chmax` (the maximum number of I/O channels), and `cach` (the cache size), all with positive correlations. Interestingly, the variable `syct` (the system cycle time) appears negatively correlated with all of the other variables, including the response variable `perf`, but these correlations are all fairly small, suggesting a weak effect.

Fig. 8.8 shows the *Spearman rank correlations* between these variables, in the same format as Fig. 8.7, which forms the basis for the "correlation trick" discussed in Chapter 9. It was shown in Chapter 5 that transformations applied to covariates can lead to improved linear regression models, and the same is true for the response variable. The correlation trick exploits the fact that Spearman rank correlations are blind to monotone transformations, while product-moment correlations are not. Thus, dramatic differences between these correlations *may* indicate that a transformation of one or both variables can highlight their relationship. The results in Figs. 8.7 and 8.8 suggest that this may be the case here: note how much stronger the negative correlations are between `syct` and all of the other variables, including `perf`.

Also, note that `syct` spans a wide range of values, from 17 to 1500, as does `perf`, from 6 to 1150. These observations suggest that a log transformation

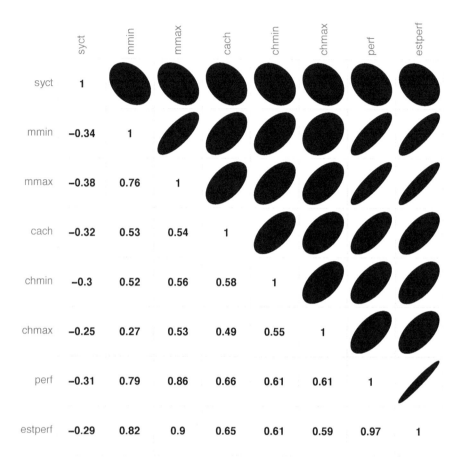

Figure 8.7: Product-moment correlations for the `cpus` variables.

may be reasonable, and the four plots shown in Fig. 8.9 provide support for this idea. The upper row shows scatterplots of `perf` versus `syct`, first on a linear scale (left plot), and then as a log-log plot where the log transformation has been applied to both variables (right plot). Comparing these plots, it is clear that `log(perf)` exhibits a much more linear relationship to `log(syct)` than the original variables do. The lower two plots are normal QQ-plots of the response variable `perf` (left plot) and the log-transformed variable `log(perf)` (right plot). Essentially all of the points in the right-hand plot fall within the 95% confidence limits shown on the plot, suggesting excellent conformance with the normal approximation for the log-transformed response variable, while the left-hand plot shows gross violations of the normality assumption for the un-transformed variable. Since least-squares linear regression models are optimal for Gaussian data, these bottom plots suggest taking `log(perf)` as our response

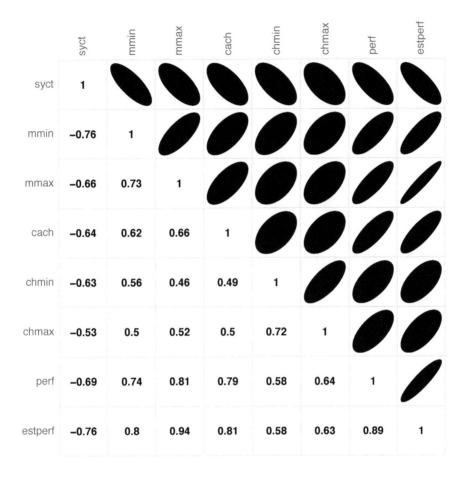

Figure 8.8: Spearman rank correlations for the `cpus` variables.

variable, and the greater linearity of the upper right-hand plot compared with the upper-left plot suggests taking `log(syct)` as a covariate instead of `syct`.

Based on these observations, the first and simplest model considered here includes only these two variables:

```
cpusModelA <- lm(I(log(perf)) ~ I(log(syct)), data = cpus)
summary(cpusModelA)

##
## Call:
## lm(formula = I(log(perf)) ~ I(log(syct)), data = cpus)
##
## Residuals:
##     Min      1Q  Median      3Q     Max
## -1.9779 -0.4859  0.0242  0.4825  2.0564
##
```

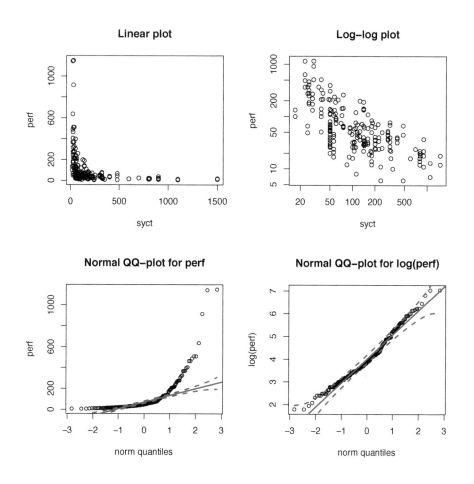

Figure 8.9: Linear and log-log plots of `perf` vs. `syct` (top row), and normal QQ-plots of `perf` and its logarithm (bottom row).

```
## Coefficients:
##              Estimate Std. Error t value Pr(>|t|)
## (Intercept)   7.40205    0.24299   30.46   <2e-16 ***
## I(log(syct)) -0.70884    0.05001  -14.17   <2e-16 ***
## ---
## Signif. codes:  0 '***' 0.001 '**' 0.01 '*' 0.05 '.' 0.1 ' ' 1
##
## Residual standard error: 0.7486 on 207 degrees of freedom
## Multiple R-squared:  0.4925,Adjusted R-squared:   0.49
## F-statistic: 200.9 on 1 and 207 DF,  p-value: < 2.2e-16
```

Based on the adjusted R-squared value obtained for this model, it appears to be somewhat predictive but not spectacularly so. In an attempt to obtain an improved model, consider adding the variables `mmax`, which exhibits the largest product-moment and Spearman rank correlations with **perf** in Figs. 8.7 and 8.8, and **cach**, which exhibits the second-highest rank correlation and the third-highest product moment correlation. It is clear from the summary results for this augmented model that these variables improve the predictions substantially:

```
cpusModelB <- lm(I(log(perf)) ~ I(log(syct)) + mmax + cach, data = cpus)
summary(cpusModelB)

##
## Call:
## lm(formula = I(log(perf)) ~ I(log(syct)) + mmax + cach, data = cpus)
##
## Residuals:
##      Min       1Q   Median       3Q      Max
## -1.43258 -0.32159  0.01254  0.30444  1.07290
##
## Coefficients:
##                Estimate Std. Error t value Pr(>|t|)
## (Intercept)   4.554e+00  2.144e-01  21.240  < 2e-16 ***
## I(log(syct)) -2.625e-01  3.896e-02  -6.739 1.59e-10 ***
## mmax          4.461e-05  3.602e-06  12.385  < 2e-16 ***
## cach          8.055e-03  9.543e-04   8.441 5.68e-15 ***
## ---
## Signif. codes:  0 '***' 0.001 '**' 0.01 '*' 0.05 '.' 0.1 ' ' 1
##
## Residual standard error: 0.4586 on 205 degrees of freedom
## Multiple R-squared:  0.8114, Adjusted R-squared:  0.8086
## F-statistic: 293.9 on 3 and 205 DF,  p-value: < 2.2e-16
```

Next, we return to the original question of whether there is a significant manufacturer effect that is not captured by these variables. To address this question, we need to construct a manufacturer variable by extracting it from the **name** text strings. To do this, first we extract these text strings as a character vector from the **cpus** data frame and examine the first few records:

```
cpuNames <- as.character(cpus$name)
cpuNames[1:5]

## [1] "ADVISOR 32/60"  "AMDAHL 470V/7"  "AMDAHL 470/7A"  "AMDAHL 470V/7B"
## [5] "AMDAHL 470V/7C"
```

It appears from this sample that the **name** variable consists of the manufacturer name followed by a computer model designation. This observation suggests splitting these name strings on spaces and retaining the first term from each split. The following *R* code accomplishes this task:

```
cpuNameSplits <- strsplit(cpuNames, split = " ")
getFirst <- function(x){x[1]}
firstTerms <- unlist(lapply(cpuNameSplits, getFirst))
```

Here, the first line applies the **strsplit** function to split the **cpuNames** strings on the space that separates the manufacturer name from the model information, returning the list **cpuNameSplits**. The second line defines a very simple custom function getFirst that returns the first element of a vector x. The third line then uses the lapply function discussed in Chapter 7 to apply getFirst to every element of the list cpuNameSplits, yielding a list, each element of which contains the computer manufacturer. The unlist function turns this list into a character vector for easier use in subsequent computations. For example, applying the table function shows who these manufacturers are and how many times each one appears in the dataset:

```
table(firstTerms)

## firstTerms
##       ADVISOR        AMDAHL        APOLLO          BASF           BTI
##             1             9             2             2             2
##     BURROUGHS        C.R.D.        CAMBEX           CDC           DEC
##             8             6             5             9             6
##            DG     FORMATION          FOUR         GOULD        HARRIS
##             7             5             1             3             7
##     HONEYWELL            HP           IBM           IPL      MAGNUSON
##            13             7            32             6             6
##     MICRODATA           NAS           NCR       NIXDORF PERKIN-ELMER
##             1            19            13             3             3
##         PRIME       SIEMENS        SPERRY       STRATUS          WANG
##             5            12            13             1             2
```

This tabulation suggests that these first terms represent unique identifiers for the manufacturers of the computers evaluated in the benchmark study. To explore the effect of including this information in our linear regression models, create an augmented data frame that includes this manufacturer variable and add this covariate as a predictor, first to the original model based on syct alone:

```
cpusPlus <- cpus
cpusPlus$mfgr <- firstTerms
cpusModelC <- lm(I(log(perf)) ~ I(log(syct)) + mfgr, data = cpusPlus)
summary(cpusModelC)

##
## Call:
## lm(formula = I(log(perf)) ~ I(log(syct)) + mfgr, data = cpusPlus)
##
## Residuals:
##      Min      1Q  Median      3Q     Max
## -1.7195 -0.3554  0.0000  0.3724  1.6262
##
## Coefficients:
##                  Estimate Std. Error t value Pr(>|t|)
## (Intercept)       9.67168    0.69352  13.946  < 2e-16 ***
## I(log(syct))     -0.90786    0.06243 -14.542  < 2e-16 ***
## mfgrAMDAHL       -0.89966    0.66569  -1.351 0.178256
## mfgrAPOLLO       -0.56906    0.76840  -0.741 0.459925
## mfgrBASF         -1.31285    0.76669  -1.712 0.088574 .
## mfgrBTI          -1.67858    0.76639  -2.190 0.029807 *
## mfgrBURROUGHS    -1.47839    0.66250  -2.232 0.026893 *
## mfgrC.R.D.       -0.80128    0.67718  -1.183 0.238285
## mfgrCAMBEX       -2.40491    0.68659  -3.503 0.000582 ***
```

```
## mfgrCDC           -1.80622    0.66151  -2.730 0.006961 **
## mfgrDEC           -0.78139    0.67686  -1.154 0.249866
## mfgrDG            -0.94245    0.66883  -1.409 0.160553
## mfgrFORMATION     -0.80635    0.69395  -1.162 0.246801
## mfgrFOUR          -1.70475    0.88330  -1.930 0.055200 .
## mfgrGOULD         -0.58654    0.72192  -0.812 0.417605
## mfgrHARRIS        -0.69243    0.66959  -1.034 0.302488
## mfgrHONEYWELL     -1.13837    0.64872  -1.755 0.081013 .
## mfgrHP            -2.00188    0.66785  -2.997 0.003111 **
## mfgrIBM           -1.13803    0.63490  -1.792 0.074759 .
## mfgrIPL           -2.00829    0.67705  -2.966 0.003429 **
## mfgrMAGNUSON      -2.22775    0.67557  -3.298 0.001177 **
## mfgrMICRODATA     -1.72155    0.88337  -1.949 0.052887 .
## mfgrNAS           -1.11859    0.64249  -1.741 0.083410 .
## mfgrNCR           -2.36041    0.65042  -3.629 0.000372 ***
## mfgrNIXDORF       -1.41702    0.72181  -1.963 0.051186 .
## mfgrPERKIN-ELMER  -0.97801    0.72251  -1.354 0.177568
## mfgrPRIME         -1.20611    0.68438  -1.762 0.079726 .
## mfgrSIEMENS       -1.58791    0.65108  -2.439 0.015714 *
## mfgrSPERRY        -1.16257    0.64868  -1.792 0.074799 .
## mfgrSTRATUS       -1.33702    0.88330  -1.514 0.131884
## mfgrWANG          -0.06109    0.76956  -0.079 0.936814
## ---
## Signif. codes:  0 '***' 0.001 '**' 0.01 '*' 0.05 '.' 0.1 ' ' 1
##
## Residual standard error: 0.6246 on 178 degrees of freedom
## Multiple R-squared:  0.6962,Adjusted R-squared:  0.645
## F-statistic:  13.6 on 30 and 178 DF,  p-value: < 2.2e-16
```

and then to the improved model that incorporates **mmax** and **cach** as predictors:

```
cpusModelD <- lm(I(log(perf)) ~ I(log(syct)) + mmax + cach + mfgr,
                    data = cpusPlus)
summary(cpusModelD)

##
## Call:
## lm(formula = I(log(perf)) ~ I(log(syct)) + mmax + cach + mfgr,
##     data = cpusPlus)
##
## Residuals:
##     Min      1Q  Median      3Q     Max
## -1.1366 -0.1813  0.0000  0.1991  0.9736
##
## Coefficients:
##                   Estimate Std. Error t value Pr(>|t|)
## (Intercept)      5.238e+00  5.239e-01   9.999  < 2e-16 ***
## I(log(syct))    -4.499e-01  4.688e-02  -9.596  < 2e-16 ***
## mmax             4.206e-05  3.919e-06  10.732  < 2e-16 ***
## cach             7.695e-03  1.015e-03   7.582 1.88e-12 ***
## mfgrAMDAHL       3.658e-02  4.880e-01   0.075   0.9403
## mfgrAPOLLO       9.684e-01  5.301e-01   1.827   0.0694 .
## mfgrBASF         2.824e-01  5.110e-01   0.553   0.5812
## mfgrBTI         -1.375e-01  5.350e-01  -0.257   0.7974
## mfgrBURROUGHS    3.546e-01  4.686e-01   0.757   0.4503
## mfgrC.R.D.       7.807e-01  4.829e-01   1.617   0.1078
## mfgrCAMBEX      -3.293e-01  4.956e-01  -0.664   0.5073
```

```
## mfgrCDC            1.569e-01  4.612e-01   0.340   0.7341
## mfgrDEC            6.987e-01  4.835e-01   1.445   0.1502
## mfgrDG             5.771e-01  4.795e-01   1.204   0.2304
## mfgrFORMATION      2.293e-01  4.971e-01   0.461   0.6452
## mfgrFOUR           4.754e-01  5.939e-01   0.800   0.4245
## mfgrGOULD          6.523e-01  4.783e-01   1.364   0.1744
## mfgrHARRIS         8.453e-01  4.807e-01   1.759   0.0804 .
## mfgrHONEYWELL      3.109e-01  4.712e-01   0.660   0.5102
## mfgrHP             1.203e-01  4.797e-01   0.251   0.8023
## mfgrIBM            3.932e-01  4.628e-01   0.850   0.3967
## mfgrIPL            1.329e-02  4.843e-01   0.027   0.9781
## mfgrMAGNUSON      -3.878e-01  4.837e-01  -0.802   0.4238
## mfgrMICRODATA      2.489e-01  5.956e-01   0.418   0.6765
## mfgrNAS            2.682e-01  4.532e-01   0.592   0.5548
## mfgrNCR           -4.044e-01  4.621e-01  -0.875   0.3827
## mfgrNIXDORF        2.295e-01  4.980e-01   0.461   0.6454
## mfgrPERKIN-ELMER  5.877e-01  5.117e-01   1.149   0.2523
## mfgrPRIME          5.920e-01  4.836e-01   1.224   0.2225
## mfgrSIEMENS        5.214e-02  4.639e-01   0.112   0.9106
## mfgrSPERRY        -5.382e-02  4.673e-01  -0.115   0.9084
## mfgrSTRATUS        5.487e-01  5.989e-01   0.916   0.3608
## mfgrWANG           1.169e+00  5.268e-01   2.220   0.0277 *
## ---
## Signif. codes:  0 '***' 0.001 '**' 0.01 '*' 0.05 '.' 0.1 ' ' 1
##
## Residual standard error: 0.3804 on 176 degrees of freedom
## Multiple R-squared:  0.8886, Adjusted R-squared:  0.8684
## F-statistic: 43.88 on 32 and 176 DF,  p-value: < 2.2e-16
```

Comparing all of these results, several things are clear. First, it does appear that adding the manufacturer improves the predictive power of these linear regression models somewhat: fairly dramatically in the case of the simpler model (i.e., cpusModelC versus cpusModelA), and at least a little for the model that includes the mmax and cach covariates. The relatively large number of distinct levels for the mfgr variable (30) implies that some of them are necessarily thin (this is clear from the table of manufacturer counts given above). This is probably responsible for the relatively large (i.e., statistically insignificant) p-values for the coefficients associated with most of these individual manufacturers.

It is possible that we could obtain a better model by re-grouping these manufacturers, but this is a topic beyond the scope of this book. The point of this example was to show how text variables that are initially not useful in building prediction models—i.e., the name variable, with one unique value for every record in the cpus data frame—can be decomposed using the text analysis tools described in this chapter to obtain a categorical variable that can be used in model-building and is at least somewhat predictive.

8.6.3 The unclaimed bank account data

The unclaimed Canadian bank account dataset from the Socrata website was introduced in Chapter 4, and it was discussed further in Sec. 8.2.3. As noted, a complication in answering questions like "What is the total unclaimed balance

for the city of Edmonton?" is the fact that Edmonton is represented in at least
32 different ways. The following example asks the broader question, "how is
the total in unclaimed balances distributed over different cities?" To address
the problem of ambiguous city representations, this analysis makes use of the
approximate matching ideas discussed in Sec. 8.5.2.

To see the scope of the problem, note that the `City` variable assumes 851
unique values; a random sample of 16 of these values illustrates their variety:

```
set.seed(3)
sample(City, size = 16)

##  [1] "EDMONTON "              "Edmonton"
##  [3] "Edmonton"               " "
##  [5] " "                      "Edmonton"
##  [7] "WINNIPEG MAN "          "EDMONTON "
##  [9] "EDMONTON,ALTA "         "Stony Plain"
## [11] " "                      "KITCHENER ONT N2B 3V9 "
## [13] " "                      "St. Albert"
## [15] "EDMONTON,AB "           "UNKNOWN,UNKNOWN "
```

As noted in Sec. 8.2.3, some variation of "Edmonton" is the most frequently
ocurring `City` value, followed by different variations on "UNKNOWN." This
sample of values also reveals case differences (e.g., "Edmonton" vs. "EDMON-
TON") and that some city names consist of two words (e.g., "Stony Plain").
To see how total unclaimed balances are distributed across different cities, we
need a *standardized* list of cities where, for example, "Edmonton" has a single
representation. Then, all of the variations seen here can be mapped into this
standard representation using the approximate string matching ideas discussed
in Sec. 8.5.2. As a first step in this direction, we do three things. First, convert
the `City` variable to lower case:

```
lowerCity <- tolower(City)
```

Second, to deal with the blank `City` entries (i.e., the single spaces in the above
list), we convert them to the word "missing":

```
blankIndex <- which(lowerCity == " ")
lowerCity[blankIndex] <- "missing"
```

Finally, the third preprocessing step is to replace all of the variations of
"unknown" with the single word "missing." To do this, we first identify these
variations using the **grep** function to search for terms containing the "unk"
substring:

```
unkIndex <- grep("unk", lowerCity)
table(lowerCity[unkIndex])

##
##      unknow alta m1m 1m1           unknown ????           unknown ab
##                       1                     121                   46
```

```
##             unknown ab              unknown alta              unknown bc
##                     87                        40                        6
##             unknown bc              unknown deu              unknown gbr
##                      4                         1                        1
##       unknown unknown  unknown unknown unknown              unknown usa
##                     97                         1                        1
##       unknown,unknown          unknown,unknwon              unknwon ab
##                   1070                         1                        1
```

Since all of these text strings do seem to indicate missing `City` entries, we
construct the final cleaned city vector by re-assigning all of these values to the
"missing" term:

```
lowerCity[unkIndex] <- "missing"
```

Taken together, these changes reduce the number of unique city entries slightly
(to 835), but as the following results illustrate, further preprocessing is needed.
    The key step in meeting our objective of cleaning up the `City` text entries
is to decide what terms to include in our standardized city name list. One way
to approach this task is to look at the terms that occur most frequently in this
text data, since these terms should include the cities of greatest interest. As
a first step in this direction, construct the document-feature matrix from the
`lowerCity` text vector, using the **ngrams** argument to include multiple-word
city names like "Stony Plain" seen in the initial random sample. The following
R code does this, allowing up to three-word city names, and displays the 20 of
these names that occur most frequently:

```
firstCityDFM <- dfm(lowerCity, ngrams = 1:3, concatenator = " ")
topfeatures(firstCityDFM, n = 20)
```

```
##       edmonton         missing              ab                  ,           , ab
##           5834            4988            1434               1387          1024
##    edmonton , edmonton  , ab             can    edmonton can            st
##            855             751             338                290           284
##         albert            park       sherwood sherwood park          alta
##            276             250             245                243           237
##    edmonton ab               .           st  .        . albert   st . albert
##            220             193             177                176           176
```

Several things are clear from these results. First, the most frequent two cities
("edmonton" and "missing") are both direct results of our preprocessing that
eliminated case effects and merged all of the different missing city designations.
Also, the fourth-ranked "city" is the punctuation mark "comma," and commas
and periods appear as parts of other city names, including alternative designations for "edmonton." In fact, our preprocessed city list also includes numbers
and other symbols, as we can see by looking at the first ten entries from a
tabulation of these text strings:

```
lowerTable <- table(lowerCity)
lowerTable[1:10]
```

```
## lowerCity
##           #03-151                    *                    ,
##              1                       3                   65
##           ,edmonton              ,germany             ,greece
##              1                       1                    1
##           ,state            ,uk  06400 porvoo,finland
##              1                       1                    1
##     1 thornton crt nw
##              1
```

As in several of the previous examples considered earlier in this chapter, these extraneous details can be removed by first tokenizing the text, constructing a document-feature matrix, and examining the top 40 features:

```
cityTokens <- tokens(lowerCity,
                     remove_numbers = TRUE,
                     remove_punct = TRUE,
                     remove_symbols = TRUE)
secondCityDFM <- dfm(cityTokens, ngrams = 1:3, concatenator = " ")
topfeatures(secondCityDFM, n = 40)
```

```
##       edmonton          missing            ab        edmonton ab
##           5834             4988          1434                971
##            can      edmonton can            st             albert
##            338              290           284                276
##      st albert             park      sherwood      sherwood park
##            273              250           245                243
##           alta               bc         grove             spruce
##            237              158           110                109
##    spruce grove    edmonton alta     vancouver            calgary
##            109               98            95                 91
##    vancouver bc            leduc           usa            park ab
##             87               82            72                 64
##          plain sherwood park ab         stony        stony plain
##             59               58            57                 57
##       albert ab     st albert ab            on         calgary ab
##             57               57            51                 48
##       grove ab  spruce grove ab          fort           montreal
##             29               29            28                 23
##            ont          toronto    winterburn            alberta
##             18               18            17                 17
```

Since these features include terms like "ab" that are obviously not valid city names, along with multiple representations for individual cities like "edmonton," we cannot use the names of these features directly as our standardized city list, but they do provide a useful basis for manually constructing such a list. In particular, a careful examination of these features suggests the following starting point for a standardized city list:

```
stdCities <- c("edmonton", "missing", "st albert", "sherwood park",
               "spruce grove", "vancouver", "calgary", "leduc",
               "stony plain", "montreal", "toronto")
```

Given this standardized list **stdCities** of valid city names and the vector **lowerCity** of pre-processed city names, the approximate string matching ap-

proaches discussed in Sec. 8.5.2 can be applied to construct `validCity`, a vector
of standardized city names that best match the corresponding `lowerCity` entries. Based on the results presented in that discussion, the Jaro-Winkler string
distance was used as the basis for this approximate matching, with the optional
parameter $p = 0.1$. Recall that this measure emphasizes common characters
that are similarly positioned in both the original string and the reference string,
with added emphasis on matches in the first four letters. Since this measure
was developed for name and address matching, it seems a good choice here.

The simplest way to apply these ideas is to use the `amatch` function with the
Jaro-Winkler distance measure. Starting with the default `maxDist` argument,
we obtain the following results:

```
validIndexA <- amatch(lowerCity, stdCities, method = "jw", p = 0.1)
validCityA <- stdCities[validIndexA]
table(validCityA, useNA = "ifany")

## validCityA
##        calgary        edmonton           leduc         missing        montreal
##             81            5796              74            4988              16
## sherwood park    spruce grove       st albert     stony plain         toronto
##            244             107             270              59              18
##      vancouver            <NA>
##             80            1083
```

These results emphasize two points. First, recall from the discussion in Sec. 8.5.2
that when the `amatch` function cannot find an acceptable match, it returns
the missing value "NA," which causes the `validCityA` value to also appear
as missing. Probably a better alternative, illustrated next, is to replace these
values with "other" to indicate that the city listed in the original data did not
match any of the values on our standardized city list. The second point is that,
for the examples discussed in Sec. 8.5.2, the default `maxDist` value (0.1) for
the `amatch` function was too stringent. In response to these two observations,
consider the following variation of the matching results presented above:

```
validIndexB <- amatch(lowerCity, stdCities, method = "jw", p = 0.1, maxDist = 0.2)
validCityB <- stdCities[validIndexB]
validCityB[which(is.na(validCityB))] <- "other"
table(validCityB, useNA = "ifany")

## validCityB
##        calgary        edmonton           leduc         missing        montreal
##             90            5812              80            4988              23
##          other   sherwood park    spruce grove       st albert     stony plain
##           1034             245             108             273              60
##        toronto       vancouver
##             18              85
```

Based on these tabulations, it appears that increasing `maxDist` from its default
value 0.1 to the less stringent matching value 0.2 has improved the quality of our
results. Before accepting this conclusion, however, it is important to examine
the results more carefully to make sure these matches are reasonable.

Since the "other" category consists of the `lowerCity` names that cannot be matched to our standardized list, we would like it to be—to modify a quote from Einstein—"as small as possible but no smaller" (what he said was that explanations should be "as simple as possible but no simpler"). For both of the `maxDist` values considered here, this category is large enough that it is not feasible to look at all of the `lowerCity` values mapped into it, but it is feasible to look at the most frequent of these values. For the default `maxDist` values, the 20 most frequent "other" values are:

```
##
##                        ,              alta                 ab
##              65                        25                  17
##      morinville ab             v1v 1v1     montreal west qc
##              8                         8                   7
##          nisku ab  north vancouver bc        victoria bc
##              7                         7                   7
##   winterburn alta          ardrossan ab   edmonton alberta
##              7                         6                   6
##        inconnue              m1m 1m1        duffield,ab
##              6                         6                   5
##        enoch,ab      fort mcmurray ab      wetaskiwin ab
##              5                         5                   5
##   winterburn,ab             gunn,ab
##              5                         4
```

This list includes several entries that should probably be re-assigned as "missing," including the comma (the most frequent term here) and abbreviations like "ab." Also, there are other text strings like "montreal west qc," "north vancouver bc," and "edmonton alberta" that should be assigned to the standardized city names "montreal," "vancouver," and "edmonton," respectively. The fact that these matches were not made here suggests that, as in the example considered in Sec. 8.5.2, the default `maxDist` value of 0.1 for the Jaro-Winkler measure is too stringent. Relaxing this value to 0.2, the top 20 "other" cities are:

```
##
##                        ,              alta                 ab
##              65                        25                  17
##      morinville ab             v1v 1v1        nisku ab
##              8                         8                   7
## north vancouver bc          victoria bc   winterburn alta
##              7                         7                   7
##      ardrossan ab             inconnue        m1m 1m1
##              6                         6                   6
##      duffield,ab              enoch,ab   fort mcmurray ab
##              5                         5                   5
##     wetaskiwin ab          winterburn,ab        gunn,ab
##              5                         5                   4
## medicine hat alta          morinville,ab
##              4                         4
```

These results are slightly better, having correctly assigned "montreal west qc" to "montreal" and "edmonton alberta" to "edmonton," but this "other" list still includes "north vancouver bc." Unfortunately, this mismatch is an undesirable consequence of the "local matching" behavior of the Jaro-Winkler measure: even though the string "north vancouver bc" *contains* the standard city name "vancouver" as a substring, the Jaro-Winkler measure is downweighting this match because this common term does not occur in the first part of "north vancouver bc." Specifically, the string distance here exceeds our matching threshold:

```
stringdist("north vancouver bc", "vancouver", method = "jw", p = 0.1)
```

```
## [1] 0.3333333
```

This observation suggests relaxing our `maxDist` threshold further, but this relaxation comes at a price. Specifically, consider the consequences of increasing this argument to 0.4, giving the following results:

```
validIndexC <- amatch(lowerCity, stdCities, method = "jw", p = 0.1, maxDist = 0.4)
validCityC <- stdCities[validIndexC]
validCityC[which(is.na(validCityC))] <- "other"
table(validCityC, useNA = "ifany")
```

```
## validCityC
##       calgary       edmonton         leduc        missing      montreal
##           151           5853           102           5009            76
##         other sherwood park  spruce grove     st albert  stony plain
##           595            265           135           333           109
##       toronto      vancouver
##            60            128
```

Here, our "other" category has been reduced in size by about 40%, which seems highly desirable, but if we examine the `City` values assigned to the standard city name "vancouver," we see a number of incorrect assignments:

```
## vancouver
##           vancouver bc              north vancouver bc
##                   57                               7
##              victoria bc                    inconnue
##                    7                               6
##           vancouver bc                 vancouver, bc
##                    5                               5
##           vancouver,bc            lancaster park ab
##                    4                               3
##      lancaster park,ab                 vancouver ab
##                    3                               3
##         vancouver b.c.                      cooper
##                    2                               1
##             coquitlem                   evansburg
##                    1                               1
##         evansville usa               france,france
##                    1                               1
##           lacombe ab                   lacombe,ab
##                    1                               1
##       manchester i gbr  north vancouver bc v7l 4s3
##                    1                               1
```

In particular, although we have correctly assigned "north vancouver bc" to the valid name "vancouver," we have also assigned a number of obvious mismatches, including "france,france," "evansville usa", and "manchester 1 gbr".

Returning to the problem that motivated this cleaning of the `City` variable—characterizing the total unclaimed balance by city—note that this kind of computation is common in database applications. To illustrate this point, the following discussion uses the `sqldf` package discussed in Chapter 4 to compute these summaries for each of the standardized city name assignments just described, using a SQL "group by" query. The first step is to create a data frame with one column for each city assignment and the unclaimed account balance for each record in the original unclaimed bank account data frame. First, however, it is useful to replace the missing assignment indicators in `validCityA` with "other," for consistency with the other assignments. The following *R* code accomplishes both of these steps:

```
validCityA[which(is.na(validCityA))] <- "other"
cityFrame <- data.frame(CityA = validCityA, CityB = validCityB,
                CityC = validCityC, Balance = unclaimedFrame$Balance)
```

Given this data frame, we can compute the total balance for each city defined by `validCityA` using the following SQL query, executed via the `sqldf` function:

```
queryA <- "select CityA as City, sum(Balance) as TotalA
                from cityFrame group by CityA"
sumA <- sqldf(queryA)
head(sumA)

##        City     TotalA
## 1   calgary    80132.39
## 2  edmonton 3091580.59
## 3     leduc   21364.10
## 4   missing 2682617.47
## 5  montreal    4058.12
## 6     other  759312.03
```

The result of this query is a data frame with one row for each city from our standardized set and two columns: the first gives the city, and the second gives the sum of the balance values for all of the original records assigned to this city. Analogous queries for the other two city assignments give the corresponding balance totals for these cases:

```
queryB <- "select CityB as City, sum(Balance) as TotalB
                from cityFrame group by CityB"
sumB <- sqldf(queryB)
queryC <- "select CityC as City, sum(Balance) as TotalC
                from cityFrame group by CityC"
sumC <- sqldf(queryC)
```

The individual summary data frames for each case can be merged into a single summary data frame using database join operations, also implemented as SQL queries via `sqldf`:

```
queryAB <- "select sumA.City, TotalA, TotalB
                   from sumA inner join sumB on
                   sumA.City = sumB.City"
sumAB <- sqldf(queryAB)
queryABC <- "select sumAB.City, TotalA, TotalB, TotalC
                    from sumAB inner join sumC on
                    sumAB.City = sumC.City"
sumABC <- sqldf(queryABC)
```

Here, the first query merges the individual summaries for the `validCityA` and `validCityB` assignments, and the second query merges the results of this first query with the `validCityC` balance summary. The final result contains the balance totals for all three of these city assignments:

```
sumABC
```

```
##              City     TotalA       TotalB       TotalC
## 1        calgary   80132.39     82119.73   182524.82
## 2       edmonton 3091580.59 3104640.21 3120354.93
## 3          leduc   21364.10     23686.20    32082.89
## 4        missing 2682617.47 2682617.47 2690746.28
## 5       montreal    4058.12      4291.43    24133.60
## 6          other  759312.03   739803.60   474336.74
## 7  sherwood park   76242.39     76303.67    82697.82
## 8   spruce grove   25295.77     26191.97    38886.50
## 9      st albert  179967.07   180264.14   212722.74
## 10   stony plain   34160.25     34303.41    57656.51
## 11       toronto   14222.64     14222.64    38670.52
## 12     vancouver   71130.91     71639.26    85270.38
```

Recall that the three valid city assignments summarized here were made with `maxDist` values of 0.1 (the default), 0.2, and 0.4, allowing an increasingly relaxed match condition for these assignments in going from A to B to C. As a consequence, fewer records were assigned to the "other" group, and this is reflected in the monotonic decrease in the balance totals for this `City` value seen here. In contrast, all of the other city assignments except "missing" increased in record count, reflected in the corresponding increase in balance totals in going from `TotalA` to `TotalB` to `TotalC`. Comparing the numbers across these city assignments, it appears that the largest unclaimed balance total is from Edmonton, followed by "missing" and then the undifferentiated mix "other," with "St. Albert" ranking fourth. Note that while the balance totals for the top-ranked cities Edmonton and "missing" do not depend strongly on which of the three city assignments we consider, some of the other city totals do. In particular, note the very large differences between `TotalB` and `TotalC` for the cities of Calgary, Montreal, and Toronto.

This example makes several important points. First, it is possible to clean up text data using the approximate matching ideas discussed in Sec. 8.5.2, although doing this successfully requires careful attention to detail. In particular, Sec. 8.5.2 noted the importance of choosing the appropriate approximate matching method: although they are not shown, the default OSA method gives terrible results here, as it did there. Also, both that discussion and this one illustrated the impact of different choices of the `maxDist` parameter. From both of these results, it appears that the default value for this argument to the `amatch` function is too small, imposing an approximate match condition that is too stringent. Conversely, if this argument is made too large, we begin to see a large number of false matches: the results presented here suggest that the value 0.4 is too large. Thus, although it would require more careful investigation to say this with certainty, it appears that the value 0.2 gives the best results here, suggesting that the `TotalB` results are probably most representative of the true breakdown of unclaimed bank balances by city. Finally, a third point that was not explored here was the degree of improvement possible by expanding our standard set of cities in making these assignments. In general, we can expect increasing the number of candidate cities to have two effects: first, it should reduce the size of the "other" set, and second, it should reduce the erroneous assignments of these cities to other valid cities (e.g., including "victoria" in the set should cause "victoria bc" to be assigned to this value rather than to "vancouver," as in the results for `maxDist = 0.4` shown above).

8.7 Exercises

1: **Note:** this exercise is essential, as it generates the data needed for all of the other exercises in this chapter. Specifically, this exercise asks you to download a text data file from the Internet and save it for use:

    1a. Using the `download.file` function discussed in Chapter 4, and the `paste` function to join the URL and the file name, separated by a forward slash ("/"), download the text file `Concrete_Readme.txt` from the following URL into a file named `concreteMetadata.txt`:

        `https://archive.ics.uci.edu/ml/`
            `machine-learning-databases/concrete/compressive`

    1b. Use the `readLines` function introduced in Chapter 4 to read the text file `concreteMetadata.txt` into a character vector: how long is this character vector? Use the `head` function to show the first six records.

2: Using the contents of the `concreteMetadata.txt` file downloaded in Exercise 1, answer the following questions:

    2a. Construct Tukey's five-number summary of the number of characters in each record of the text file;

    2b. Compute the average number of characters in each record.

3: The text data file downloaded in Exercise 1 describes a concrete compressive strength dataset that is explored further in Chapter 10, and it was used in published studies evaluating artificial neural networks as prediction models. Using the `grep` function discussed in Section 8.2.2, determine the following:

3a. How many records in the file `concreteMetadata.txt` contain the term "neural," all in lower case?

3b. How many records in this file contain the term "neural," with arbitrary case (e.g., "neural," "Neural," or "NEURAL")?

3c. What are the specific records identified in (3b)?

4: In working with text data, we often encounter text vectors that are too long to display on a single line. For example, this frequently occurs if we merge all of the elements of a single text vector, like the contents of the `concreteMetadata.txt` file. Such merges can be useful if we wish to characterize multiple documents, each composed of several text strings (e.g., if we wanted to analyze the metadata for several different files). This exercise introduces two useful tools for, first, merging multi-component text vectors into a single long text vector, and, second, for displaying inconveniently long strings.

4a. The `paste` function applied to a character vector with a non-`NULL` value for the optional `collapse` argument returns a single text string consisting of the components of the character vector separated by the `collapse` value. Use this function to merge the contents of `concreteMetadata.txt` into a single string called `collapsedText`, with the original vector elements separated by single spaces. How many characters are in this string?

4b. The `strwrap` function wraps a long character string into single-line components. Use this function to display `collapsedText`. How many lines of text does this function generate?

5: The `strsplit` function was introduced in Section 8.2.5 as a useful way of breaking long text strings into shorter components. An important practical aspect of using this function is that the `split` argument is a regular expression, and the point of this exercise is to give an illustration of the practical significance of this fact.

5a. Use the `strsplit` function to split the long text string `collapsedText` constructed in Exercise 4 into components, specifying `split = ''.''` and use the `unlist` function to convert the list returned by the `strsplit` function into a character vector. How long is this character vector? Use the `strwrap` function to list its first element.

5b. Repeat (5a), but with the optional argument `fixed = TRUE` in the call to `strsplit`. How long is this character vector? Use the `strwrap` function to list its first element.

6: The character sets discussed in Section 8.3.2 can be used to extract records containing specific text combinations. For example, e-mail addresses and certain other terms contain embedded hyphens ("-"), as do some names of persons or places. This exercise asks you to use character sets in

regular expressions with the **grep** function to find all records from the
concreteMetadata that contain any of the following combinations:

6a. a lower-case letter, followed by a hyphen, followed by a lower-case
letter;

6b. a lower-case letter, followed by a hyphen, followed by an upper-case
letter;

6c. an upper-case letter, followed by a hyphen, followed by a lower-case
letter;

6d. an upper-case letter, followed by a hyphen, followed by an upper-case
letter.

In all cases, list the elements of the **concreteMetadata** that contain the
indicated character combinations.

7: Exercise (3c) used the **grep** function to search for the term "neural" in
the **concreteMetadata** text vector, returning all records that contain this
term. Alternative terms that sometimes arise in the neural network literature include "neuron" or compound terms like "neuro-fuzzy." Using the
grouping metacharacters "(" and ")" and the "|" alternation metacharac-
ter, modify the search used in Exercise (3c) to return all records containing
either "neura" or "neuro," without regard to case. List all records containing either of these terms.

8: The following exercise asks you to apply the basic text analysis procedures from the **quanteda** package to the merged text data constructed in
Exercise 4. Specifically:

8a. Use the default options of the **tokens** function to convert the vector **collapsedText** from Exercise 4 into a bag-of-words object called
tokensA, apply the **dfm** function to this result to construct a document-
feature matrix where the features are words, and display the top 30
of these features.

8b. Convert **collapsedText** to a lower-case text string and create a bag-
of-words object called **tokensB**, using the optional arguments of the
tokens function to remove numbers, punctuation marks, and symbols. Apply the **dfm** function to this result and display the top 30
features of the resulting document-feature matrix. How do these results differ from those in (8a)?

8c. Apply the **tokens_remove** function to remove the English stopwords
from **tokensB** to obtain the object **tokensC**. Apply the **dfm** function to this result and display the top 30 features of the resulting
document-feature matrix. Which of these three results gives you the
clearest idea of what the original document is about?

9: The `tokens_ngrams` function in the `quanteda` package can be applied to tokenized text to construct n-grams of arbitrary order: setting the optional argument `n = 1` gives back the original tokenized text, setting `n = 2` (the default) returns bigrams (i.e., sequences of two successive words), while setting `n = 3` returns trigrams (i.e., sequences of three successive words). This exercise asks you to construct the bigram and trigram characterizations from the fully normalized text data used in Exercise (8c) and compare the results. Specifically:

    9a. Construct the bigrams from the `tokensC` bag-of-words from Exercise (8c), generate its associated document-feature matrix, and list the 30 most frequent features.

    9b. Repeat (9a) for trigrams: of the three results—the frequent words from Exercise (8a), the bigrams from (9a), or the trigrams—which one gives the clearest and most succinct summary of what the original metadata document is about?

10: As illustrated in several of the examples presented in this chapter, *wordclouds* can be extremely useful in providing a simple graphical representation of results like those obtained in Exercises 8 and 9. This exercise asks you to construct the following four wordclouds, each in its own plot:

    10a. A wordcloud of the top 10 terms from Exercise (8a), with the title "Raw text";

    10b. A wordcloud of the top 10 terms from Exercise (8c), with the title "Normalized text";

    10c. A wordcloud of the top 10 terms from Exercise (9a), with the title "Bigrams";

    10d. A wordcloud of the top 10 terms from Exercise (9b), with the title "Trigrams".

Note: the default scaling for the `wordcloud` function in the `wordcloud` package is too large for some of these results to display completely. See the help files for the optional `wordcloud` argument `scaling` and reduce the upper scaling limit as necessary to make these terms fit. Also, to obtain repeatable results (i.e., with deterministic rather than random word placement), specify `random.order = FALSE` in the `wordcloud` calls.

Chapter 9

Exploratory Data Analysis: A Second Look

As noted in Chapter 1 and discussed further in Chapter 3, the purpose of exploratory data analysis is to find potentially important patterns or structure in a dataset. This can take the form of individual observations or groups of observations that are unusual in some important respect (e.g., outliers), or the identification of important—and possibly unexpected—relationships between variables in a dataset. The focus of Chapter 3 was primarily on graphical visualization tools and simple descriptive statistics, with a minimum of mathematics. The purpose of this chapter is twofold: first, to introduce some more advanced exploratory data analysis tools to address important practical issues not covered in Chapter 3, and, second, to provide additional technical background to help us both decide when to use one technique instead of another, and understand the results we obtain from some of these modeling and analysis tools. Much of the material presented in this chapter is based on using random variables as an approximate description of real-world data.

The idea of using random variables as a mathematical description of the uncertainty and natural variability found in real-world measurement data was introduced in Chapter 3. The primary advantage of doing this is that it allows us to develop mathematical tools to answer questions like the following:

1. What is the "typical value" of variable X?

2. How much variation should we expect to see around this typical value?

3. We have measurements of two variables, X and Y: do these variables move together, in opposite directions, or are they essentially unrelated?

The purpose of this chapter is to introduce some of the computational tools available to answer questions like this and some of the ideas on which they are based. As before, mathematics is kept to a minimum, but it is important to introduce enough background to allow us to select the appropriate tool for each case and to understand the results generated by this tool. For a more detailed treatment of these topics, refer to *Exploring Data* [58], the references cited there, or the references cited in the discussions that follow.

9.1 An example: repeated measurements

Many of the key ideas presented in this chapter can be understood by considering the following situation: we have a sequence of N measurements $\{x_k\}$ of some variable x, taken under what should be "the same conditions" (e.g., within a narrow span of time, or from a group of N similar data sources). To analyze this data, we begin by constructing a simple mathematical model that assumes there is a *correct but unknown value* x_0 for this variable, observed with *additive random errors* $\{e_k\}$. That is, our observed data sequence may be written as:

$$x_k = x_0 + e_k, \tag{9.1}$$

for $k = 1, 2, \ldots, N$. One of the important practical consequences of using random variables to describe the uncertainty and natural variability in observed data is that anything we compute from this random data also becomes a random variable. The distribution of this random variable depends on both that assumed for our data and the mathematical form of the characterization we compute from it. This point is developed in detail in Sec. 9.1.2, along with several others closely related to it. This development necessarily uses some mathematics—nothing beyond algebra, but a modest amount of it—so a brief non-mathematical summary is given first, in Sec. 9.1.1. This summary covers the key ideas and results from the more detailed discussion in Sec. 9.1.2, to serve either as an overview for those not wishing to go through the mathematical details, or as an introductory overview for those who do.

9.1.1 Summary and practical implications

The example considered here is an important one for three reasons. First, it provides a (relatively) simple demonstration of the point made in the previous paragraph that, if we adopt a random variable model to describe the uncertainty in our data, everything we compute from that data becomes a random variable. Second, this example serves as a prototype for the much larger class of problems where we are attempting to compute a data characterization from a collection of N data observations for which we believe this characterization is applicable to the complete data collection. Examples include the mean of repeated data observations discussed in detail in Sec. 9.1.2, other descriptive statistics like the median or the standard deviation, along with more general problems like estimating linear regression model parameters from the observations in a dataset. Finally, third and most important of all, some of the specific working assumptions and results presented in Sec. 9.1.2 apply at least approximately to a much broader class of data analysis problems.

More specifically, most data characterizations we encounter in practice are, first, *unbiased*, and, second, *asymptotically normal* [58, p. 249]. The first of these characteristics means that the quantity we are attempting to compute from an uncertain dataset will be "correct, on average." In terms of the expected values introduced in Sec. 9.1.2, the expected value of the random quantity we compute from the data will be the correct value, assuming our assumed random variable model is correct. The second characteristic—asymptotic normality—means that, if we compute it from a sufficiently large dataset, our data characterization will exhibit an approximately Gaussian distribution, with a variance that gets smaller as our dataset gets larger. These ideas will be used in later sections of this chapter to provide the basis for specific uncertainty assessments for many of the important data characterizations described here.

9.1.2 The gory details

As noted in the introductory paragraph in Sec. 9.1, once we adopt a random variable model to describe the uncertainty in our data sequence $\{x_k\}$, anything we compute from this sequence also becomes a random variable, characterized by a probability distribution. Perhaps the simplest way to see this point is to consider the average $\bar{x}$ computed from this data sequence:

$$\bar{x} = \frac{1}{N} \sum_{k=1}^{N} x_k. \tag{9.2}$$

Substituting the measurement model assumed in Eq. (9.1) into this equation gives the following result:

$$\begin{aligned}
\bar{x} &= \frac{1}{N} \sum_{k=1}^{N} (x_0 + e_k), \\
&= \frac{1}{N} \sum_{k=1}^{N} x_0 + \frac{1}{N} \sum_{k=1}^{N} e_k, \\
&= x_0 + \frac{1}{N} \sum_{k=1}^{N} e_k. \tag{9.3}
\end{aligned}$$

This result tells us that our computed data summary $\bar{x}$ is composed of two components: the first is the correct but unknown value x_0 contained in our measurement sequence, and the second is the *average of the measurement errors*. We can say much more about this result, but to do so we need to introduce two things: a specific probability distribution for the error sequence $\{e_k\}$, and the concept of *expected values*.

As noted in Chapter 3, the most popular random variable model for numerical data is the Gaussian distribution, characterized by two parameters: a *mean value* μ, and a *standard deviation* σ. The density function that defines this probability distribution was given in Chapter 3 and is repeated here for convenience:

$$p(x) = \frac{1}{\sqrt{2\pi}\sigma} \exp\left\{-\frac{1}{2}\left(\frac{x-\mu}{\sigma}\right)^2\right\}. \tag{9.4}$$

The practical reason for introducing a specific probability distribution is that it allows us to compute *expected characterizations* of our data, assuming this probability distribution has been chosen reasonably enough to represent a good approximation. That is, given an assumed probability model like the Gaussian distribution defined in Eq. (9.4), we can ask and answer questions like, "What value do we expect to see for $\bar{x}$?" and "How much variation should we see around this expected value?"

To answer these questions, we need the concept of an *expected value* for a random variable x, or more generally, for some function $f(x)$ of this random variable. For a random variable defined by a probability density function $p(x)$, the *expected value of x*, $E\{x\}$, is defined as:

$$E\{x\} = \int_{-\infty}^{\infty} x p(x) dx. \tag{9.5}$$

For those without any exposure to calculus, this equation won't mean much, but don't worry: the key point here is that, if we model some variable x as a random variable with probability density function $p(x)$ like that defined in Eq. (9.4), it is possible to compute the expected value $E\{x\}$ from the parameters defining this distribution. As a specific example, for the Gaussian distribution, this value is:

$$E\{x\} = \mu. \tag{9.6}$$

As noted, expected values can also be defined for functions of a random variable x, and a particularly important example of this is the variance, which is defined as the expected value of squared deviations of x from its expected value $E\{x\}$. Again for the Gaussian distribution, this value is:

$$\text{var}\{x\} = E\{(x - E\{x\})^2\} = E\{(x - \mu)^2\} = \sigma^2. \tag{9.7}$$

Finally, three other details that will be useful in the following discussion are, first, that the expected value of a *determinisitic constant c*—i.e., any quantity that does not depend on our random variable—is simply that constant; second, that the expected value of a sum of variables is simply the sum of their individual expectations; and third, that the expected value of cx where c is a deterministic constant and x is a random variable is simply c times the expected value of x. Mathematically, if u, v, and x are random variables and c is a deterministic constant, these results are:

$$\begin{aligned} E\{c\} &= c \\ E\{u+v\} &= E\{u\} + E\{v\}, \\ E\{cx\} &= cE\{x\}. \end{aligned} \tag{9.8}$$

Returning to our repeated measurement problem, if we assume our measurement errors are well approximated by a Gaussian distribution with mean μ and standard deviation σ, it follows from Eqs. (9.3) and (9.8) that:

$$
\begin{aligned}
E\{\bar{x}\} &= E\left\{x_0 + \frac{1}{N}\sum_{k=1}^{N} e_k\right\}, \\
&= E\{x_0\} + E\left\{\frac{1}{N}\sum_{k=1}^{N} e_k\right\}, \\
&= x_0 + \frac{1}{N}E\left\{\sum_{k=1}^{N} e_k\right\}, \\
&= x_0 + \frac{1}{N}\sum_{k=1}^{N} E\{e_k\}, \\
&= x_0 + \frac{1}{N}\sum_{k=1}^{N} \mu, \\
&= x_0 + \mu. \tag{9.9}
\end{aligned}
$$

Walking through this sequence of steps, the first line simply substitutes Eq. (9.3) into the definition of the expected value, and the second line follows from the second equation in Eq. (9.8). The third line follows, first, from the fact that the expected value of the deterministic constant x_0 is simply x_0 (this value is unknown but it is constant and does not depend on the errors $\{e_k\}$), and, second, from the third equation in Eq. (9.8). The fourth line again follows from the additivity of expectations—i.e., the second equation in Eq. (9.8)— and the fifth line follows from Eq. (9.6), applicable because we have assumed a Gaussian distribution for the measurement errors $\{e_k\}$ here. Finally, the last line in Eq. (9.9) follows from the fact that the average of a sequence of N numbers, all with the same value μ, is simply that value.

What does this result mean in practical terms? It means that, under our working assumptions of a Gaussian additive error model, the expected value of the *computed mean* $\bar{x}$—something we calculate directly from the observed data using Eq. (9.2)—is equal to the sum of the correct but unknown value x_0, plus the mean of the assumed error distribution.

As noted, the Gaussian error assumption made here is extremely popular, probably the most widely assumed error distribution there is. Another extremely popular assumption for measurement errors is that the measurements are *unbiased*, meaning that the expected value of these errors is $\mu = 0$. Under this assumption, it follows that the expected value $E\{\bar{x}\}$ of the average of the repeated measurement values $\{x_k\}$ is the correct but unknown value x_0. In practical terms, this means that, *on average, the mean value $\bar{x}$ is correct*.

But wait: the same thing is true of the individual data observations $\{x_k\}$. Specifically, it follows from the above assumptions—with a lot less algebra—that for a single unbiased measurement (i.e., $\mu = 0$):

$$E\{x_k\} = E\{x_0 + e_k\} = E\{x_0\} + E\{e_k\} = x_0 + \mu = x_0. \tag{9.10}$$

What, then, is the advantage of computing the average? The answer to this question comes when we look at the variance—or, equivalently, the standard deviation—of the computed mean and compare it with that for the individual observations. For the individual observations, the variance is:

$$
\begin{aligned}
\mathrm{var}\{x_k\} &= E\{(x_k - E\{x_k\})^2\}, \\
&= E\{(x_0 + e_k - x_0)^2\}, \\
&= E\{e_k^2\}, \\
&= E\{(e_k - \mu)^2\} \text{ (since } \mu = 0 \text{ for unbiased measurements)}, \\
&= \sigma^2. \tag{9.11}
\end{aligned}
$$

In words, then, the variance of each individual observation is simply equal to the variance assumed for the measurement errors. This result is a direct consequence of the unbiased, additive error model: the randomness inherent in the data comes from that assumed for the measurement errors.

For the average $\bar{x}$ of the observed measurements, we obtain a different result:

$$
\begin{aligned}
\mathrm{var}\{\bar{x}\} &= E\{(\bar{x} - E\{\bar{x}\})^2\}, \\
&= E\left\{\left(x_0 + \frac{1}{N}\sum_{k=1}^{N} e_k - x_0\right)\right\}, \\
&= E\left\{\left(\frac{1}{N}\sum_{k=1}^{N} e_k\right)^2\right\}, \\
&= \frac{1}{N^2}E\left\{\left(\sum_{k=1}^{N} e_k\right)^2\right\}. \tag{9.12}
\end{aligned}
$$

We can simplify this result by invoking the following algebra trick:

$$\left(\sum_{k=1}^{N} e_k\right)^2 = \sum_{j=1}^{N}\sum_{k=1}^{N} e_j e_k. \tag{9.13}$$

Substituting this result into the last line of Eq. (9.12) gives:

$$
\begin{aligned}
\mathrm{var}\{\bar{x}\} &= \frac{1}{N^2}E\left\{\sum_{j=1}^{N}\sum_{k=1}^{N} e_j e_k\right\}, \\
&= \frac{1}{N^2}\sum_{j=1}^{N}\sum_{k=1}^{N} E\{e_j e_k\}. \tag{9.14}
\end{aligned}
$$

To simplify this result, we need to make one more assumption, also extremely popular and discussed further in Sec. 9.6.1. Specifically, this result simplifies considerably if we assume that the individual errors e_j and e_k are *statistically independent* for $j \neq k$. More specifically, we assume that, although all measurement errors are approximated by a zero-mean Gaussian distribution with standard deviation σ here (or, equivalently, variance σ^2), the individual errors e_k and e_j may be viewed as independent random draws from this Gaussian distribution. One consequence, discussed further in Sec. 9.6.1, is that if e_k and e_j are statistically independent, then:

$$E\{e_j e_k\} = \begin{cases} E\{e_j\}E\{e_k\} & \text{if } j \neq k, \\ E\{e_j^2\} & \text{if } j = k, \end{cases}$$

$$= \begin{cases} 0 & \text{if } j \neq k, \\ \sigma^2 & \text{if } j = k. \end{cases} \tag{9.15}$$

Substituting this result into Eq. (9.16) then gives:

$$\text{var}\{\bar{x}\} = \frac{1}{N^2} \sum_{k=1}^{N} \sigma^2 \text{ (only terms with } j = k \text{ contribute)},$$

$$= \frac{1}{N^2} \cdot N\sigma^2,$$

$$= \frac{\sigma^2}{N}. \tag{9.16}$$

Comparing this result with that from Eq. (9.11) for $\text{var}\{x_k\}$*, we see that the variance of the average is smaller by a factor of* N *than the variance of the individual observations.*

This last result is practically important for a number of reasons. First, it tells us that averaging a collection of measurements of what should be the same thing improves the precision of our result, increasingly as the size N of the dataset grows. Further, under the assumptions made here—i.e., a statistically independent sequence of zero-mean Gaussian measurement errors, all with standard deviation σ—the mean $\bar{x}$ can be shown to have a Gaussian distribution with mean x_0 and standard deviation $\sigma/\sqrt{N}$. This mean and standard deviation follow from the results just presented, but the fact that $\bar{x}$ is itself Gaussian does not; further discussion of this result is beyond the scope of this book, but details can be found in *Exploring Data* [58, p. 243]. A more general version of this result is the *Central Limit Theorem*, which says that, with some important restrictions that are satisfied by the vast majority of probability distributions, the average of any sequence of random numbers with mean μ and standard deviation σ exhibits an approximately Gaussian distribution with mean μ and standard deviation $\sigma/\sqrt{N}$ for sufficiently large N, *regardless of the original data distribution* [58, p. 246]. Even more importantly, this behavior—called *asymptotic normality*—turns out to hold for almost all data characterizations that are in wide use, including means, standard deviations, correlation coefficients, the parameters of linear regression models, and much more.

9.2 Confidence intervals and significance

As emphasized in Sec. 9.1, if we adopt the random variable model as a mathematical description of the uncertainty and natural variability inherent in our data, everything we compute from that data becomes a random variable. Thus, questions like, "What are the values of the linear regression model coefficients that are most consistent with the data?" or "Are these coefficients significantly different from zero?" all become questions about random variables. An advantage of this random variable formulation is that we can use the machinery of probability theory to quantify the uncertainty inherent in our computed results. The following discussions introduce two of the most useful of these random variable characterizations, in general terms: *confidence intervals* in Sec. 9.2.3, and *p-values* in Sec. 9.2.4. Specific applications of these ideas will then be presented in subsequent sections of this chapter in connection with uncertainty and significance assessments of other data characterizations (e.g., confidence intervals for the odds ratio in Sec. 9.3.3 and for correlation coefficients in Secs. 9.6.1 and 9.6.2). Because confidence intervals are closely related to the quantiles of a distribution, this topic is introduced in Sec. 9.2.2 to provide important background. Even before this discussion, however, Sec. 9.2.1 briefly discusses the important distinction between *data characterizations* that we compute directly from our data observations, and *model characterizations* that describe the probability models we use to approximate those data observations.

9.2.1 Probability models versus data

In this book, the terms *model, data model,* or *probability model* refer to *mathematical expressions* developed as *approximate descriptions* of the behavior of observed data collections. As a specific example, the *linear regression models* discussed in Chapter 5 postulate an approximate linear relationship between a numerical response variable and one or more numerical predictor variables. Similarly, a probability model attempts to characterize the irregular variations seen in a sequence of data values in terms of a probabiity distribution like the Gaussian distribution discussed in Chapter 3. Two crucial features of these models are, first, that they are *approximations* rather than exact descriptions of our data, and, second, the reason we use them is that they allow us to say more about how we expect our data to behave than we could without them. The way we accomplish this second task is to, first, construct a data model that we believe reasonably approximates our data, and second, use the data model to compute characterizations that cannot be obtained from the data values alone.

As a specific example, if we have a sequence of numerical data observations, we can easily calculate the minimum and maximum observed values, but we cannot say how likely we would be to observe, from a new batch of data collected under similar conditions, values smaller than the observed minimum or larger than the observed maximum. If we fit a probability distribution to our data—e.g., a Gaussian distribution based on the mean and standard deviation computed from the data—we can answer these questions, *based on the assumption that our probability model is adequate.* As a specific example, recall from the discussion in Chapter 3 that the 18-week chick weight values in the ChickWeight data frame from the *R* datasets package appeared to be well approximated by a Gaussian distribution. The minimum, maximum, mean, and standard deviation are easily computed from this data sequence:

```
wts18 <- ChickWeight$weight[which(ChickWeight$Time == 18)]
min(wts18)

## [1] 72

max(wts18)

## [1] 332

mean(wts18)

## [1] 190.1915

sd(wts18)

## [1] 57.39476
```

If we make the assumption that these data values are reasonably well approximated by a Gaussian distribution, we can obtain a *specific* Gaussian distribution by specifying its mean and standard deviation to be those of the data, listed above. From this specific data model, we can answer the questions posed above: if we collected a larger set of 18-week chick weight data, raised under the same conditions, we can estimate the probability of observing a value smaller than 72 as 0.02 and the probability of observing a value larger than 332 as 0.007.

The adequacy assumption for the probability model is extremely important, since this assumption can be violated in at least two important ways: first, it is possible that our assumed mathematical model has the wrong form, and, second, it is possible that the data characteristics we use to construct the best approximate model of this type are inaccurate. For the simple Gaussian probability model, the first violation corresponds to the Gaussian distribution not being an adequate description of the data variation, and the second violation corresponds to the mean and standard deviation we use to fit this Gaussian data model being wrong (e.g., most of the data values are reasonably well approximated by the Gaussian distribution, but there are a few extreme outliers that make the estimated standard deviation much too large).

This distinction—between data characterizations and model characterizations used to describe the data—is an important source of confusion among those new to data modeling, especially using probability models. An important example is the difference between the *average of the data values*, something we compute by adding these values and dividing by the number of them we have, and the *expected value* or *mean* of the distribution we use to approximate those data values. If we assume a random variable model to approximate our data, the arithmetic average becomes a random variable, as discussed in Sec. 9.1, while the mean of that distribution is an exact quantity we can compute from the parameters of the distribution. In fact, this distinction carries over to other important characterizations like standard deviation or the quantiles discussed in the next section, that can be computed—in different ways—from either the data values or the assumed distribution. In cases where our assumed distribution is a good approximation of our data values and we have a large enough data sample, there should be little difference in these computed values. The most important point, however, is that assuming a distribution allows us to compute quantities like the confidence intervals described in Sec. 9.2.3 that cannot be computed without making these distributional assumptions.

9.2.2 Quantiles of a distribution

It was noted in the introductory discussion of the boxplot that the basis for this graphical data characterization is five quantiles of the data: the sample minimum, the lower quartile, the median, the upper quartile, and the sample maximum. These values can be computed from any numerical data sequence in R using the `quantile` function by specifying only the data vector. Thus, for the 18-week chick weight data from the previous example, these values are:

```
wts18 <- ChickWeight$weight[which(ChickWeight$Time == 18)]
quantile(wts18)
```

```
##     0%    25%    50%    75%   100%
##   72.0  152.5  187.0  230.5  332.0
```

Like the mean and the standard deviation, quantiles can be computed either from the observed data, as in this example, or from an assumed data distribution, as discussed below. If our assumed data distribution adequately approximates our data and for sufficiently large data samples, these values should be approximately the same, but there are some important exceptions. For now, note that a useful optional argument for the `quantile` function is `probs`, which can be used to give us a more complete picture of how our data sample is distributed over its range of possible values. As a specific example, the *deciles* of the 18-week chick weight data are given by:

```
quantile(wts18, probs = seq(0, 1, 0.1))
```

```
##     0%    10%    20%    30%    40%    50%    60%    70%    80%    90%   100%
##   72.0  116.8  146.4  159.4  178.0  187.0  203.6  216.6  233.6  261.4  332.0
```

This characterization provides reference values to tell us about the center of the data distribution (for example, the 50% quantile is the sample median) and about extreme values in the tail of the data distribution (e.g., the range of the lowest 10% of the values and the highest 10% of the values).

As discussed in Chapter 3, the cumulative distribution function $P(z)$ for a continuous random variable (e.g., Gaussian) tells us the probability of observing a data sample x_k with value less than any specified value z, i.e.:

$$P(z) = \text{Probability}\{x_k < z\}. \tag{9.17}$$

In fact, this was the function used to compute the probability of observing an 18-week chick weight smaller than 72 in the example discussed in Sec. 9.2.1. Specifically, these computations were:

```
minWt <- min(wts18)
maxWt <- max(wts18)
meanWt <- mean(wts18)
sdWt <- sd(wts18)
pMin <- pnorm(q = minWt, mean = meanWt, sd = sdWt)
pMin

## [1] 0.01973403
```

The *quantile function* $Q(x)$ is essentially the inverse of $P(z)$, telling us what value of z satisfies the condition $P(z) = x$:

$$Q(x) = \{z \text{ such that } P(z) = x\}. \tag{9.18}$$

This function can be used to answer questions like, "where is the center of the distribution?" or "how far out into the tails of the distribution do I need to go for the probability of seeing more extreme values to be less than, say, 5%?" The center of the distribution is typically taken as the median: there is a 50% probability of observing data samples larger than this value, and a 50% probability of observing data samples smaller than this value. For symmetric distributions like the Gaussian, the median of the distribution is the same as the mean of the distribution, but for asymmetric distributions this is no longer the case.

As noted, if we have a large enough data sample and an assumed distribution that adequately approximates that sample, the quantiles of this distribution should agree reasonably well with those computed from the data using the quantile function. For the 18-week chick weight example, we can make this comparison explicitly, comparing the quantile results presented earlier with the distribution quantiles computed from the qnorm function. Here, we specify the mean and standard deviation of the distribution, and the range of probabilities of interest are specified with the p argument:

```
qnorm(mean = meanWt, sd = sdWt, p = seq(0, 1, 0.1))

##  [1]     -Inf 116.6371 141.8868 160.0936 175.6507 190.1915 204.7323
##  [8] 220.2893 238.4961 263.7458      Inf
```

If we compare these results for the 10% through the 90% quantiles, we find that they are within 1% of the values computed from the data given earlier. The striking difference is in the 0% and 100% quantiles, which have the values $-\infty$ and ∞ for the Gaussian distribution, corresponding to the fact that this distribution exhibits a finite probability of exhibiting *any* real value, although this probability becomes extremely small if we move far enough out into the tails of the distribution. In contrast, the 0% and 100% *data quantiles* returned by the quantile function correspond to the sample minimum and maximum, respectively, and these values are finite.

R supports many different distributions, several of which are discussed briefly later in this chapter, and for most of these distributions, this support includes the following four functions:

1. a density function like `dnorm` that computes the values of the probability density $p(x)$ for a range of data values x, given any required distribution parameters (e.g., the mean and standard deviation for the Gaussian data);

2. a cumulative distribution function like `pnorm` that computes the cumulative probability $P(x)$ for this distribution, given the same parameters;

3. a quantile function like `qnorm` that computes the quantiles of the distribution for a given set of probabilities and the distribution parameters;

4. a random number generator like `rnorm` that generates random samples from the distribution, given the number of samples desired and the distribution parameters.

Often, the distribution parameters have default values for a standard form of the distribution. For example, the default mean is 0 and the default standard deviation is 1 for the functions `dnorm`, `pnorm`, `qnorm`, and `rnorm`. The main reason for introducing quantile functions like `qnorm` here is that they form the basis for computing the confidence intervals discussed next.

9.2.3 Confidence intervals

It has been noted repeatedly, first, that the random variable model provides an extremely useful and popular description of uncertainty in real-world data, and, second, that once we adopt it, everything we compute from that real-world data inherits a probability distribution. This point was discussed in some detail for the average of a data sequence in Sec. 9.1, where an explicit distribution was developed for this case, under the assumption of Gaussian data. A further consequence of this second point is that there is inherent uncertainty around anything we compute from our data, and the key idea of the following discussion is that it is useful to quantify this uncertainty. Again considering the average as a specific example, note that the average we compute from our data approximates the mean of the distribution we assume to describe the data, and this approximation becomes more and more accurate the more data observations we have to base it on. The idea of a confidence interval for this case—i.e., the average of a sequence of observations—is that it defines a range where this distribution mean will lie with a specified probability.

This idea is illustrated in Fig. 9.1 for the important case of the *standard Gaussian distribution* or *standard normal distribution*, which has zero mean and unit standard deviation. As the labels on this plot indicate, if we draw samples from this distribution, there is a 95% probability that they will fall into the region between the two vertical lines in the plot, with a 2.5% probability of falling to the left of the left-most of these lines, and a 2.5% probability of falling

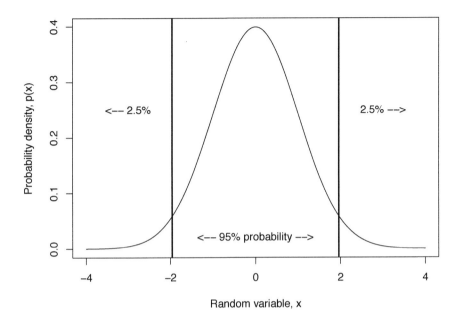

Figure 9.1: Probabilities of observing Gaussian random samples in three parts of the distribution.

to the right of the right-most line. The positions of these lines are given by the qnorm function described in Sec. 9.2.2, i.e.:

```
qnorm(p = 0.025, mean = 0, sd = 1)

## [1] -1.959964

qnorm(p = 0.975, mean = 0, sd = 1)

## [1] 1.959964
```

The utility of this idea is that we can use it to quantify the uncertainty in quantities like the arithmetic average that we compute from real data. Specifically, if we assume a Gaussian distribution for our data—as in the case of the 18-week chick weight data considered earlier—we can construct a confidence interval that tells us where the *distribution mean* lies, with specified probability. A derivation of this result involves a little algebra, which is presented in the optional section below, but the main result looks like this: the distribution mean lies in the interval $[\mu_-, \mu_+]$ with probability $1 - \alpha$, where α is typically taken as either 0.05 (for 95% confidence intervals) or 0.01 (for 99% confidence intervals). Specifically, the $1 - \alpha$ confidence interval for the mean μ of a Gaussian distribution describing a sequence of N approximately Gaussian data samples with mean $\bar{x}$ is:

$$CI_{1-\alpha} = [\bar{x} - z_{1-\alpha/2}\sigma/\sqrt{N}, \bar{x} + z_{1-\alpha/2}\sigma/\sqrt{N}], \tag{9.19}$$

where $z_{1-\alpha/2}$ is the $1 - \alpha/2$ quantile of the normal distribution. For the case of 95% confidence intervals, $\alpha = 0.05$ and $z_{1-\alpha/2}$ is the 0.975 quantile listed above, corresponding to the right-most line in Fig. 9.1. To actually use this result, we need the standard deviation σ for the Gaussian distribution, but like the distribution mean μ, this value is not known. For a long time, it was assumed that σ could simply be replaced by the standard estimator $\hat{\sigma}$ computed from the data, but it was shown in 1908 by a mathematician publishing under the pseudonym "Student" that the following ratio exhibited what is now known as *Student's t-distribution with $N - 1$ degrees of freedom* [47, Ch. 28], discussed briefly in Sec. 9.5.2:

$$t = \frac{\sqrt{N}(\bar{x} - \mu)}{\hat{\sigma}}, \tag{9.20}$$

where $\hat{\sigma}$ is the estimated standard deviation, given by:

$$\hat{\sigma} = \sqrt{\frac{1}{N - 1} \sum_{k=1}^{N} (x_k - \bar{x})^2}. \tag{9.21}$$

Thus, the confidence interval in Eq. (9.19) should be replaced by:

$$CI_{1-\alpha} = [\bar{x} - t_{N-1,1-\alpha/2}\sigma/\sqrt{N}, \bar{x} + t_{N-1,1-\alpha/2}\sigma/\sqrt{N}], \tag{9.22}$$

where $t_{N-1,1-\alpha/2}$ is the $1 - \alpha/2$ quantile of the Student's t-distribution with $N - 1$ degrees of freedom.

As noted in Sec. 9.5.2, the Student's t-distribution has heavier tails than the Gaussian distribution, but approaches this distribution in the limit of infinite degrees of freedom. The practical consequence is that the actual confidence intervals for the mean are wider than those given by Eq. 9.19, but this difference becomes smaller with increasing sample size N. For example, consider the 18-week chick weight data considered earlier: the mean of the 47 data observations is 190.2, the standard deviation is 57.39, so Eq. (9.19) would give a lower 95% confidence limit of 173.8 and an upper limit of 206.6. For comparison, using the Student's t-distribution with 46 degrees of freedom yields a lower 95% confidence limit of 173.3 and an upper limit of 207. Note that both of these numbers differ by less than 1% between the exact result based on the Student's t-distribution and that based on the Gaussian assumption for the mean: since the Gaussian *data distribution* assumed for both results is only an approximation, anyway, this difference is not really important here.

One of the other points made in Sec. 9.1 was that, *for most data charac-terizations we are likely to compute, the distribution of this characterization is approximately Gaussian, provided we have enough data.* This result is called *asymptotic normality* and the example just presented is a specific illustration, but its practical importance lies in two facts. First, asymptotic normality holds for almost all of the data characterizations we encounter in practice: there are exceptions, but they are rare, and even these exceptions can sometimes be overcome easily, as in the case of the odds ratio discussed in Sec. 9.3.3. The second reason for the practical importance of asymptotic normality is that it provides a *much* easier way of deriving confidence intervals when it does hold. In particular, if our data characterization is *unbiased,* meaning that it is "correct on average" and satisfies asymptotic normality, all we need to compute confidence intervals is an expression for the standard deviation of our estimator. Several examples presented in later sections of this chapter provide specific illustrations of this point.

The algebra

(The following discussion derives the form of the confidence interval shown in Eq. (9.19): readers not interested in these details can skip this discussion and simply take results like this one and other confidence intervals presented in this chapter as given. The derivation is included here for those wanting a more detailed view of where these results come from.)

The starting point of this observation is that, as shown in Sec. 9.1, the mean $\bar{x}$ of a sequence $\{x_k\}$ of N Gaussian data observations with mean μ and variance σ^2 is another Gaussian distribution, also with mean μ but with smaller variance σ^2/N. Two useful characteristics of the Gaussian distribution are that:

1. if z is Gaussian with mean a and variance b^2, then $z - a$ is Gaussian with mean 0 and variance b^2;

2. if z is Gaussian with mean 0 and variance b^2, then z/b is Gaussian with mean 0 and variance 1.

From these two results, it follows that, first, $\bar{x} - \mu$ has a Gaussian distribution with mean zero and variance σ^2/N, and, second, that the following ratio has a Gaussian distribution with mean zero and variance 1:

$$r = \frac{\bar{x} - \mu}{\sigma/\sqrt{N}} = \frac{\sqrt{N}(\bar{x} - \mu)}{\sigma}. \tag{9.23}$$

From this result, we can calculate the probability that r lies in tails of the distribution like those indicated in Fig. 9.1. Specifically:

$$
\begin{aligned}
\text{Prob}\,(r < z_{\alpha/2}) &= \alpha/2, \\
\text{Prob}\,(r > z_{1-\alpha/2}) &= 1 - \text{Prob}\,(r < z_{1-\alpha/2}), \\
&= 1 - (1 - \alpha/2) = \alpha/2. \tag{9.24}
\end{aligned}
$$

Since only one of these conditions can hold, the probability of either of them holding is simply the sum of the individual probabilities, so we have:

$$\text{Prob}\,(r < z_{\alpha/2} \text{ or } r > z_{1-\alpha/2}) = \alpha. \tag{9.25}$$

This result can be simplified by noting, first, that $z_{\alpha/2} = -z_{1-\alpha/2}$. This observation is a consequence of the symmetry of the standard Gaussian distribution about its mean value zero. Also, since $z_{\alpha/2} < 0$ for the small values of alpha considered here (i.e., $\alpha = 0.05$ is the most common choice in practice, followed by $\alpha = 0.01$), it is more convenient to work with the positive value $z_{1-\alpha/2}$. Based on these observations, we can re-write Eq. (9.25) as:

$$\begin{aligned} \text{Prob}\,(r < z_{\alpha/2} \text{ or } r > z_{1-\alpha/2}) &= \text{Prob}\,(r < -z_{1-\alpha/2} \text{ or } r > z_{1-\alpha/2}) \\ &= \text{Prob}\,(|r| > z_{1-\alpha/2}) = \alpha. \quad (9.26) \end{aligned}$$

From this result, it follows that:

$$\text{Prob}\,(-z_{1-\alpha/2} < r < z_{1-\alpha/2}) = 1 - \text{Prob}\,(|r| > z_{1-\alpha/2}) = 1 - \alpha. \quad (9.27)$$

Substituting the definition of r from Eq. (9.23) into this expression and re-arranging it yields an expression for the probability associated wtih the unknown distribution mean μ, based on $\bar{x}$ and σ:

$$\begin{aligned} 1 - \alpha &= \text{Prob}\left(-z_{1-\alpha/2} < \frac{\bar{x} - \mu}{\sigma/\sqrt{N}} < z_{1-\alpha/2}\right) \\ &= \text{Prob}\left(-z_{1-\alpha/2}\sigma/\sqrt{N} < \bar{x} - \mu < z_{1-\alpha/2}\sigma/\sqrt{N}\right) \\ &= \text{Prob}\left(-z_{1-\alpha/2}\sigma/\sqrt{N} - \bar{x} < -\mu < z_{1-\alpha/2}\sigma/\sqrt{N} - \bar{x}\right) \\ &= \text{Prob}\left(\bar{x} - z_{1-\alpha/2}\sigma/\sqrt{N} < \mu < \bar{x} + z_{1-\alpha/2}\sigma/\sqrt{N}\right). \quad (9.28) \end{aligned}$$

The confidence interval expression given in Eq. (9.19) is simply a re-expression of this result: the unknown value μ lies in the interval in this last expression with probability $1 - \alpha$.

9.2.4 Statistical significance and p-values

The concept of *statistical significance* is closely related to the confidence intervals just discussed, and *p-values* are the standard numerical measure of statistical significance. Perhaps the easiest way to introduce the concept of statistical significance is to look at a common example where it arises. Recall from Chapter 5 that the linear regression modeling procedure `lm` returns estimates of model parameters, along with a number of related quantities, including p-values associated with these coefficients. As a specific example, consider the linear regression model that predicts the heating gas consumption (`Gas`) from both the outside temperature (`Temp`) and the binary insulation indicator (`Insul`), including an interaction term between these two variables. The code to fit and characterize this model looks like this:

```
whiteModel <- lm(Gas ~ Temp * Insul, data = whiteside)
summary(whiteModel)

##
## Call:
```

```
## lm(formula = Gas ~ Temp * Insul, data = whiteside)
##
## Residuals:
##       Min       1Q   Median       3Q      Max
## -0.97802 -0.18011  0.03757  0.20930  0.63803
##
## Coefficients:
##                 Estimate Std. Error t value Pr(>|t|)
## (Intercept)      6.85383    0.13596  50.409  < 2e-16 ***
## Temp            -0.39324    0.02249 -17.487  < 2e-16 ***
## InsulAfter      -2.12998    0.18009 -11.827 2.32e-16 ***
## Temp:InsulAfter  0.11530    0.03211   3.591 0.000731 ***
## ---
## Signif. codes:  0 '***' 0.001 '**' 0.01 '*' 0.05 '.' 0.1 ' ' 1
##
## Residual standard error: 0.323 on 52 degrees of freedom
## Multiple R-squared:  0.9277,Adjusted R-squared:  0.9235
## F-statistic: 222.3 on 3 and 52 DF,  p-value: < 2.2e-16
```

The numerical results in the "Coefficients" section of this summary include the coefficient values themselves (the column labelled "Estimate"), the *standard error*, which is an estimate of the standard deviation of the coefficient value, the associated t-value, analogous to that defined in Eq. (9.20) for the mean, and a p-value discussed in detail below (the column labelled "$Pr(> |t|)$"). The model coefficients are obtained using the least squares fitting procedure described in Chapter 5, and the standard errors are obtained by exploiting the special structure of the ordinary least squares solution for the linear regression problem. The t-value is the ratio of the estimated coefficient to the standard error, something that may be verified by computing these ratios and comparing them with the t-values listed in the summary above: there are slight differences due to round-off errors (the displayed results are rounded to five digits, but the t-values displayed are computed from more precise coefficient and standard error values). The p-values given in the last column are the probabilities from the Student's t-distribution of observing a sample larger in absolute value than the t-value listed. The degrees of freedom for this t-distribution is given in the summary display (52) and corresponds to the number of data observations minus the number of estimated parameters.

These p-values have the following interpretation: if they are smaller than a specified threshold (typically, 5%), they provide evidence that the estimated parameters are large enough in magnitude to be regarded as *statistically significant*. Statistical significance implies that the probability of obtaining the observed result under a *reference condition* called the *null hypothesis* is too small for this result to have occurred by random chance. In the case of linear regression model coefficients, the null hypothesis is that "the true value of this parameter is zero," which would mean that "this parameter does not substantially improve the model's predictive ability." The fact that all four of the p-values shown here are extremely small (in particular, *much* smaller than the standard 5% threshold $\alpha = 0.05$) gives us a strong basis for rejecting this null hypothesis, and thus including all four terms in our model.

For the case of linear regression model parameters, the connection between statistical significance and confidence intervals is this: if a particular coefficient is statistically significant at the level α (i.e., if the p-value is less than α), this means that the value zero does not belong to the $1 - \alpha$ confidence interval. To see that this is the case here, note that the general form of the $1 - \alpha$ confidence interval for a model coefficient a_i based on an estimate $\hat{a}_i$ with standard error SE_i for a model with ν degrees of freedom is analogous to Eq. (9.22) for the mean,

$$\text{CI}_{1-\alpha} = [\hat{a}_i - t_{\nu, 1-\alpha/2} SE_i, \hat{a}_i + t_{\nu, 1-\alpha/2} SE_i]. \tag{9.29}$$

For $\alpha = 0.05$ and $\nu = 52$, we can use this equation to obtain the following lower 95% confidence limits for these model parameters:

```
##    (Intercept)            Temp      InsulAfter Temp:InsulAfter
##     6.58099603     -0.43836236     -2.49135850     0.05086618
```

and the corresponding upper 95% confidence limits:

```
##    (Intercept)            Temp      InsulAfter Temp:InsulAfter
##      7.1266594      -0.3481153      -1.7685976      0.1797416
```

Note that the critical value zero does not fall into any of these four confidence intervals, reflecting the fact that the p-values for these parameters listed above are all smaller than $\alpha = 0.05$ here.

More generally, the concepts of statistical significance and p-values are associated with *statistical hypothesis tests*, where we have a hypothesis of interest (e.g., "this model coefficient is large enough in magnitude to provide evidence that this parameter should be included in the moel") that we want to evaluate. To make this evaluation, we proceed as in this example, by first specifying a null hypothesis that corresponds to the case where our hypothesis *does not hold* (e.g., "this parameter does not belong in the model, so it's correct parameter value is zero"). Then, we compute the probability that the null hypothesis is true (this is our p-value), and if this probability is sufficiently small (traditionally, less than 5% or less than 1%), we reject the null hypothesis and take this result as support for our hypothesis (i.e., "the four estimated coefficients in this model have extremely small p-values, supporting our hypothesis that these terms should be included in our model).

In some applications (e.g., medical publications), p-values are expected (indeed, journals may require them as a condition for publication), but they have at least two major limitations. First, they are less informative than confidence intervals, which give us both a quantitative assessment of the variability of our computed results and an assessment of statistical significance. Specifically, the null hypothesis we would use to obtain p-values generally corresponds to a specific value of our data characterization (e.g., a model parameter value of zero), so we can assess statistical significance by looking to see whether this specific value falls in our confidence interval or not. The second limitation is that p-values are not always conveniently available. As a particularly important case, many of the predictive model types discussed in Chapter 10 do not provide convenient characterizations of model parameter statistics and their associated *p*-values: indeed, extremely complex models like random forests do not even provide parameter estimates. This limitation motivates the alternative model characterization tools introduced in Chapter 10.

For these reasons, this chapter places more emphasis on confidence intervals than on *p*-values. Particularly important examples are the use of confidence intervals to characterize binary variables discussed in Sec. 9.3 and as an adjunct to correlation analysis, discussed in Sec. 9.6.

9.3 Characterizing a binary variable

Binary variables are those that can only assume one of two values, like the `Insul` variable in the `whiteside` data frame from the `MASS` package discussed many times in earlier chapters of this book. As noted in Chapter 5, even if these variables are categorical, like the `Insul` variable with values "Before" and "After," it is a simple matter to re-express them as equivalent numerical variables with values 0 and 1. This means that binary variables are in some sense "intermediate" between categorical variables and (discrete) numerical variables like the integer-valued count data considered in Sec. 9.4. Thus, binary variables are amenable to characterization by both techniques like the contingency tables discussed in Sec. 9.7.1 for categorical variables, and arithmetic characterizations like means and standard deviations discussed in Chapter 3 for numerical variables. Neither of these classes of techniques exploit the unique character of binary variables, however, and better insights into these variables can often be obtained using techniques that do exploit this unique character. The following discussions introduce, first, a useful probability model for binary data (the binomial distribution, introduced in Sec. 9.3.1), second, confidence intervals on binary probabilities that exploit the properties of this distribution (in Sec. 9.3.2), and, third, the *odds ratio* (in Sec. 9.3.3), a specialized measure of association between binary variables like those introduced in Secs. 9.6 and 9.7.

9.3.1 The binomial distribution

Many problems involving random binary outcomes can be formulated as some variation of the following *urn problem* [58, Sec. 3.3.1]:

> An urn contains N balls, with M being white and the other $N - M$ being black. If n balls are drawn from the urn with replacement (i.e., the ball is withdrawn, its color is noted, and it is returned to the urn, which is then stirred or shaken to re-mix the balls), what is the probability that at least k of these n balls are white?

If $p = M/N$ is the fraction of white balls in the urn, the distribution of the number k of white balls drawn in n successive draws with replacement is the *binomial distribution*, a discrete probability distribution defined by:

$$P\{k \text{ white balls in } n \text{ succssive draws}\} = \binom{n}{k} p^k (1-p)^{n-k}, \qquad (9.30)$$

where the *binomial coefficient* is defined as:

$$\binom{n}{k} = \frac{n!}{k!(n-k)!}, \qquad (9.31)$$

and $n!$ is the factorial function:

$$n! = 1 \cdot 2 \cdot 3 \cdots n. \qquad (9.32)$$

Two important points should be noted about the binomial distribution. First, it is a *discrete distribution*, like the count data distributions discussed in Sec. 9.4. That is, in contrast to cases like the continuous Gaussian distribution introduced in Chapter 3, the binomial distribution only assigns probabilities to integer values k. This distinction is important because, as we will see in Sec. 9.4, discrete distributions behave differently from continuous distributions in certain important respects (e.g., discrete random variables can exhibit *ties* or repeated observations of the same data value, while continuously distributed random variables cannot). The second important point is that the range of the integer k in Eq. (9.30) is restricted to lie between $k = 0$ (i.e., "the sample of n balls contained only black balls and no white balls") and $k = n$ (i.e., "the sample contained all white balls").

A reasonably detailed discussion of this distribution is given by Collett [15, Sec. 2.1], but the following observations will suffice for our purposes here. First, while the binomial distribution represents the most popular probability model for binary data, it is not always a reasonable choice since, as Agresti notes, "there is no guarantee that successive binary observations are independent or identical" [2, p. 6]. For example, if the population being sampled is small and fixed, it may be more appropriate to assume *sampling without replacement*, implying that the population shrinks by one with each random sample drawn and the distribution of the remaining sample pool changes as a result; in these cases, the *hypergeometric distribution* is a better choice [58, Sec. 3.3.2]. Second, the mean and variance of the binomial distribution are given by [58, p. 119]:

$$\begin{aligned} E\{k\} &= np, \\ \text{var}\{k\} &= np(1-p). \end{aligned} \qquad (9.33)$$

Note that the mean result implies that the expected number of white balls in our sample of n balls is simply the fraction of white balls in the urn (p) times the size of our sample (n). Also, the variance expression implies that the precision of this estimate depends strongly on p: in particular, if this fraction of white balls in the urn is either zero or 1, the variance is zero, reflecting the fact that there is no uncertainty. That is, if there are no white balls in the urn, all samples will contain only black balls, with no uncertainty; similarly, if our urn contains only white balls, so will our sample, again with no uncertainty. If our urn contains an equal mix of white and black balls ($p = 0.5$), the variance has its maximum possible value ($n/4$), reflecting the fact that the uncertainty of our expected outcome is greatest in this case. Finally, for large n, the binomial distribution becomes approximately Gaussian, a useful characteristic of a number of non-Gaussian distributions that become approximately Gaussian in important limiting cases. In fact, this approximate normality of the binomial distribution forms the basis of the popular classical binomial confidence intervals discussed next.

9.3.2 Binomial confidence intervals

In a typical data analysis problem that makes use of the binomial distribution, we have a collection of n observations, each exhibiting one of two possible values, and we are interested in drawing conclusions about the fraction of these observations that assume one of these two values. One example, considered below, is the case where we want to know how this fraction varies over the different levels of a possibly related categorical variable. Regardless of the actual levels of this binary variable and their interpretations, we can always represent one of these values as the binary response $b_k = 1$ and the other one as the binary response $b_k = 0$, for $k = 1, 2, \ldots, n$. Given this representation, we can estimate the fraction of the first of these levels (commonly termed the "probability of success") as:

$$\hat{p} = \frac{n_1}{n}, \tag{9.34}$$

where n_1 is the number of records k for which $b_k = 1$ and n is the total number of records. This computation is extremely easy, and it gives us a number, but as the following example illustrates, this may not be enough information if we want to compare this fraction across different scenarios.

Specifically, consider the following question: is there evidence in the Australian vehicle insurance data to suggest that different vehicle types exhibit different risks of having an accident? To examine this question, we consider the binary insurance claim indicator variable `clm` in the `dataCar` data frame from the `insuranceData` package, coded as 1 if the policy filed a claim for accident damage, and 0 if they did not. The other variable of interest here is `veh_body`, a categorical variable with 13 unique values. The estimated claim probabilities $\{\hat{p}_i\}$ for each of these vehicle body types can be estimated from Eq. (9.34) in a number of different ways in R, and one of these is:

```
nVector <- table(dataCar$veh_body)
cTable <- table(dataCar$veh_body, dataCar$clm)
n1Vector <- cTable[, 2]
pHat <- n1Vector/nVector
```

Here, the first line counts the number of times each unique `veh_body` level occurs in the data, while the second line generates a *contingency table* (see Sec. 9.7.1 for a detailed discussion of contingency tables) that gives the number of records of

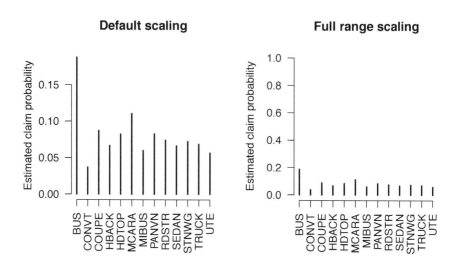

Figure 9.2: Two plots of estimated claim probability versus vehicle body type.

each vehicle body type for the two possible values of clm. The second column of this table gives the number of records for which clm = 1, corresponding to the variable n_1 in Eq. 9.34, and the last line of code computes $\hat{p}_i$ for each veh_body value using this equation.

Two plots of $\hat{p}_i$ vs. i are shown in Fig. 9.2 based on these results. The left-hand plot uses R's default scaling, which gives the clearest possible view of the differences between these individual claim probability estimates, while the right-hand plot uses the complete range of possible claim probabilities, from 0 to 1. These two plots give very different views of the differences between these claim probabilities, and it is not obvious which—if either—of these views is more representative. In particular, these estimated probabilities and their associated vehicle body types are:

```
##
##        BUS        CONVT       COUPE       HBACK       HDTOP       MCARA
## 0.18750000 0.03703704 0.08717949 0.06682527 0.08233059 0.11023622
##       MIBUS        PANVN       RDSTR       SEDAN       STNWG       TRUCK
## 0.05997211 0.08244681 0.07407407 0.06638780 0.07213579 0.06857143
##         UTE
## 0.05669429
```

By themselves, these numerical results—emphasized by the left-hand plot in Fig. 9.2—suggest that "BUS" is by far the most dangerous of the vehicle body types considered here, while "CONVT" is the safest, but is this really an accurate reflection of what the available data is suggesting, given the very small numbers of records we have for some of these body types? Putting confidence intervals around our estimates can be very useful in answering these questions.

It was noted in Sec. 9.3.1 that, for large n, the binomial distribution becomes approximately Gaussian, and Collett gives a reasonably detailed discussion of how this fact can be used to develop simple confidence intervals for the probability estimate $\hat{p}$ given in Eq. 9.34 [15, pp. 26–27]. In particular, substituting $\hat{p}$ from Eq. 9.34 into the variance expression for the binomial distribution in Eq. (9.33) and taking the square root gives the following standard error estimate for $\hat{p}$:

$$se(\hat{p}) = \sqrt{\frac{\hat{p}(1-\hat{p})}{n}}. \tag{9.35}$$

Under the approximate normality assumption, we then obtain the classical confidence interval for the probability p of a positive response:

$$p \in [\hat{p} - z_{1-\alpha/2}se(\hat{p}), \hat{p} + z_{1-\alpha/2}se(\hat{p})] \tag{9.36}$$

where α is the desired confidence limit, and $z_{1-\alpha/2}$ is the Gaussian quantile associated with the upper tail probability $\alpha/2$ discussed in Sec. 9.2.3.

Despite the historical popularity of this estimator, which dates back to Laplace in the early nineteenth century, Agresti notes that "it performs poorly unless n is very large," citing the results of a detailed study made in 2001 [11]. As a simple correction, Agresti notes that much better performance can be obtained by adding $z_{\alpha/2}^2/2$ to each count (i.e., the number n_1 of positive responses and the number n_0 of negative responses) before applying Eq. (9.34) to estimate the probability of positive response and Eq. (9.35) to estimate the standard errors. These replacements lead to the following modified estimators:

$$\tilde{p} = \frac{n_1 + z_{1-\alpha/2}^2/2}{n + z_{1-\alpha/2}^2}$$

$$se(\tilde{p}) = \sqrt{\frac{\tilde{p}(1-\tilde{p})}{n + z_{1-\alpha/2}^2}},$$

$$p \in [\tilde{p} - z_{1-\alpha/2}se(\tilde{p}), \tilde{p} + z_{1-\alpha/2}se(\tilde{p})]. \tag{9.37}$$

Note that this modification changes both the center of the confidence interval and its limits, and these modifications are greatest for small sample sizes (i.e., small n) and/or small numbers of positive responses (i.e., small n_1).

These modified estimates and confidence intervals are available from the R package PropCIs with the function addz2ci, which is called with the number n_1 of positive responses as the argument x and the total number of observations n, as the argument n. The other required argument for this function is conf.level, corresponding to $1 - \alpha$. Even with this size-corrected estimator, it is possible to obtain probability estimates that are negative or greater than 1, although this is rarer than with the uncorrected classical estimator; since these estimates are infeasible, another feature of the `addz2ci` function is that it replaces negative estimates with zero and replaces estimates larger than 1 with 1.

A simple function that calls `addz2ci` to generate size-corrected probability estimates and their confidence intervals and plot the results is shown below:

```
binCIplot <- function(n1Vector, nVector, cLevel = 0.95,
                       output = FALSE, pchEst = 16,
                       pchLo = 2, pchHi = 6, yLims = NULL, ...){
  #
  library(PropCIs)
  nPts <- length(n1Vector)
  pFrame <- NULL
  for (i in 1:nPts){
    estSum <- addz2ci(n1Vector[i], nVector[i], cLevel)
    upFrame <- data.frame(n1 = n1Vector[i], n = nVector[i],
                          est = estSum$estimate,
                          loCI = estSum$conf.int[1],
                          upCI = estSum$conf.int[2])
    pFrame <- rbind.data.frame(pFrame, upFrame)
  }
  #
  if (is.null(yLims)){
    yMin <- min(pFrame$loCI)
    yMax <- max(pFrame$upCI)
    yLims <- c(yMin, yMax)
  }
  plot(pFrame$est, ylim = yLims, pch = pchEst, ...)
  points(pFrame$loCI, pch = pchLo)
  points(pFrame$upCI, pch = pchHi)
  #
  if (output){
    return(pFrame)
  }
}
```

The two required arguments for this function are the vectors `n1Vector` of positive response counts n_1 and `nVector` of total observations n for each case. The function calls `addz2ci` to compute size-corrected estimates of the probability of positive response $\tilde{p}$ and its upper and lower confidence limits defined in Eq. (9.37) for each case. The primary result of this function is to generate a plot showing these probability estimates and their confidence limits. The optional argument `cLevel` is the desired confidence limit $1 - \alpha$ that must be passed to the `addz2ci` function via the `conf.level` argument. The optional argument `output` causes the function to return a data frame with the numbers of positive responses and total observations, the size-corrected probability estimates, and

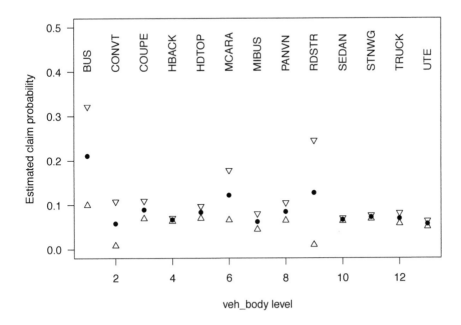

Figure 9.3: Estimated claim probabilities and their 95% confidence limits for each veh_body level from the dataCars data frame.

their upper and lower confidence limits. The remaining optional parameters (pchEst, pchLo, pchHi, and yLims) control the point shapes for the three components of the plot and the y-axis limits. Note that the default value of yLims is NULL; in this case, the y-axis limits are chosen from the minimum value of the lower confidence limits to be plotted and the maximum of the upper confidence limits. Finally, the ... mechanism can be used to pass optional arguments to the plot function used to generate the plot.

As an example, Fig. 9.3 shows the result generated for the veh_body claim probabilities, like the two plots shown in Fig. 9.2, but in a different format and using the addz2ci function to compute 95% confidence intervals around the size-corrected probability estimates. Specifically, this plot was generated using the function binCIplot listed above, called as:

```
pFrame <- binCIplot(n1Vector, nVector, las = 1,
                    xlab = "veh_body level",
                    ylab = "Estimated claim probability",
                    yLims = c(0, 0.5), output = TRUE)
text(seq(1,13,1), 0.4, names(nVector), srt = 90, adj = 0)
```

By specifying output as TRUE here, the binCIplot function returns the following summary data frame:

```
pFrame
```

```
##             n1      n        est          loCI         upCI
## BUS          9     48 0.21065629 0.099654553 0.32165803
## CONVT        3     81 0.05799911 0.008262025 0.10773620
## COUPE       68    780 0.08920264 0.069248470 0.10915681
## HBACK     1264  18915 0.06691323 0.063352671 0.07047378
## HDTOP      130   1579 0.08334425 0.069727595 0.09696090
## MCARA       14    127 0.12167955 0.065663728 0.17769537
## MIBUS       43    717 0.06231707 0.044670525 0.07996362
## PANVN       62    752 0.08456896 0.064733103 0.10440483
## RDSTR        2     27 0.12712529 0.009561774 0.24468881
## SEDAN     1476  22233 0.06646271 0.063188802 0.06973662
## STNWG     1173  16261 0.07223684 0.068258323 0.07621535
## TRUCK      120   1750 0.06951639 0.057613543 0.08141924
## UTE        260   4586 0.05706531 0.050354482 0.06377614
```

Returning to the question of least and most likely vehicle types to file a claim, we can see that the addition of confidence limits to the plots changes some of our perceptions quite substantially. For example, although the estimated claim probability is still the smallest for the "CONVT" veh_body value, this estimate is only based on 81 records, so its confidence limits are fairly wide, enough that they contain the confidence intervals for all veh_body values that appear more than 1000 times in the dataset (these are "HBACK," "HDTOP," "SEDAN," "STNWG," "TRUCK," and "UTE"). Similarly, although the classical probability estimate for "BUS" is much higher than that for "RDSTR" (0.188 vs. 0.074), it is clear from their wide confidence intervals in Fig. 9.3 that these differences may not be as great as they appear because they are both based on very small numbers of records (48 for "BUS" and 27 for "RDSTR"). In fact, the confidence intervals for "RDSTR" are wide enough to completely contain those for all other veh_body types except "BUS" and "CONVT."

Binary variables arise frequently in practice, and we often want to know something about how the probability of one of the two responses varies with one or more other variables. One way of approaching this problem is through the use of logistic regression models and other binary classifiers discussed in Chapter 10, but the binomial confidence intervals described here can be extremely useful in preliminary exploratory analysis, giving us a clear indication of how these probabilities vary over the levels of a single categorical variable. If this categorical variable only has two levels, a better approach is to use the *odds ratio* described in the next section.

9.3.3   Odds ratios

The previous discussions on binary variables have focused on the probability p that one of the two values occurs (e.g., the "probability of success"), a quantity that must lie between 0 and 1. So long as $p < 1$, these probabilities may also be described in terms of *odds*, defined as [2, p. 44]:

$$\Omega = \frac{p}{1-p}.$$ (9.38)

Note that Ω can assume any non-negative value, and if we view p as representing a probability of success (however defined), $\Omega > 1$ implies that success is more likely than failure (i.e., $p > 1-p$), while $\Omega < 1$ implies that failure is more likely than success. Also, note that given Ω, we can recover the probability p by the following inverse transformation:

$$p = \frac{\Omega}{\Omega + 1}.$$ (9.39)

The primary reason for introducing the odds here is that this quantity forms the basis for the *odds ratio*, which provides an extremely useful characterization of the relationship between two binary variables. Specifically, given two probabilities, p_1 and p_2, describing the probability of the same type of success in two different settings, the odds ratio is defined as:

$$OR = \frac{\Omega_1}{\Omega_2} = \frac{p_1/(1-p_1)}{p_2/(1-p_2)} = \frac{p_1(1-p_2)}{p_2(1-p_1)}.$$ (9.40)

To compute the odds ratio from data samples, we substitute the classical probability estimators for p_1 and p_2. This is most easily done from a 2×2 contingency table that tells us, in each of the two settings considered, the number of successes and the number of failures. A more detailed discussion of contingency tables is given in Sec. 9.7.1, but for the case of two binary variables, these tables represent an array of the following four numbers:

1. n_{11} gives the number of successes in the first setting;

2. n_{12} gives the number of failures in the first setting;

3. n_{21} gives the number of successes in the second setting;

4. n_{22} gives the number of failures in the second setting.

Note that this array describes the joint behavior of two binary variables: one that codes "success" or "failure," and another that codes "first setting" or "second setting." As examples presented below will demonstrate, these numbers are easily computed in R with the built-in `table` function.

Given these four numbers, we can estimate the odds Ω_1 as:

$$\hat{\Omega}_1 = \frac{n_{11}/(n_{11}+n_{12})}{n_{12}/(n_{11}+n_{12})} = \frac{n_{11}}{n_{12}},$$ (9.41)

with a similar expression for Ω_2. The odds ratio is then easily computed:

$$\widehat{OR} = \frac{\hat{\Omega}_1}{\hat{\Omega}_2} = \frac{n_{11}/n_{12}}{n_{21}/n_{22}} = \frac{n_{11}n_{22}}{n_{12}n_{21}}.$$ (9.42)

While the final expression here looks more elegant than the "ratio of ratios" before it, this ratio of ratios is actually better behaved numerically for large datasets, where products like $n_{11}n_{22}$ or $n_{12}n_{21}$ can cause numerical overflows.

One advantage of the odds ratio is that if we scale either the rows or the columns of the underlying 2×2 contingency table by a constant factor, the odds ratio remains unchanged. Thus, the odds ratio represents a size-independent measure of association between the two variables defining this contingency table. Another advantage of the odds ratio is that it is easy to compute confidence intervals for $\widehat{OR}$, which do depend on the absolute values of the numbers n_{ij} from which the odds ratio is computed.

It was noted in Sec. 9.1 that most data characterizations we encounter in practice exhibit asymptotic normality, and this is true of the odds ratio estimator $\widehat{OR}$ defined in Eq. 9.42. It turns out, however, that the log of this estimator also exhibits asymptotic normality, and it approaches this asymptotic limit faster than the basic odds ratio estimator does [2, p. 71]. In practical terms, this means that for a given set of counts n_{ij}, we can obtain more accurate confidence intervals for $\log \widehat{OR}$ than for $\widehat{OR}$ itself. Further, since the log transformation is invertible, we can construct confidence intervals for the log of the odds ratio and then convert them back to confidence intervals for the odds ratio itself. The function ORproc described below implements this strategy, using the following results. First, the standard error S_L for the log odds ratio is given by the following function of the contingency table counts n_{ij} [2, p. 71]:

$$S_L = \sqrt{\frac{1}{n_{11}} + \frac{1}{n_{12}} + \frac{1}{n_{21}} + \frac{1}{n_{22}}}. \tag{9.43}$$

It follows immediately from this result that the $1 - \alpha$ confidence interval for the log of the odds ratio is:

$$\mathrm{CI}_{1-\alpha}^{L} = [\log \widehat{OR} - z_{1-\alpha/2}S_L, \log \widehat{OR} + z_{1-\alpha/2}S_L]. \tag{9.44}$$

The inverse of the log transform is the exponential function $f(x) = e^x$, which is monotonically increasing, meaning that it preserves numerical order: if $x < y$, then $e^x < e^y$. As a practical consequence, we can apply this inverse transformation to the confidence interval for the log of the odds ratio to obtain a confidence interval for the odds ratio itself. This result is given by:

$$
\begin{aligned}
\mathrm{CI}_{1-\alpha}^{OR} &= [\exp(\log \widehat{OR} - z_{1-\alpha/2}S_L), \exp(\log \widehat{OR} + z_{1-\alpha/2}S_L)] \\
&= [e^{-z_{1-\alpha/2}S_L}\widehat{OR}, e^{z_{1-\alpha/2}S_L}\widehat{OR}]. \tag{9.45}
\end{aligned}
$$

An important consequence of this approach to constructing confidence intervals is that the result is not symmetric: the estimated odds ratio is guaranteed to lie within the confidence interval, and the confidence interval becomes narrower as the numbers n_{ij} increase, but the estimated odds ratio does not lie in the center of the interval.

To illustrate the utility of the odds ratio and the computations involved, consider the following example based on the Pima Indians diabetes dataset discussed in Chapter 3. Recall from that discussion that several variables exhibit physiologically impossible zero values (e.g., diastolic blood pressure, serum insulin, and triceps skinfold thickness) that appear to code missing data. Further, this dataset includes a binary diagnosis field `diabetes` taking values "neg" and "pos," and it has become a standard benchmark for evaluating binary classifiers that attempt to predict this diagnosis from other variables in the dataset, an application considered further in Chapter 10. The variable with the highest fraction of missing data is `insulin`, which is zero for 48.7% of the records. One possible approach to dealing with this problem is to simply omit those records with zero `insulin` values, but this is potentially dangerous as it can lead to seriously biased results if the records with missing `insulin` values differ systematically from the rest of the data. A particularly important case would be missing values that were strongly associated with the `diabetes` variable. The odds ratio and its confidence interval allow us to examine this question.

The basis for this examination is the following R function, which computes the odds ratio estimator $\widehat{OR}$ and its confidence interval at the level `cLevel`:

```
ORproc <- function(tbl, cLevel = 0.95){
  #
  n11 <- tbl[1,1]
  n12 <- tbl[1,2]
  n21 <- tbl[2,1]
  n22 <- tbl[2,2]
  #
  OR <- (n11/n12) * (n22/n21)
  #
  sigmaLog <- sqrt(1/n11 + 1/n12 + 1/n21 + 1/n22)
  alpha <- 1 - cLevel
  zalpha2 <- qnorm(1 - alpha/2)
  logOR <- log(OR)
  logLo <- logOR - zalpha2 * sigmaLog
  logHi <- logOR + zalpha2 * sigmaLog
  loCI <- exp(logLo)
  upCI <- exp(logHi)
  #
  outFrame <- data.frame(OR = OR, confLevel = cLevel,
                     loCI = loCI, upCI = upCI)
  return(outFrame)
}
```

The one required argument for this function is the 2×2 contingency table `tbl`, easily constructed using R's `table` function:

```
zeroInsulin <- as.numeric(PimaIndiansDiabetes$insulin == 0)
diabetic <- as.numeric(PimaIndiansDiabetes$diabetes == "pos")
cTable <- table(zeroInsulin, diabetic)
cTable

##              diabetic
```

```
## zeroInsulin   0    1
##             0 264 130
##             1 236 138
```

It is not obvious from this table whether there is an association and, if so, how strong it is. Applying the `ORproc` function with the default `cLevel` argument yields the odds ratio estimate and its 95% confidence interval, which allows us to answer this question:

```
ORproc(cTable)

##         OR confLevel     loCI     upCI
## 1 1.187484      0.95 0.8823442 1.598149
```

The critical value for the odds ratio is 1: if there is no association between the two binary variables from which the odds ratio is computed, it should have the value 1. If positive values of one variable are more likely to occur with positive values of the other variable, the odds ratio will be larger than 1, and if positive values of one variable are more likely to occur with negative values of the other, the odds ratio will be smaller than 1. Here, positive values of `zeroInsulin` correspond to missing insulin values, recorded as zeros, and positive values of `diabetic` correspond to a positive diagnosis for diabetes. By itself, the odds ratio estimate $\widehat{OR} \simeq 1.187$ suggests a slight positive association—i.e., that `insulin` is somewhat more likely to be missing for diabetic patients—but the confidence intervals make it clear that this result is not significant. Specifically, since the critical value 1 lies within the 95% confidence interval for this odds ratio estimate, we cannot conclude there is an association between missing `insulin` values and diabetic diagnosis, in either direction.

The key point of this discussion has been to define the odds ratio and demonstrate its utility in exploring the potential relationship between binary variables. It is particularly useful in cases like this one where we are interested in determining whether missing data values appear to be systematic, raising the possibility of biased results.

9.4   Characterizing count data

Numerical variables come in two basic types: *continuous-valued* variables like those considered in Sec. 9.5, and *discrete-valued* variables considered in the following discussions. The typical discrete-valued variable represents the *count* of some quantity, like the population of a city, the number of claims filed by an insurance policyholder, or the number of times a hospital patient has been readmitted. This distinction is important for two main reasons: first, probability-based techniques developed to deal with these two cases are fundamentally different, and, second, data characterizations that are very reasonable for continuous-valued numerical data sometimes fail spectacularly for discrete-valued numerical data, a point discussed in detail in Sec. 9.5.4.

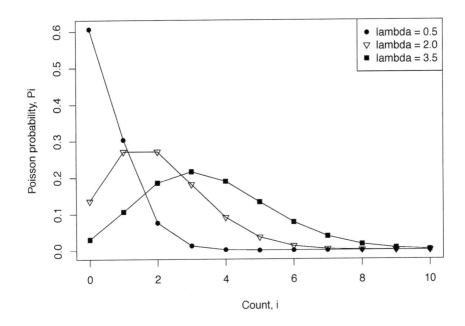

Figure 9.4: Three Poisson distribution examples: $\lambda = 0.5, 2.0, 3.5$.

9.4.1 The Poisson distribution and rare events

In much the same way that the Gaussian distribution is the most popular approximation applied to continuous-valued data, the *Poisson distribution* is the distribution most widely used to approximate count data. This distribution assigns a probability P_k to each integer value $i = 0, 1, 2, \ldots$ that is given by:

$$P_i = \frac{e^{-\lambda}\lambda^i}{i!}, \tag{9.46}$$

where $i! = 1 \cdot 2 \cdots i$ for $i \geq 1$ and $0! = 1$ by convention. Here, λ is a positive constant that represents the single parameter defining the Poisson distribution and determines all distributional characteristics. In fact, both the mean and variance of the Poisson distribution are equal to this value:

$$
\begin{aligned}
E\{i\} &= \lambda, \\
\mathrm{var}\{i\} &= \lambda. \tag{9.47}
\end{aligned}
$$

This feature of the Poisson distribution is one of its most important characteristics, a point discussed further in Sec. 9.4.2.

Fig. 9.4 shows a plot of P_i versus i for $i = 0, 1, \ldots, 10$, for three values of λ, indicated in the legend in the upper right of the plot. In each case, lines are overlaid on the points representing the P_i values as an aid to seeing how P_i behaves as a function of i. For $\lambda = 0.5$, these probabilities decay monotonically with increasing i, a characteristic of the Poisson distribution for all $0 < \lambda < 1$. For $\lambda > 1$, the Poisson probabilities increase to a maximum and gradually decay to zero, as illustrated by the results for $\lambda = 2.0$ (open triangles in Fig. 9.4) and $\lambda = 3.5$ (the solid squares). These probability curves were generated with the dpois function, which returns the values $\{P_i\}$ defined in Eq. (9.46), given the required arguments x, the sequence $\{i\}$ of count values, and lambda, the distribution parameter λ. A closely related R procedure is ppois, which returns the cumulative distribution function, defined as the probability $F(q)$ that observed counts are at most q:

$$F(q) = \text{Prob}\{i \leq q\} = \sum_{i=0}^{q} P_i. \tag{9.48}$$

Other support functions in R for the Poisson distribution include the quantile function qpois that returns the smallest integer q such that $\text{Prob}\{i \leq q\} \geq p$, given the probability threshold p and the distribution parameter lambda, and rpois, which returns a sequence of Poisson random numbers, given the sequence length argument n and the parameter lambda.

The Poisson distribution is often used to describe counts of rare events, perhaps most famously in the paper published by L. von Bortkiewicz in 1898 with the title *"Das Gesetz der kleinen Zalen"* ("The law of small numbers") [71] where he used the Poisson distribution to describe the number of deaths by horse or mule kicks in the Prussian army corps. Versions of this dataset are available in R from the **prussian** data frame in the **pscl** package, and the **HorseKicks** and **VonBort** data frames in the **vcd** package. The **VonBort** data frame gives the individual counts for all of the Prussian army corps considered by von Bortkiewicz, which are mostly zero, spanning a narrow range of values; also, the mean and variance of these counts are approximately equal:

```
table(VonBort$deaths)

##
##   0   1   2   3   4
## 144  91  32  11   2

mean(VonBort$deaths)

## [1] 0.7

var(VonBort$deaths)

## [1] 0.762724
```

These last two results suggest that the Poisson distribution may be a reasonable approximation for these count values, but the *Poissonness plot* introduced in Sec. 9.4.3 gives a better picture of the degree of agreement.

9.4.2 Alternative count distributions

A limitation of the Poisson distribution is that, as discussed in Sec. 9.4.1, it is completely characterized by a single parameter that determines both its mean and its variance or standard deviation. Real count data commonly exhibits *overdispersion*, where the variance is greater than the mean, inconsistent with the Poisson distribution. As a specific example, consider the variable pregnant from the PimaIndiansDiabetes data frame included in the mlbench R package of benchmark datasets. This variable counts the number of times each of the adult female patients characterized in this dataset has been pregnant, varying from 0 to 17. For this count variable, we see that the variance is substantially larger than the mean:

```
mean(PimaIndiansDiabetes$pregnant)

## [1] 3.845052

var(PimaIndiansDiabetes$pregnant)

## [1] 11.35406
```

Here, this difference is large enough to make us question the validity of the Poisson distributional assumption, a question examined further in Sec. 9.4.3, where the Poissonness plot is introduced as an informal graphical assessment tool for this working hypothesis.

Given the possibility that the Poisson distribution is not adequate, it is important to have more flexible alternatives. The most popular alternative is the *negative binomial distribution*, more flexible because it is defined by two parameters, k and π [2, p. 31]:

$$P_i = \frac{(i + k - 1)!}{i!(k-1)!}\pi^i(1-\pi)^k, \qquad (9.49)$$

for $i = 0, 1, 2, \ldots$. The parameter π lies between 0 and 1 and the parameter k is typically a positive integer (positive non-integer values of k are possible, but this requires the factorials in Eq. (9.49) to be replaced with gamma functions [4]). For this distribution, the mean and variance are:

$$
\begin{aligned}
E\{i\} &= \frac{k\pi}{1-\pi}, \\
\mathrm{var}\{i\} &= \frac{k\pi}{(1-\pi)^2} = \frac{E\{i\}}{1-\pi}.
\end{aligned}
\qquad (9.50)
$$

Thus, the variance of the binomial distribution is greater than the mean—reflecting overdispersion—by an amount that depends on the parameter π.

Like the Poisson distribution discussed in Sec. 9.4.1, the negative binomial distribution is supported in R by a family of four functions. The dnbinom function computes probabilities defined in Eq. (9.49), given the required arguments x corresponding to the count i, size, corresponding to the k parameter, and prob, corresponding to the parameter π. The `pnbinom` function returns the negative binomial cumulative probabilities $F(i)$ analogous to those defined in Eq. (9.48), given the quantile `q` and the distribution parameters `size` (k) and `prob` (π). Similarly, `qnbinom` returns the smallest integer q such that $\text{Prob}\{i \le q\} \ge p$, given the probability threshold `p` and the distribution parameters `size` and `prob`. The function `rnbinom` returns a vector of `n` negative binomial random samples, given `n` and the distribution parameters.

Finally, it is important to note that other alternatives to the Poisson distribution are available for count data. Detailed discussions are beyond the scope of this book, but two important special cases are *zero-inflated count distributions* and *long-tailed distributions* like the *Zipf* distribution, both discussed briefly in *Exploring Data* [58, Ch. 3]. Zero-inflated count distributions are *mixture distributions* like those discussed briefly in Sec. 9.5.2 and well supported in R in the `pscl` package. The basic idea is to regard the observed data sequence as arising from two sources: with some probability π, the observation is drawn from a standard Poisson or negative binomial distribtion, but with probability $1 - \pi$, the value is exactly zero. The net result is that the observed data sequence has more zeros than would be expected from the simpler distribution; in the case of a zero-inflated Poisson distribution, the variance is larger than the mean by an amount that depends on this probability π, representing another way to model overdispersed count data. For a good introduction to zero-inflated count distributions and related ideas like using them in regression modeling, refer to the `pscl` package vignette by Zeileis, Kleiber, and Jackman [85]. Finally, the Zipf distribution represents a very different alternative, appropriate to modeling *long-tailed data* like counts of book sales on Amazon. There, some books (e.g., *New York Times* bestsellers) have very large weekly or monthly sales numbers, while the vast majority have very small sales numbers (including zero) for most weeks or months (e.g., Alfred Lawson's *Born Again* discussed in Chapter 8). Brief discussions of the Zipf distribution and its generalization, the *Zipf-Mandelbrot distribution*, are given in *Exploring Data* [58, Sec. 3.7], and the family is supported in R by the functions `dzipfman`, `pzipfman`, `qzipfman`, and `rzipfman` in the `tolerance` package, analogous to the corresponding functions described earlier for the Poisson and negative binomial distributions.

9.4.3 Discrete distribution plots

The normal QQ-plot was introduced in Chapter 2 for continuous-valued numerical data, giving us an informal assessment of the Gaussian approximation and highlighting specific disagreements with this distributional assumption. For count data, the *Poissonness plot* provides an informal graphical test of the hypothesis that it approximately obeys the Poisson distribution introduced in Sec. 9.4.1. This diagnostic tool is available through the `distplot` function in the `vcd` package, along with the corresponding *negative binomialness plot* for informally assessing the adequacy of the negative binomial distribution.

Fig. 9.5 shows a Poissonness plot for the von Bortkiewicz horse kick data discussed earlier. Recall from the tabulation of this data presented earlier that

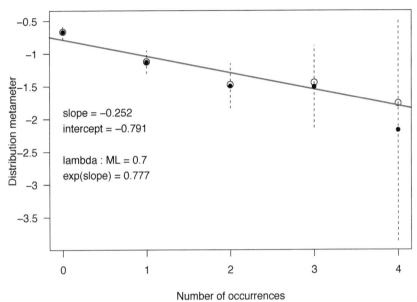

Figure 9.5: Poissonness plot of `deaths` from the `VonBort` data frame.

the number of deaths in any year and any Prussian army corps varied from 0 to 4, so the x-axis for this plot ranges from 0 to 4. The Poissonness plot displays a derived quantity called the *distribution metameter* against the range of observed counts and, if the Poisson approximation is reasonable, these points should fall approximately on a line. This distribution metameter is derived from the log of the Poisson probabilities defined in Eq. (9.46); specifically:

$$\ln P_i = -\lambda + i \ln \lambda - \ln i! \qquad (9.51)$$

If we observe the count i a total of n_i times in a sequence of N data observations, the fraction n_i/N should be approximately equal to P_i, and this idea forms the basis for the Poissonness plot. Specifically, the distribution metameter for the Poisson distribution is defined as:

$$\phi(n_i) = \ln(n_i/N) + \ln i! = \ln n_i - \ln N + \ln i!, \qquad (9.52)$$

which is easily computed from the available data values $\{n_i\}$. The key to the Poissonness plot lies in the observation that if $n_i/N \simeq P_i$, then:

$$\phi(n_i) \simeq (\ln \lambda)i - \lambda. \qquad (9.53)$$

Thus, analogous to the QQ-plot for continuous distributions, if the counts $\{n_i\}$ conform approximately to a Poisson distribution, a plot of $\phi(n_i)$ versus i should approximately conform to a straight line with slope $\ln \lambda$ and intercept $-\lambda$.

The Poissonness plot in Fig. 9.5 was constructed from the von Bortkiewicz horse kick data frame `VonBort` from the `vcd` package using the `distplot` function from the same package. By default, `distplot` generates a Poissonness plot with a reference line and 95% confidence intervals around each count value i. The open circles in this plot represent the distribution metameter $\phi(n_i)$, the dashed lines represent the confidence intervals around each of these points, and the solid circles represent the centers of these confidence intervals. If the Poisson distribution approximates the count data well, these points should fall close enough to the reference line that it intersects most or all of the confidence intervals. These conditions are met in Fig. 9.5, supporting the argument that the Poisson distribution is a reasonable approximation to the horse kick data. (Note that Fig. 9.5 looks very different from Fig. 8.2 shown in *Exploring Data* [58, p. 335] for this same dataset: the difference lies in how the counts are grouped. In Fig. 9.5, we are looking at the ungrouped count data for all years and all army corps; the Poissonness plot presented in *Exploring Data* was based on total yearly counts for each of the 20 years included in the dataset. The two plots are very different in appearance, but both provide evidence in support of the Poisson distribution as a reasonable approximation for these two different sets of counts, both derived from the von Bortkiewicz data.)

Fig. 9.6 shows another Poissonness plot, this one for the variable `pregnant` from the `PimaIndiansDiabetes` data frame in the `mlbench` package. Recall from Sec. 9.4.2 that this count data sequence exhibits substantial overdispersion and therefore seems poorly approximated by the Poisson distribution. In Fig. 9.6, the counts range from 0 to 17, and the 95% confidence intervals around the distribution metameter points for counts less than about 10 are so narrow as to be indistinguishable from the data points. Thus, the reference line in this plot lies outside the majority of these confidence intervals, casting further doubt on the appropriateness of the Poisson distribution here. The points that appear most consistent are those for large counts, where the number of data observations is small enough that the confidence intervals are extremely wide.

Fig. 9.7 shows a *negative binomialness plot* for the `pregnant` variable, also generated by the `distplot` function from the `vcd` package. These plots are obtained by specifying the optional `type` argument as "nbinomial." Like the Poissonness plot, the basis for the negative binomialness plot is a distribution metamater constructed from the negative binomial distribution such that plotting this metameter against the observed count values i should give an approximate straight line. (In the interest of keeping this discussion to a manageable length and avoiding more mathematics than necessary, a detailed discussion of this metameter is not given here; for more details, refer to *Exploring Data* [58, Sec. 8.4.2].) Comparing Figs. 9.6 and 9.7 suggests that the negative binomial distribution provides a better approximation than the Poisson distribution does here. In particular, note that the points representing smaller counts generally lie closer to the line in Fig. 9.7 than they do in Fig. 9.6.

Poissonness plot

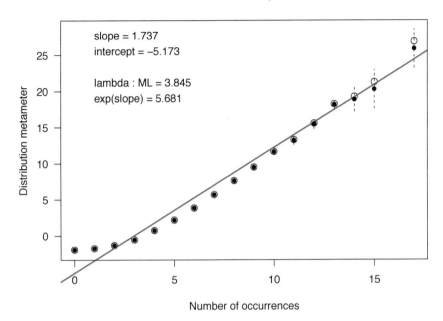

Figure 9.6: Poissonness plot of the **pregnant** variable from the PimaIndiansDiabetes data frame.

9.5 Continuous distributions

The popular Gaussian distribution was introduced in Chapter 3 and a number of its characteristics were discussed there. It was also noted that, despite its popularity, this distribution is not always a good approximation for the behavior of real data, and Sec. 9.5.1 considers this point in more detail. By highlighting where the Gaussian assumption can fail, this discussion lays the foundation for Sec. 9.5.2, which introduces a number of important non-Gaussian distributions that can be much better approximations for certain data sources (e.g., variables that can only assume positive values or variables that are restricted to bounded ranges). In addition, the normal QQ-plot introduced in Chapter 3 can be used to highlight specific deviations from normality that may be extremely useful in suggesting alternative distributions, and this point is illustrated in Sec. 9.5.3. Finally, the important distinction between continuously distributed and discrete random variables noted earlier is revisited in Sec. 9.5.4 in connection with the problems of *ties* in data sequences (i.e., multiple observations of exactly the same value) and the problem of *implosion* that this can cause for extremely outlier-resistant data characterizations like the median and the MAD scale estimator.

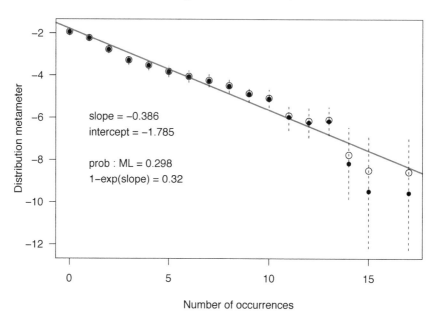

Negative binomialness plot

Figure 9.7: Negative binomialness plot of the **pregnant** variable from the PimaIndiansDiabetes data frame.

9.5.1 Limitations of the Gaussian distribution

The popularity of the Gaussian distribution as an approximate description of numerical data variables has been noted repeatedly, and in many cases, this approximation is quite reasonable. As a specific example, we saw this was the case with the chick weight data discussed in Chapter 3. There are, however, cases where the Gaussian distribution assumption is untenable, including the following types of numerical data:

1. strongly asymmetric data distributions like those often seen for variables that can only assume positive values;

2. variables that are restricted to a bounded range (e.g., probabilities or fractions that must lie between 0 and 1);

3. variables exhibiting "heavy-tailed behavior" like the *glint noise* example discussed below;

4. *multi-modal* data distributions like the Old Faithful geyser eruption data discussed below.

These are not the only situations where the Gaussian distribution does not adequately describe numerical data—numerical variables with discrete distributions like the count data discussed in Sec. 9.4 represent another broad class—but the cases listed here occur frequently enough in practice to merit discussion.

For the case of variables that can assume only positive values, the Gaussian distribution is inapplicable on theoretical grounds since, for any mean and standard deviation, there is nonzero probability of drawing a negative value from the distribution. If the mean is substantially larger than the standard deviation, this probability can be small enough to ignore, but in cases where the data variability is on the same order as the mean, the distribution is necessarily asymmetric, and this asymmetry must be taken into account. This point is illustrated in Fig. 9.8, which shows two density plots: the one on the left is for the *gamma distribution* discussed in Sec. 9.5.2, and the one on the right is for the Gaussian distribution. In both cases, the mean of the distribution is 2 and the standard deviation is $\sqrt{2}$ (approximately 1.414), and vertical lines are included in both plots at the mean (the solid line) and at the mean plus or minus one standard deviation (the dashed lines). In the case of the gamma distribution on the left, negative deviations of more than about 1.4 standard deviations are not possible, since this would result in a negative value which has zero probability for the gamma distribution. For the Gaussian distribution with the same mean and standard deviation, however, such negative values have a probability of 0.079. In cases where x must be strictly positive, this behavior of the Gaussian distribution is unacceptable, forcing us to consider alternatives like the gamma distribution shown here.

The inappropriateness of the Gaussian distribution for variables that must occupy a strictly bounded range, such as probabilities or fractions, follows from the same basic reasoning: for any choice of mean and standard deviation, there is nonzero probability that Gaussian samples will fall outside these limits. Again, if the variation seen in the data is small enough that values near the limits are essentially never seen, the Gaussian approximation may not be too bad in practice. In cases where we expect to observe values near these limits, however, the probability of the Gaussian distribution violating them is large enough that we need to consider other distributions. Probably the most popular distribution appropriate to these situations is the *beta distribution* discussed in Sec. 9.5.2, which is defined on a bounded range and thus respects whatever limits we impose. The beta distribution is described by two shape parameters—in addition to the range limits—and is extremely flexible, including the uniform distribution as a special case, along with both symmetric and asymmetric distributions on the specified data range.

The problem of outliers in numerical data was discussed in some detail in Chapter 3, as was the often ineffective "three-sigma edit rule" for detecting them. As noted, the basis for this procedure was the observation that the probability of observing values more than three standard deviations from the mean of the Gaussian distribution is only about 0.3%. Even in the absence of outliers, we often see more points than this this far out into the tails of the distribution. Such cases are called "heavy-tailed" and probably the best

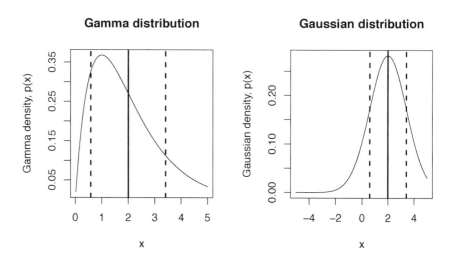

Figure 9.8: Comparison of two distributions with mean 2 and variance 2: gamma (left) and Gaussian (right).

known case—and one of the most extreme cases—is the infinite-variance *Cauchy distribution* discussed in Sec. 9.5.2. A simple application that leads to the Cauchy distribution as an approximate data distribution is that of *glint noise* in radar signals, arising from the fact that the intensity of the radar return signal is approximately proportional to the tangent of the angle between the outgoing radar beam and the surface from which it reflects. If this angle is uniformly distributed over the range from $-\pi/2$ to $\pi/2$, the return signal intensity obeys a Cauchy distribution, based on the geometric arguments given briefly by Johnson, Kotz, and Balakrishnan [46, p. 320]. This distribution is also a member of the *Student's t family* discussed in Sec. 9.5.2.

Finally, the last case listed above where the Gaussian distribution is inappropriate is that of *multi-modal* data. This behavior is illustrated by the geyser data frame from the MASS package, which describes 299 eruptions of the Old Faithful geyser in Yellowstone National Park, Wyoming, giving the duration of each eruption and the waiting time between successive eruptions observed in August 1985. Fig. 9.9 presents two views of each variable: the top two plots are scatterplots of the individual values of these variables, while the bottom plots are the corresponding density estimates. Note that both density plots

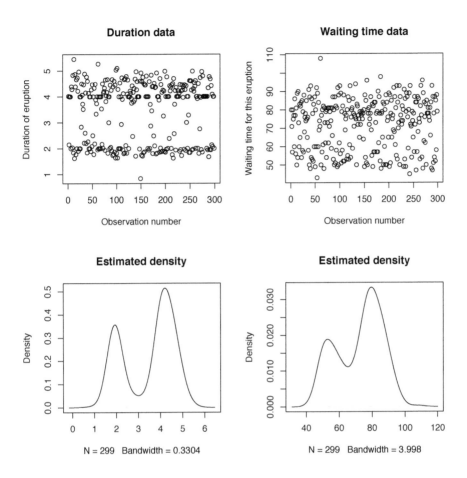

Figure 9.9: Data (top) and estimated densities (bottom) for the `geyser` variables `duration` (left) and `waiting` (right).

show two pronounced peaks, indicating *bimodal* behavior, a feature reflected in the "gaps" seen in the raw data scatterplots, especially that for the `duration` variable (top left). Often, multi-modal distributions arise when multiple data sources are involved. For example, comprehensive automobile insurance policies typically cover a variety of losses, including small ones like broken windshields with repair costs of a few hundred dollars and much larger ones like car theft, where replacement costs are generally one to two orders of magnitude greater. In such cases, by combining "the same" response variable (e.g., losses in dollars) from different generation mechanisms (e.g., broken windshields and stolen cars), substantial distributional heterogeneity can result, leading to multi-modal data distributions. Clearly, the unimodal, symmetric Gaussian distribution cannot adequately approximate the data distribution in these cases.

9.5.2 Some alternatives to the Gaussian distribution

Although Gauss laid the foundations in the early 1800s for what has become arguably *the* standard probability distribution and its enormous body of associated theory and methods, the distribution was derived as an approximation to a binomial distribution by de Moivre in 1733 [46, p. 85]. In the nearly 300 years since then, a huge body of statistical theory has developed, and many, many distributions have been proposed and investigated: in their two-volume treatment of continuous univariate distributions, Johnson, Kotz, and Balakrishnan list over 100 named distributions in their indices [46, 47]. Clearly, nothing like a comprehensive or detailed treatment is possible here, but it is important to introduce a few of the better known continuous distributions, to show that most of the data characteristics discussed in Sec. 9.5.1 are well supported by reasonably standard probability distributions. Specifically, the following discussion introduces the following three distributions to address the first three of the Gaussian limitations described there:

1. the *gamma distribution*, an asymmetric distribution appropriate to positive numerical data;

2. the *beta distribution*, an extremely flexible distribution appropriate to bounded numerical data;

3. the *Student's t-distribution*, a family of heavy-tailed, symmetric data distributions.

While the beta distribution can exhibit a very restricted form of bimodal behavior, more general multimodal behavior is most easily described using *mixture distributions*, an idea discussed briefly at the end of this section.

The standard form of the gamma distribution is defined by the following density function, for $x > 0$ [46, p. 337]:

$$p(x) = \frac{x^{\alpha-1}e^{-x}}{\Gamma(\alpha)}, \tag{9.54}$$

where α is a constant commonly called the *shape parameter* and $\Gamma()$ denotes the *gamma function* [4]. If α is an integer greater than 1, the gamma function can be expressed in terms of the better known *factorial function:*

$$\Gamma(\alpha) = (\alpha - 1)! = 1 \cdot 2 \cdots (\alpha - 2) \cdot (\alpha - 1). \tag{9.55}$$

The key point here is that the gamma distribution dates back to Laplace in 1836 [46, p. 343], so like the Gaussian distribution, there is considerable theoretical and practical support for it. For example, the mean and variance of this distribution are both equal to the shape parameter α. In R, the density function $p(x)$ defined in Eq. (9.54) can be computed with the dgamma function, the cumulative distribution function can be computed with the pgamma function, quantiles can be obtained with the qgamma function, and gamma-distributed random samples can be generated with the rgamma function.

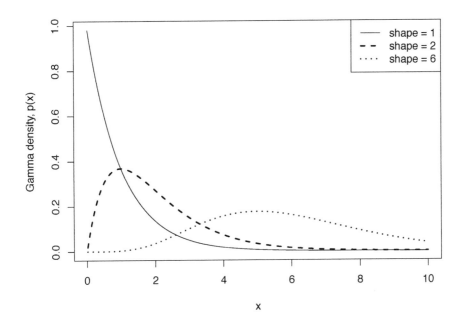

Figure 9.10: Comparison of three gamma densities.

Fig. 9.10 shows three gamma density functions with different shape parameters, illustrating its range of behavior. These curves were generated with the R function dgamma, specifying the required argument x as a vector of 500 values, evenly spaced from 0.02 to 10, and the required argument shape (α in Eq. (9.54)) as indicated in the plot. The solid curve corresponds to $\alpha = 1$, a special case also known as the *exponential distribution* because its density reduces to $p(x) = e^{-x}$. In fact, this special case exhibits a number of special mathematical properties, enough so that Johnson, Kotz, and Balakrishnan devote an entire chapter to it [46, Ch. 19], separate from the one devoted to the gamma distribution. The key point here is that this distribution is monotonically decaying from its maximum as x increases. The dashed curve in Fig. 9.10 corresponds to $\alpha = 2$, exhibiting a maximum at $x = 1$, a mean value of 2, and a standard deviation of $\sqrt{2}$ (approximately 1.414). As discussed in Sec. 9.5.1, it is possible for a random variable with this distribution to exhibit values many standard deviations above the mean, but no more than about 1.4 standard deviations below the mean (for this example, the probability of observing a sample 1.4 standard deviations or more above the mean is about 15%). Finally, the third example has $\alpha = 6$, exhibiting a more symmetric shape with a single peak or mode at $x = 5$. In fact, this distribution looks approximately Gaussian, illustrating the fact that the

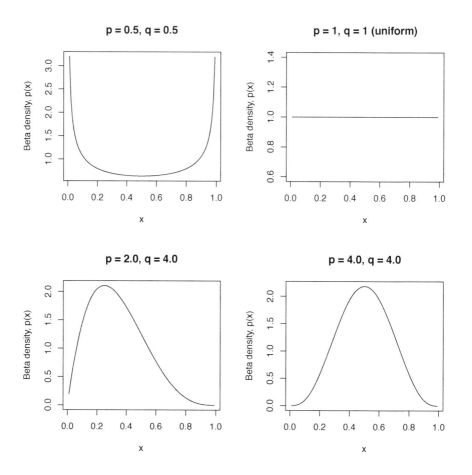

Figure 9.11: Four special cases of the beta distribution.

gamma distribution does become approximately Gaussian as α becomes large. Several of these points are revisited in Sec. 9.5.3 in connection with the use of quantile-quantile plots to assess the reasonableness of various distributions, including both the Gaussian and the gamma.

The second non-Gaussian distribution considered here is the beta distribution, with four examples shown in Fig. 9.11. This distribution is defined for $0 \le x \le 1$ by the density function [47, p. 210]:

$$p(x) = \frac{x^{p-1}(1-x)^{q-1}}{B(p,q)}, \tag{9.56}$$

where p and q are two strictly positive shape parameters and $B(p,q)$ is the *beta function* [4, p. 142], closely related to the gamma function $\Gamma()$. As with the gamma distribution, R provides extensive support for the beta distribution through the functions dbeta for the density defined in Eq. (9.56), pbeta for the cumulative distribution function, qbeta for the quantiles of the distribution, and rbeta to generate random samples from it. The shape parameters are specified by the arguments shape1 and shape2, and they make the beta distribution extremely flexible, as seen in the four examples in Fig. 9.11. The upper left plot shows the density for $p = q = 1/2$, and it illustrates the point noted earlier that the beta distribution is flexible enough to include some very special forms of bimodal behavior: here, even though the mean of this distribution is 0.5, there is very little probability of observing samples drawn from this distribution near the mean value. Instead, samples are more likely to be concentrated near the extremes. The upper right plot shows the *uniform distribution*, a special case obtained by setting $p = q = 1$ and leading to the constant density $p(x) = 1$ for $0 \le x \le 1$. Intuitively, this distribution expresses the idea that "any value in the admissible range is equally likely," a model that is sometimes very useful. The lower left plot shows the beta distribution for $p = 2$ and $q = 4$, resulting in a density that looks somewhat like the gamma distribution with $\alpha = 2$ plotted in Fig. 9.10. The key difference is that there is no finite upper limit on the values of the random samples we can draw from the gamma distribution, but for the beta distribution, samples cannot take values larger than 1. Finally, the lower right plot shows the beta distribution for $p = q = 4$, giving a symmetric distribution that looks somewhat like the Gaussian distribution, but again restricted to the range of x values between 0 and 1.

The third alternative considered here is the *Student's t-distribution*, characterized by a parameter d called the *degrees of freedom*. This density is defined for all x from $-\infty$ to $+\infty$ by [47, p. 363]:

$$p(x) = \frac{1}{\sqrt{d}B(1/2, d/2)} \left(1 + \frac{x^2}{d}\right)^{-(d+1)/2}. \tag{9.57}$$

In R, this density function is computed with the dt function, the cumulative distribution function is computed with the pt function, quantiles are computed with the qt function, and random samples are generated with the rt function. This distribution approaches a Gaussian limit as $d \to \infty$, but the variance is infinite unless $d > 2$, implying extremely heavy-tailed behavior: that is, for small values of the d parameter, the standard deviation is undefined, and the probability of observing values very far from the center of the distribution (at zero) is substantial. This point is illustrated in Fig. 9.12, which shows the shapes of the density functions (upper two plots) and samples drawn from each distribution (lower two plots) for the two extreme limits of this distribution. Specifically, the left-hand plots show the results for $d = 1$, corresponding to the Cauchy distribution used to model glint noise discussed briefly in Sec. 9.5.1, while the right-hand plot shows the Gaussian limit obtained as $d \to \infty$. The primary difference in the distributions is that the Cauchy density function (top left) decays *much* more slowly as $x \to \pm\infty$ than the Gaussian density does. This difference means that random samples far from the center of the distribution are much more likely from the Cauchy distribution than they are from the Gaussian

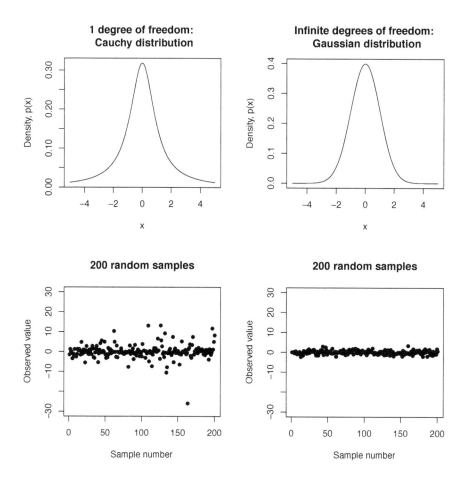

Figure 9.12: Densities and random samples for Student's *t*-distributions.

distribution, a point illustrated by the random samples generated using the rt function shown in the bottom two plots. In particular, note the point at ~ -27 at approximately sample number 160 in the lower left plot, and the generally much wider scatter seen in this plot relative to the Gaussian sample on the right.

The last form of strongly non-Gaussian behavior discussed in Sec. 9.5.1 was multi-modality, illustrated by the pronounced bimodal distribution seen for the Old Faithful geyser data. While it is true that the beta distribution can exhibit a special form of bimodal behavior if the shape parameters p and q are both less than 1, this behavior does not match that seen in the estimated densities for this example. It is not difficult to construct random variable models that do exhibit this behavior, using *mixture distributions*. The basic idea is to characterize the data not with a single distribution, but rather with a collection of several distributions, one for each mode seen in the data. In the case of the

Old Faithful data, we can use two Gaussian distributions, one to describe the lower peak in the density, and the other to describe the upper peak in the data. Essentially, we assume each observation has some probability p of being drawn from the first distribution, and probability $1 - p$ of being drawn from the second distribution. By appropriately selecting this probability and the means and standard deviations of the two Gaussian component distributions, we can construct a reasonably good approximation of the Old Faithful geyser duration and waiting time data. Although a detailed introduction to mixture distributions is beyond the scope of this book, in favorable cases like the Old Faithful waiting time data, it is easy to fit these distributions using the tools available in the mixtools R package. Specifically, the normalmixEM function fits k individual Gaussian distributions to a data sequence, returning estimates of the individual means (mu) and standard deviations (sigma), along with the estimated probability that a sample is drawn from each one (lambda). By default, $k = 2$, although this parameter is one of the arguments that can be specified in the function call. Loading the mixtools library to make this function available and applying it to the waiting time data gives the following results:

```
library(mixtools)
wait <- geyser$waiting
mixmdl <- normalmixEM(wait)

## number of iterations= 48

summary(mixmdl)

## summary of normalmixEM object:
##               comp 1     comp 2
## lambda    0.307595   0.692405
## mu        54.202705  80.360354
## sigma      4.952040   7.507601
## loglik at estimate:   -1157.542
```

In practical terms, we may interpret these results as meaning that, with approximately 30.8% probability, the waiting time has a mean value of just under an hour (54.2 minutes), with a standard deviation of about 5 minutes, but with a probability of 69.2%, the expected waiting time is closer to an hour and a half (80.4 minutes), with a standard deviation of about 7.5 minutes. Histograms and Gaussian density curves for these components are shown in the left-hand plot in Fig. 9.13.

Letting $g(x; \mu, \sigma)$ denote the Gaussian density with mean μ and standard deviation σ, these component densities can be combined into an overall density estimate by forming the following weighted sum:

$$p(x) = 0.308g(x; 54.203, 4.952) + 0.692g(x; 80.360, 7.508). \qquad (9.58)$$

This density is shown as the heavy dotted curve in the right-hand plot in Fig. 9.13, where the solid line corresponds to the density estimate for the Old Faithful waiting time data included in the discussion of this dataset in Sec. 9.5.1.

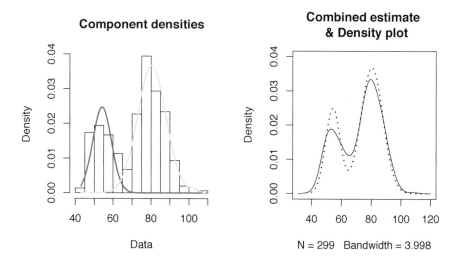

Figure 9.13: Component densities for the Old Faithful waiting time data (left), and mixture density (dotted curve) overlaid with `density` result (right).

The primary points here are, first, that the mixture distribution approach just described gives a result that is in reasonable agreement with what we obtain by applying the `density` function directly to the data, and, second, that this mixture density was easy to obtain using the `mixtools` package in *R*. The advantage of the mixture density approach in this case is that it gives us a relatively simple quantitative description of the data, as opposed to the curve we obtain from the `density` function, which provides a useful visual display but no basis for subsequent computations or quantitative interpretation like the verbal one given above. For a more complete introduction to fitting mixture densities to data, refer to the vignette by Benaglia *et al.* accompanying the `mixtools` package [7].

9.5.3   The `qqPlot` function revisited

The quantile-quantile plot function `qqPlot` from the `car` package was introduced in Chapter 3 as a useful tool, both for assessing conformance of data to the Gaussian distribution and for identifying a variety of systematic deviations from this assumption and a number of other important data features (e.g., outliers and inliers). By default, the `qqPlot` function generates Gaussian or

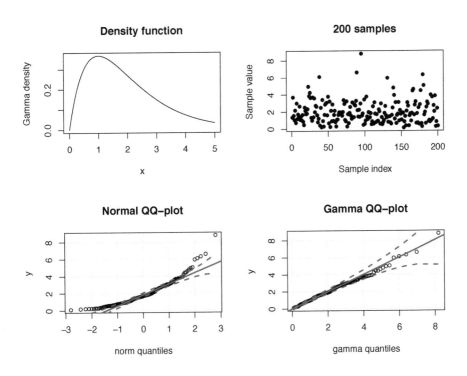

Figure 9.14: Gamma density, samples, and normal and gamma QQ-plots.

normal QQ-plots, but it can also generate QQ-plots for other reference distributions. Specifically, the `distribution` argument (which may be conveniently abbreviated as `dist`) specifies the root name for the comparison distribution, and its default value `norm` specifies the normal distribution. It is, however, valid to specify any distribution available in our *R* installation that has associated quantile and density functions, including all of those discussed in Sec. 9.5.2. In cases like the beta or gamma distribution where shape parameters or other associated parameters must be specified, these values must be included in the call to the `qqPlot` function, as the following examples illustrate. The main point is that the `qqPlot` function can be used to assess the reasonableness of other continuous reference distributions besides the Gaussian.

Fig. 9.14 summarizes an example built around the gamma distribution introduced in Sec. 9.5.2 with shape parameter $\alpha = 2$. The upper left plot shows this density function $p(x)$ for $0 \leq x \leq 5$, generated by specifying the `dgamma` function with a vector `x` of values spanning this range and `shape = 2`. The upper right plot shows 200 random samples drawn from this distribution to give an idea what the data values themselves look like. The lower two plots are QQ-plots, the one in the lower left the default Gaussian QQ-plot, while the one on the right uses the gamma distribution with the correct shape parameter as the reference distribution. This plot was generated with the following R code:

```
set.seed(33)
y <- rgamma(n = 200, shape = 2)
qqPlot(y, dist = "gamma", shape = 2, main = "Gamma QQ-plot")
```

Note that the normal QQ-plot on the left shows clear evidence of the asymmetry characteristic of the gamma distribution, and it also suggests the possibility of upper tail outliers. In contrast, the data points cluster reasonably closely around the reference line in the gamma QQ-plot on the right, and essentially all of these points fall within the 95% confidence intervals around this line. These results suggest, correctly, that the gamma distribution with shape parameter $\alpha = 2$ is a very reasonable distribution to describe this data sequence.

To conclude this discussion of non-Gaussian QQ-plots and their use, consider the `infant.mortality` variable from the `UN` data frame in the `car` package, giving the United Nations estimates of infant mortality rate for 207 nations in 1998, although some observations are missing. Fig. 9.15 shows four characterizations of these data values. The upper left plot shows the default normal QQ-plot generated by the `qqPlot` function, which is strongly suggestive of distributional asymmetry, an impression confirmed by the estimated density in the upper right plot. Since this estimated distribution is similar in shape to the gamma distribution with $\alpha = 2$ considered in the previous example, the lower left QQ-plot is constructed for this reference distribution, exactly as before. Note that the upper tail falls well within the 95% confidence limits around the reference line here, and while the lower tail does fall a bit above the reference line and its confidence limits, the agreement is *much* better than in the normal QQ-plot shown above. Decreasing the shape parameter from $\alpha = 2$ to $\alpha = 1$ results in a monotonically decaying density which seems at odds with the shape of the density plot shown in the upper right, but the resulting QQ-plot shown in the lower right for this case appears to do a better job of fitting the lower tail of the distribution, although at the expense of a somewhat poorer fit to the upper tail. The point of this example is that the `qqPlot` function can tell us, with a moderate amount of trial-and-error exploration, both where Gaussian assumptions appear unwarranted and what type of distribution might be a more reasonable alternative. In this case, these results suggest that if we need to develop a predictive model for infant mortality, we are probably better off considering something like the gamma GLM model class introduced in Chapter 10 than a standard linear regresison model, which is best suited to approximately Gaussian response variables.

9.5.4   The problems of ties and implosion

It has been noted several times that ties—i.e., duplicated data values—are not possible in Gaussian random variables; indeed, ties are not possible for *any* continuously distributed random variable. Ties do occur in real data, however, for a number of reasons. One is that data observations are not infinitely precise, so they are sometimes only recorded with limited precision to reflect this fact.

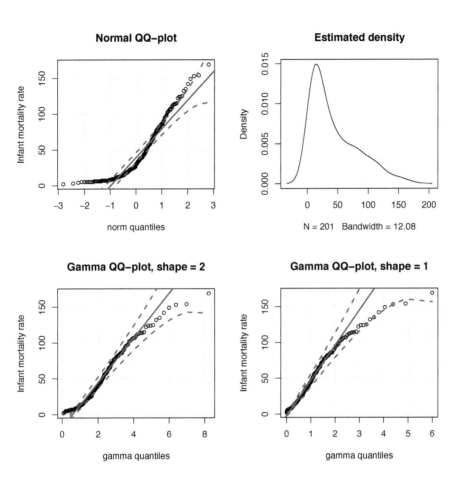

Figure 9.15: UN infant mortality data: three QQ-plots and a density estimate.

Indeed, this practice has been suggested as desirable in releasing certain kinds of data to the public [45]. For example, applying *R's* `table` function to the `income` variable from the `Chile` data frame in the `car` package gives:

```
##
##   2500   7500  15000  35000  75000 125000 200000
##    160    494    768    747    269     88     76
```

In some cases, the presence of repeated values like this makes little or no difference, but if ties occur frequently enough in a data sequence, they can cause a number of otherwise very useful data characterizations to fail completely.

An example that illustrates this point is the `claimcst0` variable from the `dataCar` data frame in the `insuranceData` package. This data frame characterizes 67,856 Australian automobile insurance policies, and the `claimcst0` variable is the numerical value of the loss associated with each policy. Since policies only exhibit losses when a claim is filed after an accident, and since most policies do not file claims (only 6.8% of the records in `dataCar` filed claims), almost all of the recorded `claimcst0` values are zero. Because some claims can be very large (e.g., over 90% of the nonzero losses are under 5000 Australian dollars, but the largest claim is ten times this amount), we might be tempted to apply outlier-resistant characterizations like the median and the MAD scale estimator, but both of these characterizations fail completely in this example. Specifically, the mean and the median give us the following characterizations of `claimcst0`:

```
mean(dataCar$claimcst0)
```

```
## [1] 137.2702
```

```
median(dataCar$claimcst0)
```

```
## [1] 0
```

While it might be reasonably argued that the median is giving us a more representative characterization of "the typical value" of this variable, the fact that this result is completely independent of the values of the nonzero portion of the data is a problem in many applications. This difficulty extends to the MAD scale estimator, where it is much more serious:

```
sd(dataCar$claimcst0)
```

```
## [1] 1056.298
```

```
mad(dataCar$claimcst0)
```

```
## [1] 0
```

The fact that the MAD scale estimator has the value zero here, independent of all of the nonzero data values, reflects the problem of *implosion* that affects highly outlier-resistant data characterizations like the median, the MAD scale estimator, and many others. As a specific example, note that this effect means that if we apply the Hampel identifier discussed in Chapter 3 for outlier detection, it will declare *all* nonzero values to be outliers in this case.

The basic problem that leads to implosion is that if more than 50% of the values in a data sequence are the same, the MAD scale estimator will be zero, *independent of all of the values in the data sequence.* Thus, the MAD scale estimate will remain zero if we add an arbitrary constant to every element of the data sequence, multiply every element by an arbitrary number, or apply transformations like the square root or the square to the data sequence. Other data characterizations that behave badly in the face of frequently repeated data values include those based on quantiles like the interquartile distance discussed in Sec. 9.2.2. As a specific example, note that the quartiles computed from the `claimcst0` variable using the `quantile` function are:

```
quantile(dataCar$claimcst0, probs = seq(0, 1, 0.25))

##     0%    25%    50%    75%    100%
##   0.00   0.00   0.00   0.00 55922.13
```

Thus, since both the lower quartile (25% quantile) and the upper quartile (75% quantile) are zero for this example, the interquartile distance is also zero, again independent of the value of all of the nonzero losses.

9.6 Associations between numerical variables

The linear regression models introduced in Chapter 5 represent a detailed description of the relationship between two or more continuous-valued numerical variables. These models are closely related to the *correlation coefficient* discussed further in Sec. 9.6.1, which provides a simple numerical measure of the extent to which two variables "move together," "move in opposition," or "move independently." In fact, the standard *product-moment correlation coefficient* described in Sec. 9.6.1 measures the degree of *linear association* between two numerical variables: it is possible for two variables to be related through a monotonic but highly nonlinear relationship and this may not be detected with the product-moment correlation coefficient. An alternative that can detect these monotone relationships is *Spearman's rank correlation measure*, introduced in Sec. 9.6.2, which forms the basis for the "correlation trick" described in Sec. 9.6.3 to see whether nonlinear data transformations applied to one or both variables can highlight nonlinear relations between them. Another characteristic of the Spearman rank correlation measure is its greater resistance to outliers, a topic considered in Sec. 9.6.5, where other outlier-resistant association measures are considered. Finally, Sec. 9.6.6 considers the problem of detecting *multivariate outliers* using these association measures.

9.6.1 Product-moment correlations

The *product-moment correlation coefficient* was developed by Karl Pearson at the end of the nineteenth century, and for that reason it is sometimes called the *Pearson correlation coefficient*. This measure remains extremely popular over a century later because it has a number of extremely useful characteristics, including the following three:

1. ease of computation: this measure is easily programmed in any language that supports basic arithmetic operations and is available as a built-in function in any software environment that supports basic statistical computations, including Microsoft *Excel*, R, *Python*, and many others;

2. ease of interpretation: the product-moment correlation coefficient is a measure of *linear association*, quantifying the tendency for one variable to vary linearly with another;

3. completeness for jointly Gaussian data: if two variables, x and y, have a *joint Gaussian distribution* (a concept discussed briefly below), they are completely characterized by their individual means and standard deviations and their product-moment correlation coefficient.

The following discussion expands on all three of these advantages and offers some simple illustrations of the application of the product-moment correlation coefficient. Despite its popularity, however, there are situations where this correlation measure performs poorly, and the following sections discuss alternatives developed to address some of these limitations.

For two random variables, x and y, the product-moment correlation ρ_{xy} is defined in terms of expected values, like the mean and standard deviation. Specifically:

$$\rho_{xy} = \frac{E\{(x - E\{x\})(y - E\{y\})\}}{\sigma_x \sigma_y}, \tag{9.59}$$

where σ_x and σ_y are the standard deviations of x and y, respectively, e.g.:

$$\sigma_x = \sqrt{E\{(x - E\{x\})^2\}}, \tag{9.60}$$

with an analogous expression for σ_y. As with the mean and standard deviation, the usual approach to estimating the product-moment correlation coefficient is to replace these expectations with the corresponding averages. For two observed data sequences $\{x_k\}$ and $\{y_k\}$, each of length N, this leads to an expression like the following:

$$\hat{\rho}_{xy} = \frac{\sum_{k=1}^{N} (x_k - \bar{x})(y_k - \bar{y})}{\sqrt{\sum_{k=1}^{N} (x_k - \bar{x})^2 \sum_{k=1}^{N} (y_k - \bar{y})^2}}, \tag{9.61}$$

where $\bar{x}$ and $\bar{y}$ are the averages of these data sequences. As noted, the simplicity of these computations has led to the extremely wide availability of built-in procedures for this correlation measure.

In R, this procedure is the function `cor` that computes product-moment correlations, along with several other numerical association measures like the Spearman rank correlation discussed in Sec. 9.6.2. As a simple example, applying this function to the `Temp` and `Gas` variables from the `whiteside` data frame in the `MASS` package gives the following result:

```
cor(whiteside$Temp, whiteside$Gas)
```

```
## [1] -0.6832545
```

As noted above, besides ease of computation, simplicity of interpretation is another advantage of the product-moment correlation coefficient, as this example illustrates. In particular, variables x and y are said to be *linearly associated* if they can be represented by the following equation:

$$y = ax + b, \tag{9.62}$$

for some constants a and b. Further, the correlation coefficient has the value $\rho_{xy} = 1$ if this relation holds for some strictly positive constant a, and it has the value $\rho_{xy} = -1$ if this relation holds for some strictly negative constant a. In fact, these values represent the extreme limits of the correlation coefficient, which must always lie between -1 and $+1$. Intermediate values of ρ_{xy} thus provide evidence of *approximate linear relations* between x and y if it lies near these extreme limits, and $\rho_{xy} = 0$ implies *statistical independence* if x and y are jointly Gaussian random variables, a point discussed below. For the `whiteside` data, the relatively large negative value of the estimated correlation coefficient is consistent with the approximately linear relationship seen in the data, with `Temp` decreasing as `Gas` increases, indicating a negative slope paramter a. This point is emphasized in Fig. 9.16, which shows a scatterplot of these two variables, along with their correlation coefficient and their best-fit linear model, obtained with the `lm` function discussed in Chapter 5 as:

```
whitesideLinearModel <- lm(Gas ~ Temp, data = whiteside)
coef(whitesideLinearModel)

## (Intercept)        Temp
##   5.4861933  -0.2902082
```

In fact, there is a close relationship between the slope of this linear regression model and the product-moment correlation coefficient. Specifically, the estimated slope $\hat{a}$ of a linear regression model may be expressed as [58, p. 201]:

$$\hat{a} = \hat{\rho}_{xy}\frac{\hat{\sigma}_y}{\hat{\sigma}_x}, \tag{9.63}$$

where $\hat{\sigma}_x$ and $\hat{\sigma}_y$ represent the standard deviations computed from the x and y variables (i.e., `Temp` and `Gas`, respectively, in this example).

The third advantage of the product-moment correlation coefficient noted above is its completeness as a characterization of the relationship between two jointly Gaussian random variables. A detailed discussion of this point requires more mathematics than is reasonable to include here, but the essential idea is the following: if two variables, x and y, exhibit a *joint Gaussian distribution*, this means that they are completely characterized by their individual means, μ_x and μ_y, standard deviations, σ_x and σ_y, and their correlation coefficient, ρ_{xy}. For other distributions, this complete characterization no longer holds, which is one reason the Gaussian distribution is so popular historically. In fact, a slightly more complex version of this result extends to multivariate Gaussian distributions involving an arbitrary number of components, a point discussed very briefly in Sec. 9.6.4.

It is also true that *for any distribution*, if the variables x and y are statistically independent, their correlation coefficient is zero. This follows from the fact that, if x and y are statistically independent, the following condition holds:

$$E\{(x - E\{x\})(y - E\{y\})\} = E\{x - E\{x\}\}E\{y - E\{y\}\}$$
$$= (E\{x\} - E\{E\{x\}\})(E\{y\} - E\{E\{y\}\})$$

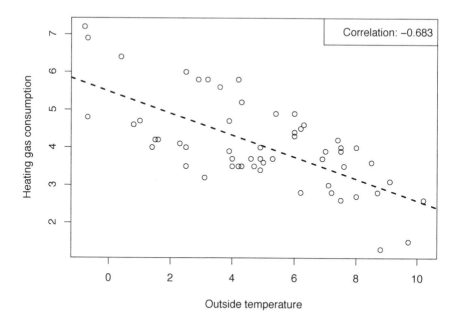

Figure 9.16: Scatterplot of Gas vs. Temp from the whiteside data frame, with linear regression line and correlation coefficient.

$$= (E\{x\} - E\{x\})(E\{y\} - E\{y\})$$
$$= 0. \tag{9.64}$$

Substituting this result into Eq. (9.59), it follows that if x and y are independent, $\rho_{xy} = 0$, regardless of the form of their joint distribution. *The converse result also holds for the Gaussian and binomial distributions, but only for these distributions.* Specifically, if x and y have a Gaussian joint distribution or are binary random variables, $\rho_{xy} = 0$ implies statistical independence, *but this result does not hold for other distributions.*

Since numerical variables are often assumed to at least approximately satisfy a joint Gaussian distribution, zero correlation is often taken as an approximate indication of statistical independence between x and y. Further, because the product-moment correlation estimator $\hat{\rho}_{xy}$ exhibits asymptotic normality under fairly general conditions—like most data characterizations we commonly encounter—it is easy to construct approximate confidence intervals for this case. Specifically, for large enough samples, under the assumption of statistical independence, $\hat{\rho}_{xy}$ exhibits an approximate Gaussian distribution with zero mean and variance $1/N$ where N is the sample size. Under these assumptions, then— i.e., that the variables are approximately normally distributed and our sample size is large enough for the asymptotic approximatation to be reasonable—we can interpret ρ_{xy} as evidence that x and y are related (i.e., not statistically independent) if $|\hat{\rho}_{xy}|$ is large enough. In particular, we can reject the statistical independence hypothesis at the level $1 - \alpha$ if $|\hat{\rho}_{xy}| > z_{\alpha/2}/\sqrt{N}$. For the common case of 95% confidence (i.e., $\alpha = 0.05$), this condition becomes $|\rho| > 1.96/\sqrt{N}$. For a more detailed—and more mathematical—discussion of these ideas, refer to *Exploring Data* [58, Sec. 10.4].

9.6.2 Spearman's rank correlation measure

In cases where there is a strong relationship between two variables that is *not linear*, the product-moment correlation coefficient may be very small. As a specific example, consider the `brain` and `body` variables in the `Animals2` data frame from the `robustbase` package, where the product-moment correlation coefficient is:

```
cor(Animals2$body, Animals2$brain)
```

```
## [1] 0.05166368
```

Note that here the number of observations N is 65, leading to a 95% confidence interval for independence of ± 0.243. Thus, if we adopt the traditional interpretation for the correlation coefficient based on an approximate Gaussian interpretation, we would conclude that these variables are unrelated. In fact, there *is* an approximate relationship between these variables, but it is not a linear one.

The *Spearman rank correlation coefficient* measures the degree to which large values of one variable tend to occur with large values of another. This is accomplished by replacing the data values with their *ranks* and computing the product-moment correlation coefficient between these ranks. Specifically, given the data sequences $\{x_k\}$ and $\{y_k\}$, the corresponding ranks $\{R_x(k)\}$ and $\{R_y(k)\}$ are the indices $\{i\}$ of the original data values in their rank-ordered lists. That is, to determine $R_x(k)$, the rank of x_k, we first re-order the data values from smallest to largest as:

$$x_{(1)} \le x_{(2)} \le \cdots x_{(N-1)} \le x_{(N)}, \tag{9.65}$$

where $x_{(1)}$ denotes the smallest data value in the original list and $x_{(N)}$ represents the largest. The rank $R_x(k)$ of x_k is the position of this specific data value in the above list: if x_k is the smallest value, then $R_x(k) = 1$, if it is the second-smallest value, $R_x(k) = 2$, and so forth. The Spearman rank correlation coefficient is then computed by applying Eq. (9.61) to the ranks $\{R_x(k)\}$ and $\{R_y(k)\}$. This result simplifies somewhat because both the mean and variance of the ranks can be computed knowing the sample size N alone, without knowledge of the data values themselves. Specifically, since the ranks take each value from 1 to N exactly once, it follows that the means of both $\{R_x(k)\}$ and $\{R_y(k)\}$ are given by:

$$\bar{R}_x = \bar{R}_y = \frac{1}{N} \sum_{k=1}^{N} k = \frac{1}{N} \frac{N(N+1)}{2} = \frac{N+1}{2}. \tag{9.66}$$

Similarly, it can be shown [58, p. 495] that the variances are given by:

$$\mathrm{var}\{R_x(k)\} = \mathrm{var}\{R_y(k)\} = \frac{N^2 - 1}{12}. \tag{9.67}$$

After these simplifications, the Spearman rank correlation coefficient is given by:

$$\hat{\rho}_{xy}^S = \frac{12}{N(N^2 - 1)} \sum_{k=1}^{N} \left[R_x(k) - \frac{N+1}{2} \right] \left[R_y(k) - \frac{N+1}{2} \right]. \tag{9.68}$$

Although the Spearman rank correlation is not as well known or as widely available as the product-moment correlation, it is easily computed.

In R, the Spearman rank correlation measure is available as an option in the cor function, by specifying method = ''spearman'', over-riding the default (method = ''pearson''). Applying this correlation measure to the **brain** and **body** variables from the **Animals2** data frame gives very different results:

```
cor(Animals2$body, Animals2$brain, method = "spearman")
```

```
## [1] 0.9241982
```

Here, we see strong evidence for a relationship between these variables, in contrast to the product-moment correlation results that suggested no relationship. The difference is that the Spearman rank correlation is a measure of *monotone association* between the variables x and y instead of a linear association measure. That is, the Spearman rank correlation has its maximum possible value of 1 if there exists an *increasing function* $f(x)$ such that:

$$y_k = f(x_k), \tag{9.69}$$

for $k = 1, 2, \ldots, N$. The linear function $f(x) = ax + b$ in Eq. (9.62) is increasing if the slope parameter a is positive, so it follows that if the product-moment correlation has the value $+1$, so does the Spearman rank correlation, but if $f(x)$ is the increasing function $f(x) = \log x$, the Spearman rank correlation still has the value $+1$, but the product-moment correlation will be less than this maximum value, possibly a lot less. Similarly, the Spearman rank correlation will have its minimum possible value of -1 if $\{x_k\}$ and $\{y_k\}$ are related by Eq. (9.69) for some *decreasing function* $f(x)$; again, the linear function $f(x) = ax + b$ in Eq. (9.62) is decreasing if the slope parameter a is negative, so if the product-moment correlation has the value -1, so does the Spearman rank correlation. These observations form the basis for the correlation trick discussed in the next section.

Another important difference between the Spearman rank correlation coefficient and the more familiar product-moment correlation coefficient is that the Spearman measure is much less sensitive to outliers, a point discussed further in Sec. 9.6.5.

9.6.3 The correlation trick

Because the Spearman rank correlation coefficient is based on ranks, this monotone association measure is invariant under the application of arbitrary increasing functions. Specifically, an increasing function is one for which the following condition is satisfied:

$$x \leq y \Rightarrow f(x) \leq f(y), \tag{9.70}$$

and if we apply this condition to the re-ordered data sequence in Eq. (9.65) we see that:

$$x_{(1)} \leq \ldots \leq x_{(N)} \Rightarrow f(x_{(1)}) \leq \cdots \leq f(x_{(N)}), \tag{9.71}$$

which means that the ranks $\{R_{f(x)}(k)\}$ of the transformed sequence $\{f(x_k)\}$ are exactly the same as the ranks $\{R_x(k)\}$ of the original data sequence. This means that if we apply an arbitrary increasing transformation to our original data sequences—e.g., log, square root, etc.—it has no effect at all on the Spearman rank correlation coefficient.

This invariance does *not* hold for the product-moment correlation coefficient, as we can see in Fig. 9.17, which shows two views of the `body` and `brain` variables from the `Animals2` data frame in the `robustbase` package. The plot on the left shows the original data variables, and the extremely small product-moment correlation coefficient for these variables is shown in the upper right of the plot. The right-hand plot shows the result of applying log transformations to both the `body` and `brain` variables, giving us a very different view of the dataset. In particular, this plot suggests a strong linear relationship between the transformed variables, aside from the presence of a small cluster of outliers below the main point cloud. The correlation coefficient between these transformed variables is shown in the upper left of this plot, and this value is well outside the 95% confidence limits (± 0.243) for statistical independence.

The *correlation trick* exploits the fact that the Spearman rank correlation is invariant to monotone transformations like the logarithm, while the product-moment correlation coefficient is not. The idea is that, in cases like the `whiteside` data where an approximate linear relationship exists between two variables—and these variables are free of outliers—the product-moment and Spearman rank correlations should be approximately equal. In fact, for this case, we have the following results:

```
cor(whiteside$Temp, whiteside$Gas)

## [1] -0.6832545

cor(whiteside$Temp, whiteside$Gas, method = "spearman")

## [1] -0.6196817
```

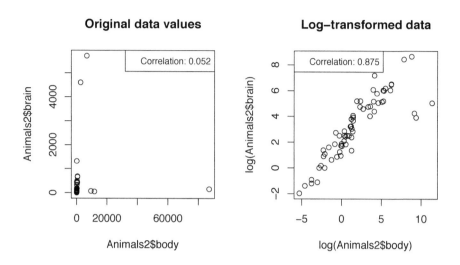

Figure 9.17: Two brain-body scatterplots from the `Animals2` data frame: the raw data (left) and the log-transformed data (right).

These values are not identical, but they are similar, in marked contrast to the case for the `Animals2` data, repeated here for convenience:

```
cor(Animals2$body, Animals2$brain)

## [1] 0.05166368

cor(Animals2$body, Animals2$brain, method = "spearman")

## [1] 0.9241982
```

The idea behind the correlation trick is that in cases like this—where the product-moment correlation coefficient is small and the Spearman rank correlation is large—an approximate nonlinear relationship often exists between the variables. In such cases, applying a monotone transformation like the log or square root to one or both variables can often reveal structure that was not apparent in the original view of the data. The difference in the appearance of the left- and right-hand plots in Fig. 9.17 provides a dramatic illustration of this idea. We see analogous differences if we apply these transformations in fitting linear regression models, as the following comparison demonstrates.

If we attempt to predict brain weight from body weight using the original, un-transformed data in a linear regression model, we obtain terrible results (note the slightly negative adjusted R-squared value):

```
brainBodyModelA <- lm(brain ~ body, data = Animals2)
summary(brainBodyModelA)

##
## Call:
## lm(formula = brain ~ body, data = Animals2)
##
## Residuals:
##    Min    1Q Median    3Q    Max
## -486.3 -262.4 -249.3 -109.7 5417.0
##
## Coefficients:
##              Estimate Std. Error t value Pr(>|t|)
## (Intercept) 2.663e+02  1.152e+02   2.313   0.024 *
## body        4.304e-03  1.048e-02   0.411   0.683
## ---
## Signif. codes:  0 '***' 0.001 '**' 0.01 '*' 0.05 '.' 0.1 ' ' 1
##
## Residual standard error: 915.1 on 63 degrees of freedom
## Multiple R-squared:  0.002669,Adjusted R-squared:  -0.01316
## F-statistic: 0.1686 on 1 and 63 DF,  p-value: 0.6827
```

Applying log transformations to both variables gives a much better model:

```
brainBodyModelB <- lm(I(log(brain)) ~ I(log(body)), data = Animals2)
summary(brainBodyModelB)

##
## Call:
## lm(formula = I(log(brain)) ~ I(log(body)), data = Animals2)
##
## Residuals:
##     Min      1Q  Median      3Q     Max
## -3.8592 -0.5075  0.1550  0.6410  2.5724
##
## Coefficients:
##              Estimate Std. Error t value Pr(>|t|)
## (Intercept)   2.17169    0.16203   13.40  <2e-16 ***
## I(log(body))  0.59152    0.04117   14.37  <2e-16 ***
## ---
## Signif. codes:  0 '***' 0.001 '**' 0.01 '*' 0.05 '.' 0.1 ' ' 1
##
## Residual standard error: 1.172 on 63 degrees of freedom
## Multiple R-squared:  0.7662,Adjusted R-squared:  0.7625
## F-statistic: 206.4 on 1 and 63 DF,  p-value: < 2.2e-16
```

(Note the use of the "as-is" function I() to protect the log transformations on both sides of the model formula; see Chapter 5 for a discussion of this.)

A thorough treatment of monotone transformations is beyond the scope of this book, but the following points are worth noting:

1. monotone transformations like log and square root are invertible, meaning that no information is lost if we need to transform the data, perform our analysis, and then transform back to the original representation to interpret or explain our results (note, however, that useful transformations can dramatically change the appearance of our data, as seen in Fig. 9.17);

2. transformations like those considered here are only applicable to numerical data, and even then there may be constraints on where they can be used (e.g., log transformations only work for strictly positive data, while square roots can be used for data with zeros and positive values);

3. as a general rule, the log transformation is most useful when the range of the data values is very wide (e.g., spanning several orders of magnitude).

For those interested in a detailed discussion of transformations, this topic is treated at length in *Exploring Data* [58, Ch. 12].

9.6.4 Correlation matrices and correlation plots

If we have a collection of more than two numerical variables, we can sometimes gain useful insight into the relationships between them by examining their *correlation matrix*, which is a square array whose ij-element gives the correlation coefficient between the variables x_i and x_j. Further, since $\rho_{xy} = \rho_{yx}$ for all numerical variables x and y, this matrix is symmetric: its ij element and its ji element are the same. Also, the diagonal elements (ii) represent the correlation of x_i with itself, which is always 1. In fact, this general structure may be seen in the two-variable case, although the real utility of correlation matrices arises when we are considering more than two numerical variables together. The `cor` function can be applied to two numerical variables x and y, as in the previous examples, but it can also be applied to numerical matrices or data frames, returning the correlation matrix in the format just described that relates all of these variables. Applying this function to the numerical variables `Temp` and `Gas` from the whiteside data frame gives the following result:

```
cor(whiteside[, c("Temp", "Gas")])

##              Temp         Gas
## Temp    1.0000000  -0.6832545
## Gas    -0.6832545   1.0000000
```

Here, there are only two off-diagonal elements in this correlation matrix: M_{12} represents the correlation ρ_{xy} between `Temp` and `Gas`, while M_{21} represents the correlation between `Gas` and `Temp`, which is the same. As noted in the above discussion, both diagonal elements have the value 1, representing the correlations of `Temp` and `Gas` with themselves.

As a more useful example, consider the correlation matrix constructed from the six variables `Price`, `MPG.city`, `MPG.highway`, `EngineSize`, `Horsepower`, and `Weight` from the `Cars93` data frame in the `MASS` package. This correlation matrix is easily constructed as:

```
keepVars <- c("Price", "MPG.city", "MPG.highway", "EngineSize",
                      "Horsepower", "Weight")
corMat <- cor(Cars93[, keepVars])
corMat
```

```
##                  Price   MPG.city MPG.highway EngineSize Horsepower
## Price        1.0000000 -0.5945622  -0.5606804  0.5974254  0.7882176
## MPG.city    -0.5945622  1.0000000   0.9439358 -0.7100032 -0.6726362
## MPG.highway -0.5606804  0.9439358   1.0000000 -0.6267946 -0.6190437
## EngineSize   0.5974254 -0.7100032  -0.6267946  1.0000000  0.7321197
## Horsepower   0.7882176 -0.6726362  -0.6190437  0.7321197  1.0000000
## Weight       0.6471790 -0.8431385  -0.8106581  0.8450753  0.7387975
##                 Weight
## Price        0.6471790
## MPG.city    -0.8431385
## MPG.highway -0.8106581
## EngineSize   0.8450753
## Horsepower   0.7387975
## Weight       1.0000000
```

As in the previous example, the values M_{ij} on the diagonal of this matrix are all 1, and the off-diagonal elements are symmetric (i.e., $M_{ji} = M_{ij}$), giving the correlation coefficient between the variable associated with each row and column. Thus, the elements M_{23} and M_{32} both list the value 0.9439358, corresponding to the correlation coefficient between MPG.city and MPG.highway. This large positive value implies that these two variables track strongly together, consistent with our intuition that these variables should be similar but not identical. In contrast, the M_{62} and M_{26} elements of this matrix show that Weight and MPG.city have a strong negative correlation: the heavier the car, the poorer the gas mileage, in general.

While we can learn much by examining the numbers in this matrix, there are enough of them even in this 6×6 correlation matrix that this task quickly becomes tedious and error prone. A useful alternative is to convert these numbers into a plot that allows us to see these relationships more easily. The corrplot package is an extremely flexible graphical tool for displaying correlation matrices in many different formats. The example shown in Fig. 9.18 illustrates this point and was constructed using the following R code:

```
library(corrplot)
corrplot(corMat, method = "ellipse", type = "upper", tl.pos = "diag",
           cl.pos = "n", col = "black")
corrplot(corMat, method = "number", type = "lower",
           tl.pos = "n", cl.pos = "n", add = TRUE, col = "black")
```

The first line here loads the corrplot package, while the second line generates the upper triangle (type = ''upper'') of the square array shown in Fig. 9.18, representing the correlation values graphically as ellipses (method = ''ellipse''). The specification tl.pos = ''diag'' is somewhat confusing, but it actually causes the variable names to appear across the top and to the left side of the complete array. The argument cl.pos = ''n'' suppresses the color bar (mapping colors to numbers) that appears by default when this plot array is

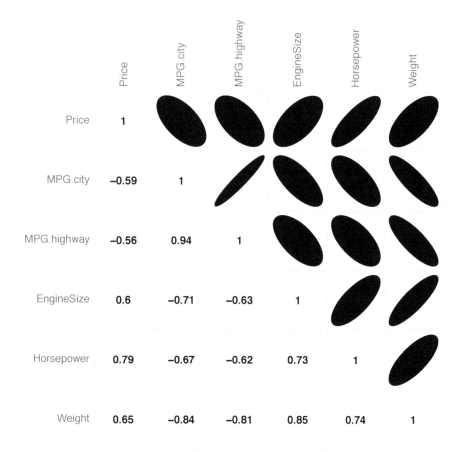

Figure 9.18: Correlation matrix from selected `Cars93` numerical variables.

generated, and `col` = ''`black`'' generates a black-and-white plot. (Note that a color version of this plot is shown in Fig. 9.28 in the second section of color figures at the end of this chapter.) The last line of code here adds numerical correlation values (`add` = `TRUE`) to the lower triangle of this array (`method` = ''`number`'' and `type` = ''`lower`''). The specification `tl.pos` = ''`n`'' suppresses the placement of variable names along the diagonal here, `cl.pos` = ''`n`'' suppresses the addition of a color bar at the bottom of the plot, and `col` = ''`black`'' again gives a black-and-white plot.

We can see many relationships quickly from the plot shown in Fig. 9.18. Specifically, ellipses that are narrow and elongated indicate strong correlations between the variables involved, positive if the ellipse slopes from lower left to upper right and negative if it slopes from the upper left to the lower right. Thus, from the second row of this plot array, labelled `MPG.city`, we see both of the relationships noted above: this variable exhibits a strong positive correlation with MPG.highway and a somewhat weaker but still strong negative correlation with Weight. Rounder, less elongated ellipses indicate weaker associations, as in the negative correlations seen here between both mileage variables and Price in the top row of this array, and the positive correlations between Price and both EngineSize and Weight. Having the numerical values for these correlations in the lower triangle of this array allows us to quickly quantify these results.

Before leaving this discussion, two points should be noted. First, as mentioned in passing earlier, the correlation matrix is most informative in the case of jointly Gaussian random variables, or data variables that are reasonably well approximated by such a distribution. In particular, a collection of p jointly Gaussian random variables $\{x_i\}$ is completely characterized by its p mean values $\mu_i = E\{x_i\}$, variances $\sigma_i^2 = E\{(x_i - \mu_i)^2\}$ (or, equivalently, standard deviations σ_i), and the matrix of correlations ρ_{ij} between these variables. An especially important result is that, for two jointly Gaussian variables, x_i and x_j, they are uncorrelated (i.e., $\rho_{ij} = 0$) if and only if they are statistically independent. The fact that statistically independent variables are uncorrelated holds for all data distributions, as noted in Sec. 9.6.1, but the converse does not: it is possible for non-Gaussian variables to be uncorrelated but not statistically independent.

Finally, the second important point to note is that all of the graphical machinery just described can also be applied to alternative correlation measures like the Spearman rank correlation coefficient introduced in Sec. 9.6.2, or the robust correlation measures described next.

9.6.5 Robust correlations

The impact of severe outliers on linear regression modeling and the utility of outlier-resistant robust linear regression procedures was demonstrated in Chapter 5 with a modification of the whiteside data frame. Recall that the motivation for this example was the use of "extreme numerical values" to represent missing data, an unfortunate practice that can have serious adverse consequences. Since linear regression modeling and correlation analysis are extremely closely related, it is not surprising that the modified whiteside data frame considered in Chapter 5 yields correlation estimates as extreme and as obviously incorrect as the linear regression results presented there. Recall that this modified data frame, whitesideMod, changed one observed value of the Temp variable and the corresponding value of the Gas variable from their nominal values (in the range from −0.8 to 10.2 for Temp and from 1.3 to 7.2 for the Gas variable) to the missing value indicator 9999. For the product-moment correlation coefficient discussed in Sec. 9.6.1, this modification has the following impact:

```
cor(whiteside$Temp, whiteside$Gas)

## [1] -0.6832545

cor(whitesideMod$Temp, whitesideMod$Gas)

## [1] 0.9999963
```

As with the linear regression model considered in Chapter 5, the impact of the outlying data point introduced by this modification is catastrophic: the correlation changes from a significant negative value (i.e., the value computed from the original `whiteside` data frame has a larger magnitude than $1.96/\sqrt{N}$, which is approximately 0.261916) to an essentially perfect positive correlation.

It was noted in passing in Sec. 9.6.2 that the Spearman rank correlation coefficient exhibits better outlier resistance than the product-moment correlation coefficient, a point illustrated nicely by this example:

```
cor(whiteside$Temp, whiteside$Gas, method = "spearman")

## [1] -0.6196817

cor(whitesideMod$Temp, whitesideMod$Gas, method = "spearman")

## [1] -0.5455435
```

Because correlations play a key role in many different data characterizations (see, for example, the discussion of principal component analysis in Sec. 9.8) and because their outlier sensitivity is so extreme, considerable statistical research has been devoted to developing outlier-resistant alternatives. (For a good introduction to this topic, refer to the book by Rousseeuw and Leroy [62].) One of these alternatives to the standard correlation estimator is the *minimum covariance determinant (MCD) estimator*, available as the `covMcd` function in the `robustbase` package. Applying this function to the original and modified `whiteside` data examples just considered, we obtain the following results:

```
library(robustbase)
MCD0 <- covMcd(whiteside[, c("Temp", "Gas")], cor = TRUE)
MCD0$cor

##              Temp        Gas
## Temp   1.0000000 -0.4866247
## Gas   -0.4866247  1.0000000

MCD <- covMcd(whitesideMod[, c("Temp", "Gas")], cor = TRUE)
MCD$cor

##              Temp        Gas
## Temp   1.0000000 -0.4933154
## Gas   -0.4933154  1.0000000
```

Note that while these correlation estimates are substantially smaller in magnitude than those obtained using either the product-moment or the Spearman rank correlation estimators, the impact of the extreme outlier pair in the `whitesideMod` dataset is almost entirely negligible here.

The primary points of this discussion are, first, that the standard correlation estimator is extremely sensitive to the presence of outliers in the data, and, second, that effective alternatives are available. One important application of these outlier-resistant correlation estimators is the detection of multivariate outliers, discussed next.

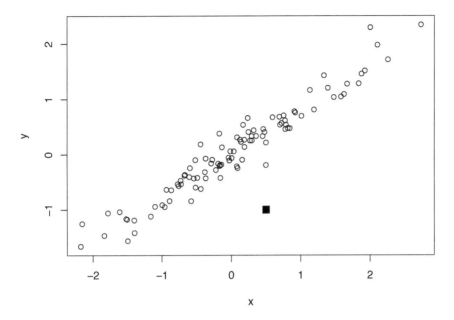

Figure 9.19: Plot of simulated Gaussian dataset with a single bivariate outlier.

9.6.6 Multivariate outliers

A characteristic of the univariate outliers discussed in Chapter 3 is that they are necessarily extreme values in the dataset: for a data observation to be inconsistent with the majority of the other data values, it must be either much larger or much smaller than these other values. In the case of *multivariate outliers*, this is no longer true: there, it is enough that the outlying points violate the relationship seen between the majority of other data values. This point is illustrated in Fig. 9.19, which shows two simulated Gaussian random variables, contaminated with a single outlier (the point marked with the large square). This outlying point lies clearly outside the main cluster of data values that correspond to a highly correlated pair of Gaussian random variables, but it is not extreme with respect to either x or y.

The standard basis for detecting multivariate outliers is correlation analysis, based on the idea that point clouds like the one shown in Fig. 9.19 are approximately elliptical for correlated Gaussian random variables. Thus, points that are "unusual with respect to the majority of the data" like the square marked in Fig. 9.19 represent those that lie outside of these elliptical point clusters. This idea allows us to define a distance measure that tells us how far points are from the center of the ellipse—the general vicinity where we expect to find most of

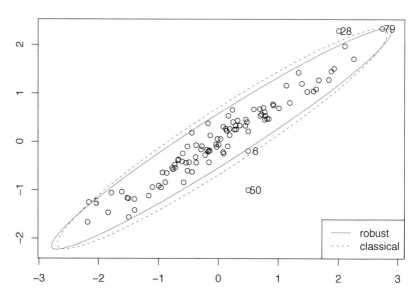

Figure 9.20: Classical (dashed) and robust (solid) Mahalanobis outlier detection plots for the contaminated Gaussian data from Fig. 9.19.

the data points—in a way that accounts for the size and shape of the ellipse. The mechanics of this approach are discussed below, but the idea is illustrated in Fig. 9.20, which shows the same data points as in Fig. 9.19, but with two added ellipses, corresponding to classical and outlier-resistant estimates of the ellipses that should contain the correlated Gaussian data with 97.5% probability. In this case, both ellipses are quite similar, so both the classical and outlier-resistant approaches yield very similar results for this example, but this is not always the case: the `aircraft` example discussed below illustrates this point. Here, however, both methods clearly identify the modified data point marked with the square in Fig. 9.19 as an outlier (the point labelled 50 in Fig. 9.20). Essentially, these methods represent extensions of the three-sigma edit rule or a robust extension like the Hampel identifier to higher dimensions.

To see how this works, first consider the following quadratic reformulation of the three-sigma edit rule for outlier detection introduced in Chapter 3. In its slightly more general form with a threshold parameter t, this rule declares the observation x_k to be an outlier if:

$$|x_k - \bar{x}| > t\sigma_x, \qquad (9.72)$$

where $\bar{x}$ is the mean of the data values $\{x_k\}$ and σ_x is the standard deviation of these data values. This condition may be re-written as:

$$\frac{|x_k - \bar{x}|}{\sigma_x} > t, \tag{9.73}$$

which may be further re-written as:

$$d(x_k) = \frac{(x_k - \bar{x})^2}{\sigma_x^2} > t^2. \tag{9.74}$$

Finally, note that the probability density associated with the value x_k is simply related to $d(x_k)$ via:

$$p(x_k) = \frac{1}{\sqrt{2\pi}\sigma_x} e^{-d(x_k)/2}. \tag{9.75}$$

The point of this derivation is to show, first, that the three-sigma edit rule can be written in terms of the function $d(x_k)$, and, second, that this function has a simple relationship to the Gaussian probability distribution.

In fact, both of these observations carry over to the multivariate case where $\mathbf{x}_k$ is a p-dimensional vector. Specifically, define the *Mahalanobis distance* $d(\mathbf{x}_k)$ as:

$$d(\mathbf{x}_k) = (\mathbf{x}_k - \bar{\mathbf{x}})^T \mathbf{S}^{-1}(\mathbf{x}_k - \bar{\mathbf{x}}), \tag{9.76}$$

where $\bar{\mathbf{x}}$ is the mean of the N multivariate observations $\{\mathbf{x}_k\}$ in the dataset and $\mathbf{S}$ is the $p \times p$ covariance matrix characterizing these multivariate observations. As in the univariate case presented above, the Gaussian probability density associated with this observation can be written simply in terms of the Mahalanobis distance:

$$p(\mathbf{x}_k) = \frac{1}{(2\pi)^{p/2}|\mathbf{S}|^{1/2}} e^{-d(\mathbf{x}_k)/2}, \tag{9.77}$$

where $|\mathbf{S}|$ denotes the determinant of the matrix $\mathbf{S}$. The key point here is that the Mahalanobis distance defines "how far out" the point $\mathbf{x}_k$ lies into the tails of the multivariate Gaussian probability distribution assumed for the nominal data. Thus, specifying a threshold for the Mahalanobis distance in the multivariate setting is analogous to applying the three-sigma edit rule in the univariate setting; in fact, this strategy reduces to the three-sigma edit rule when $p = 1$.

Like the three-sigma edit rule, the idea behind using the Mahalanobis distance for outlier detection is to define a threshold value t and declare the multivariate data point $\mathbf{x}_k$ to be an outlier if $d(\mathbf{x}_k) > t$. By appealing to the connection with the multivariate Gaussian distribution just noted, we can choose t to correspond to a value that is only likely to be exceeded with some specified small probability if our data sample conforms well to a multivariate Gaussian distribution. As noted in the discussion of Fig. 9.20, the ellipses shown there were chosen large enough to contain Gaussian data points with probability 97.5%. Thus, in the sample of 100 points considered in that example, we would expect

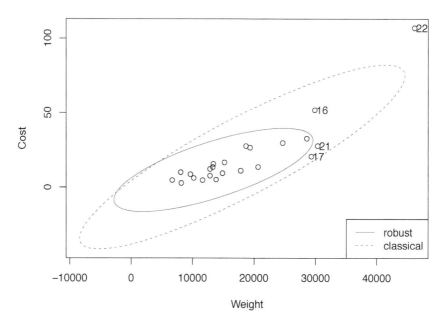

Figure 9.21: Classical (dashed) and robust (solid) Mahalanobis outlier detection plots for the `aircraft` Weight (X3) and Cost (Y) data.

to see two or three nominal points declared as marginal outliers, which is consistent with what we see in the plot (specifically, the points labelled 5, 28, and 79 lie just barely outside the dashed curve representing the classical Mahalanobis distance ellipse). Assuming their influence on the estimated covariance matrix **S** is not too great, any genuine outliers in the data should lie well outside this ellipse, which is exactly what we see for the outlying point labelled 50.

The covariance matrix **S** represents a multivariate extension of the univariate variance, and as discussed in Sec. 9.6.1, it is intimately related to the correlation matrix. Consequently, this matrix is extremely sensitive to outliers in the data, so the classical Mahalanobis-based approach to multivariate outlier detection often fails, like its univariate counterpart the three-sigma edit rule. One solution is to replace the standard covariance estimator with an outlier-resistant alternative like the MCD estimator discussed in Sec. 9.6.5. The robust ellipse (solid curve) shown in Fig. 9.20 was constructed exactly this way using the function `tolEllipsePlot` from the `robustbase` package.

Fig. 9.21 presents another example of multivariate outlier detection using this function, but here the choice of method makes a much greater difference. Specifically, this plot was constructed from the `Weight` (X3) and `Cost` (Y) variables from the `aircraft` data frame from the `robustbase` package, representing data from the US Office of Naval Research characterizing 23 aircraft. Here, both the classical Mahalanobis ellipse (the larger, dashed curve) and the MCD-based robust ellipse (the smaller, solid curve) identify case 22 as an outlier, but this is the only outlier detected with the classical characterization. The robust ellipse is much smaller and it identifies three other aircraft (nos. 16, 17, and 21) as outliers in their relationship between cost and weight.

9.7 Associations between categorical variables

Ordered categorical variables like those found frequently in survey data (e.g., "strongly disagree," "disagree," "no opinion," "agree," or "strongly agree") can be converted to numerical ranks and the Spearman rank correlation measure described in Sec. 9.6.2 can be applied to characterize their association. More typically, we encounter *un-ordered* or *nominal* variables where ranks cannot be defined. It may still be true, however, that two different categorical variables are strongly associated: certain levels of one variable appear frequently together with certain levels of another variable and rarely with other values of this variable. The following discussions present three ways of assessing the relationship between categorical variables: *contingency tables*, which form the basis for the mosaic plots introduced in Chapter 2, are considered in Sec. 9.7.1; the *chi-square measure* is a numerical measure of association between categorical variables introduced in Sec. 9.7.2; and *Goodman and Kruskal's tau measure* is another, less widely known measure of association between categorical variables that has certain important practical advantages discussed in Sec. 9.7.3.

9.7.1 Contingency tables

The special case of the 2×2 contingency table was introduced in Sec. 9.3.3 in connection with the odds ratio as a measure of association between binary variables. More generally, an $I \times J$ *contingency table* describes the joint behavior of two categorical variables, x and y, with I and J distinct levels, respectively. Specifically, an $I \times J$ contingency table is a rectangular array of counts n_{ij} for $i = 1, 2, \ldots, I$ and $j = 1, 2, \ldots, J$ that tell us how many records exhibit the ith value of x together with the jth value of y. Contingency tables represent the most complete description of the relationship between two categorical variables, but by themselves, these tables are generally not easy to interpret. In particular, although there are cases where we can clearly see the relationships between categorical variables from a contingency table, these cases tend to be very special, corresponding either to complete predictability or complete independence.

The following examples from the `Soils` data frame in the `car` package illustrate this point, which contains 48 observations of 14 variables, including five categorical (factor) variables. Of these, the `Gp` variable has $I = 12$ levels and the `Contour` variable has $J = 3$ levels. The 12×3 contingency table for this pair of variables is easily generated with R's built-in `table` function, which provides a clear view of the relationship between these variables:

```
table(Soils$Gp, Soils$Contour)

##
##       Depression Slope Top
##  D0            4     0   0
##  D1            4     0   0
##  D3            4     0   0
##  D6            4     0   0
##  S0            0     4   0
##  S1            0     4   0
##  S3            0     4   0
##  S6            0     4   0
##  T0            0     0   4
##  T1            0     0   4
##  T3            0     0   4
##  T6            0     0   4
```

It is clear from this table that the value of `Gp` completely determines the value of `Contour`. In fact, this dataset contains data collected on the basis of a *designed experiment*, in which four soil samples are taken from areas characterized by each of the three possible values of `Contour`, and the first letter of the `Gp` value assigned to each sample indicates which `Contour` value is associated with that sample. Similarly, the second character of each `Gp` value codes the `Depth` value, with each of the four possible values of this variable equally represented in the dataset, a relationship that is clearly evident in the contingency table for `Depth` and `Gp` (not shown). In fact, such complete predictability—e.g., of `Contour` or `Depth` from the sample indicator `Gp`—is typical of designed experiments.

Another typical characteristic of designed experiments is complete independence between experimental conditions like `Contour` and `Depth` in this example. Independence means that knowing the value of one variable tells us nothing about the value of the other, corresponding to the counts n_{ij} in the contingency table being independent of both i and j. The contingency table for the `Contour` and `Depth` variables from the `Soils` data frame provides an example:

```
table(Soils$Contour, Soils$Depth)

##
##              0-10 10-30 30-60 60-90
##  Depression     4     4     4     4
##  Slope          4     4     4     4
##  Top            4     4     4     4
```

As a more typical example, consider the `msleep` data frame from the `ggplot2` package, which gives 11 characterizations of 83 mammals, including the name of each animal and four other character variables. Two of these character variables are `order`, the fourth level of the biological classification hierarchy below kingdom, phylum, and class, and `vore`, indicating what each animal eats (e.g., herbivore, insectivore, etc.). The complete relationship between these two variables is represented by the following 19×5 contingency table:

```
table(msleep$order, msleep$vore, useNA = "ifany")

##
##
##                     carni herbi insecti omni <NA>
##   Afrosoricida          0     0       0    1    0
##   Artiodactyla          0     5       0    1    0
##   Carnivora            12     0       0    0    0
##   Cetacea               3     0       0    0    0
##   Chiroptera            0     0       2    0    0
##   Cingulata             1     0       1    0    0
##   Didelphimorphia       1     0       0    1    0
##   Diprotodontia         0     1       0    0    1
##   Erinaceomorpha        0     0       0    1    1
##   Hyracoidea            0     2       0    0    1
##   Lagomorpha            0     1       0    0    0
##   Monotremata           0     0       1    0    0
##   Perissodactyla        0     3       0    0    0
##   Pilosa                0     1       0    0    0
##   Primates              1     1       0   10    0
##   Proboscidea           0     2       0    0    0
##   Rodentia              1    16       0    2    3
##   Scandentia            0     0       0    1    0
##   Soricomorpha          0     0       1    3    1
```

The structure of this contingency table is more complicated than the previous two examples, and while some relationships are apparent—for example, all members in the dataset from the order Carnivora are, not surprisingly, carnivores—others are more complex. In particular, note that the members of order Rodentia represented here include one carnivore, 16 herbivores, two omnivores, and three others with missing vore values. This last observation is important: to fully characterize the relationship between two categorical variables, it is essential to allow for the possibility of missing data, treating "missing" as a distinct level, a point discussed in Sec. 9.7.2.

The key point of this last example was to emphasize the point made earlier that, despite the completeness of this characterization, contingency tables are difficult to interpret succinctly. This is particularly true for categorical variables involving many levels since the number of entries in an $I \times J$ contingency table is IJ: the table characterizing order and vore just described contains 95 numbers. These observations motivate the numerical association measures between categorical variables described in the next two sections, which can be especially useful if we are attempting to discover relationships between many pairs of categorical variables.

9.7.2 The chi-squared measure and Cramér's V

The best known measure of association between categorical variables is the *chi-squared measure*, introduced by the statistician Karl Pearson at the end of the nineteenth century. This measure is a number computed from the entries of a contingency table, and sufficiently large values of this number provide evidence against the null hypothesis that the two variables are independent.

The chi-squared measure can be defined in several different but algebraically equivalent ways. The following definition is based on probability estimates computed from the entries $\{n_{ij}\}$ of the contingency table constructed for the categorical variables x with I levels and y with J levels:

$$p_{ij} = n_{ij}/N,$$

$$p_{i\cdot} = \sum_{j=1}^{J} p_{ij},$$

$$p_{\cdot j} = \sum_{i=1}^{I} p_{ij}, \tag{9.78}$$

where $N = IJ$. Note that p_{ij} is the classical estimate of the probability of observing the ith value of x together with the jth value of y, and $p_{i\cdot}$ and $p_{\cdot j}$ are the estimated *marginal probabilities* of observing the ith value of x for any value of y, and of observing the jth value of y for any value of x. In terms of these probability estimates, the chi-squared measure is given by [31, p. 9]:

$$\chi^2 = N \sum_{i=1}^{I} \sum_{j=1}^{J} \frac{(p_{ij} - p_{i\cdot} p_{\cdot j})^2}{p_{i\cdot} p_{\cdot j}}. \tag{9.79}$$

If x and y are regarded as random categorical variables with joint probability p_{ij} of observing the ith value of x with the jth value of y, these variables are *statistically independent* if $p_{ij} = p_{i\cdot} p_{\cdot j}$ for all i and j. Note that under these conditions, the numerator in the double sum in Eq. (9.79) is always zero, yielding the result $\chi^2 = 0$. For large samples, χ^2 has a chi-square distribution with $N - I - J + 1$ degrees of freedom [2, p. 79], and this forms the basis for the classical independence test. That is, if the computed value of χ^2 has probability smaller than $1 - \alpha$ under the chi-squared distribution with $N - I - J + 1$ degrees of freedom, we reject the independence hypothesis, concluding that there is a relationship between the variables x and y.

Although the chi-squared measure is well established, by itself it suffers from two inherent limitations. The first is that its magnitude increases with increasing N, making it difficult to interpret by itself. One solution to this difficulty is to normalize χ^2 so that it spans the range from 0 to 1, with the value 0 suggesting statistical independence and the value 1 suggesting strong dependence. Several normalizations have been proposed, and probably the two most common are [2, p. 112] the *contingency coefficient* proposed by Karl Pearson in 1904:

$$C = \sqrt{\frac{\chi^2}{\chi^2 + N}}, \tag{9.80}$$

and *Cramér's V*, the square root of a normalized measure proposed by Cramér in 1946:

$$V = \sqrt{\frac{\chi^2}{N \min(I - 1, J - 1)}}. \tag{9.81}$$

Both of these characterizations are returned by the assocstats function from the vcd package, along with the χ^2 measure itself and its associated p-values under the chi-squared distribution.

The second limitation of the chi-squared measure is noted by Goodman and Kruskal in their discussion of it [31, p. 10]:

> The fact that an excellent test of independence may be based on χ^2 does not mean that χ^2, or some simple function of it, is an appropriate *measure* of degree of association.

This observation provided motivation for their development of alternative association measures like that discussed in Sec. 9.7.3.

Before proceeding to this discussion, however, it is instructive to consider the χ^2 values resulting from the three contingency table examples discussed in Sec. 9.7.1. The function assocstats noted above is called with the contingency table computed by the table function, so the result for the first of these examples is:

```
tbl <- table(Soils$Gp, Soils$Contour)
assocstats(tbl)

##                      X^2 df    P(> X^2)
## Likelihood Ratio 105.47 22 7.0521e-13
## Pearson           96.00 22 3.2000e-11
##
## Phi-Coefficient    : NA
## Contingency Coeff.: 0.816
## Cramer's V         : 1
```

The first part of this result lists the χ^2 value computed two different ways (for a discussion, refer to the discussion in Agresti's book [2, Sec. 3.2.1]). The result listed as "Pearson" is the χ^2 measure defined in Eq. (9.79), and this entry also gives the degrees of freedom for the χ^2 distribution (here, df = $12 \times 3 - 12 - 3 + 1 = 22$) and the p-value associated with the χ^2 value under this distribution. Since this p-value is much less than the standard threshold $\alpha = 0.05$, we reject independence for this example, as we should. The other three values included in these results are: (1) the *phi coefficient*, missing here since it is applicable only to 2×2 contingency tables, where it is equal to $\sqrt{\chi^2/N}$ [2, p. 112]; (2) the contingency coefficient C defined in Eq. (9.80); and (3) Cramér's V defined in Eq. (9.81). The fact that Cramér's V has its maximum possible value of 1 here correctly reflects the extremely strong association between these categorical variables: Gp completely determines Contour.

The second example, involving the variables Countour and Depth, gives:

```
tbl <- table(Soils$Contour, Soils$Depth)
assocstats(tbl)

##                  X^2 df P(> X^2)
## Likelihood Ratio   0  6        1
## Pearson            0  6        1
```

```
##
## Phi-Coefficient   : NA
## Contingency Coeff.: 0
## Cramer's V        : 0
```

As noted in the preceeding discussion of this example, these variables are statistically independent since n_{ij} is independent of both i and j, consistent with the zero value of χ^2 seen here, the complete lack of statistical significance (i.e., the maximum possible p-value of 1), and the zero values for both C and V.

The third example is more interesting, since the relationship between the two variables is not obvious from the contingency table itself. Here, we have:

```
tbl <- table(msleep$order, msleep$vore, useNA = "ifany")
assocstats(tbl)

##                    X^2 df   P(> X^2)
## Likelihood Ratio 155.34 72 4.6264e-08
## Pearson          191.98 72 7.4052e-13
##
## Phi-Coefficient   : NA
## Contingency Coeff.: 0.836
## Cramer's V        : 0.76
```

Here, the χ^2 value is large and the associated p-value is small enough to exhibit high statistical significance. This fact, along with the large values for both C and V, suggests a strong relationship between order and vore, consistent with the conclusions drawn from the examination of this contingency table given in Sec. 9.7.1. As we will see in the next section, however, the Goodman-Kruskal τ association measure gives us greater insight into all three of these examples.

Before proceeding to that discussion, however, it is important to say something about the treatment of missing values in this analysis. In particular, if we do not include "missing" as a level in our analysis, we obtain a different contingency table and different values for the χ^2 measure, its associated degrees of freedom and p-value, and the derived quantities C and V:

```
tbl <- table(msleep$order, msleep$vore)
assocstats(tbl)

##                    X^2 df   P(> X^2)
## Likelihood Ratio 139.22 54 1.8616e-09
## Pearson          166.34 54 2.2549e-13
##
## Phi-Coefficient   : NA
## Contingency Coeff.: 0.828
## Cramer's V        : 0.854
```

Here, we see that the χ^2 value is smaller if we do not include the missing level, as is the degrees of freedom associated with this value, but the p-values are actually more significant in this case, while the C value is slightly smaller and Cramér's V is somewhat larger. In fact, there are other cases where we cannot obtain values for these statistics without including the missing level. A case in point is the relationship between genus and vore from the msleep data frame:

```
tbl <- table(msleep$genus, msleep$vore)
assocstats(tbl)

##                      X^2  df P(> X^2)
## Likelihood Ratio 188.65 228  0.97314
## Pearson             NaN 228     NaN
##
## Phi-Coefficient    : NA
## Contingency Coeff.: NaN
## Cramer's V         : NaN
```

This example will be explored further in Sec. 9.7.3, but, here, the key point is the importance of accounting for the possibility of missing values in examining the relationship between categorical variables.

9.7.3 Goodman and Kruskal's tau measure

As Goodman and Kruskal observed in the quote cited above from their book [31, p. 10], the popular χ^2 measure of independence has its limitations as a measure of association, providing part of the motivation for the authors to seek alternatives. Another motivation was to develop measures of association between categorical variables that were easier to interpret. Two of the measures Goodman and Kruskal (their λ and τ measures) proposed in their book are discussed by Agresti, motivated by the idea that they may be interpreted as describing the fraction of the variation of the categorical variable y that can be explained given the variable x. Specifically, this family of association measures has the following general form [2, p. 56]:

$$A(x, y) = \frac{V(y) - E\{V(y|x)\}}{V(y)}, \tag{9.82}$$

where $V(y)$ represents a measure of the variation of y and $V(y|x)$ represents the conditional variation of y given x. Since this conditional variation depends on x, the association measure $A(x, y)$ is based on its expected value $E\{V(y|x)\}$, which averages out this x-dependence. To compute A, we need definitions of the variation function $V(y)$ and its associated conditional variation $V(y|x)$, and two different choices lead to Goodman and Kruskal's λ and τ measures. This discussion focuses on their τ measure, which forms the basis for the GoodmanKruskal R package.

Before describing Goodman and Kruskal's τ measure in detail, it is important to note two points about the general class of association measures defined in Eq. (9.82). First, the expected conditional variation $E\{V(y|x)\}$ measures the residual variation of y that is not due to variation in x. Thus, if y is completely predictable from x, there is no residual variation, so $E\{V(y|x)\} = 0$, implying $A(x, y) = 1$. Conversely, if x is unrelated to y, knowledge of x is of no help in predicting y and $V(y|x) = V(y)$ for all x, implying $A(x, y) = 0$. For all other cases between these two extremes, $A(x, y)$ assumes a value between 0 and 1.

The second important point to note about association measures defined by Eq. (9.82) is that it is *asymmetric*: $A(x, y) \neq A(y, x)$, in general. Two important exceptions are the case of two binary variables ($I = J = 2$), where $A(x, y) = A(y, x)$, and statistical independence, where $A(x, y) = A(y, x) = 0$ if x and y are statistically independent. In more general cases, however, this asymmetry arises from the fact that $A(x, y)$ essentially measures the utility of x in predicting y and, as the following examples demonstrate, the utility of y in predicting x may be very different.

For Goodman and Kruskal's τ measure, the two terms appearing in Eq. (9.82) are given by [2, pp. 68–69]:

$$V(y) \quad = \quad 1 - \sum_{j=1}^{J} p_{\cdot j}^2,$$

$$E\{V(y|x)\} \quad = \quad 1 - \sum_{i=1}^{I} \sum_{j=1}^{J} \frac{p_{ij}^2}{p_{i\cdot}}, \tag{9.83}$$

where p_{ij}, $p_{i\cdot}$, and $p_{\cdot j}$ are the estimated individual and marginal probabilities defined in Eq. (9.78). Substituting these expressions into Eq. (9.82) and performing some messy algebraic simplification leads to the following expression for Goodman and Kruskal's τ measure:

$$\tau(x, y) = \frac{\sum_{i=1}^{I} \sum_{j=1}^{J} \left(\frac{p_{ij}^2 - p_{i\cdot}^2 \cdot p_{\cdot j}^2}{p_{\cdot j}} \right)}{1 - \sum_{j=1}^{J} p_{\cdot j}^2}. \tag{9.84}$$

As noted in the above discussion, if x and y are statistically independent, then $\tau(x, y) = \tau(y, x) = 0$. In general—as in the case of the numerical correlation measures discussed in Sec. 9.6—the converse is not true, but it is for Goodman and Kruskal's τ measure. That is, $\tau(x, y) = 0$ if and only if x and y are statistically independent, implying $\tau(y, x) = 0$ as well. This equivalence between zero association and statistical independence does not hold for Goodman and Kruskal's λ measure [2, p. 69], which partially motivates our focus on the τ measure here.

To see the general behavior of the Goodman-Kruskal τ measure, including the extent and potential utility of its asymmetry, it is useful to consider the three contingency table examples discussed in Secs. 9.7.1 and 9.7.2. Recall that the first of these examples considered the relationship between the variables Gp and Contour from the Soils data frame in the car package. Goodman and Kruskal's τ measure is implemented with the GKtau function in the GoodmanKruskal R package, and applying this function to these two variables gives the following result:

```
library(GoodmanKruskal)
GKtau(Soils$Gp, Soils$Contour)

##       xName        yName Nx Ny tauxy tauyx
## 1 Soils$Gp Soils$Contour 12  3     1 0.182
```

To clearly establish which variable is x and which is y, the GKtau function returns xName and yName, along with the numbers of unique values, Nx and Ny for each. This is followed by the values of $\tau(x,y)$ and $\tau(y,x)$, and this example illustrates the asymmetry of the τ measure noted above. Specifically, since the Countour value is completely predictable from Gp (indeed, it is coded as the first letter of the group name, as noted in Sec. 9.7.1), $\tau(x,y) = 1$ for this case. Conversely, since each Countour value appears in four different groups, Gp is only weakly predictable from Contour, which is responsible for the small $\tau(y,x)$ value seen here. Recall that the Cramér's V value for this case was $V = 1$, correctly indicating the presence of a very strong relationship between these variables, but giving no indication of its inherent asymmetry.

The second contingency table example considered the relationship between the variables Contour and Depth in the Soils data frame. Recall that these variables were independently set experimental conditions, so both the χ^2 value and the normalized Cramér's V value were zero. Applying the GKtau function to this case, we see that $\tau(x,y)$ and $\tau(y,x)$ are both zero, reflecting this statistical independence:

```
GKtau(Soils$Contour, Soils$Depth)

##            xName         yName Nx Ny tauxy tauyx
## 1 Soils$Contour Soils$Depth  3  4     0     0
```

The third example examined the relationship between the variables order and vore from the msleep data frame in the ggplot2 graphics package, where order describes a broad class of animals and vore describes what they eat. Recall that the Cramér's V value computed from these variables was large, suggesting a strong relationship between these two variables. As in the first example, the Goodman-Kruskal τ measure gives us a more detailed view of this relationship:

```
GKtau(msleep$order, msleep$vore)

##           xName         yName Nx Ny tauxy tauyx
## 1 msleep$order msleep$vore  19  5  0.62 0.223
```

In particular, note that order is more predictive of vore than the other way around, suggesting that the biological classification order groups together similar enough animals that many of them eat the same types of food. Another contributing factor to this asymmetry is that because there are 19 orders and only 5 vore values (including missing), we cannot predict order from vore nearly as well.

In the discussion of χ^2 and Cramér's V for this example, it was noted that the handling of the missing vore values was important in quantifying the relationship between these two variables. The same conclusion holds for Goodman and Kruskal's τ measure, as the following example illustrates. By default, the GKtau function includes missing values as a level if it appears in either the x or the y data, specified by the optional includeNA argument which has the default value "ifany." Specifying this argument as "no" causes records with missing x or y values to be ignored, giving us a slightly different result:

```
GKtau(msleep$order, msleep$vore, includeNA = "no")

##            xName        yName Nx Ny tauxy tauyx
## 1 msleep$order msleep$vore 19  4 0.733 0.229
```

Here, the treatment of missing vore values makes little difference in our ability to predict order from vore—it is about equally poor in either case—but omitting the records with missing vore values does offer non-negligible improvement in our ability to predict vore from order.

Correlation matrices were introduced in Sec. 9.6.4 as a way of simultaneously visualizing relationships between several numerical variables. These symmetric matrices are easily computed with the cor function, and they can be visualized with the corrplot package. Similarly, association matrices between several categorical variables can be computed using the GKtauDataframe function from the GoodmanKruskal package. As an example, note that the msleep data frame includes five character-valued categorical variables—name, genus, vore, order, and conservation—and apply the GKtauDataframe function to the portion of the msleep data frame that contains these variables. The following R code does this and displays the results:

```
keepVars <- c("name", "genus", "vore", "order", "conservation")
dfSub <- as.data.frame(msleep[, keepVars])
GKmat <- GKtauDataframe(dfSub)
GKmat

##                 name  genus  vore  order conservation
## name          83.000  1.000 1.000  1.000        1.000
## genus          0.927 77.000 1.000  1.000        0.968
## vore           0.049  0.055 5.000  0.223        0.072
## order          0.220  0.237 0.620 19.000        0.312
## conservation   0.073  0.077 0.101  0.101        7.000
## attr(,"class")
## [1] "GKtauMatrix"
```

(Note that msleep is a tibble, an extension of the R data frame supported by the tibble package; the function GKtauDataframe does not work with tibbles, but tibbles are easily converted to data frames with the as.data.frame function, giving the desired result.)

It is important to say something about the structure of the matrix GKmat. First, note that the large diagonal elements contain the numbers of unique values for each variable, while the off-diagonal elements contain the values of $\tau(x,y)$ and $\tau(y,x)$. Specifically, reading across each row, the off-diagonal elements give the value of $\tau(x,y)$ where x is the variable associated with the row (i.e., the variable name to the left) and y is the variable associated with the column (i.e., the variable name at the top). A clearer picture of these results is provided by the plot in Fig. 9.29 in the second group of color figures, included at the end of this chapter. This plot was generated by simply applying the generic plot function:

```
plot(GKmat)
```

The reason this approach works is that the `GKtauDataframe` function returns an object of class "GKtauMatrix," as indicated above when `GKmat` is displayed, and a `plot` method has been developed for this object class.

The plot in Fig. 9.29 presents a graphical representation of this association matrix, where each diagonal element includes text of the form "K = xx," where "xx" denotes the number of unique values for the variable associated with both the row and the column. Thus, for example, the variable `name` has $K = 83$ unique values, corresponding to one for each record in the dataset (i.e., each record characterizes a species of animal, and `name` identifies that species). The off-diagonal elements in each row give numerical and graphical representations of the value of the Goodman-Kruskal association measure $\tau(x, y)$ where x represents the variable associated with the row (i.e., the name to the left of the plot array) and y corresponds to the variable associated with the row (i.e., the name above the plot). In these plots, highly elongated ellipses correspond to associations $\tau(x, y)$ near the maximum value of 1, while nearly circular ellipses correspond to associations $\tau(x, y)$ that are near zero. As a specific example, for the fourth row, x corresponds to the variable `order`, which is only slightly predictive of `name`, `genus`, and `conservation`, but reasonably predictive of `vore`, as we have seen in the previous results. Comparing this with the row above, where x represents `vore`, we see that this variable has essentially no ability to predict `name`, `genus`, or `conservation`, and is only slightly predictive of `order`, again consistent with the results we have seen before.

The first row of this array represents a particularly important special case. The `msleep` data frame has 83 rows, and the `name` variable has a unique value for each row. In practical terms, this means that knowing the `name` value uniquely determines a row in the data frame, which uniquely determines the value of *all* of the variables. This is the reason that all of the entries in this first row are diagonal lines, corresponding to the most extreme elongation possible for an ellipsoid and indicating perfect predictability, $\tau(x, y) = 1$. *This is an important point: whenever x has one unique value for every record in a dataset, $\tau(x, y) = 1$ for all other variables y in the dataset.* It is also important to note that the asymmetry of Goodman and Kruskal's τ measure means that the converse does not hold: $\tau(y, x)$ is *not* generally equal to 1 for these cases. This point is clearly illustrated by looking at the *column* labelled `name` in Fig. 9.29: because there are almost as many distinct `genus` values as `name` values (i.e., 77 vs. 83), the association $\tau(y, x)$ for this case is quite large (0.93), but the other three of these associations are much smaller.

Also, while x exhibiting one unique value for each record is a *sufficient condition* for $\tau(x, y) = 1$, it is *not a necessary condition*. This point is illustrated by the `genus` row in Fig. 9.29: while the associations between the 77-level `genus` variable and both `name` and `conservation` are quite large, they are less than

1. In contrast, `genus` is perfectly predictive of both `vore` and `order`, the latter being a consequence of the fact that `genus` is a lower level of the biological classification hierarchy, so each genus is associated with a unique order.

Finally, it is worth emphasizing that Goodman and Kruskal's τ measure can be an extremely useful exploratory data analysis tool for discovering directed relationships between categorical variables, particularly in cases where there are many of them with names that are not overly descriptive. In particular, if x and y are two cryptically named variables and y is in fact a grouped or cleaned version of x, it follows that $\tau(x, y) = 1$ but $\tau(y, x) < 1$, exactly as with the `genus`/`order` pair considered here. If we did not collect or name these variables ourselves, but are given them as part of an analysis dataset, we may be completely unaware of these important relationships. The Goodman-Kruskal τ measure gives us a way of uncovering these relationships and using this knowledge to guide data interpretation and/or predictive model-building.

9.8 Principal component analysis (PCA)

Principal component analysis (PCA) is an analysis method for multivariate numerical datasets that seeks to define new variables called *projections* or *principal components* that are linear combinations of the original variables, satisfying the following two constraints:

1. Each principal component is uncorrelated with all of the others;

2. The variance of the principal component is as large as possible, subject to the first constraint.

Mathematically, if we have a dataset of N observations of each of p numerical variables $\{x_1, x_2, \ldots, x_p\}$, each of our principal components will be a vector of N values, defined by an expression of the form:

$$z_k = a_1 x_{1k} + a_2 x_{2k} + \cdots + a_p x_{pk}, \qquad (9.85)$$

where x_{ik} is the kth value of the ith variable x_i. To obtain the first principal component, we choose the p numerical coefficients $\{a_i\}$ so that the variance of z_k is as large as possible, subject to the following normalization constraint:

$$\sum_{i=1}^{p} a_i = 1. \qquad (9.86)$$

The notation gets a little messy here, but to distinguish the distinct principal components, assume we have solved the computational problem just described (there are R procedures that do this, discussed below), and denote the first principal component as $z_k^{(1)}$, defined by Eq. (9.85) with coefficients $\{a_i^{(1)}\}$. To obtain the second principal component, we repeat the optimization process used to find the first principal component, but subject to the additional *orthogonality* constraint to guarantee that these components are uncorrelated:

$$\sum_{k=1}^{N} z_k^{(1)} z_k^{(2)} = 0. \tag{9.87}$$

Again, procedures are available to compute the principal components for a given numerical dataset, but before discussing these, it is important to say something about where this approach may be useful and where caution is required.

Typically, principal component analysis is used as a means of *dimension reduction:* we have a dataset with an inconveniently large number of numerical variables, and we wish to work with a smaller subset for purposes of either data visualization or building predictive models. For example, while graphical visualization of a dataset with many numerical variables poses serious difficulties, a scatterplot of the first two principal components can sometimes reveal significant structure in the underlying data. PCA can provide useful solutions to modeling and visualization problems if many of the variables in the original dataset are highly correlated and are qualitatively similar enough that the linear combination can be at least approximately interpreted. Such problems are frequently encountered in chemical composition analysis based on spectroscopic data, where the individual variables in the dataset represent the absorption of electromagnetic radiation at different wavelengths: responses at nearby wavelengths are typically highly correlated, but in a highly data-dependent fashion. The R package `chemometrics` has been developed to deal with problems of this type, and the vignette by Garcia and Filzmoser provides a fairly thorough discussion of its use and the underlying mathematical ideas on which it is based [28]. There are, however, several important practical details that can strongly influence the utility of the results. First, PCA results depend strongly on data scaling: if the original data variables exhibit wildly different ranges of values, the large variables will completely dominate the PCA results. For this reason, data normalization is typically applied prior to PCA, usually by subtracting the mean of each variable and dividing these differences by their standard deviation to obtain a set of variables that all have zero mean and unit variance. Alternatively, PCA can be applied to correlation matrices computed from the original data. Second, because PCA is essentially a variance decomposition, it is strongly sensitive to outliers, which has led to the development of a variety of robust PCA algorithms, a topic discussed briefly below. Finally, like correlation analysis, PCA is appropriate to numerical data and requires special extensions to handle the common problem of mixed datasets with both numerical and categorical variables, another topic discussed briefly later in this section.

As noted, R provides a number of built-in functions to compute principal components. The basic two are `princomp`, discussed by Venables and Ripley [70, pp. 302–305], and `prcomp`, both included in the `stats` package in the basic R distribution. There are a number of differences in detail between these two procedures, both in terms of the calling arguments and the nature and structure of what is returned. Possibly the most important difference is that these two procedures use different algorithms to solve the problem of computing the coefficients $\{a_i^{(j)}\}$ of the principal components, commonly called the *loadings*. According to the help files available for `princomp`, the algorithm used by the `prcomp` is preferred. Another implementation of the basic PCA procedure is available as the function `classPC` in the `robustbase` package, which also supports *robust principal component analysis*, discussed briefly below. First, however, it is useful to present a representative example to show how PCA works in a case where the results are relatively easy to interpret. The basis for this example is the `crabs` data frame from the `MASS` package, used by Venables and Ripley to illustrate the `princomp` procedure [70, pp. 302–305]. Because the `princomp` help files actually recommend the `prcomp` procedure instead, it is used here, with the recommended centering and scaling options (centering is the default option, but scaling is not). The general structure of the `crabs` data frame can be obtained with the `str` function:

```
str(crabs)

## 'data.frame':  200 obs. of  8 variables:
##  $ sp   : Factor w/ 2 levels "B","O": 1 1 1 1 1 1 1 1 1 1 ...
##  $ sex  : Factor w/ 2 levels "F","M": 2 2 2 2 2 2 2 2 2 2 ...
##  $ index: int  1 2 3 4 5 6 7 8 9 10 ...
##  $ FL   : num  8.1 8.8 9.2 9.6 9.8 10.8 11.1 11.6 11.8 11.8 ...
##  $ RW   : num  6.7 7.7 7.8 7.9 8 9 9.9 9.1 9.6 10.5 ...
##  $ CL   : num  16.1 18.1 19 20.1 20.3 23 23.8 24.5 24.2 25.2 ...
##  $ CW   : num  19 20.8 22.4 23.1 23 26.5 27.1 28.4 27.8 29.3 ...
##  $ BD   : num  7 7.4 7.7 8.2 8.2 9.8 9.8 10.4 9.7 10.3 ...
```

Here, the variable `sp` indicates the crab's species, designated as "B" for blue and "O" for orange, and `sex` indicates its sex. The numerical variables of interest here are the last five columns of the data frame, which give sizes in millimeters of the frontal lobe (`FL`), the rear width (`RW`), the carapace length (`CL`), the carapace width (`CW`), and the body depth (`BD`). Applying the `prcomp` function to these five columns of the data frame gives a result that may be summarized as:

```
prinComp <- prcomp(crabs[, 4:8], center = TRUE, scale = TRUE)
summary(prinComp)

## Importance of components:
##                           PC1     PC2     PC3     PC4     PC5
## Standard deviation     2.1883 0.38947 0.21595 0.10552 0.04137
## Proportion of Variance 0.9578 0.03034 0.00933 0.00223 0.00034
## Cumulative Proportion  0.9578 0.98810 0.99743 0.99966 1.00000
```

Because we have started with five numerical variables, it is possible to construct five principal components that satisfy the orthogonality condition in Eq. (9.87), and the above summary gives the standard deviation of each of these five principal components, along with the proportion of the variance explained by each component. The coefficients $\{a_i^{(j)}\}$ from Eq. (9.85) that define each principal component (i.e., the loadings) are returned in the `rotation` element of the object returned by `prcomp`. Here, these loadings are:

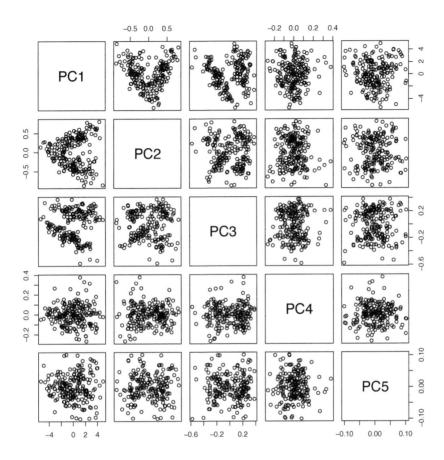

Figure 9.22: Scatterplot summary of the five `crabs` principal components.

```
prinComp$rotation

##          PC1        PC2         PC3          PC4         PC5
## FL 0.4520437  0.1375813  0.53076841  0.696923372  0.09649156
## RW 0.4280774 -0.8981307 -0.01197915 -0.083703203 -0.05441759
## CL 0.4531910  0.2682381 -0.30968155 -0.001444633 -0.79168267
## CW 0.4511127  0.1805959 -0.65256956  0.089187816  0.57452672
## BD 0.4511336  0.2643219  0.44316103 -0.706636423  0.17574331
```

The principal components themselves, commonly called the *scores*, are returned in the x element of `prinComp`.

Fig. 9.22 is a scatterplot of these five principal components, constructed by first converting the x element of the `prinComp` object returned by the `prcomp` function to a data frame and then applying the generic `plot` function:

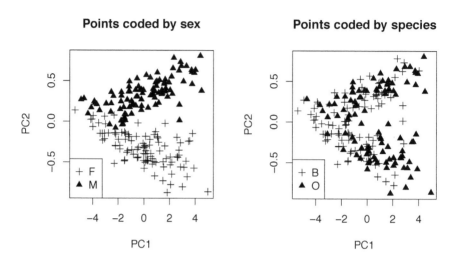

Figure 9.23: Scatterplot of PC2 vs. PC1, coded by **sex** (left) and **sp** (right).

```
pcaDF <- as.data.frame(prinComp$x)
plot(pcaDF)
```

In the first row of this array, the scatterplots between the first principal component (PC1) and the second and third (PC2 and PC3) show the most interesting structure, while the rest look more like structureless blobs. This is typical: the first few principal components generally have the greatest "information content," while the latter ones are dominated by "noise." For this reason, the following discussion emphasizes these first two scatterplots, taking PC1 as the x-axis and PC2 or PC3 as the y-axis (i.e., the top two plots in the PC1 column in Fig. 9.22).

The two scatterplots in Fig. 9.23 show the second principal component from the **crabs** data (PC2) plotted against the first (PC1), constructed by plotting the corresponding components of the **x** list element returned by the **prcomp** function. These plots differ only in their point coding: on the left, points are marked with solid triangles if **sex** is "M" and as plus signs if **sex** is "F," while on the right, solid triangles correspond to the "O" value of the species variable **sp** and plus signs correspond to the value "B." Comparing these plots, the one on the left shows a reasonably complete separation of the two parts of the

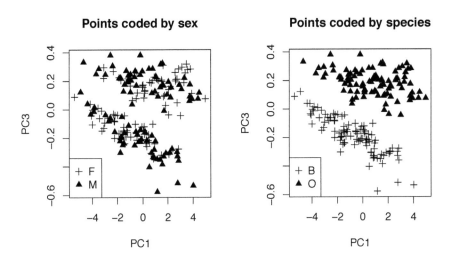

Figure 9.24: Scatterplot of PC3 vs. PC1, coded by **sex** (left) and **sp** (right).

sideways V-shaped cloud into a lower leg of female crabs and an upper leg of male crabs. In contrast, the right-hand plot shows a complete mix of species in the two visually distinct portions of this scatterplot. Thus, it appears that the first two principal components code different relationships for male and female crabs between the size variables from which these vectors were constructed.

Fig. 9.24 shows the corresponding scatterplots of the third principal component (PC3) versus the first (PC1), again based on the x list element returned by the `prcomp` function, in the same format and coding as Fig. 9.23. Again, we see two reasonably well-separated point clouds. On the left, where the points are coded by the `sex` variable, the male and female point shapes are about equally distributed between these point clouds, like the species distribution in the right-hand plot in Fig. 9.23. In contrast, the coding by species on the right separates these point clouds perfectly, suggesting that the first and third principal components together code different relationships between species and the five size variables in the `crabs` dataset.

Taken together, Figs. 9.23 and 9.24 illustrate a case where PCA illuminates significant and physically interpretable structure in the dataset. This example thus demonstrates that principal component analysis can be a useful tool for highlighting the structure of relationships between numerical variables in a dataset, but it is important to be aware that PCA results are not always as clear or readily interpretable as in this example. In particular, three potential difficulties are, first, that the inherent relationships between variables may be complex enough that PCA results are not particularly illuminating, second, that because it is a variance-based data decomposition, the basic PCA approach is highly sensitive to outliers in the data, and, third, that we are often faced with the presence of categorical variables in our dataset. The following discussions briefly consider the second and third of these points: robust extensions of PCA, and extensions to mixed data involving both numerical and categorical variables.

Since both PCA and correlation analysis are based on estimated variances, their outlier sensitivities are similar. A number of examples illustrating this sensitivity are discussed in the paper by Hubert *et al.* [42], clearly illustrating the limitations of classical PCA and the need for robust alternatives. Indeed, this outlier sensitivity is not surprising since standard PCA can be based on standard correlation matrices, as noted in the above discussion. Thus, one possible approach to outlier-resistant PCA is to apply standard methods to robust correlation matrices, computed using the methods described in Sec. 9.6.5. In addition, other outlier-resistant PCA methods have been developed and some of these are available in the *R* packages `pcaPP` and `rrcov`. A detailed treatment of these ideas is beyond the scope of this book, but for a reasonably detailed (but necessarily mathematical) discussion of the essential ideas behind robust PCA, refer to the paper by Hubert *et al.* [42].

One possible approach for extending PCA to mixed datasets with both numerical and categorical variables is to adopt the dummy variable scaling used to include categorical predictors in linear regression models, discussed in Chapter 5. A practical difficulty with this idea is that the dummy numerical variables created to represent the different levels of a categorical variable necessarily exhibit strong negative correlations, since only one of these variables can take the value 1, and all others must be 0, for any data record k. To see this point, create a new categorical variable `crabType` for the `crabs` dataset to represent both their sex and species, with levels "FB," "MB," "FO," and "MO." Next, manually construct the dummy variables required to represent `crabType`, taking the "FB" level as our base class. To do this, we define the following binary variables for the other three non-base levels:

```
sex <- crabs$sex
sp <- crabs$sp
crabTypeMB <- as.numeric((sex == "M") & (sp == "B"))
crabTypeFO <- as.numeric((sex == "F") & (sp == "O"))
crabTypeMO <- as.numeric((sex == "M") & (sp == "O"))
```

To see the correlations between these variables, combine them into a data frame and apply the `cor` function:

```
dummyDF <- data.frame(crabTypeMB = crabTypeMB,
                      crabTypeFO = crabTypeFO,
                      crabTypeMO = crabTypeMO)
```

```
cor(dummyDF)

##            crabTypeMB crabTypeFO crabTypeMO
## crabTypeMB  1.0000000 -0.3333333 -0.3333333
## crabTypeFO -0.3333333  1.0000000 -0.3333333
## crabTypeMO -0.3333333 -0.3333333  1.0000000
```

Next, to see the effects of these negative correlations on our principal components results, create a new data frame containing the numerical variables from the crabs dataset, along with these three dummy variables and apply the prcomp function to the results:

```
crabs2 <- crabs[, 4:8]
crabs2 <- cbind.data.frame(crabs2, dummyDF)
prinComp2 <- prcomp(crabs2, center = TRUE, scale = TRUE)
```

Since the crabs2 data frame has eight numerical variables, the prcomp procedure constructs eight principal components, and Fig. 9.25 is an array of scatterplots between all pairs of these principal components, like Fig. 9.22 for the original PCA results. The details of these plot arrays are very different: for example, note the pronounced cluster structure in *all* of the scatterplots involving components PC2, PC3, and PC4 in Fig. 9.25. Both because of the greater number of principal components arising from the dummy variables required to represent our new categorical variable, and because of the much greater degree of clustering seen in these scatterplots, this second set of PCA results is more difficult to interpret and use. This difficulty has led to the development of extensions of the basic PCA approach to mixed datasets (i.e., those containing both numerical and categorical data). One implementation available in R is the PCAmixdata package described in the paper by Chavent *et al.* [13].

This example also provides a clear illustration of the need for data scaling when constructing principal components. Specifically, Fig. 9.26 shows the results obtained by calling prcomp with center = FALSE and scale = FALSE with the crabs2 data frame. Comparing this scatterplot with Fig. 9.25, constructed using the recommended centering and scaling discussed earlier, we see very little similarity between them. In fact, the scatterplots between PC1 and PC2 in Fig. 9.26 look very similar to the corresponding plots in the original scatterplot matrix shown in Fig. 9.22, while that between PC3 and PC8 in Fig. 9.26 looks quite similar to that between PC1 and PC3 in Fig. 9.22. As in Fig. 9.25, the scatterplots shown in Fig. 9.26 show more cluster structure in scatterplots involving higher-order principal components than the original results shown in Fig. 9.22, but the extremely narrow, well-separated clusters seen in many of the scatterplots in Fig. 9.25 are not seen in Fig. 9.26, reflecting the influence of centering and scaling the data.

A detailed discussion of principal component analysis and its use in finding structure in high-dimensional numerical data is beyond the scope of this book, but because PCA is a popular tool, a few final comments are in order. First, it is worth emphasizing that, like the standard implementation of correlation

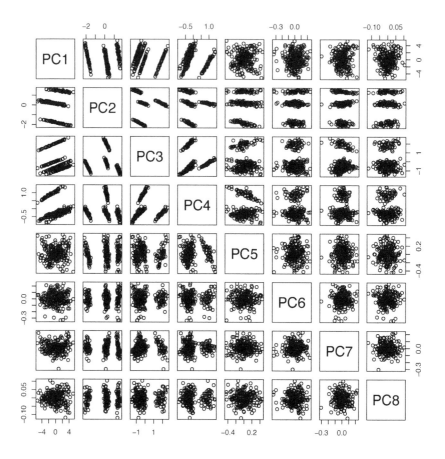

Figure 9.25: Scatterplots of the eight augmented `crabs` principal components.

analysis discussed in Sec. 9.6.1, the standard implementation of PCA is only applicable to collections of *numerical* variables. Extensions like those described by Chavent *et al.* [13] have been developed, but as the comparison of Figs. 9.22 and 9.25 presented above illustrates, principal component analysis for mixed data is inherently more complicated than that for purely numerical data. Second, as the comparison between Figs. 9.25 and 9.26 illustrates, data scaling is an important pre-processing step prior to constructing principal components: failure to perform this preprocessing step can profoundly change both the nature and the utility of PCA results. Third, again like correlation analysis, outliers can profoundly influence PCA results, motivating the development of robust PCA algorithms like those implemented in the R packages `pcaPP` and `rrcov` discussed briefly above. Finally, it is worth concluding with the following caution, offered by Venables and Ripley in their brief discussion of PCA [70, p. 304]:

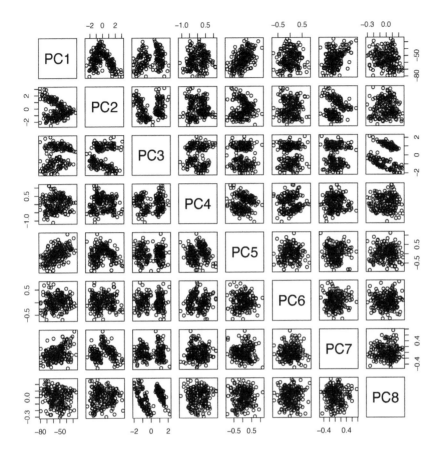

Figure 9.26: Scatterplot summary of the eight augmented `crabs` principal components, computed without centering or scaling.

A *warning:* principal component analysis will reveal the gross features of the data, which may already be known, and is often best applied to residuals after known structure has been removed. As we discovered in Figure 4.13 on page 96, animals come in varying sizes and two sexes!

9.9 Working with date variables

The following discussion of *date variables* is included in this chapter not because they are commonly characterized using probability-based tools like the other data types considered here, but rather because these variables can exhibit at least two very different forms of behavior. In particular, date variables are typically represented externally as character strings for human interpretation, but represented internally as numbers to allow simple computations like the number of days between two different dates. Further, several different external representations are in wide use, leading to the possibilities of errors or inconsistencies if one format is assumed but another is actually used (e.g., European dates typically put the day first, followed by the month, while the standard in the U.S. is to put the month first, followed by the day). The following discussion briefly considers date representations and a few of the useful conversions and computations that are possible for date variables in R.

Base R supports a `Date` object class, together with a number of useful tools for working with these objects. As a simple example, consider:

```
dayInYear <- seq(1, 365, 1)
dates <- as.Date("2017-01-01") + dayInYear - 1
dates[1:10]

##  [1] "2017-01-01" "2017-01-02" "2017-01-03" "2017-01-04" "2017-01-05"
##  [6] "2017-01-06" "2017-01-07" "2017-01-08" "2017-01-09" "2017-01-10"

dates[360:365]

##  [1] "2017-12-26" "2017-12-27" "2017-12-28" "2017-12-29" "2017-12-30"
##  [6] "2017-12-31"
```

Here, the first line creates the vector of numbers from 1 to 365, indexing the days of any non-leap year. The `as.Date()` call in the second line converts the character string "2017-01-01" into an R object of class `Date`, which allows us to perform simple arithmetic operations, exploiting the numerical character of date variables. The effect of this addition—of the numbers 0 through 364—is to create all of the dates of the year 2017, as the last two lines illustrate. Similarly, subtraction of date variables is possible, as are a number of other arithmetic operations on dates:

```
as.Date("2017-05-23") - as.Date("2017-02-19")

## Time difference of 93 days

min(dates)

## [1] "2017-01-01"

max(dates)

## [1] "2017-12-31"

median(dates)

## [1] "2017-07-02"
```

We cannot push these date operations too far, however: while we can compute the median of a sequence of dates, the `quantile` function fails for date variables.

Beyond these simple arithmetic operations, base R supports a variety of other operations on `Date` variables. Three examples are the functions `weekdays`, `months`, and `quarters`, which convert dates into character strings representing days of the week, months in the year, and quarters of the year. To illustrate these functions, consider the following data frame construction:

```
scheduleFrame <- data.frame(dayInYear = dayInYear, date = dates)
scheduleFrame$dayOfWeek <- weekdays(scheduleFrame$date)
scheduleFrame$month <- months(scheduleFrame$date)
scheduleFrame$quarter <- quarters(scheduleFrame$date)
str(scheduleFrame)

## 'data.frame': 365 obs. of  5 variables:
## $ dayInYear: num  1 2 3 4 5 6 7 8 9 10 ...
## $ date     : Date, format: "2017-01-01" "2017-01-02" ...
## $ dayOfWeek: chr  "Sunday" "Monday" "Tuesday" "Wednesday" ...
## $ month    : chr  "January" "January" "January" "January" ...
## $ quarter  : chr  "Q1" "Q1" "Q1" "Q1" ...
```

Note that when we incorporate a `Date` variable into a data frame, the `Date` class is retained. This is important, because if the date variable were converted to a character string, the date manipulation functions used here would fail. An important case where the `Date` class is *not* preserved is that where we write a data frame containing date variables to a CSV file and then later read it back. Specifically, note that `date` is converted from a `Date` variable to a factor variable when we perform this sequence of operations:

```
write.csv(scheduleFrame, "scheduleFrame.csv", row.names = FALSE)
scheduleFrame2 <- read.csv("scheduleFrame.csv")
str(scheduleFrame2, vec.len = 2)

## 'data.frame': 365 obs. of  5 variables:
## $ dayInYear: int  1 2 3 4 5 ...
## $ date     : Factor w/ 365 levels "2017-01-01","2017-01-02",..: 1 2 3 4 5 ...
## $ dayOfWeek: Factor w/ 7 levels "Friday","Monday",..: 4 2 6 7 5 ...
## $ month    : Factor w/ 12 levels "April","August",..: 5 5 5 5 5 ...
## $ quarter  : Factor w/ 4 levels "Q1","Q2","Q3",..: 1 1 1 1 1 ...
```

More generally, much can be done with date variables in R, either using the `Date` class available in base R, or with other, more flexible date manipulation packages like `chron` or `lubridate`. A detailed discussion of this topic is beyond the scope of this book, but a very useful introduction to some of the intricacies of working with date variables is available in the help file for R's `as.Date` function. Another useful source of information is the discussion by Crawley in *The R Book* [16, pp. 89–96].

9.10 Exercises

1: It was noted at the end of Section 9.1.1 and emphasized again in Section 9.2.3 that most data characterizations we encounter in practice are asymptotically normal, meaning that if they are computed from a large enough sample, quantities like means, medians, or regression coefficients exhibit an approximately Gaussian distribution. This exercise asks you to use the qqPlot function from the car package to explore this. Specifically:

1a. Initialize the R random number system with a seed value of 79 and using a simple for loop, create a list with 100 gamma-distributed random samples, each of size 200 with shape parameter shape = 3. Using the lapply function, create a vector gammaMeans of the means of these 100 sequences and construct its normal QQ-plot. Does asymptotic normality appear to hold here?

1b. Repeat (1a) for 100 random samples from a Student t-distribution with one degree of freedom (the Cauchy distribution), generating and characterizing the vector CauchyMeans. Does asymptotic normality appear to hold here?

1c. Repeat (1b) but for the median instead of the mean, generating and characterizing the vector CauchyMedians. Does asymptotic normality appear to hold here?

2: Section 9.9 introduced the Date object class in R and showed some of the computations that are possible with these objects. This exercise asks you to create and work with some of these objects. Specifically:

2a. In the U.S., one of the most commonly used character formats for the date March 4, 2000 is "03/04/2000." Use the as.Date function to convert this character string into a date with the default format option: does the resulting Date object correspond to March 4, 2000? Verify this conclusion by computing the number of days between the current date (use the Sys.Date() function) and this Date object: how many years does this number of days correspond to?

2b. Using the format argument of the as.Date function, define startDate as the Date object corresponding to the character string "01/01/2000" and endDate as that corresponding to the string "01-jan-2020". (Hint: see help(strptime) for details about specifying the format argument.) Compute the number of days between these dates.

2c. Using the seq function, create a vector of dates called dateSeq with one date for each day between startDate and endDate. Set the random number generator seed to 39 and use the sample function to create a vector dateSample of 40,000 random dates in this range. Note that you must use the sample function to draw a random sample with replacement, causing some dates to be repeated: use the table function to tabulate the number of times each date occurs in this sequence. What is the range of these repetition counts?

3: The Greene data frame in the car package characterizes appeals decisions for refugee claimants who were previously rejected by the Canadian Immigration and Refugee Board.

3a. One potentially important categorical variable in this dataset is `judge`, the identity of the judge assigned to hear the appeal. Use the `table` function to construct the contingency table of `decision` by `judge` and display the results. From this contingency table, construct the vectors `n1Vector` of positive decision counts for each judge, and `nVector` of the total number of appeals heard by each judge. From these, compute and display the classical estimates of the probability of a positive decision for each judge.

3b. Create the function `binCIplot` listed in Section 9.3.2 and use it to generate a plot of the size-adjusted binomial confidence intervals from the `addz2ci` function in the `PropCIs` package. Does this plot suggest that the judge assignment has a strong influence on the appeals decision, a weak influence, or no significant influence?

4: The odds ratio was introduced in Section 9.3.3 as a measure of association between two binary variables: an odds ratio near 1 implies the two variables are independent, while very large or very small odds ratios imply a strong association between these variables. Construct the function `ORproc` listed in Section 9.3.3 and use it to examine the relationship between these binary variables from the `Greene` data frame in the `car` package introduced in Exercise 3:

4a. The variable `decision` is a binary indicator of whether the appeal was granted or not, and `language` indicates whether the case was heard in English or in French. Do these variables appear to be related?

4b. The variable `rater` is an opinion made by an independent observer concerning whether the appeal should be granted or not: is there evidence of agreement between `rater` and `decision`?

5: It was noted in Exercise 2 that sampling with replacement from a set results in repeated values. The number of repetitions of each value will depend on both the size of the set and the size of the sample: if both are large, this distribution of counts will often be approximately Poisson. This exercise asks you to explore this hypothesis for the tabulation of repeated date values constructed in Exercise (2c). Specifically:

5a. Using the `truehist` function from the `MASS` package, construct a histogram of the counts from this tabulation (hint: the R table object constructed in Exercise 2 must be converted to a numeric vector). Does the shape of this plot suggest a Poisson distribution?

5b. One of the hallmarks of the Poisson distribution is the fact that the mean is equal to the variance. Does this condition appear to hold, at least approximately, for the counts from the date tabulation in (5a)?

5c. Section 9.4.3 introduced the Poissonness plot as an informal graphical test of the Poisson hypothesis. Use the `distplot` function from the vcd package to construct a Poissonness plot for the date counts considered here: does this plot support the Poisson hypothesis here?

6: The Leinhardt data frame in the car package contains infant mortality rates per 1000 live births for 105 countries. This variable, infant, contains missing values but spans a range of positive values and exhibits pronounced asymmetry, making the Gaussian distributional assumption questionable. This exercise asks you to consider the gamma distribution discussed in Section 9.5.2 as an alternative:

6a. The gamma distribution is characterized by two parameters: the shape parameter α introduced in Section 9.5.2 that determines the overall shape of the distribution, and a *scale* parameter β that determines the range of the data values. The mean and variance of the gamma distribution are related to these parameters by the following expressions:

$$
\begin{aligned}
\text{mean} &= \alpha\beta, \\
\text{variance} &= \alpha\beta^2.
\end{aligned}
$$

One way of fitting a gamma distribution to data is to use the *method of moments*, matching the observed mean and variance with their computed values, μ and σ^2, respectively, and solving for the parameters α and β. Doing this yields the following parameter estimates:

$$
\hat{\alpha} = \frac{\mu^2}{\sigma^2},
$$

$$
\hat{\beta} = \frac{\sigma^2}{\mu}.
$$

Use these estimators to obtain the distribution parameters alphaHat and betaHat that best fit the infant variable.

6b. Using the parameter estimates obtained in (6a), construct a side-by-side plot array with: on the left, a histogram constructed from the truehist function from the MASS package, overlaid with a heavy solid line showing the estimated gamma density; on the right, a gamma QQ-plot using the shape parameter determined from (6a). Do these plots suggest the gamma distribution is a reasonable approximation for this data sequence?

7: As noted at the end of Section 9.6.4, the correlation matrix display introduced there for the product-moment correlation coefficient can also be applied to the Spearman rank correlation measure introduced in Section 9.6.2 and used in the correlation trick described in Section 9.6.3. This exercise uses the corrplot function in the corrplot package to create a graphical multivariate extension of the correlation trick, creating a display that allows us to compare product-moment correlations and Spearman rank correlations for a collection of several numerical variables. The basis for this exercise is the subset of numerical variables included in the `msleep` data frame from the `ggplot2` package. Specifically, using columns 6 through 11 of this data frame, first generate the product-moment correlation matrix as an object called `corMatPM` and the Spearman rank correlation matrix as an object called `corMatSP`. To handle the missing data in this data frame, use the option `use = ''complete.obs''` in computing these correlations. Next, use the `corrplot` function to create a single plot array showing both correlations. Specifically, using code similar to that shown in the example at the end of Section 9.6.4, plot the product-moment correlations in the upper portion of the plot array, but use the optional argument `addCoef.col = ''black''` to overlay numerical correlation values on the plot. Finally, using code similar to that used to add the lower portion to the plot in the example in Section 9.6.4, display the Spearman rank correlations in the lower portion of the plot, again with the correlations represented as ellipses overlaid with numbers. Which two variable pairs show the greatest differences between the two correlations?

8: It was shown in Section 9.6.3 that the relationship between brain weight and body weight from the `Animals2` dataset in the `robustbase` package was monotone but not linear, and both of these variables are included in the `msleep` dataset from the `ggplot2` package considered in Exercise 7. This exercise asks you to construct a data frame `sleepDF` from the columns of the `msleep` data frame that includes the untransformed values for `sleep_rem`, `sleep_cycle`, and `awake`, along with variables `logBrain` and `logBody` obtained by applying the log transformation to the `msleep` variables `brainwt` and `bodywt`. From this data frame, first compute the product-moment correlation matrix `corMatPM` analogous to that in Exercise 7 and then compute the robust correlation matrix `corMatRob` using the `covMcd` function discussed in Section 9.6.5. From these correlation matrices, construct a `corrplot` display in the same format as in Exercise 7, with the product-moment correlations shown in the upper portion of the display and the robust correlation estimates shown in the lower portion. Which variable is associated with the largest number of correlation sign changes between the product-moment correlation and the robust correlation estimate?

9: The elliptical contour plots used for bivariate outlier detection in Section 9.6.6 based on the MCD covariance estimator can be extremely effective in the case of two variables, but something different is needed for detecting multivariate outliers involving more than two variables. One useful technique is the *distance-distance plot*, which computes both the classical Mahalanobis distance and the robust alternative based on the MCD covariance estimator. Given these Mahalanobis distances, the distance-distance plot shows the robust distance as the y-axis versus the classical distance as the x-axis. This plot can be generated from the `plot` method for objects of class `mcd` returned by the `covMcd` function in the `robustbase` package. In addition to the basic plot, the resulting display also includes reference lines to help identify unusual points with respect to either the classical Mahalanobis distance (a vertical line) or the robust distance (a horizontal line). Points falling outside either or both of these limits are marked with a number indicating the corresponding record from the data frame used to compute the MCD covariance estimator. Using the data frame `sleepDF` constructed in Exercise 8, generate the corresponding distance-distance plot. Which species from the `msleep` data frame are the two most extreme outliers with respect to the robust distance?

10: The `happy` data frame from the `GGally` package contains 10 variables taken from the General Social Survey, an annual survey of American demographic characteristics, opinions, and attitudes, between the years 1972 and 2006. This dataset contains 51,020 records and the variables included are related to happiness; six of these variables are categorical. Create a data frame from `happy` that contains only these categorical variables and construct a plot of the Goodman-Kruskal association measure between all of these variables like that shown for the `msleep` example in Section 9.7.3. Which other variable is the best predictor of happiness? Which variable is best predicted from happiness? Are these associations strong or weak?

Color Figures, Group 2

Figure 9.27: Colored wordcloud of frequent words from chapter 34 (Epilogue) of Lawson's novel *Born Again*.

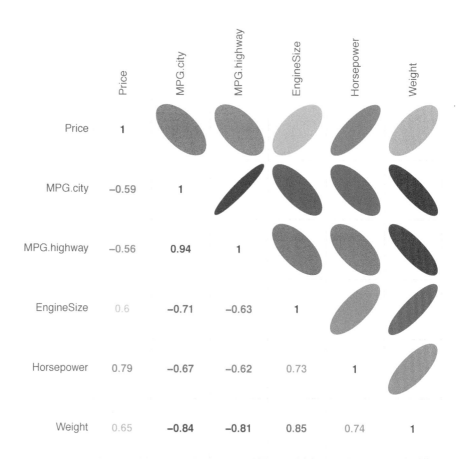

Figure 9.28: Correlation matrix from selected `Cars93` numerical variables.

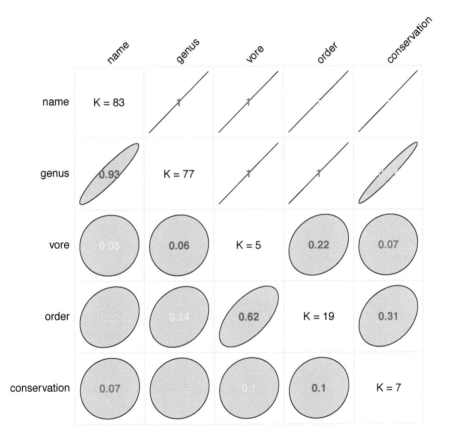

Figure 9.29: Goodman-Kruskal τ association matrix for the categorical variables in the `msleep` data frame.

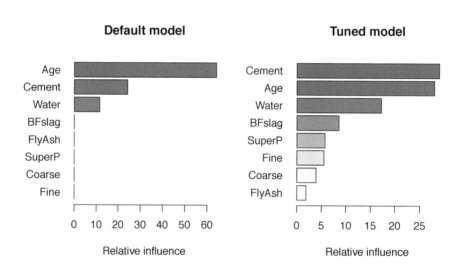

Figure 9.30: Relative importances from the `gbm` models described in Sec. 10.5.3: the default model (left) and the tuned model (right).

Chapter 10

More General Predictive Models

Chapter 5 introduced the predictive modeling problem through linear regression models, the simplest and historically most popular approach to building predictive models. This approach is not always the best, however, for two reasons. First, linear regression models are not always applicable; the problem of binary classification is a case in point: there, we are attempting to predict the probability that a binary response variable will assume one of its two possible values. If we approach this problem using the linear regression methods described in Chapter 5, we are likely to get invalid predictions: predicted probabilities that are either negative or greater than 1, neither of which is possible. The second limitation of linear regression models is that, increasingly, even when these models can be built and yield valid predictions, members of the newer class of *machine learning models* give much better predictions. Although this chapter cannot give a complete introduction to this model class, it does offer a gentle introduction and some suggestions on where to go to learn more.

10.1 A predictive modeling overview

Because predictive models come in many different types, and because the types of predictive models that are applicable in a given application depend strongly on a number of key problem details, it is useful to begin with a high-level overview of the nature and objectives of the predictive modeling problem. In general terms, the objective of this problem is to build a mathematical model that predicts a *response variable* or *target* from one or more *predictor variables* or *covariates*. The following discussion is divided into two parts: Sec. 10.1.1 begins with a high-level overview of the objectives of the predictive modeling problem, establishing notation and terminology that will be used throughout the rest of this chapter; Sec. 10.1.2 then outlines the basic sequence of steps involved in building predictive models.

10.1.1   The predictive modeling problem

The predictive modeling problem considered here starts with a dataset $\mathcal{D}$ containing N records, including a response variable y and p predictor variables $x_1, x_2, \ldots, x_p$. Given this dataset, the objective of the predictive modeling problem is to construct a mathematical function f that maps the values of the predictor variables for each record into a prediction of the response variable for that record:

$$\hat{y}_k = f(x_{1k}, x_{2k}, \ldots, x_{pk}). \tag{10.1}$$

Here, $\hat{y}_k$ is the model's prediction of the kth observation y_k of the response variable from the dataset $\mathcal{D}$, and x_{ik} denotes the kth observation of the ith prediction variable x_i from the dataset. One of the key requirements of a good predictive model is that it generate accurate predictions, meaning:

$$\hat{y}_k \simeq y_k, \text{ for } k = 1, 2, \ldots, N. \tag{10.2}$$

Here, the symbol "$\simeq$" means "is approximately equal to" and, as in the linear regression problem discussed in Chapter 5, this requirement can be interpreted in different ways, leading to different solutions. In the case of continuous response variables, the least squares criterion used in fitting linear regression models is a popular choice. That is, we interpret Eq. (10.2) to mean that the sum of squared errors should be minimized:

$$\text{minimize } J = \sum_{k=1}^{N} (\hat{y}_k - y_k)^2. \tag{10.3}$$

The details of how we accomplish this minimization depend on the mathematical form of the function f in Eq. (10.1); in the predictive modeling approaches considered here, this choice amounts to the selection of an R function that finds the best element of a particular class of functions f.

The simplest illustration of this idea is the class of linear regression models considered in Chapter 5, where the form of the function f is:

$$f(x_{1k}, x_{2k}, \ldots, x_{pk}) = \sum_{i=1}^{p} a_i x_{ik}, \tag{10.4}$$

where $\{a_i\}$ is a set of p unknown constants, whose values are determined by minimizing the *performance metric* J defined in Eq. (10.3). As discussed in Chapter 5, the standard R procedure for solving this problem is `lm`, but other procedures are also available that optimize slightly different performance metrics (e.g., the `lmrob` function in the `robustbase` package).

An important aspect of the linear regression problem is that it is appropriate to *numerical* response variables, ideally with approximately Gaussian, continuously distributed response variables. There are important applications where this assumption does not hold, and in these cases, linear regression models may be totally inappropriate. One particularly important case is that of a *binary target*, where the response variable y_k can take only one of two values. Often, these *binary classification problems* are formulated in terms of models that predict the probability that y_k assumes one of these values, but, even here, the fact that this probability must lie between 0 and 1 means that linear regression models are inappropriate. An important modification of the linear regression problem that handles this case is the class of *logistic regression models* discussed in Sec. 10.2, which guarantees that these probability limits are satisfied. The logistic regression model belongs to the larger class of *generalized linear models* discussed in Sec. 10.2.4, which can be used to fit response variables that must be positive (e.g., gamma-GLM models) or that are discrete-valued count variables (e.g., Poisson-GLM models).

The basic formulation outlined here—select a class of prediction functions f and choose the best member of this class for our dataset by minimizing a performance metric over all members of this class—is applicable to many different predictive modeling problems. This chapter cannot begin to consider all of the model structures that are in wide use, but it does provide an introduction to a few useful examples, including *decision tree models* in Sec. 10.3, *MOB models* that combine decision trees and either linear regression models or generalized linear models in Sec. 10.4, and the broader class of *machine learning models* discussed briefly in Sec. 10.5, including *random forest models* introduced in Sec. 10.5.2, and *boosted tree models* introduced in Sec. 10.5.3, both extensions of the decision tree models discussed in Sec. 10.3. These models often yield more accurate predictions than linear regression models or simple decision tree models, but this improved accuracy comes at the price of greater model complexity and a difficulty in understanding how they achieve this performance. Two useful tools for "peeking inside these black boxes" are *partial dependence plots* and *variable importance measures*, both introduced in Sec. 10.6.

10.1.2 The model-building process

The basic process of building a predictive model consists of these steps:

1. Select the target variable y to be predicted (as noted in Sec. 10.1.1, the type of this variable determines the model structures available in Step 2);

2. Choose a model structure, fixing the form of the function f in Eq. (10.1);

3. Select the predictor variables $x_1, x_2, \ldots, x_p$;

4. Obtain the dataset $\mathcal{D}$ that contains the target and predictor variables;

5. Select the fitting procedure and fit the model (often the easiest step);

6. Evaluate the results (see Chapter 5 or Sec. 10.2.3);

7. Iterate the above steps as necessary to obtain a satisfactory model.

The iteration noted at the end of this process is a key aspect of model-building: typically, one pass through Steps 1 to 6 does not yield the predictive model we want, either because the model performance is unacceptable, or because we believe that we can do better, possibly by considering other model structures in Step 2 or alternative predictor variables in Step 3. Also, since these variables must be present in our dataset $\mathcal{D}$, the selection of additional predictor variables in Step 3 may require the generation of a new dataset. Also, note that the evaluation procedures used in Step 6 depend, like the choice of model structure, on the target variable type. In particular, very different performance measures are typically used for regression-like problems with numerical target variables and binary classification problems; this point is discussed in detail in Sec. 10.2.3.

10.2   Binary classification and logistic regression

Binary classification models or *binary classifiers* are predictive models whose response variable is binary, assuming only two possible values (e.g., "policyholder filed claim" or "policyholder did not file claim"). Typically, these binary classification models predict not the response variable itself, but rather the *probability of a positive response*, where one of the two possible response values has been designated "positive." Probably the binary classifier with the longest history and the most completely developed underlying theory is *logistic regression*, an extension of the linear regression models discussed in Chapter 5, and a member of the larger class of *generalized linear models* introduced in Sec. 10.2.4.

10.2.1   Basic logistic regression formulation

A more extensive discussion of these ideas is given in Sec. 10.2.4, but the essential basis for all generalized linear models (GLMs), including the logistic regression model, is these three elements:

1. A probability distribution that is appropriate to the response variable y;

2. A *linear prediction equation* like the linear regression models considered in Chapter 5;

3. A *link function* that relates the linear prediction equation to the expected value of y.

For the specific case of logistic regression, the appropriate probability distribution is the binomial distribution discussed in Chapter 9, characterized by a single probability parameter p that any observed response is positive. In the logistic regression problem, the objective is to build a model that predicts this probability from other, related variables. Thus, if our binary response variable is an indicator of whether or not an insurance client filed a claim, we might be interested in predicting the probability that a policyholder will file a claim (we take "filed a claim" as our positive response) from various potentially relevant characteristics (e.g., driver age, vehicle age and type, prior driving record, etc.).

If we attempted to construct a linear regression model like those discussed in Chapter 5 to predict the probability of a positive response from these other characteristics, there is nothing in our model to restrict these predictions to the range of valid probabilities, which must lie between 0 and 1. We can overcome this difficulty, however, by predicting not the probability p itself, but applying the *logit link function*, defined as:

$$\text{logit}(p) = \log\left(\frac{p}{1-p}\right). \tag{10.5}$$

The effect of this function is to map probabilities in the range $0 < p < 1$ into the entire set of real numbers, $-\infty < \text{logit}(p) < \infty$. Combining these three components—the binomial distribution for our binary responses, the linear prediction equation, and the logit link function—yields the mathematical form of the basic logistic regression model:

$$\log\left(\frac{\hat{p}_k}{1-\hat{p}_k}\right) = a_0 + \sum_{i=1}^{m} a_i x_{i,k}. \tag{10.6}$$

Here, as in the linear regression models discussed in Chapter 5, the coefficient a_0 represents the intercept term in the model, $\{a_i\}$ for $i = 1, 2, \ldots, m$ is the set of coefficients for the m covariates x_1 through x_m, and $x_{i,k}$ represents the kth observation of the variable x_i. Given the observed responses $\{y_k\}$ and the values $\{x_{i,k}\}$ for these covariates, the objective of the logistic regression problem is to determine the values of the coefficients a_i for $i = 0, 1, 2, \ldots, m$ that yield the most accurate predicted probabilities $\hat{p}_k$ of positive response for each record k.

Substantial effort has gone into developing reliable methods for solving this model-fitting problem, and the basic implementation in R is the glm function, described and illustrated in Sec. 10.2.2. Before proceeding to that discussion, however, it is important to note that the logit function is invertible, leading to an explicit formula for the predicted probability $\hat{p}_k$ in terms of the data values and the model parameters. Specifically, if $z = \text{logit}(p)$, the inverse relationship is:

$$p = \text{logit}^{-1}(z) = \frac{1}{1+e^{-z}}. \tag{10.7}$$

Applying this result to Eq. (10.6) yields the following formula for the predicted probabilities:

$$\hat{p}_k = \left[1 + \exp\left\{-a_0 - \sum_{i=1}^{m} a_i x_{i,k}\right\}\right]^{-1}. \tag{10.8}$$

The point of this result is to show that there is a simple, explicit expression available for computing the predicted probabilities $\{\hat{p}_k\}$ of positive response for every record in our dataset, given the observed covariate values $\{x_{i,k}\}$ and the model parameters $\{a_i\}$. In practice, however, we don't need to implement this prediction expression ouselves because R's generic predict function has a built-in method for the logistic regression models returned by the glm function, exactly as in the case of linear regression models discussed in Chapter 5. This function is discussed further in the example that follows.

10.2.2 Fitting logistic regression models

To provide a concrete illustration of how logistic regression models are fit— and to serve as a basis for comparison with other binary classifiers discussed in this chapter—consider the problem of predicting the diabetic diagnosis of the female patients characterized in the data frame `PimaIndiansDiabetes` from the `mlbench` *R* package. Applying the `BasicSummary` function described in Chapter 3 gives a general overview of the contents of this data frame:

```
BasicSummary(PimaIndiansDiabetes)

##    variable    type levels topLevel topCount topFrac missFreq missFrac
## 1  pregnant numeric     17        1      135   0.176        0        0
## 2   glucose numeric    136       99       17   0.022        0        0
## 3  pressure numeric     47       70       57   0.074        0        0
## 4    triceps numeric    51        0      227   0.296        0        0
## 5   insulin numeric    186        0      374   0.487        0        0
## 6      mass numeric    248       32       13   0.017        0        0
## 7  pedigree numeric    517    0.254        6   0.008        0        0
## 8       age numeric     52       22       72   0.094        0        0
## 9  diabetes  factor      2      neg      500   0.651        0        0
```

As discussed previously, physiologically infeasible zero values dominate the variables `triceps` and `insulin`, representing nearly 30% and 50% of the total records, respectively. For these reasons, the binary classifiers considered here omit both of these variables from the covariate set. Also, the performance measures discussed in Sec. 10.2.3 require the binary response to be coded numerically, as 0 or 1, so the original response variable `diabetes`, a factor with two levels coded as "neg" and "pos," is replaced with a binary variable, with 0 corresponding to "neg" and 1 corresponding to "pos." Finally, for the reasons discussed in Chapter 5, it is important to partition the dataset into separate training and validation subsets. Because the versions of this example considered here compare a relatively small collection of cases, each on an equal footing, the data frame is partitioned into approximately equal-sized training and validation subsets, where all models are fit to the training subset and primary evaluations are made using the validation subset. The following *R* code generates the subsets `PimaTrain` and `PimaValid` used here:

```
keepVars <- c("pregnant", "glucose", "pressure", "mass", "pedigree", "age")
PimaSub <- PimaIndiansDiabetes[, keepVars]
diabetesFlag <- ifelse(PimaIndiansDiabetes$diabetes == "neg", 0, 1)
PimaSub$diabetes <- diabetesFlag
TVflag <- TVHsplit(PimaSub, split = c(0.5, 0.5), labels = c("T", "V"))
PimaTrain <- PimaSub[which(TVflag == "T"), ]
PimaValid <- PimaSub[which(TVflag == "V"), ]
```

The mechanics of fitting a logistic regression model in *R* are nearly identical to those of fitting a linear regression model to predict a numerical target variable. The primary difference is that the generalized linear model function `glm` is used instead of the linear regression modeling function `lm`. Specifically, the following call builds a logistic regression model that predicts the binary response `diabetes` from all other variables in the training dataset:

```
logisticFull <- glm(diabetes ~ ., data = PimaTrain, family = "binomial")
```

As in the linear regression models discussed in Chapter 5, the `glm` function supports the formula interface, so the modelling objective "predict `diabetes` from everything else" is specified here as `diabetes ~ .`, with the `data` argument used to specify the source of all of these variables. The one major difference between this example and the linear regression examples considered in Chapter 5 is the `family` argument here, which specifies that the binomial distribution is appropriate to the binary response variable `diabetes`. Like linear regression models, a summary of the model coefficients and their associated statistics can be obtained with the `summary` function:

```
summary(logisticFull)

##
## Call:
## glm(formula = diabetes ~ ., family = "binomial", data = PimaTrain)
##
## Deviance Residuals:
##     Min       1Q   Median       3Q      Max
## -2.6874  -0.6947  -0.3504   0.6676   3.0097
##
## Coefficients:
##               Estimate Std. Error z value Pr(>|z|)
## (Intercept) -9.364797   1.132591  -8.268  < 2e-16 ***
## pregnant     0.078015   0.044793   1.742   0.0816 .
## glucose      0.036704   0.004932   7.443 9.87e-14 ***
## pressure    -0.016226   0.008599  -1.887   0.0592 .
## mass         0.104031   0.022120   4.703 2.56e-06 ***
## pedigree     0.723745   0.463455   1.562   0.1184
## age          0.033482   0.013797   2.427   0.0152 *
## ---
## Signif. codes:  0 '***' 0.001 '**' 0.01 '*' 0.05 '.' 0.1 ' ' 1
##
## (Dispersion parameter for binomial family taken to be 1)
##
##     Null deviance: 483.93  on 377  degrees of freedom
## Residual deviance: 338.36  on 371  degrees of freedom
## AIC: 352.36
##
## Number of Fisher Scoring iterations: 5
```

Looking at the p-values associated with these predictor variables, it appears that the most important terms in this model are the intercept, the variable `glucose`, and the variable `mass`. To provide a basis for comparing model performance using the measures described in Sec. 10.2.3, it is useful to consider two other models. The first is the logistic regression model that includes only these most significant terms, and the motivation for considering this model is to see how much performance we lose by omitting the less significant variables, resulting in a simpler model that is easier to interpret. The following *R* code fits this model and gives the corresponding summary information:

```
logisticRef <- glm(diabetes ~ glucose + mass, data = PimaTrain,
                               family = "binomial")
summary(logisticRef)

##
## Call:
## glm(formula = diabetes ~ glucose + mass, family = "binomial",
##     data = PimaTrain)
##
## Deviance Residuals:
##     Min       1Q   Median       3Q      Max
## -2.2252  -0.7464  -0.4062   0.7183   3.1485
##
## Coefficients:
##               Estimate Std. Error z value Pr(>|z|)
## (Intercept) -8.509286   0.963348  -8.833  < 2e-16 ***
## glucose      0.038312   0.004762   8.046 8.56e-16 ***
## mass         0.091274   0.020572   4.437 9.13e-06 ***
## ---
## Signif. codes:  0 '***' 0.001 '**' 0.01 '*' 0.05 '.' 0.1 ' ' 1
##
## (Dispersion parameter for binomial family taken to be 1)
##
##     Null deviance: 483.93  on 377  degrees of freedom
## Residual deviance: 358.76  on 375  degrees of freedom
## AIC: 364.76
##
## Number of Fisher Scoring iterations: 5
```

Comparing the coefficients in these models, we see that the `glucose` coefficient in the simpler model is larger by a factor of six than that in the original model, while the `mass` coefficient is larger by a factor of nine. These differences suggest the models are very different, making the question of differences in their prediction performance particularly interesting. Also, note that the intercept term is automatically included in this model, even though it is not explicitly included in the model formula here. As in the case of linear regression models, it is possible to construct an intercept-only logistic regression model, serving as a "minimal performance benchmark" for subsequent comparisons:

```
logisticNull <- glm(diabetes ~ 1, data = PimaTrain, family = "binomial")
summary(logisticNull)

##
## Call:
## glm(formula = diabetes ~ 1, family = "binomial", data = PimaTrain)
##
## Deviance Residuals:
##     Min       1Q   Median       3Q      Max
## -0.9093  -0.9093  -0.9093   1.4716   1.4716
##
## Coefficients:
```

```
##             Estimate Std. Error z value Pr(>|z|)
## (Intercept)  -0.6694     0.1087  -6.159 7.3e-10 ***
## ---
## Signif. codes:  0 '***' 0.001 '**' 0.01 '*' 0.05 '.' 0.1 ' ' 1
##
## (Dispersion parameter for binomial family taken to be 1)
##
##     Null deviance: 483.93  on 377  degrees of freedom
## Residual deviance: 483.93  on 377  degrees of freedom
## AIC: 485.93
##
## Number of Fisher Scoring iterations: 4
```

Here, note that the only parameter in this model (i.e., the intercept) is an order of magnitude smaller than that seen in the other two models.

An explicit expression for the predicted probabilities $\hat{p}_k$ of positive response (here, "patient is diabetic") was given at the end of Sec. 10.2.1, where it was noted that these predictions are available in R from the generic **predict** function. Use of this function for logistic regression models is similar to that for linear regression models, but with one extremely important difference: to obtain predicted probabilities, it is necessary to specify the optional argument **type = ''response''**. The following R code generates both the training-set and validation-set predictions for all three of the models described above, which will form the basis for the performance comparisons discussed next:

```
pHatFullT <- predict(logisticFull, newdata = PimaTrain, type = "response")
pHatFullV <- predict(logisticFull, newdata = PimaValid, type = "response")
pHatRefT <- predict(logisticRef, newdata = PimaTrain, type = "response")
pHatRefV <- predict(logisticRef, newdata = PimaValid, type = "response")
pHatNullT <- predict(logisticNull, newdata = PimaTrain, type = "response")
pHatNullV <- predict(logisticNull, newdata = PimaValid, type = "response")
```

10.2.3 Evaluating binary classifier performance

Many different performance metrics have been developed for binary classifiers, and a number of these are available in the R package **MLmetrics**, which also includes a number of performance measures for regression-type prediction models. Some of these binary performance measures are appropriate to classifiers that assign an explicit binary prediction b_k to each record that corresponds to one of the two possible values for y_k, but the two measures considered here are suitable for binary classifiers that predict the probability p_k of positive response for each record, like the logistic regression models just discussed. The first and better known of these measures is based on the *Receiver Operating Characteristic (ROC) curve*, called the *AUC*, an abbreviation for "area under the ROC curve." The second, not nearly as well known, is based on the square root of the *Brier score*, originally developed for evaluating the quality of weather forecasts. For reasons explained in the following discussions, these measures exhibit some fundamentally different characteristics, so it is very useful to consider the two of them together.

ROC curves and the AUC

The *Receiver Operating Characteristic (ROC) curve* is a plot based on the following important binary classifier characteristics [2, p. 228]:

1. *sensitivity* is the probability that a positive response is correctly classified;

2. *specificity* is the probability that a negative response is correctly classified.

For a binary classifer like those considered here that estimates the probability of a positive response, these classifications must be based on a *threshold probability* p_0 such that a positive response is assigned to the kth data observation if $\hat{p}_k > p_0$ and a negative response is assigned if $\hat{p}_k \leq p_0$. As a consequence, the sensitivity and specificity depend on this threshold probability, i.e.:

$$
\begin{aligned}
\text{Sensitivity} &= P\{\hat{p}_k > p_0 | y_k = 1\}, \\
\text{Specificity} &= P\{\hat{p}_k \leq p_0 | y_k = 0\}.
\end{aligned}
\tag{10.9}
$$

Note that as we vary the threshold probability p_0, the sensitivity varies from 0 when $p_0 = 1$ to 1 when $p_0 = 0$, while the specificity varies in the opposite direction, from 1 when $p_0 = 1$ to 0 when $p_0 = 0$. The ROC curve summarizes this trade-off, plotting Sensitivity versus $1-$Specificity as the detection threshold p_0 varies from 0 to 1.

Two example ROC curves are shown in Fig. 10.1, both based on the logistic regression model `logisticFull` for the Pima Indians diabetes dataset described in Sec. 10.2.2. The ROC curve on the left is based on the training set predictions, while that on the right is based on the validation set predictions. These curves were generated using the `roc` function from the `pROC` package, which returns an object of class "roc" that can be plotted and used to compute the AUC values discussed below. For the training data predictions, this object is created as:

```
library(pROC)
ROCobjectT <- roc(PimaTrain$diabetes, pHatFullT)
```

Note that the x-axes on these curves are not "1 - `Specificity`" as suggested in the above discussion, but rather `Specificity` with these x-axes *reversed* so that the `Specificity` value 1 appears at the left end of the plot, decreasing towards zero as we move to the right end of the plot. This point is important because these ROC curves exhibit the concave shape traditionally seen in these plots, lying above the diagonal reference line (ROC curves with traditionally ordered `Specificity` axes are equivalent but look very different [38, p. 316]). This diagonal reference line corresponds to a classifier based on *random guessing*, and ROC curves like the two shown here that lie above this reference line correspond to classifiers that exhibit better performance than random guessing. (Note that if the curve were convex, always lying below this reference line, this suggests the response labels have been reversed in building the classifier since we can obtain better results by replacing y_k with $1 - y_k$.)

A useful numerical characterization of classifier accuracy is the *area under the ROC curve* or *AUC*. The idea here is that a perfect classifier would exhibit both

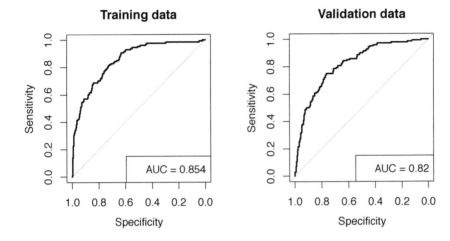

Figure 10.1: ROC curves for the logistic regression model `logisticFull` for the Pima Indians diabetes data: training data predictions (left) and validation data predictions (right).

sensitivity and specificity values of 1 for all threshold probabilities p_0, yielding an ROC curve that would form a right triangle with the random guessing reference line. Such perfect classification is generally not achievable in practice, but the area under this ideal classifier ROC curve would be 1, the maximum possible value. In contrast, the area under the random guessing line is 0.5, while curves like those shown in Fig. 10.1 exhibit intermediate areas. This area can be computed from the data used to generate the ROC curve, and it is shown in the lower right of the two plots in Fig. 10.1. This value may be obtained by applying the `auc` function from the `pROC` package to objects of class "roc" returned by the `roc` function, or it may be obtained directly from the observed responses and prediction probabilities using the `AUC` function from the `MLmetrics` package:

```
library(MLmetrics)
AUC(pHatFullT, PimaTrain$diabetes)
```

```
## [1] 0.853875
```

```
AUC(pHatFullV, PimaValid$diabetes)
```

```
## [1] 0.8195429
```

Like the goodness-of-fit measures described in Chapter 5 for linear regression models, the quality assessment provided by the AUC value for binary classifiers is susceptible to overfitting. For this reason, it is useful to adopt the same general strategy advocated there, of fitting our model to the training dataset and evaluating its performance against a similar but not identical validation dataset. The fact that the AUC value is slightly lower for the validation set predictions than for the training set predictions is consistent with our expectations that training set-based evaluations are somewhat optimistic, possibly reflecting some degree of overfitting.

We can use ROC curves and their associated AUC values to compare the performance of different logistic regression models, or to compare the performance of different types of binary classifiers. For example, the validation set AUC values for the three logistic regression models `logisticFull`, `logisticRef`, and `logisticNull` described in Sec. 10.2.2 are:

```
AUC(pHatFullV, PimaValid$diabetes)
```

```
## [1] 0.8195429
```

```
AUC(pHatRefV, PimaValid$diabetes)
```

```
## [1] 0.7914714
```

```
AUC(pHatNullV, PimaValid$diabetes)
```

```
## [1] 0.5
```

These results suggest that we suffer only a small degradation in performance if we simplify our logistic regression model from the one that uses all available covariates (`logisticFull`) in predicting the diabetic status of the women characterized in the Pima Indians diabetes dataset to one that retains only the variables `glucose` and `mass` (`logisticRef`). Since this second model involves four fewer prediction variables, it is substantially simpler, suggesting this slight performance decline may be worth accepting for the sake of interpretability (i.e., it is easier to explain to others with limited technical backgrounds).

A thorough discussion of ROC curves, the AUC measure, and other popular performance measures for binary classifiers like those available in the `MLmetrics` package is beyond the scope of this book. A useful reference that discusses these topics in detail, including a survey of some of the available software in R, is the paper by Robin *et al.* [60].

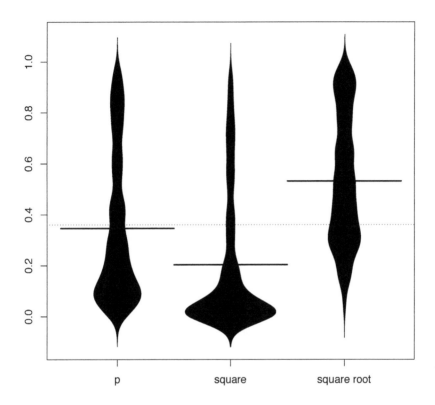

Figure 10.2: Effect of square and square root transformations on validation set positive response probabilities from the `logisticFull` model.

The Brier score and its square root

One of the disadvantages of the ROC curve and its associated AUC performance measure is that this measure is invariant under arbitrary monotone transformations of the predicted probabilities $\{\hat{p}_k\}$, so long as these transformations preserve the 0 to 1 range limit. As a specific example, note that both the square and the square root preserve this range, but these transformations profoundly change the distribution of the estimated probabilities. This point is illustrated in Fig. 10.2, which gives a beanplot summary of the distributions of `pHatFullV`, the validation set predictions from the `logisticFull` logistic regression model, its square, and its square root. Applying the `AUC` function from the `MLmetrics` package to these three variables, we see these transformations have no impact on this performance measure:

```
AUC(pHatFullV, PimaValid$diabetes)
```

```
## [1] 0.8195429
```

```
AUC(pHatFullV^2, PimaValid$diabetes)
```

```
## [1] 0.8195429
```

```
AUC(sqrt(pHatFullV), PimaValid$diabetes)
```

```
## [1] 0.8195429
```

The AUC is a well-accepted and useful measure of binary classifier performance, but these observations suggest that it does not give a complete performance assessment, motivating the need to consider other measures.

For a binary response y_k coded as 0 or 1, the *Brier score* is the average squared error between the observed responses $\{y_k\}$ and the predicted probabilities $\{\hat{p}_k\}$, i.e.:

$$B = \frac{1}{N} \sum_{k=1}^{N} (y_k - \hat{p}_k)^2. \tag{10.10}$$

This measure is a specialization of a more general measure proposed in 1950 by Brier for evaluating the quality of weather forecasts, with more than two possible response values [10]. Here, since the response y_k assumes only the values 0 or 1 and $0 \le \hat{p}i_k \le 1$ for all k, it follows that the difference $y_k - \hat{p}_k$ must lie between -1 (if $y_k = 0$ and $\hat{p}_k = 1$) and $+1$ (if $y_k = 1$ and $\hat{p}_k = 0$). As a consequence, it follows that $(y_k - \hat{p}_k)^2$ lies between 0 and 1 for all k, so the average value B must also lie between these limits. Thus, the Brier score always falls between 0 and 1, with the following interpretations:

1. $B = 1$ implies that the predicted probabilities $\{\hat{p}_k\}$ are the worst possible, in that $\hat{p}_k$ is always 0 when $y_k = 1$ and $\hat{p}_k$ is always 1 when $y_k = 0$;

2. $B = 0$ implies that the predicted probabilities are perfect: $\hat{p}_k = 1$ whenever $y_k = 1$ and $\hat{p}_k = 0$ whenever $y_k = 0$;

3. A particularly interesting reference case is that if $\hat{p}_k = 0.5$ for all k, giving $(y_k - \hat{p}_k)^2 = 0.25$ for all k, independent of y_k, then $B = 0.25$.

This last observation motivates the adoption of the *root-Brier score* advocated here. Specifically, taking the square root of the Brier score preserves its range and the interpretation of its extremes: $\sqrt{B} = 0$ still implies the worst possible disagreement between the observations $\{y_k\}$ and the predictions $\{\hat{p}_k\}$, and $\sqrt{B} = 1$ still implies perfect agreement. The advantage of this transformation is that it moves the reference case described in (3) into the middle of this range: $\sqrt{B} = 0.5$. Thus, the root-Brier score proposed here is defined as:

$$R = \sqrt{B} = \sqrt{\frac{1}{N} \sum_{k=1}^{N} (y_k - \hat{p}_k)^2}. \tag{10.11}$$

Again, the three special cases noted above for this measure are:

1. $R = 1$ implies the worst possible disagreement between observations and predictions: $\hat{p}_k = 0$ whenever $y_k = 1$ and $\hat{p}_k = 1$ whenever $y_k = 0$;

2. $R = 0$ implies perfect agreement: $\hat{p}_k = 0$ whenever $y_k = 0$ and $\hat{p}_k = 1$ whenever $y_k = 1$;

3. $\hat{p}_k = 0.5$ for all k implies $R = 0.5$.

An advantage of this reformulation of the basic Brier score is that it provides a crude interpretation guide: R values between 0 and 0.5 indicate general agreement between the predictions and the observed responses, better as R decreases towards zero, while R values between 0.5 and 1 indicate general disagreement, more severe as R approaches 1. The root-Brier score is easily implemented as a custom function in R, but an even easier way of computing it is to use the RMSE function from the MLmetrics package, with the two required arguments y_pred corresponding to $\{\hat{p}_k\}$ and y_true corresponding to $\{y_k\}$.

One advantage of both the original Brier score and the root-Brier score is that, in contrast to the AUC, these measures are *not* invariant under monotone transformations of classifier probabilities. To see this point, reconsider the example illustrated in Fig. 10.2. Applying the RMSE function from the MLmetrics package here yields:

```
RMSE(pHatFullV, PimaValid$diabetes)

## [1] 0.4068554

RMSE(pHatFullV^2, PimaValid$diabetes)

## [1] 0.443952

RMSE(sqrt(pHatFullV), PimaValid$diabetes)

## [1] 0.4404869
```

Here, the best results are achieved with the original probabilities pHatFullV from the logistic regression model, while applying either the square or the square root transformation results in a noticeable performance degradation.

Like the AUC, the root-Brier score can also be used to compare the performance of different models. Thus, for the three logistic regression models considered in Sec. 10.2.2, we have the following performance indications:

```
RMSE(pHatFullV, PimaValid$diabetes)

## [1] 0.4068554

RMSE(pHatRefV, PimaValid$diabetes)

## [1] 0.4188516

RMSE(pHatNullV, PimaValid$diabetes)

## [1] 0.4801311
```

As with the AUC comparisons, we see a slight decline in performance in going from the more complex logistic regression model `logisticFull` to the simpler one, `logisticRef`. Also, note that while the trivial intercept-only model `logisticNull` generates a constant prediction for all data records, this prediction is 0.339 for all k rather than 0.50; thus, while the root-Brier score for this case is close to the "non-informative" value 0.5, it is not exactly equal to this value. This reflects the fact that the model `logisticNull` is using the prevalence of diabetes in the dataset—the only predictor available to it—to achieve slightly better predictions than complete ignorance (i.e., "the probability is 50-50"). Note that this slight advantage of the `logisticNull` model over complete ignorance is not reflected in the AUC, which assigns both complete ignorance ($\hat{p}_k = 0.5$ for all k) and the `logisticNull` prediction ($\hat{p}_k = 0.339$) the same "random guessing" score of 0.5.

These differences between the AUC and the root-Brier score show that they are providing distinct binary classifier performance assessments, and experience suggests that these assessments can be advantageously used together. That is, rather than relying entirely on one, it is a good idea to examine both, preferring classifiers that exhibit both large AUC values (definitely greater than 0.5, the closer to 1 the better) and small root-Brier scores (definitely smaller than 0.5, the closer to 0 the better).

10.2.4 A brief introduction to glms

As noted in Sec. 10.2.1, the class of *generalized linear models* is an extension of the linear regression model class discussed in Chapter 5 that involves three components:

1. A probability distribution appropriate to the response variable y;

2. A linear prediction equation like those discussed in Chapter 5;

3. A *link function* $g(\cdot)$ that relates the linear predictor to the expected value of the response variable.

The class of generalized linear models includes both the linear regression models discussed in Chapter 5 and the logistic regression models discussed in Sec. 10.2.1, along with several others, including those appropriate to the gamma distribution for positive response variables and the Poisson distribution for count data. More specifically, in their simplest form, these generalized linear models have the following structure:

$$\text{predictor: } \eta_k = a_0 + \sum_{i=1}^{p} a_i x_{k,i},$$
$$\text{link: } g(\mu_k) = \eta_k, \tag{10.12}$$

where μ_k is the mean (expected value) of the response variable. Each of the special cases noted above corresponds to a specific distributional assumption for the response variable y, and each of these distributions is typically associated with a specific link function. The most popular of these special cases are:

1. linear regression: Gaussian distribution and identity link, $g(z) = z$;

2. logistic regression: binomial distribution and logit link, $g(z) = \mathrm{logit}(z)$;

3. gamma regression: gamma distribution and log link, $g(z) = \ln z$;

4. Poisson regression: Poisson distribution and log link, $g(z) = \ln z$.

A detailed discussion of the generalized linear model class is necessarily fairly mathematical and is therefore not presented here; for those interested in further details, the book by McCullagh and Nelder provides a very thorough treatment with many examples [55]. The key points here are that, first, the underlying statistical theory for this class of models has been well developed, and, second, effective computational procedures for fitting them to data are widely available.

In R, all four of the special cases listed above can be fit using the glm function, selecting between them via the optional family argument to specify the distribution and the link function. Specifically, the glm function is called with the following family specifications for the four model types listed above:

1. linear regression: family = gaussian;

2. logistic regression: family = binomial;

3. gamma regression: family = Gamma(link = "log");

4. Poisson regression: family = poisson.

Note that in three of these four cases, the default link function is the one listed above, corresponding to the *canonical link* for the distribution, which exhibits particularly favorable statistical characteristics when used in fitting generalized linear models [55, p. 32]. In the case of the gamma distribution, the canonical link function is the reciprocal, but the log link is much more widely used in practice. For a more complete discussion of the range of link functions available with the glm option, refer to the help documentation for the family argument, and for a discussion of some of the practical differences between these link function choices, refer to the book by McCullagh and Nelder [55].

As a partial illustration of the flexibility of the glm function and the impact of different choices of the link function, consider the problem of predicting the pregnant variable—the number of times each female adult has been pregnant— from the other variables in the Pima Indians diabetes dataset. Since pregnant is a count variable, the Poisson distribution is appropriate. The default model generated by the glm function is:

```
PimaPregnant1 <- glm(pregnant ~ ., data = PimaTrain, family = poisson)
summary(PimaPregnant1)
```

```
##
## Call:
## glm(formula = pregnant ~ ., family = poisson, data = PimaTrain)
##
## Deviance Residuals:
##     Min       1Q    Median       3Q      Max
## -4.2979  -1.0746  -0.2004   0.8266   3.4197
##
## Coefficients:
##                Estimate Std. Error z value Pr(>|z|)
## (Intercept) -0.1400070  0.1819611  -0.769 0.441636
## glucose      0.0005765  0.0009070   0.636 0.525019
## pressure    -0.0012160  0.0016680  -0.729 0.466002
## mass         0.0097653  0.0040065   2.437 0.014794 *
## pedigree    -0.0839805  0.0879954  -0.954 0.339895
## age          0.0317986  0.0021072  15.091  < 2e-16 ***
## diabetes     0.2229941  0.0642419   3.471 0.000518 ***
## ---
## Signif. codes:  0 '***' 0.001 '**' 0.01 '*' 0.05 '.' 0.1 ' ' 1
##
## (Dispersion parameter for poisson family taken to be 1)
##
##     Null deviance: 1186.30  on 377  degrees of freedom
## Residual deviance:  892.69  on 371  degrees of freedom
## AIC: 1901.5
##
## Number of Fisher Scoring iterations: 5
```

The two most significant predictors in this model are `age` and `diabetes`, the binary diabetes indicator, and `mass` is marginally significant in this model, but none of the other coefficients are.

As noted above, the default link function is $g(z) = \ln z$, but other link options that can be specified explicitly are the identity link $g(z) = z$ and the square root link $g(z) = \sqrt{z}$. Specifying the identity link yields the following, slightly different model:

```
PimaPregnant2 <- glm(pregnant ~ ., data = PimaTrain,
                                  family = poisson(link = "identity"))
summary(PimaPregnant2)

##
## Call:
## glm(formula = pregnant ~ ., family = poisson(link = "identity"),
##     data = PimaTrain)
##
## Deviance Residuals:
##     Min       1Q    Median       3Q      Max
## -4.4811  -1.1035  -0.0437   0.7972   3.2336
##
## Coefficients:
##                Estimate Std. Error z value Pr(>|z|)
## (Intercept) -2.487488   0.532204  -4.674 2.95e-06 ***
## glucose      0.002452   0.003291   0.745   0.4561
## pressure    -0.010740   0.005548  -1.936   0.0529 .
## mass         0.016743   0.009865   1.697   0.0897 .
```

```
## pedigree    -0.503436   0.230455  -2.185   0.0289 *
## age           0.191246   0.010998  17.390  < 2e-16 ***
## diabetes      0.499405   0.249230   2.004   0.0451 *
## ---
## Signif. codes:  0 '***' 0.001 '**' 0.01 '*' 0.05 '.' 0.1 ' ' 1
##
## (Dispersion parameter for poisson family taken to be 1)
##
##     Null deviance: 1186.30  on 377  degrees of freedom
## Residual deviance:  803.25  on 371  degrees of freedom
## AIC: 1812
##
## Number of Fisher Scoring iterations: 12
```

In particular, note that the only highly significant prediction variable in this model is `age`, with `diabetes` now marginally significant, along with `pedigree`, but `mass` is no longer significant at the 5% level. This second example also illustrates that the `family` specification actually calls a function from the `stats` package (typing `help(family)` brings up a detailed help page), and this function accepts an optional argument `link`. As noted above, valid link options for the `poisson` family are "log" (the default), "identity," used here, and "sqrt," used in the next example:

```
PimaPregnant3 <- glm(pregnant ~ ., data = PimaTrain,
                        family = poisson(link = "sqrt"))
summary(PimaPregnant3)

##
## Call:
## glm(formula = pregnant ~ ., family = poisson(link = "sqrt"),
##     data = PimaTrain)
##
## Deviance Residuals:
##     Min      1Q   Median      3Q     Max
## -4.4550  -1.1229  -0.0902  0.7759  3.2243
##
## Coefficients:
##               Estimate Std. Error z value Pr(>|z|)
## (Intercept)  0.3894606  0.1636580   2.380  0.01733 *
## glucose      0.0007333  0.0009158   0.801  0.42330
## pressure    -0.0022055  0.0016239  -1.358  0.17441
## mass         0.0069350  0.0036839   1.883  0.05977 .
## pedigree    -0.1088842  0.0821209  -1.326  0.18487
## age          0.0404414  0.0024507  16.502  < 2e-16 ***
## diabetes     0.1851645  0.0659919   2.806  0.00502 **
## ---
## Signif. codes:  0 '***' 0.001 '**' 0.01 '*' 0.05 '.' 0.1 ' ' 1
##
## (Dispersion parameter for poisson family taken to be 1)
##
##     Null deviance: 1186.30  on 377  degrees of freedom
## Residual deviance:  847.57  on 371  degrees of freedom
## AIC: 1856.4
##
## Number of Fisher Scoring iterations: 5
```

Comparing these three models, we see that the variable `age` is the most significant predictor in all cases, with `diabetes` always significant at the 5% level, but substantially more significant than this under both the log and square root links. The significance of the variables `pedigree` and `mass` varies strongly with the choice of link function, with both exhibiting 5% significance for the identity link, only `mass` exhibiting this level of significance for the log link, and neither variable exhibiting this significance for the square root link. A detailed discussion of considerations in selecting a link function is beyond the scope of this book, but the two key points of this example are, first, that different choices can lead to somewhat different results, and, second, that adopting the default option—while not necessarily the best choice in all cases—requires the least explanation. This is particularly true here, where the default option is the canonical link that does exhibit desirable statistical properties and is probably the most popular in practice.

To conclude this discussion, it is worth noting that R has a number of other model-fitting functions that support a wide range of other extensions of linear regression models to different distributions of response variables. That is, it was noted in Chapter 5 that the linear regression model is probably best suited to response variables with an approximately Gaussian distribution, while the discussion presented here has focused on generalized linear models that are appropriate to response variables with binomial, Poisson, or gamma distributions. The `betareg` package provides analogous support for modeling response variables that are reasonably characterized by the beta distribution introduced in Chapter 9, such as fractions or probabilities; the package vignette by Grün *et al.* provides an excellent and thorough introduction [35]. A much more extensive range of models, based on *generalized additive models*—generalized linear models that include nonparametric smoothing terms to account for nonlinear covariate effects—is provided by the `gamlss` suite of R packages, which may be viewed as sets of two, three, or four coupled generalized additive models that allow for covariate-dependent dispersion, skewness, or other shape-parameter dependences [64]. It should be emphasized that all of these modeling tools—starting with the `glm` function, moving on to the additional flexibility inherent in the `betareg` package, and from there to the extreme flexibility of the `gamlss` suite of packages—have increasingly steep learning curves as you move from simpler to more complex model types. For this reason, it is critically important to examine the results obtained from as many practial perspectives as possible before basing important decisions or future results on the predictions or interpretations of these models.

10.3   Decision tree models

Decision tree models or *recursive partitioning models* generate predictions of the target variable by first partitioning the dataset into subsets based on the values of the prediction variables, and then assigning a prediction to each of these subsets. The key characteristic of these models is that every record in the dataset is mapped into one and only one subset, which guarantees a unique prediction. The main advantage of these models is that, at least in simple cases, they are easily interpreted and explained to others. The primary disadvantage of these models is their extreme sensitivity to changes in the data: as the example presented in Sec. 10.3.2 demonstrates, fitting "the same" decision tree model to "similar" datasets can yield very different results. This sensitivity is one of the primary motivations for the development of currently popular machine learning models like random forests and boosted trees, both extensions of the decision tree models considered here that give more stable predictions (e.g., lower sensitivity to small changes in the data), but at the expense of much greater model complexity.

10.3.1 Structure and fitting of decision trees

Decision tree models can be developed to predict a variety of different response variable types, including numerical, binary, categorical, and count data. In all cases, the basic idea is as described above: given a training dataset $\mathcal{D}$ with N records, a decision rule is developed that assigns each record in the dataset to one of m *terminal nodes* or *leaves*, and each of these terminal nodes generates a predicted value of the response variable. As noted in the brief introduction to binary classifiers given in Sec. 10.2, these classifiers can generate either binary responses or the probability of a positive response, and this is true of decision tree classifiers as well. The classification trees considered here are of the second type, generating a predicted probability that the response is positive. Because they are slightly easier to explain, the following discussion begins with the case of simple regression tree models, but a classification tree example is presented in Sec. 10.3.2.

The basic approach generally used in building decision tree models is based on *binary splits* and *recursive partitioning*. The first step in this model-building approach is to split the original dataset $\mathcal{D}$ into two parts (i.e., construct a *binary partition*) such that every record k in $\mathcal{D}$ belongs to one subset (i.e., $k \in A$) or the other (i.e., $k \in B$). Given this partition, we then assign a constant prediction to each of these subsets:

$$\hat{y}_k = \begin{cases} a & \text{if } k \in A, \\ b & \text{if } k \in B. \end{cases} \tag{10.13}$$

These predicted values a and b are chosen to minimize the sum of squared prediction errors for the complete dataset, which is given by:

$$J = \sum_{k \in A} (y_k - a)^2 + \sum_{k \in B} (y_k - b)^2. \tag{10.14}$$

Note that *for a fixed partition*, this problem is easy to solve. First, note that the value chosen for a influences only the contribution of the first sum in Eq. (10.14), while the value chosen for b influences only the contribution of the second sum. This means we can choose a to minimize the first sum, without regard to the second, and we can choose b to minimize the second sum, without regard to the first. In fact, these best-fit constants are simply the means of the subsets A and B, respectively. Thus, once we have a binary partition, this first step in the regression tree problem is extremely easy to solve. The more challenging part is constructing this binary partition, a task illustrated in the simple example discussed below.

Before considering how this initial partition is obtained, note that decision trees typically do not stop after this first step. Instead, they apply *recursive partitioning*, obtaining better predictions by further splitting each of the subsets A and B into two additional subsets, $A1$ and $A2$ for the A subset and $B1$ and $B2$ for the B subset. As before, given these additional binary splits, we assign constant predictions, $a1$, $a2$, $b1$, and $b2$, respectively, to these four new subsets. Again, these predictions are easily computed, as they are simply the average of the response variable in each subset. In principle, we could continue this process until we ran out of data, but in practice, two things are commonly done to terminate this recursive partitioning process. The first is to impose a minimum size restriction on the final subsets created, corresponding to the terminal nodes or leaves of the tree. This constraint guarantees that our model-building procedure terminates before we run out of data, resulting in "shorter" trees and guaranteeing that each prediction is based on at least a specified minimum number of data samples. The second thing commonly done to restrict tree size is *pruning*, re-merging some of the deepest binary splits in the tree using a complexity-penalized measure, analogous to the adjusted R-squared measure for linear regression models. For detailed discussions of these ideas, see the discussions of pruning given by Venables and Ripley [70, pp. 257–258] or Hastie, Tibshirani, and Friedman [38, pp. 307–308].

As noted, the more difficult aspect of building tree-based models is the task of forming the binary partitions, both initially and at each subsequent stage of the recursive partitioning process. The fundamental idea is to use the values of a selected covariate to construct these binary partitions, an idea illustrated in the following sequence of decision tree models, each developed to predict the heating gas variable, `Gas`, from the `whiteside` data frame in the `MASS` package. To make this example as concrete as possible, explicit results are presented obtained with `rpart`, one of several packages available in R to fit decision trees. This package provides the most widely available support for decision tree models in R since it is a recommended package, available in almost all R installations. Other packages that build simple decision tree models include `party` and `partykit`, providing two different implementations of the `ctree` procedure based on conditional inference, and the `C5.0.default` procedure in the `C50` package, which implements the C5.0 algorithm, a decision tree algorithm popular in the machine learning community. In the interest of space, all of the decision tree examples considered here were fit using the `rpart` procedure from the `rpart` package.

The simplest possible regression tree model is one like that summarized in the left-hand plot in Fig. 10.3, where the binary partition is determined by the two values of the binary variable `Insul` in the `whiteside` dataset. This model

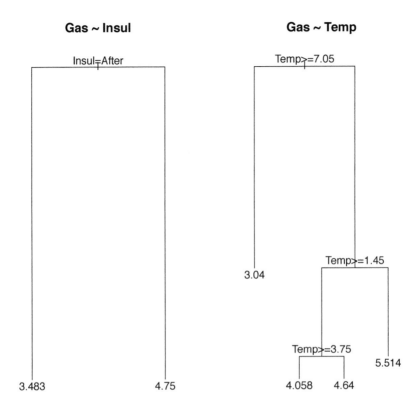

Figure 10.3: Two `rpart` decision tree models for the `whiteside` data.

was generated and plotted using the following R code:

```
rpartWhiteA <- rpart(Gas ~ Insul, data = whiteside)
plot(rpartWhiteA)
text(rpartWhiteA, xpd = TRUE, minlength = 5)
```

(Here, `xpd = TRUE` allows the text labels to extend slightly outside the plot area, avoiding truncation, and `minlength = 5` allows the complete `Insul` value "After" to be displayed.) Recall that the `Insul` variable assumes the two values "Before" and "After," and partitioning the original `whiteside` dataset into subsets based on this variable yields a binary partition with the following optimal predictions for each subset:

```
mean(whiteside$Gas[which(whiteside$Insul == "Before")])
```

```
## [1] 4.75
```

```
mean(whiteside$Gas[which(whiteside$Insul == "After")])
```

```
## [1] 3.483333
```

In the graphical representation of this regression tree model shown in the left-hand plot in Fig. 10.3, note that the top of the tree—the *root node*—is labelled as "Insul=After," indicating that the left branch of the tree should be taken for this case. This means that the corresponding record is assigned to the left-hand terminal node of the tree, with the `Gas` prediction given by 3.483, corresponding to the average for the "After" records as shown above. Conversely, if this condition is not met—i.e., if `Insul` has the value "Before"—the record is assigned to the right-hand terminal node, with the `Gas` prediction 4.75, again consistent with the mean values listed above. Note that in this case, since covariate `Insul` only assumes two possible values, recursive splitting of these partitions is not possible, so the decision tree building process terminates here.

For a numeric predictor like `Temp`, binary partitions are constructed differently. There, a threshold value is tentatively selected and all records with covariate values less than this threshold are put in one partition, while all other records are put in the other. The value of the performance measure J defined in Eq. (10.14) depends on this threshold value, so the binary partition is constructed by choosing the threshold value that minimizes J. This idea is illustrated in the right-hand plot in Fig. 10.3, showing a regression tree model that predicts `Gas` from `Temp`. The first stage in the model-building process finds that the best binary partition of the complete dataset is based on the `Temp` threshold 7.05, so the root node of this decision tree is labelled as "Temp $>=$ 7.05" in Fig. 10.3. If we meet this condition, we take the left branch of the decision tree, leading to a terminal node with the predicted `Gas` value of 3.04. This node was not split because the default minimum node size used by `rpart` to terminate the tree-building process is 20, while there are only 15 records with `Temp` values greater than or equal to 7.05. These details are not obvious from Fig. 10.3, but they are easily obtained from the `rpart` model using either the `summary` method or the `print` method, which gives a more succinct description:

```
rpartWhiteB <- rpart(Gas ~ Temp, data = whiteside)
print(rpartWhiteB)
```

```
## n= 56
##
## node), split, n, deviance, yval
##       * denotes terminal node
##
##  1) root 56 75.014290 4.071429
##    2) Temp>=7.05 15 10.476000 3.040000 *
##    3) Temp< 7.05 41 42.742440 4.448780
##      6) Temp>=1.45 34 23.310590 4.229412
```

```
##          12) Temp>=3.75 24 10.998330 4.058333 *
##          13) Temp< 3.75 10  9.924000 4.640000 *
##        7) Temp< 1.45 7  9.848571 5.514286 *
```

Each node in the tree is numbered in this summary, with the root node numbered
1, and each node is summarized by the criterion used to define the binary split,
the number of records contained in the resulting subset, the fitting measure J
(the *deviance*), and the prediction associated with that node (i.e., the mean of
the Gas values in the subset). Terminal nodes are marked with asterisks in this
display, and we see that node 2, defined by Temp $>= 7.05$, is a terminal node
with only 15 observations, with a predicted Gas value of 3.04, consistent with
the model structure shown in Fig. 10.3. For Temp values less than this threshold,
we take the right branch, which we can see from the above summary contains
41 records (node 3). This node is large enough to split, so a second binary
partitioning is performed, yielding the optimum threshold value 1.45. Records
with Temp values greater than equal to this threshold value take the left branch,
leading to node 6, representing a subset with 34 records, while records with Temp
values less than 1.45 take the right branch, leading to node 7, with 7 records.
Because this right branch is too small to split further, it becomes a terminal
node, generating a predicted Gas value of 5.514. The 34-record left branch is
large enough to split further, yielding an optimum threshold value of 3.75 and
two terminal nodes (12 and 13). With 24 records, node 12 is large enough to
split further, but this split does not improve performance enough to justify the
increased tree complexity, so it is eliminated by the pruning process.

These examples have illustrated how the decision tree building process works
for a single prediction variable, either binary or numeric, but, in general, decision trees are built based on multiple prediction variables. The example shown
in Fig. 10.4 helps to illustrate how decision trees are built in these cases. Specifically, this model predicts Gas from both Temp and Insul, obtained with the
following rpart call:

```
rpartWhiteC <- rpart(Gas ~ Temp + Insul, data = whiteside)
```

Because we have two prediction variables to consider, each stage of the recursive
partitioning process constructs the best binary split from each variable, compares their performance, and selects the best one to partition the data. Here, the
binary split of the complete dataset on Insul gives a smaller J value than the
best split on the Temp variable used in the right-hand model shown in Fig. 10.3,
so the split based on the Insul variable is chosen for the top-level data partition. In contrast to the decision tree model based on Insul alone, we can
further improve this model's prediction accuracy by splitting both of the resulting data subsets on the Temp variable, as we can see from the model structure
shown in Fig. 10.4. For the Insul = ''Before'' subset (the right branch),
this subset is only split once, yielding two terminal nodes, with predicted Gas
values that depend on both Insul and Temp. The Insul = ''After'' subset
(the left branch) is split further, ultimately yielding three terminal nodes, each
with their own Gas prediction.

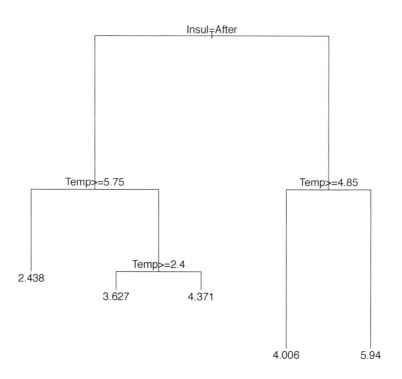

Figure 10.4: Two variable `rpart` decision tree model for the `whiteside` data.

The point of this discussion has been to give a general description of how decision tree models are built, illustrating this process with three simple regression examples based on the R procedure `rpart`. The following two examples are included to illustrate three points. First, decision tree models can be used to implement binary classifiers, representing a potentially useful alternative to logistic regression models. Second, decision tree models don't always perform as well as "classical" predictive models like those based on linear or logistic regression, but, third, decision tree models can sometimes perform substantially better than these other models. These last two observations illustrate a more general point that will be considered further in later sections of this chapter: the relative performance of different types of predictive models is strongly problem-dependent.

10.3.2 A classification tree example

Given a binary response variable coded as 0 or 1, the `rpart` decision tree building procedure constructs a binary classifier that predicts the probability of positive response. Thus, in favorable cases, decision tree classifiers may represent a reasonable alternative to logistic regression models since they are usually simpler to explain to those with limited technical backgrounds. The criteria and specific procedures for fitting classification tree models are somewhat different than those for regression tree models, but these details will not be considered here, since the general structure of these models is the same in both cases. For details on building classification tree models, refer to the discussions by Venables and Ripley [70, pp. 253–256] or Hastie, Tibshirani, and Friedman [38, pp. 308–310].

To provide a specific example to compare with the logistic regression results presented in Sec. 10.2.2, consider the problem of building the classification tree model shown in Fig. 10.5 that predicts the probability of a positive diabetes diagnosis (i.e., `diabetes = 1`) from the `glucose` and `mass` variables in the Pima Indians diabetes dataset. Since the logistic regression model based on these two predictors performed almost as well as the full logistic regression model built from all of the available predictors, it is interesting to compare the results obtained using these two predictors in a decision tree model. The R code needed to fit this classification tree model and the results look like this:

```
rpartModel <- rpart(diabetes ~ glucose + mass, data = PimaTrain)
print(rpartModel)

## n= 378
##
## node), split, n, deviance, yval
##       * denotes terminal node
##
##  1) root 378 84.6560800 0.338624300
##    2) glucose< 139.5 274 44.5547400 0.204379600
##      4) glucose< 103.5 120  6.5916670 0.058333330
##        8) mass< 38.85 103  0.9902913 0.009708738 *
##        9) mass>=38.85 17  3.8823530 0.352941200 *
##      5) glucose>=103.5 154 33.4090900 0.318181800
##       10) mass< 27.35 40  3.6000000 0.100000000 *
##       11) mass>=27.35 114 27.2368400 0.394736800
##         22) mass< 43 107 25.0467300 0.373831800
##           44) mass>=34.65 40  7.5000000 0.250000000 *
##           45) mass< 34.65 67 16.5671600 0.447761200 *
##         23) mass>=43 7  1.4285710 0.714285700 *
##    3) glucose>=139.5 104 22.1538500 0.692307700
##      6) glucose< 158.5 49 12.2449000 0.510204100
##       12) mass< 34.05 24  5.3333330 0.333333300 *
##       13) mass>=34.05 25  5.4400000 0.680000000 *
##      7) glucose>=158.5 55  6.8363640 0.854545500 *
```

A graphical representation of this model is presented in Fig. 10.5, which shows that the top-level splits in this classification tree are all based on the `glucose` variable, while all of the lower-level splits are based on the `mass` variable. One of the advantages of this model over the logistic regression models considered

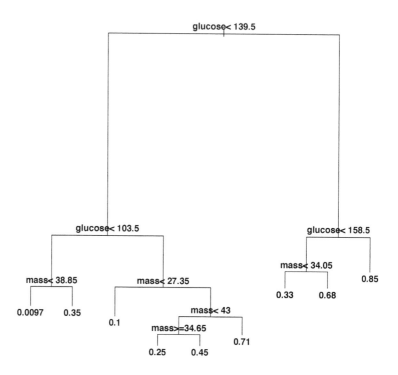

Figure 10.5: Binary classification tree for the Pima Indians diabetes dataset.

earlier is that this model yields simple rule-based statements. For example, following the left branches of this decision tree yields a characterization of the set of patients with the lowest probability of a diabetic diagnosis (0.97%): glucose < 103.5 and mass < 38.5. Similarly, following the right branches of this decision tree yields a characterization of the set of patients with the highest probability of a diabetic diagnosis (85%): glucose ≥ 158.5 and mass ≥ 34.05.

The interpretability of classification tree models is a significant advantage over logistic regression models in terms of explaining what they mean to others with limited technical backgrounds, but how accurate are the predictions they generate? Because they generate a probability of positive response, we can apply the AUC and root-Brier scores described in Sec. 10.2.3 to the validation set predictions and compare these measures with those obtained for the logistic regression model logisticRef built using the same covariates. As with the linear and logistic regression models, and many other predictive models available in R, predictions from **rpart** models can be generated using the generic **predict** function. For example, the validation set predictions for the classification tree model shown in Fig. 10.5 are generated as:

```
rpartHatV <- predict(rpartModel, newdata = PimaValid)
```

Given these predictions, we can apply the **AUC** function from the **MLmetrics** package to compute the area under the ROC curve, for both the **logisticRef** logistic regression model and the **rpart** classification tree model:

```
AUC(pHatRefV, PimaValid$diabetes)
## [1] 0.7914714
AUC(rpartHatV, PimaValid$diabetes)
## [1] 0.7522286
```

Since larger AUC values indicate better performance, we see that we are paying a non-negligible performance price for the greater interpretability of the classification tree model in this example. Conversely, as it was argued at the beginning of the discussion of the root-Brier score presented in Sec. 10.2.3 that the AUC is blind to certain types of performance limitations, it is a good idea to also compare the root-Brier scores for these models. Recall that these scores are easily computed via the **RMSE** function from the **MLmetrics** package, applied to the predicted positive response probabilities and the observed responses. These results for the logistic regression and classification tree models are:

```
RMSE(pHatRefV, PimaValid$diabetes)
## [1] 0.4188516
RMSE(rpartHatV, PimaValid$diabetes)
## [1] 0.4370589
```

Since smaller values of the root-Brier score indicate better performance, these results also suggest that we are paying a performance price for the greater interpretability of the classification tree model in this example. That said, it is important to emphasize that general performance conclusions cannot be drawn from this example: there are cases where, as here, tree-based models perform worse than classical models, but there are other cases, like the one considered next, where the opposite is true.

10.3.3 A regression tree example

A simple regression tree example based on the **whiteside** data frame from the **MASS** package was used in Sec. 10.3.1 to illustrate the structure and construction

of these models, but the following discussion introduces a more realistic example
that will form the basis for a number of performance comparisons for more
complex model structures. Specifically, this example first constructs a linear
regression model and then a regression tree model using `rpart`, comparing the
performance of these models using the validation R-squared measure introduced
in Chapter 5 and laying the foundation for performance comparisons with the
other model structures introduced in later sections of this chapter.

The basic predictive modeling problem considered here is that of predicting
the compressive strength of laboratory concrete samples on the basis of sample
characteristics, including the sample age in days and seven composition variables that describe the amounts of different components included in the concrete mix. The dataset on which this example is based is derived from the
`ConcreteCompressiveStrength` data frame from the `randomUniformForest`
package. The following code loads this package and shows its contents and
structure:

```
library(randomUniformForest)
str(ConcreteCompressiveStrength, vec.len = 3)

## 'data.frame': 1030 obs. of  9 variables:
## $ Cement                        : num  540 540 332 332 ...
## $ Blast Furnace Slag            : num  0 0 142 142 ...
## $ Fly Ash                       : num  0 0 0 0 0 0 0 0 ...
## $ Water                         : num  162 162 228 228 192 228 228 228 ...
## $ Superplasticizer              : num  2.5 2.5 0 0 0 0 0 0 ...
## $ Coarse Aggregate              : num  1040 1055 932 932 ...
## $ Fine Aggregate                : num  676 676 594 594 ...
## $ Age                           : int  28 28 270 365 360 90 365 28 ...
## $ Concrete compressive strength : num  80 61.9 40.3 41 ...
```

Note that many of the variable names in this data frame contain spaces and
several are inconveniently long, as is the name of the data frame itself. The
following R code renames the data frame and the variables to make this dataset
easier to work with:

```
concreteData <- ConcreteCompressiveStrength
colnames(concreteData) <- c("Cement", "BFslag",
         "FlyAsh", "Water", "SuperP",
         "Coarse", "Fine", "Age", "Strength")
```

The use of random data partitioning into at least training and validation subsets
was advocated in Chapter 5 to avoid overfitting in linear regression models,
and this approach is good practice for fitting all of the other model structures
considered here. The following code generates a random 50/50 split into training
and validation subsets:

```
concreteFlag <- TVHsplit(concreteData, split = c(0.5, 0.5),
                                       labels = c("T", "V"))
concreteTrain <- concreteData[which(concreteFlag == "T"), ]
concreteValid <- concreteData[which(concreteFlag == "V"), ]
```

In the following example, and all of those considered later in this chapter using this dataset, models are fit to the training subset and evaluated with respect to the validation subset.

To serve as a reference for evaluating the performance of the regression tree model considered here, we first fit a linear regression model that predicts the response variable, **Strength**, from all of the other variables in the dataset:

```
lmConcrete <- lm(Strength ~ ., data = concreteTrain)
summary(lmConcrete)

##
## Call:
## lm(formula = Strength ~ ., data = concreteTrain)
##
## Residuals:
##      Min       1Q   Median       3Q      Max
## -27.673   -5.925    0.221    6.309   34.719
##
## Coefficients:
##                Estimate Std. Error t value Pr(>|t|)
## (Intercept)  -1.525056  36.210632  -0.042    0.966
## Cement        0.118228   0.011233  10.525  < 2e-16 ***
## BFslag        0.097822   0.013922   7.027 6.52e-12 ***
## FlyAsh        0.086548   0.016619   5.208 2.74e-07 ***
## Water        -0.212892   0.053550  -3.976 7.99e-05 ***
## SuperP        0.119586   0.121881   0.981    0.327
## Coarse        0.007213   0.012844   0.562    0.575
## Fine          0.022235   0.014766   1.506    0.133
## Age           0.120035   0.007554  15.889  < 2e-16 ***
## ---
## Signif. codes:  0 '***' 0.001 '**' 0.01 '*' 0.05 '.' 0.1 ' ' 1
##
## Residual standard error: 10.03 on 531 degrees of freedom
## Multiple R-squared:  0.6366, Adjusted R-squared:  0.6312
## F-statistic: 116.3 on 8 and 531 DF,  p-value: < 2.2e-16
```

Based on the coefficients and their p-values for this model, it appears that the variables **Cement**, **BFslag**, **FlyAsh**, and **Age** all make significant, positive contributions (i.e., increasing the values of these variables is associated with increased compressive strength of the sample), while **water** exhibits a significant negative contribution (i.e., less water is associated with stronger samples).

As in the **whiteside** data example discussed in Sec. 10.3.1, fitting a regression tree model to the training set using the **rpart** procedure is simple:

```
rpartConcrete <- rpart(Strength ~ ., data = concreteTrain)
```

The structure of this model is shown in Fig. 10.6. While this model is more complex than those considered earlier, it is still relatively easy to identify the values of these variables that are associated with the weakest and strongest concrete samples. In particular, since the **Strength** predictions for this model appear as the numbers below each terminal node, it appears that, *roughly*, the concrete samples on the left side of the decision tree are weaker than those on

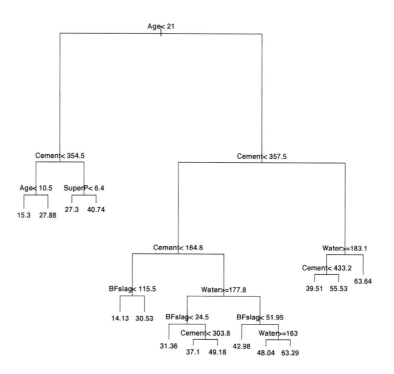

Figure 10.6: Regression tree model for the concrete compressive strength data.

the right, with those in the middle exhibiting intermediate strength. Further, the first binary split that largely defines these two sides of the tree is on Age $<$ 21, suggesting that younger samples are weaker and older samples tend to be stronger, although not uniformly so. More specifically, the left-most leaf on this tree is that for Age $<$ 10.5 and Cement $<$ 354.5, which exhibits the second-smallest predicted strength (15.3), while the right-most leaf is that for Age $\geq$ 21, Cement $\geq$ 357.5, and Water $<$ 183.1, which exhibits the highest strength prediction (63.54). Note that these results are consistent with the linear model coefficients for Age, Cement, and Water, although the regression tree model implies interactions between these variables that are not present in the linear regression model. A more detailed view of how the predictions generated by both of these models depend on these and other variables can be obtained using the *partial dependence plots* described in Sec. 10.6.1.

To compare the performance of the linear regression model and the regression tree model for this case, compute the validation R-squared measure introduced in Chapter 5 using the function **ValidationRsquared** described there:

```
lmConcreteHatV <- predict(lmConcrete, newdata = concreteValid)
ValidationRsquared(concreteValid$Strength, lmConcreteHatV)

## [1] 0.5785755

rpartConcreteHatV <- predict(rpartConcrete, newdata = concreteValid)
ValidationRsquared(concreteValid$Strength, rpartConcreteHatV)

## [1] 0.6910238
```

Recall that this measure is computed from the validation set model predictions and observed response values and it varies between 0, corresponding to terrible model performance, and 1, corresponding to a perfect fit to the validation dataset. Here, the fact that the validation set R-squared value is larger for the regression tree model than for the linear regression model implies that it is generating somewhat more accurate predictions. A more complete view of the character of the predictions generated by both models is presented in Fig. 10.7, showing the predicted versus observed response plots described in Chapter 5 for the linear regression model on the left, and for the regression tree model on the right. The detail that emerges most prominently from these plots is the discontinuous character of the regression tree predictions, reflecting the fact that each record in the dataset is assigned to one of a discrete set of terminal nodes and all records assigned to a given node are given the same predicted response.

10.4 Combining trees with regression

The regression tree models just described assign each record in the dataset to a terminal node, and a single predicted response is generated for all records assigned to that node. Classification tree models behave similarly: each record is assigned to a unique terminal node and a binary response or the probability of a positive outcome is generated for all records assigned to that node. The models considered here are more general: as in the tree models just described, each record is again assigned to a unique terminal node, but rather than a constant prediction, each terminal node generates a prediction based on either a linear regression model or a generalized linear model. These are *model-based recursive partitioning (MOB) models*, and they can be fit to a dataset using the **partykit** package in *R*, which also supports model prediction and a number of other useful characterization tools (e.g., model structure plots).

One motivation for these models comes from the observation that in many cases, the underlying dataset is heterogeneous so that relatively simple models may fit subsets of the data quite well, but these local models are quite different in structure or details for the different data subsets. While it might be possible to obtain a more complex model that adequately fits the entire dataset, this

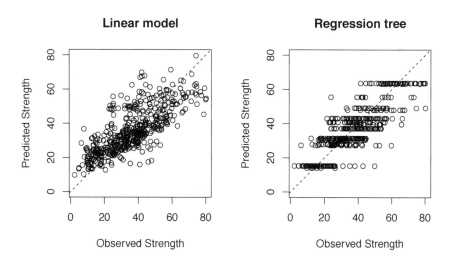

Figure 10.7: Predicted vs. observed plots for the linear regression model (left) and the regression tree model (right).

model is likely to be a lot more complex, involving interactions between the variables used to define the subsets and the covariates used to predict the target variable within each subset. As a specific example, consider the prediction of the heating gas consumption `Gas` in the `whiteside` data frame from the `MASS` package, based on the outside temperature `Temp`. Both simple linear regression models describing this relationship and more complex linear regression models involving the binary `Insul` variable were discussed in detail in Chapter 5, and it was noted that the most complex of these models, involving an interaction between `Temp` and `Insul`, could be rewritten as a pair of models, one for the subset defined by `Insul` = ``Before`` and another for the subset defined by `Insul` = ``After``. In this case, fitting an MOB model gives us this result immediately.

Specifically, the `lmtree` function from the **partykit** package fits MOB models consisting of a decision tree, with terminal nodes containing linear regression models that generate the model responses. If we suspect that a linear relationship between `Temp` and `Gas` should provide a reasonable approximation to the data, but that the details of this relationship may depend on the binary variable `Insul`, we can generate the following `lmtree` model:

```
lmTreeWhite <- lmtree(Gas ~ Temp | Insul, data = whiteside)
```

Here, the formula `Gas ~ Temp | Insul` indicates that we want to construct a decision tree that assigns every record in the dataset to a terminal node based on the `Insul` variable, and that once we reach that terminal node, we should predict `Gas` from `Temp` based on a linear regression model, based on all of the observations assigned to that node. Fig. 10.8 shows the structure of this model: the root node partitions the complete dataset into two subsets, one corresponding to the `Insul = ''Before''` subset (Node 2, with 26 observations), and the other corresponding to `Insul = ''After''` (Node 3, with 30 observations). Scatterplots of `Gas` versus `Temp` for the observations in each node are included as node summaries, overlaid with the corresponding linear regression lines. A succinct summary of this model structure can be obtained with the generic function **print**, as was the case for **rpart** models; specifically:

```
print(lmTreeWhite)

## Linear model tree
##
## Model formula:
## Gas ~ Temp | Insul
##
## Fitted party:
## [1] root
## |   [2] Insul in Before: n = 26
## |       (Intercept)        Temp
## |         6.8538277   -0.3932388
## |   [3] Insul in After: n = 30
## |       (Intercept)        Temp
## |         4.723850    -0.277935
##
## Number of inner nodes:    1
## Number of terminal nodes: 2
## Number of parameters per node: 2
## Objective function (residual sum of squares): 5.425247
```

A more complete characterization of this model can be obtained with the generic **summary** function, which gives the complete summary results for each component linear regression model, including model parameters, their standard errors and associated *p*-values, and the adjusted R-squared goodness-of-fit measures discussed in Chapter 5.

Without going into the mathematical details, the basic approach used by the `lmtree` function to fit MOB models like this one consists of the following steps, described in the **partykit** package vignette, "Parties, Models, Mobsters:

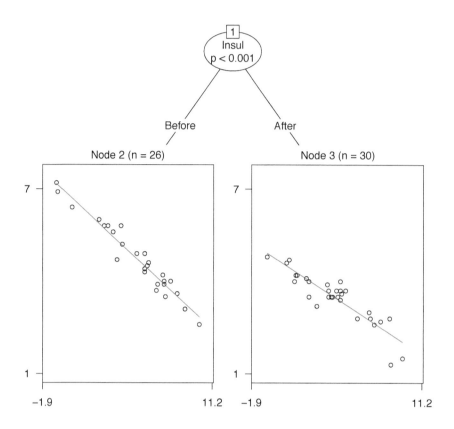

Figure 10.8: Structure of the `lmtree` model for the `whiteside` dataset.

A New Implementation of Model-Based Recursive Partitioning in R," by Zeileis and Hothorn, available in the help files for the package. Starting from the root node, containing all observations in the dataset, these steps are:

1. Fit the model once to all observations in the current node;

2. Assess whether the parameter estimates are stable with respect to every partitioning variable $z_1, \ldots, z_m$. If there is some overall instability, select the variable z_j associated with the highest parameter instability; otherwise, stop;

3. Compute the split points that locally optimize the objective function J;

4. Split the node into child nodes and repeat the procedure until some stopping criterion is met.

The **partykit** package provides code to implement this strategy, both for linear regression models via the **lmtree** function, and for the generalized linear models discussed in Sec. 10.2.4 via the **glmtree** function. For a detailed discussion of these procedures, with examples, refer to the package vignette cited above; for an even more detailed discussion of fitting MOB models, refer to the 2008 journal paper by the same authors [84].

Another example of a linear regression MOB model was that for the Boston housing dataset discussed in Chapter 6, similar to the **whiteside** example just described in that it led to the construction of readily interpretable local linear regression models for suburban and urban subsets of the overall dataset, highlighting the different relationships between the data variables inherent in these distinct data subsets.

The following example is a little more complex, motivated by linear regression and **rpart** regression tree models for the same data and achieving somewhat better performance than either one. Subsequent comparisons highlight key differences between the MOB models and these other models, which may be viewed as components of the MOB model. Specifically, this example uses the **lmtree** procedure to fit an MOB model to the concrete compressive strength data considered in Sec. 10.3.3. Since the top split in the **rpart** model was based on **Age**, the only non-composition variable, the MOB model considered here uses **Age** as the conditioning variable z to construct the tree in the model, and retains all of the composition variables as covariates in the linear regression models at the terminal nodes. The following R code fits this model and provides a detailed summary of the results:

```
lmTreeConcrete <- lmtree(Strength ~ . - Age | Age, data = concreteTrain)
summary(lmTreeConcrete)

## $`2`
##
## Call:
## lm(formula = Strength ~ . - Age)
##
## Residuals:
##     Min      1Q  Median      3Q     Max
## -15.579  -3.389   0.139   2.534  17.882
##
## Coefficients:
##               Estimate Std. Error t value Pr(>|t|)
## (Intercept) 107.590637  44.531379   2.416 0.017079 *
## Cement        0.072874   0.014076   5.177 8.35e-07 ***
## BFslag        0.024000   0.016560   1.449 0.149664
## FlyAsh        0.007426   0.019076   0.389 0.697707
## Water        -0.279498   0.071244  -3.923 0.000141 ***
## SuperP       -0.016829   0.179371  -0.094 0.925396
## Coarse       -0.038007   0.015825  -2.402 0.017734 *
## Fine         -0.027979   0.018933  -1.478 0.141883
## ---
## Signif. codes:  0 '***' 0.001 '**' 0.01 '*' 0.05 '.' 0.1 ' ' 1
##
## Residual standard error: 5.859 on 130 degrees of freedom
```

```
## Multiple R-squared:  0.7871,Adjusted R-squared:  0.7756
## F-statistic: 68.66 on 7 and 130 DF,  p-value: < 2.2e-16
##
##
## $`4`
##
## Call:
## lm(formula = Strength ~ . - Age)
##
## Residuals:
##     Min      1Q  Median      3Q     Max
## -18.5911  -3.8316  -0.5797   3.5134  28.1382
##
## Coefficients:
##             Estimate Std. Error t value Pr(>|t|)
## (Intercept) -93.03634   34.54313  -2.693  0.00753 **
## Cement        0.17312    0.01115  15.530  < 2e-16 ***
## BFslag        0.15518    0.01348  11.514  < 2e-16 ***
## FlyAsh        0.11652    0.01605   7.262 4.43e-12 ***
## Water        -0.10100    0.04734  -2.134  0.03381 *
## SuperP        0.04883    0.11492   0.425  0.67124
## Coarse        0.03368    0.01233   2.732  0.00673 **
## Fine          0.06330    0.01442   4.389 1.66e-05 ***
## ---
## Signif. codes:  0 '***' 0.001 '**' 0.01 '*' 0.05 '.' 0.1 ' ' 1
##
## Residual standard error: 6.606 on 260 degrees of freedom
## Multiple R-squared:  0.7957,Adjusted R-squared:  0.7902
## F-statistic: 144.7 on 7 and 260 DF,  p-value: < 2.2e-16
##
##
## $`5`
##
## Call:
## lm(formula = Strength ~ . - Age)
##
## Residuals:
##     Min      1Q  Median      3Q     Max
## -16.6897  -4.2789  -0.4505   3.9169  17.0642
##
## Coefficients:
##             Estimate Std. Error t value Pr(>|t|)
## (Intercept) 39.62983   57.24371   0.692 0.490023
## Cement       0.08170    0.01774   4.605 9.94e-06 ***
## BFslag       0.08754    0.02223   3.938 0.000135 ***
## FlyAsh       0.05464    0.02629   2.078 0.039718 *
## Water       -0.15112    0.09098  -1.661 0.099181 .
## SuperP       0.34659    0.20357   1.703 0.091125 .
## Coarse       0.01055    0.01970   0.535 0.593398
## Fine        -0.01074    0.02256  -0.476 0.634829
## ---
## Signif. codes:  0 '***' 0.001 '**' 0.01 '*' 0.05 '.' 0.1 ' ' 1
##
## Residual standard error: 7.065 on 126 degrees of freedom
## Multiple R-squared:  0.7174,Adjusted R-squared:  0.7017
## F-statistic: 45.7 on 7 and 126 DF,  p-value: < 2.2e-16
```

The main disadvantage of this summary is that it does not describe the `Age` splits used to define the three terminal nodes of the tree; this information is available from the generic `print` function, and the key details may be summarized as:

- Node 2: Age $\leq$ 7, subset with 138 records;

- Node 4: $7 <$ Age $\leq$ 28, subset with 268 records;

- Node 5: Age $>$ 28, subset with 134 records.

Recall from the discussion of the linear regression model that the four composition variables that were highly significant were `Cement`, `BFslag`, `FlyAsh`, and `Water`, with the first three positive and the last one negative. Here, the most significant coefficients and their magnitudes vary substantially across the three `Age` subsets generated by this model, although their signs are all the same as those in the linear regression model. Thus, for the "young concrete samples," with `Age` less than seven days, the most significant variable is `Cement`, although with a positive coefficient somewhat less than that for the linear regression model (0.1182283), and the only other highly significant coefficient is that for `Water`, which is somewhat more negative than that for the linear regression model (-0.2128923). For the intermediate age concrete samples (Node 4, between 7 and 28 days), the `Cement` coefficient becomes both much larger and much more significant, the `Water` term becomes enough smaller in magnitude to be only marginally significant, while all of the other composition variables except `SuperP` become highly significant, all with positive coefficients. Finally, the "mature concrete samples" (Node 5, older than 28 days) exhibit significant coefficients only for `Cement` and `BFslag`, both with positive coefficients roughly comparable to those in the linear model (0.1182283 for `Cement` and 0.0978217 for `BFslag`). The key point of this example is that it gives us potentially useful insight into how the influence of these different composition variables varies with the age of the concrete sample.

The validity of these insights rests on how well the MOB model predicts the concrete compressive strength relative to the linear regression model. To address this question, compare their validation R-squared measures (see Chapter 5):

```
ValidationRsquared(concreteValid$Strength, lmConcreteHatV)

## [1] 0.5785755

lmTreeConcreteHatV <- predict(lmTreeConcrete, newdata = concreteValid)
ValidationRsquared(concreteValid$Strength, lmTreeConcreteHatV)

## [1] 0.8160398
```

These results suggest the MOB model gives *much* better predictions of compressive strength. In fact, as indicated at the beginning of this discussion, the MOB model provides more accurate validation set predictions than the `rpart` model, which itself outperformed the linear regression model, with a validation R-squared value of 0.6910238. A more complete view of the performance differences between the linear regression model and the MOB model may be seen

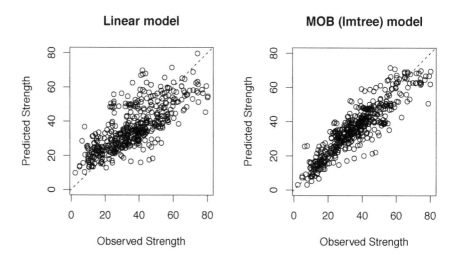

Figure 10.9: Predicted vs. observed plots for the linear regression model (left) and the `lmtree` model (right).

in Fig. 10.9, which gives validation set predicted versus observed plots for these two models. The tighter clustering of the points in the right-hand plot around the equality reference line reflects the much larger validation R-squared value for the MOB model. Also, note the absence of the horizontal bands seen in the predicted versus observed plot in Fig. 10.7 for the `rpart` model, illustrating the smoother prediction behavior of MOB models relative to regression tree models.

10.5   Introduction to machine learning models

The term *machine learning models* refers to a broad class of prediction models developed by the machine learning community, including neural networks, support vector machines, random forests, and boosted trees [38]. These models have become extremely popular because they sometimes give dramatically better predictions than models like those considered in Chapter 5 or in earlier sections of this chapter. Probably the most significant price paid for this improved performance is their much greater complexity, making their interpretation and explanation to others more difficult. Also, as will be demonstrated in Sec. 10.5.3, these methods often have important tuning parameters that can strongly influence performance.

Because machine learning is an extremely active research area, no extensive survey of these models can be given here. Instead, the following discussion focuses on two important tree-based machine learning model classes—random forests and boosted trees—both because they represent a logical extension of the simpler tree-based models discussed in Sec. 10.3, and because they are well known, widely used members of the machine learning model class. For a more thorough introduction to machine learning models, refer to the book by Hastie, Tibshirani, and Friedman [38].

10.5.1 The instability of simple tree-based models

One of the main motivations for both of the tree-based machine learning model classes considered here is the inherent instability of simple decision trees. This instability is illustrated in Fig. 10.10, which shows two classification trees, both built to predict the probability of a positive diagnosis in the Pima Indians diabetes dataset from the variables glucose, age, and mass. The left-hand model was built from the training subset PimaTrain described in Sec. 10.2.2, while the right-hand model was built from the validation subset PimaValid. In both models, the first binary split occurs on the glucose variable, and the threshold for both splits is approximately the same, but the second-level splits are based on different variables: in the left-hand model, both are based on glucose, while in the right-hand model, one split is based on age and the other is based on mass. As we move further down these two trees, their differences become more pronounced and they even have different numbers of terminal nodes (11 in the left-hand model versus 12 in the right-hand model).

Since the motivation for the random training/validation data partitioning is to create two datasets that are similar but not identical, the substantial differences in the classification tree models built from these two datasets are troubling. In particular, although it has been emphasized that one of the advantages of decision tree models is their ease of interpretation, the instability of these models demonstrated in Fig. 10.10 raises the question of how valid these interpretations are. That is, since both of these models were built from "comparable representative subsets" of the same data source, which—if either—of the different interpretations following from these models is appropriate?

One of the primary sources of this instability is the nature of the recursive partitioning process itself: once we have made a split, this choice determines the results obtained in all successive splits. Thus, even if we could obtain a better final model by modifying earlier decisions, recursive partitioning algorithms cannot do this: once a decision is made, we are stuck with it. This inher-

Training Data   **Validation Data**

Figure 10.10: Two `rpart` models for the Pima Indians diabetes data, one fit to the training dataset (left) and the other fit to the validation data (right).

ent instability has led to a number of machine learning strategies that attempt to build an *ensemble of trees* and combine their predictions to obtain a more stable final result. Two of these approaches are *random forests*, introduced in Sec. 10.5.2, and *boosted tree models*, introduced in Sec. 10.5.3.

10.5.2   Random forest models

The basic idea behind random forests is, rather than fitting a single classification or regression tree to our dataset, we fit a large number of trees, each generated from a different random subset of the data. The predictions from this "forest" are then combined, by averaging the individual predictions in the case of a regression model or a binary classifier that predicts positive response probabilities, or by a majority voting scheme for binary classifiers that predict one of the two possible responses. More specifically, the random forest algorithm generates a collection of B regression or classification trees using the following steps, repeated once for each of the B trees generated [38, p. 588]:

1. Draw a random *bootstrap* sample from the dataset of size n_{train} by sampling with replacement;

2. For each bootstrap sample, fit a regression or classification tree model, using the following steps to build an un-pruned tree:

   a. Randomly select m of the p available prediction variables;

   b. Select the best binary split based on these m variables, as in the decision tree algorithm described in Sec. 10.3.1;

   c. Use this split to partition the node into two daughter nodes;

   d. Repeat steps (a) through (c) for each daughter node, continuing until the minimum node size limit n_{min} is reached.

The resulting B un-pruned trees are used to generate predictions, which are then combined as described above.

More detailed descriptions of the random forest algorithm are given in the book by Hastie, Tibshirani, and Friedman [38, Ch. 15], in the paper by Liaw and Wiener [52], or the original paper by Breiman that proposed random forests [9]. Two key points to note here are, first, that the random forest algorithm makes use of two forms of random selection—bootstrap sampling of data subsets, and random selection of covariate subsets—to make the individual trees in the forest as independent as possible, and, second, the fact that this algorithm involves several tuning parameters (e.g., B, n_{train}, m, and n_{min}). Fortunately, random forest performance does not appear to depend strongly on these parameters, as Hastie, Tibshirani, and Friedman note [38, p. 590]:

> In our experience, random forests do remarkably well, with very little tuning required.

The example presented below seems to support this position, as it gives an excellent model using the default parameter settings.

Several random forest implementations are available in R, but the best known is probably the `randomForest` package described by Liaw and Wiener [52]. Here, this package is used to fit a random forest model to predict `Strength` from the other covariates in the `concreteTrain` dataset used in the previous regression examples. The following R code loads the `randomForest` library and calls the `randomForest` function to fit this model:

```
library(randomForest)
rfConcreteModel <- randomForest(Strength ~ ., data = concreteTrain,
                                           importance = TRUE)
```

These default settings are $B = 500$, specified by the optional argument `ntree`, an n_{train} value equal to 63.2% of the number of records in the training dataset `concreteTrain`, m is one-third of the number p of columns in this dataset, rounded down to its integer part, and n_{min} is 5. Like the other model types considered here, a method is available for the generic `predict` function for random forest models. Applying the validation R-squared function described in Chapter 5 to these predictions and comparing the results with the linear regression model, we see that this random forest model gives much better predictions:

```
ValidationRsquared(concreteValid$Strength, lmConcreteHatV)

## [1] 0.5785755

rfConcreteHatV <- predict(rfConcreteModel, newdata = concreteValid)
ValidationRsquared(concreteValid$Strength, rfConcreteHatV)

## [1] 0.8571199
```

In fact, this default random forest model gives the best prediction performance seen so far for this example. The substantially better quality of these predictions relative to the linear regression reference model is highlighted in the predicted versus observed plots for these two models shown in Fig. 10.11.

10.5.3 Boosted tree models

Like random forests, boosted tree models are based on the idea of combining predictions from many individual decision tree models, each obtained from modified versions of the original data. Many versions of this idea exist and new ones continue to evolve. Detailed descriptions are necessarily quite mathematical, as in the very thorough paper by Friedman [27], which introduced a number of important ideas, including the partial dependence plots and variable importance measures discussed briefly in Sec. 10.6. Probably the simplest of the many boosted tree implementations is the AdaBoost procedure introduced by Freund and Schapire in 1997 [25] for classification problems and used by Hastie, Tibshirani, and Friedman to introduce these ideas [38, p. 337]. Like the random forest approach described in Sec. 10.5.2, this approach constructs a collection of B classification trees, but unlike the random forest algorithm, here these trees are constructed *sequentially*, each one based on results generated by the preceeding one. The essential idea is to give increased weight at each iteration to observations that were not well predicted in the previous iteration, refitting the same data at each stage but with these modified weights, resulting in a different classification tree. In the end, these results are combined to obtain a final classifier that performs much better than any of the individual classifiers generated at each iteration. Indeed, the performance of the AdaBoost algorithm was sufficiently impressive when it was first announced that Breiman declared it "the best off-the-shelf classifier in the world" [38, p. 340].

A variety of boosted tree implementations are available in R, including the popular and flexible **gbm** package considered here. One of the optional arguments

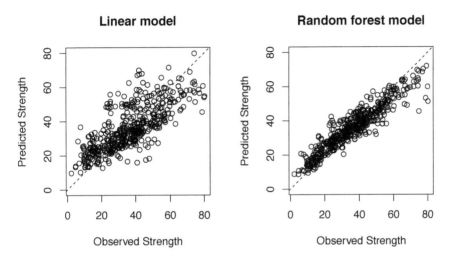

Figure 10.11: Predicted vs. observed plots for the linear regression model (left) and the `randomForest` model (right).

for the gbm function in this package is `distribution`, which can be specified as "gaussian" to optimize the squared error loss function for numerical response variables, "laplace" to optimize the sum of absolute errors, "t-dist" to optimize a loss function derived from the t-distribution discussed in Chapter 9, and several others, including "bernoulli" for binary classification problems, "multinomial" for multi-class classification problems, and "poisson" for count data. As the following example illustrates, this function can be used to build models with excellent prediction performance, although not necessarily using the default options.

The following R code fits the default `gbm` model to the concrete compressive strength dataset and compares its validation R-squared performance with the linear regression model:

```
library(gbm)
gbmConcreteModel1 <- gbm(Strength ~ ., data = concreteTrain)

## Distribution not specified, assuming gaussian ...

gbmConcreteHatV1 <- predict(gbmConcreteModel1, newdata = concreteValid,
                                                n.trees = 100)
ValidationRsquared(concreteValid$Strength, gbmConcreteHatV1)

## [1] 0.02493458

ValidationRsquared(concreteValid$Strength, lmConcreteHatV)

## [1] 0.5785755
```

The terrible prediction performance of this model is further highlighted by the left-hand plot in Fig. 10.12, which shows that the predictions are almost constant, corresponding to a validation R-squared value of zero.

The difficulty is that the default tuning parameters for the gbm method are extremely poorly suited to this particular problem. For one thing, the default interaction.depth value is 1, corresponding to a *boosted stumps model*, which does not allow for any interactions between variables. Increasing this argument to 8 gives slightly better performance, but still dramatically worse than the linear regression model for this problem. Significant trial-and-error tuning ultimately gave the following model, with even better validation set performance than the random forest model described in Sec. 10.5.2:

```
gbmConcreteModel2 <- gbm(Strength ~ ., data = concreteTrain,
                                       distribution = "gaussian",
                                       interaction.depth = 30,
                                       shrinkage = 0.05,
                                       n.trees = 1000,
                                       n.minobsinnode = 20)
```

Besides increasing the interaction.depth argument from its default value of 1 to 30, allowing much more complex trees to be built at each step, three other optional arguments have been modified. First, increasing n.minobsinnode from its default value of 5 to 20, the default value for the rpart decision trees discussed earlier, limits the complexity of the individual trees in the gbm model. To understand the impact of the other two arguments, note that the boosting algorithm may be viewed as an iterative optimization procedure, where each iteration makes a small change to the previous tree. The size of this change is controlled by the *learning rate*, corresponding to the shrinkage argument. The default shrinkage value is 0.001 but the simulation-based example described in the gbm help file uses the much larger value 0.05 adopted here, noting that "0.001 to 0.1 usually work." That example also sets the number of iterations, n.trees, to 1000, motivating this choice here. For the concrete compressive strength model, the default values n.trees = 100 and shrinkage = 0.001 do not allow this optimization procedure to find a reasonable model.

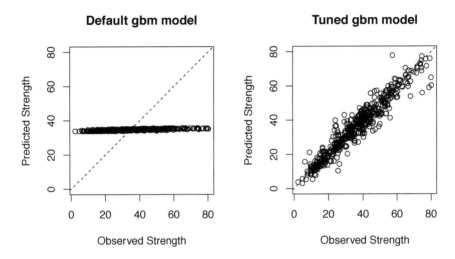

Figure 10.12: Predicted vs. observed plots for the default gbm model (left) and the tuned gbm model (right).

The right-hand plot in Fig. 10.12 shows the predicted versus observed response plot for this model, demonstrating excellent performance. In fact, the validation R-squared value shows that this model predicts even better than the random forest model described in Sec. 10.5.2:

```
gbmConcreteHatV2 <- predict(gbmConcreteModel2, newdata = concreteValid,
                                            n.trees = 1000)
ValidationRsquared(concreteValid$Strength, gbmConcreteHatV2)

## [1] 0.9192591
```

This example illustrates several important points. First, comparing the performance of the linear regression model, the rpart regression tree model, the lmtree MOB model, the random forest model, and the tuned gbm model for the concrete compressive strength prediction problem supports the argument made earlier that machine learning models can significantly outperform simpler, more traditional predictive models. Second, the performance difference between the two `gbm` models considered here provides a dramatic illustration of how sensitive model performance can be to method tuning parameters for machine learning models. This second point is extremely important because these models typically have several different parameters that can be adjusted, and the process of tuning these parameters can be reasonably time-consuming. Finally, the third point that this example illustrates only indirectly is the difficulty of model interpretation: whereas we can examine the model parameters and their statistics for linear regression models or `lmtree` models, and we can construct simple graphical representations of `rpart` decision tree models, neither random forest models nor boosted tree models lend themselves to this type of interpretation. This difficulty of interpretation motivates auxilliary tools like the partial dependence plots discussed in Sec. 10.6.1, and the variable importance measures discussed in Sec. 10.6.2.

10.6 Three practical details

The random forest and boosted tree examples presented in Secs. 10.5.2 and 10.5.3 have shown that these machine learning models can provide much more accurate predictions than simpler, better-known models like those based on linear regression or decision trees. It has been noted, however, that this improved performance comes at a price: while it is easy to interpret and explain decision tree models to those with limited technical backgrounds, and regression models are well understood by those with some classical statistics background, machine learning models like random forests or boosted trees cannot be described in terms of either simple diagrams or a small set of model parameters. As a consequence, interpretation of these model poses a practical complication that may significantly limit their use, despite their potential for much greater prediction accuracy. Partial solutions to this problem are provided by the partial dependence plots discussed in Sec. 10.6.1 and the variable importance measures discussed in Sec. 10.6.2, both proposed by Friedman as ways of interpreting boosted tree models [27].

Another important issue in building and using predictive models is the impact of categorical predictors with thin levels. If we adopt the data partitioning strategy advocated in Chapter 5 and used in most of the examples considered in this chapter, it is possible that extremely thin levels of a categorical predictor will not appear in all data partitions. If a level is absent from the training set, this level will not be incorporated in the model (e.g., no parameter will be estimated for this level in a regression model), which can cause the `predict` function to fail when we apply it to a validation or holdout dataset that has this level present. This problem and some possible solutions are discussed in Sec. 10.6.3.

10.6.1 Partial dependence plots

Partial dependence plots were described by Friedman as a useful tool to help understand the behavior of boosted tree models [27], and they provide a way to visualize the way the predictions of any model, no matter how complex, vary with the individual variables on which they are based. The essential idea is, for each prediction variable x_j, to define a grid of values, $\{x^*_{jm}\}$ for $m = 1, 2, \ldots, M$, and look at how the model's predictions vary with x_j over this range, while averaging out the dependence on all other variables. That is, note that the predicted reasponse y_k for record k in the dataset may be written as:

$$y_k = f(x_{k1}, \ldots, x_{kp}), \tag{10.15}$$

for some mathematical function $f(\cdots)$, which may be arbitrarily complicated and difficult to describe, but is defined by the end result of the model fitting procedure. The partial dependence plot is constructed by computing the predictions at each grid point, replacing all x_{jk} values with x^*_{jm} and averaging the predictions over all of the other variables:

$$\tilde{y}_{jm} = \frac{1}{N} \sum_{k=1}^{N} f(x_{k1}, \ldots, x^*_{jm}, \ldots, x_{kp}). \tag{10.16}$$

The m averaged predictions $\{\tilde{y}_{jm}\}$ are then plotted against the grid values $\{x^*_{jm}\}$ to provide a graphical assessment of how, on average, the predictions $\{\hat{y}_k\}$ depend on the variable x_j.

For linear regression models, partial dependence plots provide a graphical representation of the model parameters. Specifically, note that the predictions of a simple linear regression model are given by:

$$f(x_{k1}, \ldots, x_{kp}) = a_0 + \sum_{i=1}^{p} a_i x_{ki}. \tag{10.17}$$

Substituting this result into Eq. (10.16) gives:

$$\tilde{y}_{jm} = \frac{1}{N} \sum_{k=1}^{N} \left(a_0 + \sum_{i \neq j} a_i x_{ik} + a_j x^*_{jm} \right)$$

$$= a_0 + \sum_{i \neq j} a_i \bar{x}_i + a_j x^*_{jm}, \tag{10.18}$$

where $\bar{x}_i$ is the average value of the covariate x_i over all records in the dataset:

$$\bar{x}_i = \frac{1}{N} \sum_{k=1}^{N} x_{ik}. \tag{10.19}$$

This result means that the partial dependence plot for the variable x_j in a linear regression model is a straight line, whose slope is equal to the model parameter

```
## plotmo grid: Cement BFslag FlyAsh Water SuperP Coarse  Fine Age
##               263.45   16.1      0   185    6.4    969 781.1  28
```

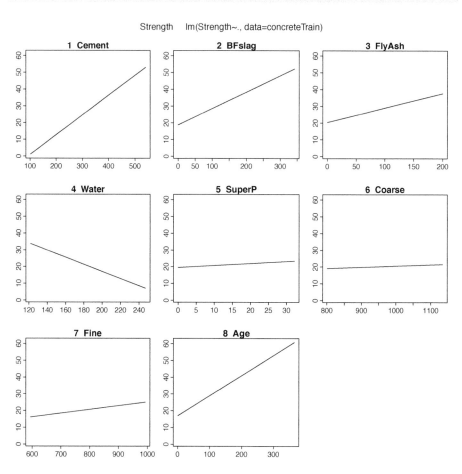

Figure 10.13: Partial dependence plot for the linear regression tree model.

a_j. Fig. 10.13 shows an approximate partial dependence plot generated by the R function plotmo described below, for the linear regression model lmConcrete that predicts concrete compressive strength from the covariates listed in this plot array. These plots are shown on common axes and are all straight lines, with positive slopes except for Water. Note the large slope magnitudes for the highly significant predictors Cement, BFslag, FlyAsh, Water, and Age.

The partial dependence plots just described can be computationally expensive: the averages in Eq. (10.16) require N predictions for each grid value x^*_{jm}, and for all partial dependence plots, we must do this for all p variables. Thus, if we use M grid points for each covariate, we need NMp predictions, which may

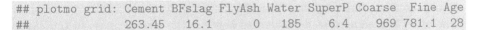

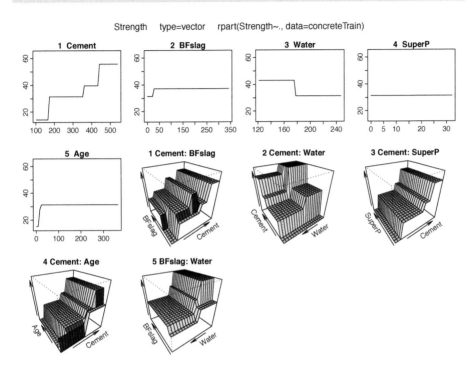

Figure 10.14: Partial dependence plot for the **rpart** regression tree model.

require significant computation time for complex models. A simpler alternative is to select a "representative value" for each covariate and use this single value instead of averaging over all values. A logical choice here is the median, $x_i^\dagger$, and this alternative is the default method for the **plotmo** function in the R package **plotmo**. The partial dependence plot in Fig. 10.13 was constructed as:

```
plotmo(lmConcrete)
```

Fig. 10.14 shows the **plotmo** results for the **rpartConcrete** regression tree model. The first five plots show the partial dependences for the individual prediction variables, exactly as in Fig. 10.13. Note that the **Cement** dependence is an increasing staircase, very roughly similar to its linear model dependence. The other two significant dependences here are the step decrease in the `Water` influence at approximately 180, and the increasing dependence on `Age` in the fifth plot that saturates at about 30 days. This plot suggests that concrete gets stronger as it matures, until it is "fully set" after about a month, with `Age` having no further influence. This interpretation, consistent with practical expectations, cannot be captured by any linear model.

The last five plots in Fig. 10.14 are *two-way interaction plots*, obtained by averaging the model predictions over all but *two* selected variables, x_j and x_ℓ, and generating a three-dimensional surface plot showing how these averaged predictions vary with these two selected covariates. As explained in Section 6 of the `plotmo` vignette "Plotting regression surfaces with plotmo" included in the `plotmo` help files, both the selection of the individual variables shown in the first set of plots and the two-way interaction plots depend on the type of the model being characterized. In the case of `rpart` models, these two-way interaction plots characterize those variables that appear in parent-child pairs in the decision tree. Here, these plots characterize the interactions between `Cement` and the four variables `BFslag`, `Water`, `SuperP`, and `Age`, and that between `BFslag` and `Water`. In the `Cement:Water` plot, we see a general decrease in strength with increasing water content that depends strongly on the `Cement` value.

Fig. 10.15 shows the default `plotmo` results for the random forest model `rfConcreteModel`. For random forests, `plotmo` shows `degree1` or main-effect plots for the ten most important variables (see Sec. 10.6.2), which includes all variables here. The dominant variations are those for `Cement`, `Water`, and `Age`. Note that the `Cement` plot actually looks quite linear, similar to the corresponding plot in Fig. 10.13 for the linear regression model. The `Water` plot looks like a smoothed version of the staircase dependence for the `rpart` model in Fig. 10.14, decreasing almost monotonically until it becomes constant at ~ 180. The `Age` dependence is also quite similar to the `rpart` plot, although with a somewhat later saturation effect (i.e., ~ 90 days versus ~ 30). Six interaction plots are shown, including one for that between `Cement` and `Water`, which looks qualitatively similar to that for the `rpart` model in Fig. 10.14, but smoother.

Fig. 10.16 shows the `plotmo` results for the default `gbm` model described in Sec. 10.5.3. For `gbm` models, single-variable (i.e., `degree1`) plots are shown for up to ten variables with relative influence greater than 1% (see Sec. 10.6.2), and two-way interaction (i.e., `degree2`) plots for the four variables with the largest relative influence. Here, only the variables `Cement`, `Water`, and `Age` meet this importance criteria and are shown, along with their two-way interaction plots. The three main effect plots are qualitatively consistent with the corresponding plots for the `rpart` regression tree model and the random forest model, but the range of these plots is much narrower than that in Figs. 10.14 for the `rpart` model or 10.15 for the random forest model, reflecting the extremely narrow range seen for these model predictions in the left side of Fig. 10.12.

Finally, Fig. 10.17 shows the corresponding `plotmo` results for the tuned `gbm` model described in Sec. 10.5.3, which gives the best validation set predictions of any of the concrete compressive strength models considered in this chapter.

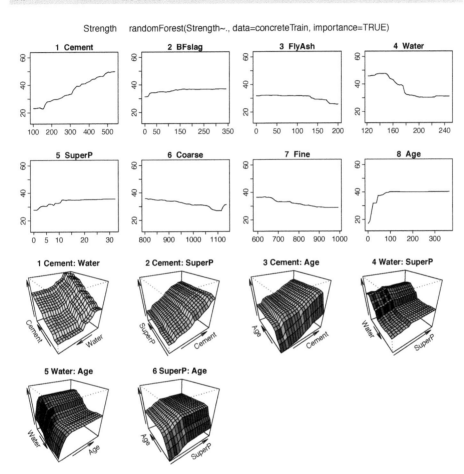

Figure 10.15: Partial dependence plot for the random forest model.

Here, all eight prediction variables meet the 1% relative importance threshold, so main effect plots are shown for all of these variables. Comparing these plots with Fig. 10.15 for the random forest model, which predicts almost as well as this gbm model, we see that the dependences of the two models on Cement, Water, and Age are qualitatively quite similar, if somewhat different in detail. Two-way interaction plots are presented between the four variables with the highest relative importance, which are the three just described, plus BFslag, corresponding to four of the five highly significant variables in the linear regression model. Again, note how similar the interaction plot between Cement and Water is to that for the random forest model in Fig. 10.15.

```
## plotmo grid:  Cement BFslag FlyAsh Water SuperP Coarse  Fine Age
##                263.45   16.1      0   185    6.4    969 781.1  28
```

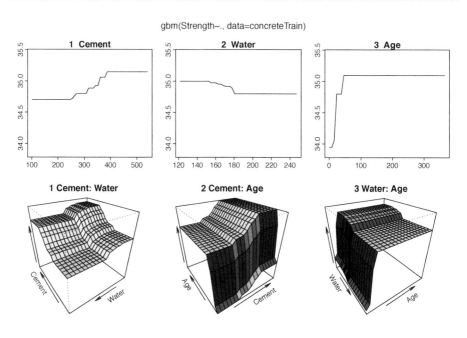

Figure 10.16: Partial dependence plot for the default `gbm` model.

The point of these examples has been to illustrate, first, that partial dependence plots may be viewed as a graphical representation of the parameters in a linear regression model, and, second, that they represent a useful extension of this idea for more complex models where the variable dependences are neither linear nor characterized by a small set of model parameters. Comparing these plots, we can begin to see why the `rpart` regression tree model, the random forest model, and the well-tuned `gbm` model give more accurate predictions of compressive strength than the linear regression model does. Specifically, these models all suggest a `Cement` dependence that is roughly linear and increasing, a `Water` dependence that is generally decreasing, but with a nonlinear saturation

```
## plotmo grid: Cement BFslag FlyAsh Water SuperP Coarse  Fine Age
##                263.45   16.1      0   185    6.4    969 781.1  28
```

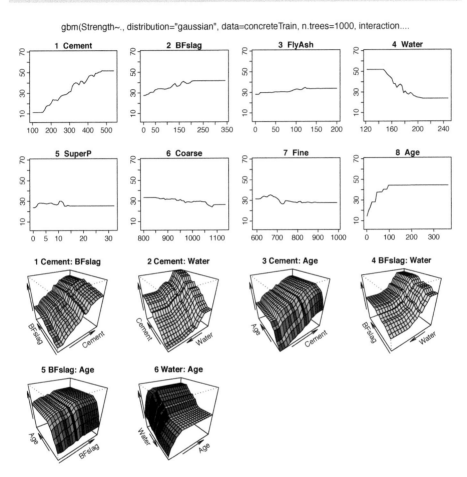

gbm(Strength~., distribution="gaussian", data=concreteTrain, n.trees=1000, interaction....

Figure 10.17: Partial dependence plot for the tuned gbm model.

effect that causes it to have no further effect beyond a value of approximately 180, and a dependence on Age that increases up to a hard saturation limit somewhere less than 100 days. In addition, all of these nonlinear models exhibit significant interactions between these variables that can only be approximated in a linear regression model if they are included explicitly: in these other models, these interactions are automatically incorporated.

10.6.2   Variable importance measures

Variable importance measures are numerical scores designed to help us answer the question of how much different covariates contribute to our predicted response. This question can be posed in several different ways, and this discussion briefly considers the following two, focusing primarily on the first:

1. How much do the covariates in a fixed model contribute to its predictions?

2. Which predictors are most important to include in building a model?

The key difference in these questions is on their focus: the first considers only those predictors that have been included in a model that has already been fit to a dataset, while the second considers the broader question of which predictors should be included in this model as we are building it. There does not appear to be general agreement on the broader question of how best to define the notion of variable importance [34], but the two versions of this question posed here have led to some potentially useful computational tools.

One potentially useful measure of variable importance is the fractional range of variation seen in the univariate partial dependence plots shown for various models in Sec. 10.6.1. Specifically, since each of these plots is shown on the same y-axis range, the relative range of variation seen in each plot gives us an idea of how much each covariate influences the predicted value of the response variable as it ranges from its minimum to its maximum value. For a linear regression model, this measure of variable importance is the magnitude of the model parameter scaled by the ratio of the range of the covariate to the range of the response variable:

$$\phi_j = \frac{\max\{\tilde{y}_{jm}\} - \min\{\tilde{y}_{jm}\}}{\max\{y_k\} - \min\{y_k\}} = |a_j| \frac{\max\{x_{jm}^*\} - \min\{x_{jm}^*\}}{\max\{y_k\} - \min\{y_k\}}. \qquad (10.20)$$

Since this measure is computed from the coefficients of a fixed model, it clearly falls into the first class of methods noted above. A potential limitation of this measure is that it takes no account of the variability of these model parameters (e.g., standard errors or p-values); for a discussion of some more complex measures that do take account of this variability, refer to the paper by Grömping [34, Sec. 3]. Another informal approach sometimes adopted in assessing variable importance in linear regression models is to rank them by statistical significance: the smaller the p-value, the greater the statistical significance of the variable.

While it would be possible to extend the ϕ_j measure described above to arbitrary model types, model-specific variable importance measures have been developed for tree-based models, and these are much more widely used. For a single regression tree, variable importance can be measured as [38, p. 368]:

1. For each predictor variable, consider all internal nodes that split on it;

2. For each of these nodes, compute the reduction in squared error between the split node and its defining data subset before the split;

3. Sum these reductions to obtain an overall measure of the variable's importance in improving the tree's prediction accuracy.

Note that under this scheme, covariates that were provided to the fitting procedure but not selected as splitting variables in the tree generated by this procedure will have zero importance. Also, note that this approach is applicable to classification tree models, although the improvement associated with splitting on each variable is measured differently (e.g., Gini measure or entropy instead of squared error). In either case, this improvement is described as an "increase in node purity" in some of the plots described below.

The **rpart** procedure in *R* returns an object of class "rpart," with the following components, illustrated by the regression tree model for the concrete compressive strength data considered in the earlier sections of this chapter:

```
names(rpartConcrete)
```
```
##  [1] "frame"     "where"     "call"
##  [4] "terms"     "cptable"   "method"
##  [7] "parms"     "control"   "functions"
## [10] "numresp"   "splits"    "variable.importance"
## [13] "y"         "ordered"
```

For this model, the variable importance measure described above has the following values for all of the covariates:

```
rpartConcrete$variable.importance
```
```
##   Cement      Age    Water    SuperP   BFslag   Coarse     Fine   FlyAsh
## 54800.93 41080.20 23296.48 19344.45 17487.01 15155.80 12747.37 10161.58
```

Although these covariates are listed in decreasing order of importance, the magnitudes of these values make them somewhat difficult to compare. These values are also returned as part of the (extremely verbose) output of the **summary** procedure for **rpart** models, but re-scaled to sum to 100, providing a clearer view of the relative importance of each covariate in contributing to the tree's prediction:

```
rawImp <- rpartConcrete$variable.importance
round(100 * rawImp/sum(rawImp))
```
```
## Cement    Age  Water SuperP BFslag Coarse   Fine FlyAsh
##     28     21     12     10      9      8      7      5
```

This approach to assessing variable importance extends to more complex tree-based models like random forests or boosted tree models, by simply averaging the variable importances computed for each tree over all trees in the model. In the case of boosted tree models generated by the **gbm** package in *R*, the **summary** function returns a summary data frame giving the relative importance of each covariate included in the **gbm** function call, normalized to sum to 100, and it generates a color-coded horizontal barplot summary of these relative influences. Two examples are shown in Fig. 9.30, in the second section of color figures at the end of Chapter 9. There, the plot on the left is the summary for the default **gbm** model built from the concrete compressive strength dataset described in Sec. 10.5.3, and the one on the right is for the optimized `gbm` model. Recall that the default model predicted very poorly, while the tuned model generated the best predictions seen for any model for this dataset: the differences in these variable importance plots show, first, that these are very different models, and, second, how the second model is able to perform so much better than the first. Specifically, note that the most important variable in the first model is `Age`, followed by `Cement`, while this ordering is reversed in the second model, where the relative importances of these variables are much more similar than in the first model. The influence of the `Water` variable is ranked third in both models, but the greatest difference lies in the influence of the other five variables, which is zero for the first model but together contribute approximately 25% of the total variable influence in the second model.

As noted, the approach of averaging the individual decision tree-based importances described above also applies to random forest models. These averaged values are returned by the function `importance`, applied to objects of class "randomForest," and the function `varImpPlot` generates a plot of these values. An example is shown in Fig. 10.18 for the random forest model of the concrete data described in Sec. 10.5.2. There, the right-hand plot shows the variable importances based on the average increase in node purity over all trees in the random forest, while the left-hand plot shows a second variable importance measure based on random permutations, described below. Note that, like the value of the `variable.importance` element of the `rpart` objects described above, these random forest importance values are not normalized, but normalized values can be obtained easily using the contents of the array returned by the `importance` function for random forest objects.

The relative variable importance values in the left-hand plot in Fig. 10.18 are based on a different approach, using random permutations [27] applied to the *out of bag (OOB) sample* created when each individual tree is fit during the process of building the random forest [38, Sec. 15.3]. Specifically, each individual tree in a random forest is fit to a random subsample of the training data, and those records in the training set that are *not* included in this random sample represent the out of bag sample. Since this OOB sample is not used in fitting the tree, it can be used in evaluating its performance. One application of this sample is to compute variable importance, in the following steps:

1. Compute the OOB sample prediction error;

2. For each covariate, x_j, apply a random permutation to all x_j values in the OOB sample and re-compute the prediction error;

3. Compute the degradation in prediction error caused by this random permuation, defined as the difference between the randomized prediction error and the original OOB prediction error;

4. Average this prediction error increase over all trees in the random forest.

The basic idea is that, if x_j contributes significantly to the model predictions, applying a random permutation should destroy its utility, causing the prediction

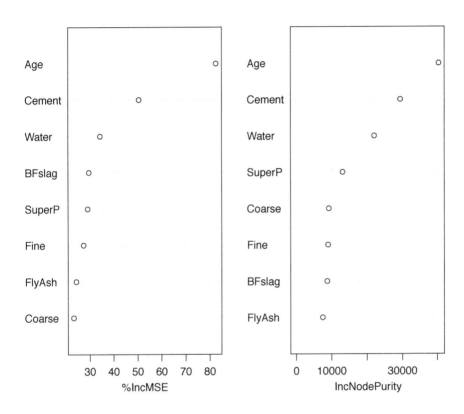

Figure 10.18: Two variable importance plots for the `rfConcreteModel` random forest model: permutation-based (left) and tree-average (right).

error computed in Step (2) to increase relative to the OOB prediction error computed in Step (1). Computing this difference and averaging it over all trees in the model thus gives a measure of each variable's importance in generating good predictions. The left-hand plot in Fig. 10.18 shows these permutation-based OOB importance measures for the random forest model `rfConcreteModel`. Note that these importance values are different from the values based on increased node purity shown in the right-hand plot, especially for the variables deemed less important under both measures. In particular, both measures rank `Age` as most important, followed by `Cement`, with `Water` third, similar to the other models considered here, but the ranks of the other five variables are substantially different in these two lists (e.g., `BFslag` is ranked fourth by the random permutation measure but seventh by the node purity measure, while `Coarse` is ranked fifth by the node purity measure but eighth by the random permutation measure). These results, together with the others presented above, emphasize that different variable importance measures, whether derived from different scores applied to the same model as here, or the same score applied to different models (e.g., the average node purity measure between random forest and boosted tree models), tend to give different results. One way to deal with this ambiguity is to look at those variables that are most important by several measures: e.g., note that Age, Cement, and Water are ranked as the top three covariates by both random forest measures, by the gbm models, and by the rpart model.

Finally, two other points are worth noting about the random permutation approach just described. The first is that, although boosted tree models do not have OOB samples associated with the individual trees, the gbm package does provide a random permutation variable importance measure, available by specifying the optional argument method = permutation.test.gbm in the summary function. The help file for the summary.gbm function notes that the idea is similar to that just described for random forests, but based on the entire training dataset instead of the (non-existent) OOB sample, and it also describes the method as "experimental."

The second point to note is that the random permutation approach can be used to obtain a variable importance measure of the second type described at the beginning of this section: a measure of the importance of a covariate in fitting the best model. As Hastie, Tibshirani, and Friedman note in connection with the random permutation OOB measure for random forests just described, this measure does not address this second question since, "if the model was refitted without the variable, other variables could be used as surrogates" [38, p. 593]. The following modification of this idea, however, does address this issue and thus represents a variable importance measure of the second type:

1. Fit a model to the training dataset and characterize its performance (e.g., validation R-squared, AUC, or root-Brier score, as appropriate);

2. For each variable x_j in the candidate covariate set, apply a random permutation to the training set observations, and re-fit the model, obtaining a new model that effectively omits x_j;

3. Characterize the performance of the model from Step (2) using the same approach (and the same validation dataset) as in Step (1);

4. Assess the importance of each variable in terms of how much the random permutation causes the overall model performance to degrade.

Note that while this approach has the advantage of being applicable to any model structure we might consider, its primary practical limitation is that it requires the model to be fit the first time and then re-fit once for each variable in the set of candidate covariates. Thus, for complicated models, this strategy may be computationally expensive.

10.6.3 Thin levels and data partitioning

The problem of thin levels in categorical data was discussed briefly in Chapter 5 in conjunction with building linear regression models with categorical predictors. One manifestation of this problem is high variability in the estimated parameters associated with these thin levels, leading to the possibility of model parameters with very large magnitudes, giving these thin levels unreasonable influence on predictions. Another, even more serious, difficulty arises when we apply the data partitioning strategy advocated in Chapter 5 to a dataset containing categorical predictors with thin levels. Specifically, if certain levels are thin enough, records listing these values are likely to be absent from the training, validation, or holdout partitions. If a level is absent from the training partition, it is not incorporated into the model (e.g., no parameter is estimated for this level in a linear regression model), which can cause the `predict` function to fail when we attempt to use it to generate predictions for the validation or holdout dataset.

As a simple example to illustrate this point, consider the distribution of levels for the `Cylinders` variable from the `Cars93` data frame from the `MASS` package:

```
table(Cars93$Cylinders)

##
##       3     4     5     6     8 rotary
##       3    49     2    31     7      1
```

Since there are only three 3-cylinder cars included in this dataset, two 5-cylinder cars, and one car with a rotary engine, it follows that there is a very high likelihood that records with `Cylinders` = 3 will be absent from one or more partitions in a T/V/H split, it is certain that records with `Cylinders` = 5 will be absent from at least one partition, and the lone rotary engine record is necessarily absent from two of these partitions.

Three possible solutions to this problem are:

1. After the T/V/H split, move any records with levels absent from the training subset into the training subset;

2. After the T/V/H split, delete any records with levels absent from the training subset from the validation and holdout subsets;

3. Re-group the categorical variable to eliminate thin levels, possibly combining thin levels into a single "other" category.

None of these solutions are ideal, but the first may be the most practical. The second approach is potentially the most problematic, since if the model generated from the training data is to be used as the basis for generating predictions on new datasets, not used in the model-building process (e.g., to predict behavior of future customers or patients, or to characterize new materials like concrete samples), we will be unable to generate predictions for cases where this thin level occurs. Finally, regrouping may be a good strategy, but this is generally an *ad*

hoc process that can be done in many different ways, and the practical consequences of different regrouping strategies are not obvious. As a consequence, if
a regrouping approach is adopted, the results should be examined carefully to
make sure that the results obtained under it are reasonable.

The problems of thin levels are most serious for categorical variables with
many different levels. For example, a potentially useful characterization of location within the U.S. is via postal zipcodes, but this represents a categorical
variable with approximately 40000 levels. Thus, even for large to very large
datasets, these many-level categorical variables may exhibit a large number of
thin levels, severely reducing the practical utility of any of the three strategies
listed above. A possible alternative in these cases is to replace the many-level
categorical variable with a numerical *score variable*, developed to reflect the
influence of these levels in a way that is easier to incorporate into predictive
models. In the case of a numerical response variable y, the simplest way of
doing this would be to define a score for each level of the categorical variable
as the average value of y over all records listing that level. That is, if c is an
L-level categorical variable, replace the categorical observations $\{c_k\}$ with score
values $\{s_k\}$ defined as:

$$c_k = \ell \implies s_k = s_\ell = \frac{1}{N_\ell} \sum_{j \in S_\ell} y_j, \qquad (10.21)$$

where ℓ indicates one of the L levels of the original categorical variable c, N_ℓ
is the number of records in the dataset for which $c_k = \ell$, and S_ℓ is the set of
records for which this is true. Alternatively, if there is some other, potentially
relevant numerical variable z for which a complete set of observations is available
(e.g., population, crime rate, average home price, or average annual income
by zipcode), these variables can be used as the basis for one or more score
variables to replace the original categorical variable c (i.e., use z_j instead of y_j
in Eq. (10.21)).

One potential difficulty with this approach is that, if a few very thin levels of c are associated with extreme values of the response variable y (or any
alternative variable z considered), the scores $\{s_k\}$ may give undue influence to
these very thin levels. One way to overcome this difficulty is to down-weight
the contributions of these levels so that levels ℓ for which N_ℓ is very small are
given less weight than those for which N_ℓ is large. For example, we can assign
size-dependent weights w_ℓ to each level ℓ and modify the scores s_ℓ defined in
Eq. (10.21) to:

$$s_\ell = w_\ell \left(\frac{1}{N_\ell} \sum_{j \in S_\ell} y_j \right) + [1 - w_\ell] \left(\frac{1}{N} \sum_{j=1}^{N} y_j \right), \qquad (10.22)$$

where N is the total number of records in the dataset. Since the standard
deviation of a sample of a numerical variable typically varies inversely with the
square root of the sample size, one logical choice for the weights $\{w_\ell\}$ is to make

them proportional to the square root of N_ℓ, e.g.:

$$w_\ell = \sqrt{\frac{N_\ell}{N}}. \tag{10.23}$$

Note that as $N_\ell \to N$, this weight approaches 1, so the weighted score s_ℓ defined in Eq. (10.22) becomes essentially the same as the original score defined in Eq. (10.21), while as $N_\ell \to 0$, this weight approaches zero and the score is replaced with that for the complete dataset.

10.7 Exercises

1: The birthwt data frame in the MASS package characterizes infant birth weight data from 1986, including both the actual birthweight (bwt) and a binary indicator (low) of low birth weight (less than 2.5 kilograms). This problem asks you to build and characterize a logistic regression model to predict this binary response; specifically:

1a. First, fit a "naive model" that predicts low from all other variables in the birthwt data frame and characterize it with the summary function. What indications do you have that something is wrong?

1b. Excluding the variable bwt, again fit a logistic regression model and characterize it using the summary function. Which variables are significant predictors of low birth weight in this model, at the 5% significance level?

1c. Use the predict function to predict the probability of low birth weight for all records in the birthwt data frame for the logistic regression model from (1b) and use the pROC package to construct and display the ROC curve for this binary classifier.

1d. Use the MLmetrics package to compute both the AUC and the root-Brier score introduced in Section 10.2.3.

2: Exercise 6 in Chapter 9 considered the Leinhardt data frame from the car package, containing infant mortality rates per 1000 live births for 105 countries. As noted, the variable infant contains missing values and does not appear well approximated by the Gaussian distribution, so the gamma distribution was proposed as a possible alternative. One of the classes of generalized linear models introduced in Section 10.2.4 is based on the gamma distribution, and these models can be fit using the glm function by specifying family = Gamma. Further, this option is actually a function that accepts the optional argument link with three valid values: "inverse" (the default), "identity," and "log." This exercise asks you to fit three gamma glm's to predict infant from the other variables in the Leinhardt data frame:

2a. Create model `gammaGLMa` with the "inverse" link, generate model predictions, and construct a plot of predicted versus observed responses. Add a heavy dashed reference line for perfect predictions;

2b. Repeat (2a) with `link = ''identity''` to build model `gammaGLMb`;

2c. Repeat (2a) with `link = ''log''` to build model `gammaGLMc`.

3: This exercise asks you to repeat the model-building process from Exercise 3, but using the outlier-resistant procedure `glmrob` from the `robustbase` package instead of the `glm` procedure used in Exercise 3. Specifically:

3a. Use the `glmrob` function to build the gamma glm model `gammaGLMaRob` that predicts `infant` from the other variables in the `Leinhardt` data frame. Re-create the predicted versus observed plot from Exercise (2a), and then add the corresponding predicted versus observed points from `gammaGLMaRob`, using a different plotting symbol to distinguish them. What primary differences do you notice between the robust and the original predictions?

3b. Repeat (3a), with `link = ''identity''`, building the `gammaGLMbRob` model.

3b. Repeat (3a), with `link = ''log''`, building the `gammaGLMcRob` model.

4: A popular alternative to the logistic regression model considered in Exercise 1 is decision tree classifiers like those constructed with the `rpart` package. This exercise asks you to repeat part of Exercise 1 with the `rpart` function from the `rpart` package to build a decision tree model:

4a. The optional argument `method` specifies the type of decision tree model `rpart` will build. Build a decision tree model that predicts the variable `low` from all other variables in `birthwt` except `bwt`, with `method = ''class''`. Show the structure and details of this model with the generic `plot` and `text` functions discussed in Section 10.3.1. (Note that setting the optional argument `xpd = TRUE` in the `text` function call prevents truncation of the text labels.)

4b. Repeat Exercise (4a) but with `method = ''anova''`. Is the structure of these models the same or different?

4c. Using the generic `predict` function with the model from (4b), generate the predicted positive response probabilities and construct the ROC curve using the `pROC` function as in Exercise (1c).

4d. Using the `MLmetrics` package, compute the AUC and root-Brier score for the `rpart` model from (4b). Does this model have a better AUC than that from Exercise (1b)? Does it have a better root-Brier score?

5: As discussed in Section 10.3, decision tree models can be built for either binary classification applications like that considered in Exercise 4, or for regression applications where the response variable is numeric. This exercise asks you to build a regression tree model using the `rpart` function in the `rpart` package to predict the variable `Price` from all other variables in the data frame `car.test.frame` in the same package. Once you have built this model, use the generic `plot` and `text` functions to show the structure of this model.

6: Many different methods have been developed for fitting decision tree models. The basic package in *R* is `rpart`, used in Exercises 4 and 5, but an alternative is the `ctree` function in the `partykit` package. This exercise asks you to essentially repeat Exercise 5, but using the `ctree` function. Specifically, fit a decision tree model using the `ctree` function and then display the structure of this model using the `plot` function (note that the `text` function is not needed here). Is this model essentially the same as that in Exercise 5 or very different?

7: The `cabbages` data frame in the `MASS` package summarizes the results of an agricultural field trial giving the vitamin C contents and head weights of 60 cabbages, grown under varying conditions. This exercise asks you to build two models to predict the vitamin C content (`VitC`): the first is a linear regression model like those discussed in Chapter 5, while the second is an MOB model like those discussed in Section 10.4. Specifically:

   7a. Using the `lm` function, fit a linear regression model that predicts `VitC` from `HeadWt` in the `cabbages` data frame. Generate a plot of predicted versus observed `VitC` values from this data frame, with a heavy dashed equality reference line. Use the `RMSE` function from the `MLmetrics` package to compute the rms error for these predictions.

   7b. Using the `lmtree` function from the `partykit` package, fit a regression tree model where each node of the tree predicts `VitC` from `HeadWt` as in (7a), but which uses the other two variables in the `cabbages` data frame as the partitioning variables to construct the decision tree. Use the generic `plot` function to show the structure of the model.

   7c. Generate a plot of predicted versus observed `VitC` values for the MOB model from (7b), with a heavy dashed equality reference line, and use the `RMSE` function from the `MLmetrics` package to compute the rms error for these predictions. Which of these two models gives the better predictions?

8: The `birthwt` data frame from the `MASS` package was used in Exercises 1 and 4 as the basis for building and comparing binary classifiers that predict the low birth weight indicator `low` from other variables in the dataset, excluding the actual birth weight variable `bwt`. This exercise and the next two consider various aspects of predicting this birth weight variable from these other variables, excluding the binary indicator `low`. Specifically:

8a. Create the data frame `birthwtMod` that excludes the `low` variable.

8b. Build a linear regression model that predicts `bwt` from all other variables in the `birthwtMod` data frame. From this model, generate a plot of predicted versus observed birth weights, with a heavy dashed equality reference line. Then, use the `RMSE` function from the `MLmetrics` package to compute the rms prediction error.

8c. Repeat (8b), except instead of a linear regression model, build a random forest model using the `randomForest` package, again generating a plot of predicted versus observed responses and computing the rms prediction error. Which model gives the better predictions?

9: For categorical predictors, the partial dependence plots discussed in Section 10.6.1 exhibit a "staircase" appearance, since the predictor only assumes a finite number of discrete values, and the partial dependence plot represents the average model prediction for each of these discrete values. This exercise illustrates this point for the two models constructed in Exercise 8. Using the `plotmo` function from the `plotmo` package:

9a. Construct the partial dependence plot for the linear model.

9b. Construct the partial dependence plot for the random forest model.

10: It was demonstrated in Section 10.5.3 that the quality of the boosted tree models generated by the `gbm` function depends very strongly on the tuning parameters used in building these models. This exercise asks you to make the following comparisons:

10a. Using the default argument values for the `gbm` function from the `gbm` package, fit a model `gbmModelA` that predicts `bwt` from all other variables in the `birthwtMod` data frame constructed in Exercise (8a). Use the generic `predict` function to generate predicted birth weights, and construct a predicted versus observed plot with a heavy dashed equality reference line. Use the `RMSE` function from the `MLmetrics` package to compute the rms prediction error. How do these values compare with those for the models from Exercise 8?

10b. Repeat Exercise (10a) but using the argument values for the best concrete compressive strength prediction model built with the `gbm` package in Section 10.5.3. Construct a predicted versus observed plot and compute the rms prediction error for this model. How does it compare with the two models constructed in Exercise 8?

Chapter 11

Keeping It All Together

The focus of this book has been on learning how to use R to analyze and interpret data in an interactive environment. It has been emphasized that if we want to share our results with others, we need to create external files containing these results that persist after our interactive R session terminates. These files can take many forms, including R code files, CSV files containing data, and graphical output saved in PNG files or other image file formats. Once we move beyond learning how to use R and trying out new features, to actually using it to produce results for others, our collection of these files usually grows rapidly. When this happens, it is easy to lose track of what we have done and how exactly we have done it. Often, we find that six months or a year after we have generated one set of results, someone wants a similar set of results, with minor variations in the data sources, the analysis approaches, method parameters, or display formats. The primary focus of this chapter is on organizing and managing this growing collection of files—data, programs, numerical results, plots, or larger documents containing some of these other results—in a way that minimizes the effort and frustration involved in providing "minor variations" on old results. Another topic covered briefly in this chapter is how to manage your R installation itself: installing R, updating packages, and updating your R version.

It is important to emphasize from the outset that there is no single "right" way to organize anything: a "good" organization of any kind of material, computer-based or otherwise, is one that works for you. What follows, then, are a few suggestions that many have found useful, offered as possibilities to include in your own organization of the electronic materials that emerge from your continuing use of R and/or other computing environments.

11.1 Managing your R installation

Everything discussed in this book depends on having R installed on your machine, and this installation process is described in Sec. 11.1.1. Also, because packages are continually updated, it is good practice—and sometimes essential—

to periodically update previously installed packages to obtain the newest version, a process that is discussed in Sec. 11.1.2. Finally, it is also a good idea to occasionally update your R installation itself, as described in Sec. 11.1.3.

11.1.1 Installing R

The standard source for all things R is CRAN, the *Comprehensive R Archive Network*, which supports a number of interlinked websites and provides versions of R for all major operating systems. The easiest way to install R is to follow the instructions for downloading the appropriate *binary distribution* on this URL:

```
https://www.R-project.org
```

The advantage of downloading and installing the binary distribution is that it is immediately executable: once the process is complete, you need only click on the R desktop icon to start an interactive R session. It is also possible to build an R installation from the source code, also available from CRAN, but this is a lot more work and may require obtaining additional software components (e.g., compilers for languages like C and *FORTRAN*). More detailed instructions on installing R are available in the document R *Installation and Administration*, which can be easily found by searching for the title on the web (it is also included with the R help files, but this isn't useful if you don't already have R installed).

Finally, it is important to note that some companies limit employee access to the Internet, for a variety of reasons. If you work for such a company and are attempting to use R at work, you may not be able to follow the instructions just given. In such cases, probably the only recourse is to talk to someone in your company's IT organization about how to get R installed on your machine. Similarly, if you can access the web but don't have the necessary permissions to install new software on your machine, it is again probably necessary to talk to someone in your company's IT organization.

11.1.2 Updating packages

Major releases of new R versions occur about twice a year, and R packages are often updated much more frequently. Thus, within a relatively short time (e.g., a few months) after you install an R package, the version you have will no longer be the most current one. Because significant efforts are made to keep R backward-compatible, code that ran earlier is likely to still run after you update your R packages; the reason new R packages are released is to distribute improvements to the R user community, fixing bugs and adding new capabilities. This provides one motivation for periodically updating your R packages.

Another reason for updating your R packages is that they typically require other R packages. Thus, if you install a new package Y to obtain some new functionality, and this package requires a package X that you already have installed, but it requires a newer version than you have, your newly installed Y package may not run correctly.

Finally, a third reason for updating your *R* installation is that "breaking changes" do sometimes occur. For example, early in the development of this book, version 0.9.8.5 of quanteda was being used for the text analysis procedures discussed in Chapter 8. When this package was updated to version 0.99.12, some of the function names and argument names were different, requiring some of the code originally included in Chapter 8 to be revised extensively.

To update *all* packages in your current *R* installation, replacing the current versions with the newest ones available, run the update.packages() command, with no optional arguments. To update a single package, specify the optional argument oldPkgs with the name of the package you want to update; for example, the following command would update just the quanteda package:

```
update.packages(oldPkgs = "quanteda")
```

For a list of packages that are currently installed on your machine but for which newer versions are available, use the old.packages function, called without optional arguments. *Note that, like the initial package installation either with the install.packages function or from the R help tab, the functions update.packages and old.packages require Internet access and the selection of a CRAN mirror site.*

11.1.3 Updating *R*

To see what version of *R* you are running, type "R.version" in your interactive *R* session: this lists the R.version variable from base *R* that gives the version of *R* you are running, along with the type of machine and operating system you are running on. This information, along with a list of the *R* packages currently loaded into your session, is also returned by the sessionInfo() function from the utils package. To see whether there is a newer version of *R* available, go to the CRAN website (https://cran.r-project.org/) and look at the version listed under "latest release." If this version is later than your R.version result, a new version is available. If so, it is probably worth following the what's new link to see what new features are available in this version of *R*.

The help files under the "R for (my operating system) FAQ" (e.g., "R for Windows FAQ" if you are running on a Microsoft Windows machine) describe the process for updating your *R* version, which they note is a matter of taste. One approach is to first uninstall the current version of *R*—which does not remove the files containing all of your packages—and then install the new version of *R*. This will create a new set of directories (i.e., with names reflecting the current version number), and it will be necessary to copy the installed packages from the original package files into this new directory. After this, the update.packages function should be run within your new *R* version to obtain the latest versions of these packages. *Conversely, if you are in the middle of a critical project, you may want to save the old version of* R *until you are sure your new version has all of the capabilities you need.*

11.2   Managing files effectively

Stephanie Winston's book, *Getting Organized* [79], begins with a cartoon of a woman talking on the phone, saying:

> I can guarantee you one thing, Sid ... it's here in our infallible files, neatly nestled somewhere between A and Z.

While her book is not concerned with organizing computer files, a few pages later she gives a self-assessment quiz with the title "How Organized Are You?" that begins with the following question [79, p. 15]:

> Does it often take you more than ten minutes to unearth a particular letter, bill, report, or other paper from your files (or piles of paper on your desk)?

Her book is concerned with organizing your life, money, and time so that the answer to such questions becomes "no." The intent of the following section is to offer some useful suggestions for achieving this objective with respect to the results and code that accumulates in files as you use R (or any other computing environment) in real-world applications.

11.2.1   Organizing directories

Files are typically organized hierarchically in directories, with upper-level directories or folders containing lower-level subdirectories or folders, which in turn contain our files. This organization can be nested as deeply as we like, but it is worth noting that the two extremes of this organization are the least useful. At one extreme, there are no subdirectories: all files are located in one main directory. The disadvantage of this organization is that, to find a particular file, we have to look at all files: this is not a difficult task if we have only a few files, but as we work, developing and saving intermediate results, R code, new data sources, PNG files containing plots, and various summary documents, our file collection is likely to grow rapidly. Once we have dozens or hundreds of files, looking through them all to find the ones we want becomes increasingly time-consuming. At the other extreme, if we build a directory structure in which every file is in its own directory, we now have to look through every directory to find the file we want, which is probably worse than putting them all in one main directory.

A useful directory structure contains subdirectories that group similar files together. To do this, however, we must decide from the outset how "similar files" should be defined. As Stephanie Winston notes, the same issue arises in organizing books on a bookshelf: if we have a biography of the author Thomas Hardy, do we group it under the heading of "biography," or do we group it together with Thomas Hardy's literary works [79, p. 153]?

In his list of "Good and Bad Practices," for reproducible research, Xie's first recommendation is to "manage all source files under the same directory" [82, p. 7]. The subject of reproducible research is introduced in Sec. 11.4, but the key point here is that Xie's advice implicitly offers one useful definition of "similar files": all *source files*, containing the data, code, and explanation required to generate a summary document like those discussed in Chapter 6. A somewhat more general organizational principle along similar lines is to organize our files by project. In the case of two projects, A and B, we would organize our directory hierarchy something like this:

1. Project A directory:

   1a. Project A data files

   1b. Project A code files

   1c. Project A summary files

   1d. Project A working notes

2. Project B directory:

   1a. Project B data files

   1b. Project B code files

   1c. Project B summary files

   1d. Project B working notes

While simple, this organization has a number of advantages. First, it is easily and obviously extendible as we work on new projects: we simply create new top-level directories for Projects C, D, and E, etc., as they come along. Second, if we know what project we need information about, the directory structure helps us find that information: the summary on Project B that we sent our boss last month should be in the "Project B summary files" subdirectory under "Project B directory." Third, it provides an obvious place to put new data, code, summary notes, or other materials as we continue to work on a project. Finally, we can also expand the lower levels of this structure, creating additional subdirectories for results specific to particular special cases of a project. For example, if we are doing business analyses by region, it might make sense to create subdirectories for each region under the "Project A summary files" directory.

One disadvantage of this project-based organization is that it implicitly assumes the code associated with each project is unique to that project, which is often not the case. In particular, as discussed in Chapter 7, developing modular code that can be used broadly across many different applications is good practice because it allows us to develop and debug the code once and re-use it many times. Thus, it should happen frequently that the code—or at least component modules of the code—is shared across projects. If we retain the strict project-oriented hierarchy described above, this implies we must put a separate copy of these common modules into each "code files" subdirectory. This, however, is bad practice, violating the DRY principle ("don't repeat yourself") advocated in Chapter 7: if the same code is to be used in several different applications, it should be put in *one* place, and all applications should retrieve it from there. This is important because it guarantees consistency between different applications using "the same" code: if multiple copies exist, some may have features and/or bugs that the others don't, which can lead to inconsistencies in interpreting or comparing these different results.

One solution to this problem—a very good solution, but one that requires substantial effort—is to organize our code into one or more *R* packages, like the ones available from CRAN. In his book, *Software for Data Analysis*, John Chambers, the principal designer of the *S* language on which *R* is based, offers the following recommendation [12, p. 79]:

> Consider organizing your programming efforts into a package early in the process. You can get along with collections of source files and other miscellany, but the package organization will usually repay some initial bother with easier to use and more trustworthy software.

As noted, what Chambers terms "some initial bother" involves substantial effort, but a very helpful book in working through this process is *R Packages* by Hadley Wickham, the author of a large and growing list of *R* packages [75].

A much simpler alternative that addresses the common code problem is to put all of our code files into a common directory, which can be retrieved using the source function as they are needed. This idea leads to the following alternative directory structure:

1. Common code directory

2. Project A directory:

    1a. Project A data files

    1b. Project A summary files

    1c. Project A working notes

3. Project B directory:

    1a. Project B data files

    1b. Project B summary files

    1c. Project B working notes

This modified has the same basic advantages listed above, but it satisfies the DRY principle, allowing the code to appear once in a common directory. Thus, any improvements or bug fixes can be made to this single copy of the source code and then used by all projects, both current and future. In fact, one of the key features of the CRAN-compliant *R* packages discussed by Chambers [12, Ch. 4] and Wickham [75] is that they have a specific directory structure, with all code source files located in one subdirectory and other package components (e.g., documentation and external data) located in other directories.

A similar approach can be applied to common data sources. That is, if several projects use some or all of the same datasets, it is a good idea to again apply the DRY principle and put *one* copy of this dataset in a common directory that is used by all of these projects. The advantages of this approach are similar to those for creating a common source code directory, and this practice can avoid some subtle data issues that may be difficult to recognize or track down. For example, it has been noted that several versions of the Pima Indians diabetes dataset are available, including one from the `mlbench` package and three others from the `MASS` package (`Pima.tr`, `Pima.tr2`, and `Pima.te`), all of which are fundamentally different. Similar issues arise frequently in research settings (e.g., Vandewalle *et al.* note that the *Lena* image, a popular benchmark for image processing studies, is available in a number of different formats, including color, black-and-white, with or without image compression [68]), and probably even more frequently with private data sources (e.g., company proprietary datasets that may be modified to add new variables, correct errors, or provide updated information). Reading data from a single source guarantees that all "similar" or "related" analyses are in fact using *the same* data. Also, this approach helps in cases where the dataset is very large: we need only a single copy rather than several, saving considerable storage space.

11.2.2 Use appropriate file extensions

Most computer file names have the general form `FileName.ext`, where `FileName` is the file name itself, and `ext` is a *file extension*, designed to help us and various software utilities know something about the file's internal format and contents. Examples of file extensions discussed in this book include:

1. `csv`, indicating a comma-separated value (CSV) file;

2. `pdf`, indicating a portable document format (PDF) file;

3. `png`, indicating a PNG image file;

4. `R`, indicating an *R* code file;

5. `rds`, indicating an RDS binary *R* object file;

6. `txt`, indicating a text file, typically ASCII-encoded;

7. `xls` or `xlsx`, indicating a Microsoft *Excel* spreadsheet file.

File extensions are not enforced, so it is possible to create files with names like `RcodeFile` with no file extension, or `MysteriousFile.zqx` with a meaningless file extension. Such names are generally undesirable, for two reasons. First, they prevent us from searching for files by type, e.g.:

```
list.files(pattern = "csv$")
```

Second, these files will not be found by software utilities designed to look for files of a specific type: e.g., most popular Microsoft programs like *Word* or *Excel* bring up a search window that shows a list of files that can be opened by that program. Since this is done by file extension, a missing or meaningless extension means that these files will not appear on this list.

11.2.3   Choose good file names

In addition to selecting appropriate file extensions, a useful step in effective file management is to select good file names that precede this extension. As with naming variables, one criterion for a good file name is that it be informative. Recall, for example, the names for the 5 variables in the `aircraft` data frame in the `robustbase` package, which are completely non-informative and must be interpreted with the aid of the associated help file:

1. `X1` is the aircraft's aspect ratio;

2. `X2` is the aircraft's lift-to-drag ratio;

3. `X3` is the aircraft's weight;

4. `X4` is the aircraft's thrust;

5. `Y` is the aircraft's cost.

Analogously, if we choose non-informative file names like `DataFile01.csv`, it is necessary to look inside the file to have any idea of what it contains; if we have dozens or hundreds of files, the search for the one or few we want can take a long time.

Like good variable names, good file names are strongly application-specific, so it is not possible to give strong guidelines for choosing file names. That said, there are some special cases where it is possible to be more specific:

1. *R* program files should end with the `.R` file extension and, ideally, contain a single function or executable procedure, and the file name should be the same as the name of the function or procedure;

2. RDS files should end with the `.rds` file extension and have the same name as the *R* object that has been saved in the file.

Data files (e.g., `.csv` or `.txt` files) should have names that succinctly describe the data they contain, while `.png` files should have names that succinctly describe the images they contain, and `.pdf` files should have abbreviated names of the documents they contain.

While it is not always applicable, an approach that is sometimes extremely useful is automatic file name generation. For example, if we have a program that fits a specific model type to the data contained in a particular CSV file and generates a predicted versus observed plot as a PNG file, it may be reasonable to have this program generate the PNG file name automatically from the name of the model, the name of the CSV file, and the fact that it is a predicted versus observed plot. A typical result might be something like:

`ConcreteCompressiveStrengthLMtreePredVsObs.png`

A potential disadvantage of this approach is that the file names can become inconveniently long, but this strategy does at least lead to informative file names that should be easy to find in a pattern-based search.

It may be obvious in retrospect, but it is worth noting just the same: a good file name should be a valid file name. For example, most software environments do not permit file names with embedded spaces, and file names with embedded special characters like $ or % may cause difficulties and should also be avoided. These observations are particularly important in the case of automatically generated file names, where these characters may be included in the file name either due to oversights or programming errors.

11.3 Document everything

As the preceeding discussion emphasized, keeping our files organized so we can quickly find what we need is important, but another important aspect of organizing our computer results is having documentation that can either inform others or remind us of important details (e.g., which variables were in a dataset, what a particular function does and what its required arguments are, or what we concluded from the analysis of a similar case 18 months ago). The following sections offer some guidelines for documenting data, source code, and results, either for the benefit of others or to remind ourselves what we did and what did and didn't work out when we generated those results.

11.3.1 Data dictionaries

The idea and importance of *data dictionaries* was introduced briefly in Chapter 3. Sadly, data dictionaries are all too often either missing or incomplete, but a good data dictionary can save numerous headaches if we ever have need to re-use a dataset, or reconstruct a "similar" one, possibly incorporating newer data or additional variables. The following discussion offers a few guidelines for developing a useful data dictionary.

A limited form of data dictionary is the metadata provided by R packages that include dataset examples. Like the R code included in packages, a CRAN-compliant package that includes datasets must include documentation describing them, typically incorporating the following three elements [75, p. 93]:

1. `Description`, a brief summary (typically one or two sentences) describing the dataset;

2. `Format`, a succinct but detailed description of the format of the dataset, giving the number of rows (records), the number of columns (fields per record), and the names and brief descriptions of each field;

3. `Source`, a literature reference or URL that identifies where the dataset was obtained, and possibly gives additional description of its contents and/or how it was constructed.

Ideally, the format section should include the units of measure for numerical variables and descriptions of the levels of categorical variables. Other extremely useful elements of a data dictionary include:

4. Published references describing the data source and prior analyses of it;

5. Summaries of missing data or other known data anomalies (e.g., outliers, inliers, externally imposed range limits, etc.);

6. Brief descriptions of any features that are likely to cause confusion or misinterpretation;

7. Distinctions between the data source and any other, similar data sources that are likely to be confused with it (e.g., variables that have been added to or removed from other, similar datasets, or record subsets that have been deemed unreliable and either removed or corrected, etc.).

Ideally, a data dictionary should be an organic document, updated every time the dataset is updated or some new unexpected feature is discovered in it. The best data dictionaries are those that document the things we wish we knew when we started working with the dataset originally but we didn't and it caused us a lot of headaches.

11.3.2   Documenting code

One of the particularly useful aspects of organizing *R* code as a CRAN-compliant package is that *R* packages have specific documentation requirements. For functions, important components of this documentation include:

1. `Description`, a short paragraph (often only one sentence) describing what the function does;

2. `Usage`, one or more text strings showing how the function is called, listing its arguments in the order in which they appear in the function definition, with default values for any optional arguments;

3. `Arguments`, a list with one entry for each argument that gives the argument name and a phrase describing its purpose;

4. `Details`, a few (often one or two) short paragraphs expanding on what the function does, how it does it, or other important but not necessarily obvious details about the function or its arguments;

5. `Value`, a brief description of what the function returns;

6. `Examples`, a brief collection of executable applications of the function that can be run to show more explicitly what it does.

R package documentation may also include references to relevant publications, or cross-references to related *R* functions. The key requirement for good function documentation is that it tells a potential new user (or reminds the developer who hasn't used it in some time) what the function does and how to use it. Thus, it should be clear what the required arguments are, what valid values for all arguments are, what the function does, and what it returns.

Another form of function documentation that is encouraged in the *R* community and which has become increasingly popular is the *vignette*, a longer document that provides broader context than user documentation like that just described. Vignettes are associated with packages and typically convey something about why the functions in the package were implemented, how they work together, and where they can be useful. Frequently, vignettes include much more detailed examples than those described above for *R* user documentation, often consisting of one or more fairly complete data stories along the lines discussed in Chapter 6. In his book on *R* package development, Wickham devotes a chapter to developing vignettes [75, Ch. 6].

11.3.3 Documenting results

Chapter 6 was devoted to the subject of developing data stories that are useful in describing our analysis results to others. An important aspect of this task was matching the data story to the audience, and the primary focus of that chapter was on data stories that focus on results and their interpretation, with only the minimum essential description of analysis methods and details. These data stories represent one extremely important form of documentation for our analysis results, but not the only one: since analyses often need to be repeated with "minor variations," it is also important to document, *for ourselves or those analysts who undertake these follow-on analyses*, key details of how we arrived at those results. These details are generally not appropriate to include in a data story to be presented to others, but it is important to retain them for future analytical reference. This implies the need for other, more detailed descriptions of a more exploratory nature that allow us to both re-create these results when needed, and to document any alternative paths we may have considered that did not lead to anything useful. This second point is important, either to prevent going down the same fruitless paths in subsequent analyses, or to document something that didn't work out in the original analysis but which might be appropriate to consider for the revised analyses. As noted in the discussion of directory organization in Sec. 11.2.1, it is probably useful to put these documents in separate subdirectories to clearly distinguish between analytical summaries to be shared with others and working notes retained for future reference.

In both cases, an extremely useful way of developing these documents is to adopt the *reproducible computing* approach described in Sec. 11.4. The advantage of this approach is that it ties summary documents—either data stories to be shared with others or working notes to be retained for future reference—to the code and data on which they are based. This approach involves using a software environment like *R Markdown* that allows us to combine verbal description with code and results generated with that code (e.g., numerical summaries or plots) all in one document. This document is then processed—in the case of R Markdown documents, via the `rmarkdown` package—to generate a document in a shareable format like a PDF file or an HTML document.

11.4   Introduction to reproducible computing

In their article on reproducible research in signal processing, Vandewalle *et al.* begin by posing the question, "Have you ever tried to reproduce results presented in a research paper?" [68]. They note, first, that this is often a very difficult task, and, second, that it is often even difficult to reproduce *their own* research results some time after the original work was completed. To address these difficulties, their paper advocates the ideas of *reproducible research* or *literate programming* that attempt to keep all necessary material together in one place, linked by software tools and practices that allow us to easily reproduce our earlier work or the work of others. The following sections present, first in Sec. 11.4.1, a brief introduction to the key ideas of generating reproducible results, and, second, in Sec. 11.4.2, an introduction to *R Markdown*, a simple environment to integrate verbal descriptions of results with the *R* code that generated them.

11.4.1   The key ideas of reproducibility

Vandewalle *et al.* define six degrees of reproducibility, from level 0 ("The results cannot be reproduced by an independent researcher"), to level 5 [68]:

> The results can be easily reproduced by an independent researcher with at most 15 minutes of user effort, requiring only standard, freely available tools (C compiler, etc.).

While specific to computer-generated results, note the similarity of this working definition of "good reproducibility" to Stephanie Winston's first organization question posed in Sec. 11.2. A useful extension of this reproduciblity criterion partitions it into two components [59, p. 22]:

1. the ease of obtaining the "raw materials" on which the results are based (i.e., software platform, detailed code descriptions, data, etc.);

2. the ease of reproducing the research results, given these raw materials.

The reason this partitioning is useful is that the first step involves the important ideas discussed earlier in this chapter and elsewhere in this book: e.g., naming data and code files so we know what is in them, providing useful data dictionaries and code documentation, and maintaining a useful file organization so we can find what we need quickly. The best way to approach the second step is more specialized, taking advantage of document-preparation software that allows us to combine text, code, and results all in one place.

One platform that allows us to do this is `Sweave`, included in the `utils` package in *R* and thus available in all *R* installations. This function maps a specified source file, typically with the extension `.Rnw`, into a LaTeX® file, which can be converted into a PDF file via the LaTeX computer typesetting software [32, 50]. This package is built on the even more flexible computer typesetting language TeX® developed by Donald Knuth [49] to support the development of documents containing arbitrarily complicated mathematical expressions. The combination of R with LaTeX through `Sweave` provides a powerful basis for creating documents that combine mathematical expressions, explanations in English (or other languages), R code, and results generated by R code.

This book was developed using the R package `knitr` [82], an even more flexible document preparation package, which allows the incorporation of all of these elements and more, including code in other programming languages. The book *Nonlinear Digital Filtering with Python* [59] was also developed using `knitr`, and it incorporates R results (for the graphics), *Python* code and results, block diagrams constructed using the LaTeX `picture` environment, and nearly 500 equations. Both of these examples illustrate that the `knitr` package is an extremely powerful environment for reproducible computing applications, but taking advantage of this flexibility does require learning to use both LaTeX and `knitr`. Good introductions to these packages are the books by Griffiths and Higham [32] and Xie [82], but these learning curves are somewhat steep and may be off-putting to those without mathematical inclinations.

Fortunately, there is a simpler alternative in the `rmarkdown` package, which allows you to create documents in the simpler R Markdown format, again allowing you to incorporate both explanatory text and R code and results, but without the necessity of learning either LaTeX or `knitr`. The following section provides a brief introduction to this document preparation environment.

11.4.2 Using R Markdown

In his book on developing R packages, Wickham advocates using R Markdown for developing vignettes. The process consists of the following sequence of steps:

1. Create an R Markdown file with the `.Rmd` extension;

2. Run the `render` function from the `rmarkdown` package:

   a. `render(RmdFileName, html_document())` generates an HTML file;

   b. `render(RmdFileName, pdf_document())` generates a PDF file

The first step can be accomplished with the `file.create` and `file.edit` functions used to create R source code files described in Chapter 7. The second step requires that the `rmarkdown` package has been installed and loaded with the `library` function, and it may require the installation of certain other external packages (e.g., creating PDF files uses LaTeX under the hood, so a version of this package must be available, possibly with some additional features).

The R Markdown file uses an extension of the markup language `Markdown`, intended as a human-readable environment for creating HTML pages. Essentially, an R Markdown document consists of these three elements:

1. Ordinary text;

2. Special markup symbols that control the format of the rendered document;

3. R code, either simple inline expressions in the text, or larger code blocks.

Wickham gives a very useful introduction to R Markdown, with a number of examples [75, Ch. 6], and much additional information is available from the Internet by searching under the query "R Markdown." A very simple R Markdown file looks like this:

```
# R Markdown example

An R Markdown file can include both text (like this) and *R* code
like the following, either inline like this (the **mtcars** data
frame has `r nrow(mtcars)` rows), or in larger blocks like this:

```{r}
str(mtcars)
```
```

The first line here includes the formatting symbol "#," which defines a section heading, making the text large, rendered in boldface. Lower-level headings (e.g., subheadings) can be created by using multiple # symbols, and the "*" symbols in the first line of text causes the letter "R" to be rendered in italics. Similarly, the double asterisk in the next line cause the name "mtcars" to be rendered in boldface. The string `r nrow(mtcars)` in the next line executes the R code nrow(mtcars) and includes the result as inline text, and the last three lines of this file cause R to execute the command str(mtcars), showing this command and displaying the results.

The key point of this example is that R Markdown is much easier to learn than LaTeX but it provides an extremely flexible basis for creating documents that combine R code and explanatory text. Wickham argues that "You should be able to learn the basics in under 15 minutes" [75, p. 62], and it probably does represent the easiest way to adopt the philosophy of reproducible computing.

Bibliography

[1] P. Adriaans and D. Zantinge. *Data Mining*. Addison-Wesley, 1996.

[2] A. Agresti. *Categorical Data Analysis*. Wiley, New York, NY, USA, 2nd edition, 2002.

[3] F.J. Anscombe. Graphs in statistical analysis. *The American Statistician*, 27:17–21, 1973.

[4] R.A. Askey and R. Roy. Gamma function. In F.W.J. Olver, D.W. Lozier, R.F. Boisvert, and C.W. Clark, editors, *NIST Handbook of Mathematical Functions*, chapter 5, pages 135–147. Cambridge University Press, Cambridge, UK, 2010.

[5] V. Barnett and T. Lewis. *Outliers in Statistical Data*. Wiley, 2nd edition, 1984.

[6] D.A. Belsley, E. Kuh, and R.E. Welsh. *Regression Diagnostics*. Wiley, 1980.

[7] T. Benaglia, D. Chauveau, D.R. Hunter, and D. Young. mixtools: An R package for analyzing finite mixture models. *Journal of Statistical Software*, 32(6):1–29, 2009.

[8] J. Breault. Data mining diabetic databases: Are rough sets a useful addition? In *Proceedings of the 33rd Symposium on the Interface, Computing Science and Statistics*. Fairfax, VA, USA, 2001.

[9] L. Breiman. Random forests. *Machine Learning*, 45(1):5–32, 2001.

[10] G.W. Brier. Verification of forecasts expressed in terms of probability. *Monthly Weather Review*, 78(1):1–3, 1950.

[11] L.D. Brown, T.T. Cai, and A. DasGupta. Interval estimation for a binomial proportion. *Statistical Science*, 16:101–133, 2001.

[12] J.M. Chambers. *Software for Data Analysis*. Springer, New York, NY, USA, 2008.

[13] M. Chavent, V. Kuentz-Simonet, A. Labenne, and J. Saracco. Multivariate analysis of mixed data: The PCAmixdata R package. 2014.

[14] R. Colburn. *Using SQL*. Que, 1999.

[15] D. Collett. *Modelling Binary Data*. Chapman and Hall/CRC, Boca Raton, FL, USA, 2nd edition, 2003.

[16] M.J. Crawley. *The R Book*. Wiley, New York, NY, USA, 2002.

[17] C.J. Date. *An Introduction to Database Systems*. Addison-Wesley, 7th edition, 2000.

[18] L. Davies and U. Gather. The identification of multiple outliers. *Journal of the American Statistical Association*, 88:782–792, 1993.

[19] P. de Jong and G.Z. Heller. *Generalized Linear Models for Insurance Data*. Cambridge University Press, New York, 2008.

[20] D. DesJardins. Paper 169: Outliers, inliers and just plain liars—new eda+ (eda plus) techniques for understanding data. In *Proceedings SAS User's Group International Conference, SUGI26*. Cary, NC, USA, 2001.

[21] P. Diaconis. Theories of data analysis: From magical thinking through classical statistics. In D.C. Hoaglin, F. Mosteller, and J.W. Tukey, editors, *Exploring Data Tables, Trends, and Shapes*, chapter 1. Wiley, 1985.

[22] N. Draper and H. Smith. *Applied Regression Analysis*. Wiley, 2nd edition, 1981.

[23] J. Duckett. *HTML & CSS*. Wiley, 2011.

[24] P. Ein-Dor and J. Feldmesser. Attributes of the performance of central processing units: a relative performance prediction model. *Communications of the ACM*, 30:308–317, 1987.

[25] Y. Freund and R. Schapire. A decision-theoretic generalization of online learning and an application to boosting. *Journal of Computer and Systems Sciences*, 55:119–139, 1997.

[26] J. Friedl. *Mastering Regular Expressions*. O'Reilly, 3rd edition, 2006.

[27] J.H. Friedman. Greedy function approximation: A gradient boosting machine. *Annals of Statistics*, 29(5):1189–1232, 2001.

[28] H. Garcia and P. Filzmoser. Multivariate statistical analysis using the R package chemometrics. 2017.

[29] M. Gardner. *Fads and Fallacies in the Name of Science*. Dover, 1957.

[30] M. Gladwell. *Outliers*. Little, Brown and Company, 2008.

[31] L.A. Goodman and W.H. Kruskal. *Measures of Association for Cross Classifications.* Springer-Verlag, New York, NY, USA, 1979.

[32] D.F. Griffiths and D.J. Higham. *Learning LaTeX.* SIAM, Philadelphia, PA, USA, 2nd edition, 2016.

[33] J.R. Groff and P.N. Weinberg. *SQL: The Complete Reference.* McGraw-Hill, 2nd edition, 2002.

[34] U. Grömping. Variable importance assessment in regression: Linear regression versus random forest. *The American Statistician*, 63(4):308–319, 2009.

[35] B. Grün, I. Kosmidis, and A. Zeileis. Extended beta regression in R: Shaken, stirred, mixed, and partitioned. *Journal of Statistical Software*, 48(11):1–25, 2012.

[36] W. Härdle. *Applied Nonparametric Regression.* Cambridge University Press, 1990.

[37] D. Harrison and D.L. Rubinfeld. Hedonic prices and the demand for clean air. *Journal of Environmental Economics and Management*, 5:81–102, 1978.

[38] T. Hastie, R. Tibshirani, and J. Friedman. *The Elements of Statistical Learning.* Springer, 2nd edition, 2009.

[39] H.V. Henderson and P.F. Vellemen. Building multiple regression models interactively. *Biometrics*, 37(2):391–411, 1981.

[40] L.D. Henry, Jr. *Zig-Zag-and-Swirl.* University of Iowa Press, 1991.

[41] S.T. Herbst and R. Herbst. *The New Food Lover's Companion.* Barrons, New York, NY, USA, 4th edition, 2007.

[42] M. Hubert, P.J. Rousseeuw, and K. VandenBranden. ROBPCA: A new approach to robust principal component analysis. *Technometrics*, 47(1):64–79, 2005.

[43] A. Hunt and D. Thomas. *The Pragmatic Programmer.* Addison-Wesley, 1999.

[44] N. Iliinsky and J. Steele. *Designing Data Visualizations.* O'Reilly, 2011.

[45] H. Jacob. *Using Published Data: Errors and Remedies.* Sage Publications, 1984.

[46] N.L. Johnson, S. Kotz, and N. Balakrishnan. *Continuous Univariate Distributions*, volume 1. Wiley, New York, NY, USA, 2nd edition, 1994.

[47] N.L. Johnson, S. Kotz, and N. Balakrishnan. *Continuous Univariate Distributions*, volume 2. Wiley, New York, NY, USA, 2nd edition, 1995.

[48] G. Klambauer. *Mathematical Analysis*. Marcel Dekker, 1975.

[49] D.E. Knuth. *The TEXbook*. Addison-Wesley, Boca Raton, FL, USA, 1994.

[50] L. Lamport. *LATEX: A Document Preparation System, User's Guide and Reference Manual*. Addison-Wesley, Reading, MA, USA, 2nd edition, 1994.

[51] M. Lewis. *Moneyball*. W.W. Norton and Company, New York, 2004.

[52] A. Liaw and M. Wiener. Classification and regression by random forest. *R News*, 2/3:18–22, 2002.

[53] R.J.A. Little and D.B. Rubin. *Statistical Analysis with Missing Data*. Wiley, 2nd edition, 2002.

[54] M. Livio. *The Golden Ratio*. Broadway Books, New York, 2002.

[55] P. McCullagh and J.A. Nelder. *Generalized Linear Models*. Chapman and Hall, New York, NY, USA, 2nd edition, 1989.

[56] P. Murrell. *Introduction to Data Technologies*. Chapman & Hall/CRC, Boca Raton, FL, 2009.

[57] P. Murrell. *R Graphics*. Chapman & Hall/CRC, Boca Raton, FL, 2nd edition, 2011.

[58] R.K. Pearson. *Exploring Data in Engineering, the Sciences, and Medicine*. Oxford University Press, New York, 2011.

[59] R.K. Pearson and M. Gabbouj. *Nonlinear Digital Filtering with Python*. CRC Press, Boca Raton, FL, USA, 2016.

[60] X. Robin, N. Turck, A. Hainard, N. Tiberti, F. Lisacek, J.-C. Sanchez, and M. Müller. pROC: an open-source package for R and S+ to analyze and compare ROC curves. *BMC Bioinformatics*, 12(77):1–8, 2011.

[61] S. Rose, D. Engel, N. Cramer, and W. Cowley. Automatic keyword extraction from individual documents. In M.W. Berry and J. Kogan, editors, *Text Mining: Applications and Theory*, chapter 1, pages 3–20. Wiley, 2010.

[62] P.J. Rousseeuw and A.M. Leroy. *Robust Regression and Outlier Detection*. Wiley, New York, NY, USA, 1987.

[63] D.A. Simovici and C. Djerba. *Mathematical Tools for Data Mining*. Springer, 2008.

[64] D.M. Stasinopoulos and R.A. Rigby. Generalized additive models for location, scale, and shape (GAMLSS) in R. *Journal of Statistical Software*, 23(7):1–25, 2007.

[65] E.R. Tufte. *Visual Explanations*. Graphics Press, Cheshire, CT, USA, 1997.

[66] R.F. van der Lans. *The SQL Guide to SQLite*. Lulu, 2009.

[67] M.P.J. van der Loo. The stringdist package for approximate string matching. *The R Journal*, 6(1):111–122, 2014.

[68] P. Vandewalle, J. Kovačević, and M. Vetterli. Reproducible research in signal processing. *IEEE Signal Processing Magazine*, 26(3):37–47, 2010.

[69] P.F. Velleman and D.C. Hoaglin. Data analysis. In D.C. Hoaglin and D.S. Moore, editors, *Perspectives on Contemporary Statistics*, number 21 in MAA Notes, chapter 2. Mathematical Association of America, 1991.

[70] W. Venables and B. Ripley. *Modern Applied Statistics with S*. Springer-Verlag, New York, NY, USA, 2002.

[71] L. von Bortkiewicz. *Das Gesetz der kleinen Zahlen*. Teubner, Leipzig, 1898.

[72] D.A. Weintraub. *Is Pluto a Planet?* Princeton University Press, Princeton, New Jersey, 2007.

[73] H. Wickham. *ggplot2*. Springer, 2009.

[74] H. Wickham. *Advanced R*. CRC Press, 2015.

[75] H. Wickham. *R Packages*. O'Reilly, Sebastapol, CA, USA, 2015.

[76] L. Wilkinson. *The Grammar of Graphics*. Springer, 2nd edition, 2005.

[77] W. Willard. *HTML: A Beginner's Guide*. McGraw-Hill, 2009.

[78] W.E. Winkler. Problems with inliers, working paper no. 22. In *Conference of European Statisticians*. Prague, Czech Republic, 1997.

[79] S. Winston. *Getting Organized*. Warner Books, New York, NY, USA, 1978.

[80] R. Winterowd. *The Contemporary Writer*. Harcourt Brace Jovanovich, 2nd edition, 1981.

[81] A. Wright. *Glut: Mastering Information through the Ages*. Joseph Henry Press, 2007.

[82] Y. Xie. *Dynamic Documents with R and knitr*. CRC Press, Boca Raton, FL, USA, 2014.

[83] N. Yau. *Data Points*. Wiley, 2013.

[84] A. Zeileis, T. Hothorn, and T. Hornik. Model-based recursive partitioning. *Journal of Computational and Graphical Statistics*, 17(2):492–514, 2008.

[85] A. Zeileis, C. Kleiber, and S. Jackman. Regression models for count data in R. *Journal of Statistical Software*, 27(8), 2008.

Index

T - #0093 - 111024 - C562 - 234/156/26 - PB - 9780367571566 - Gloss Lamination